SOLAR PHOTOSPHERE:
STRUCTURE, CONVECTION AND MAGNETIC FIELDS

INTERNATIONAL ASTRONOMICAL UNION

UNION ASTRONOMIQUE INTERNATIONALE

SOLAR PHOTOSPHERE: STRUCTURE, CONVECTION AND MAGNETIC FIELDS

PROCEEDINGS OF THE 138TH SYMPOSIUM OF THE
INTERNATIONAL ASTRONOMICAL UNION,
HELD IN KIEV, U.S.S.R., MAY 15–20, 1989

EDITED BY

J. O. STENFLO

Institute of Astronomy, Zurich, Switzerland

KLUWER ACADEMIC PUBLISHERS
DORDRECHT / BOSTON / LONDON

Library of Congress Cataloging in Publication Data

Solar photosphere : structure, convection, and magnetic fields :
  symposium no. 138 held in Kiev, USSR, May 15-20, 1989 / edited by
  J.O. Stenflo.
      p.   cm.
  At head of title: International Astronomical Union, Union
astonomique internationale.
    ISBN 978-0-7923-0530-9   ISBN 978-94-009-1061-4 (eBook)
    DOI 10.1007/978-94-009-1061-4

  1. Solar photosphere--Congresses.   I. Stenflo, Jan Olof.
II. ·International Astronomical Union.
QB528.S65   1989
523.7'4--dc20   89-27781

ISBN 978-0-7923-0530-9

Published on behalf of
the International Astronomical Union
by
Kluwer Academic Publishers, P.O. Box 17, 3300 AA Dordrecht, The Netherlands.

Kluwer Academic Publishers incorporates
the publishing programmes of
D. Reidel, Martinus Nijhoff, Dr W. Junk and MTP Press.

Sold and distributed in the U.S.A. and Canada
by Kluwer Academic Publishers,
101 Philip Drive, Norwell, MA 02061, U.S.A.

In all other countries, sold and distributed
by Kluwer Academic Publishers Group,
P.O. Box 322, 3300 AH Dordrecht, The Netherlands.

Printed on acid-free paper

TABLE OF CONTENTS

FOREWORD

Solar and stellar photospheres constitute the layers most accessible to observations, forming the interface between the interior and the outside of the stars. The solar atmosphere is a rich physics laboratory, in which the whole spectrum of radiative, dynamical, and magnetic processes that tranfer energy into space can be observed. As the fundamental processes take place on very small spatial scales, we need high-resolution observations to explore them. On the other hand the small-scale processes act together to form global properties of the sun, which have their origins in the solar interior. The rapid advances in observational techniques and theoretical modelling over the past decade made it very timely to bring together scientists from east and west to the first IAU Symposium on this topic.

The physics of the photosphere involves complicated interactions between magnetic fields, convection, waves, and radiation. During the past decade our understanding of these generally small-scale structures and processes has been dramatically advanced. New instrumentations, on ground and in space, have given us new means to study the granular convection. Diagnostic methods in Stokes polarimetry have allowed us to go beyond the limitations of spatial resolution to explore the structure and dynamics of the subarcsec magnetic structures. Extensive numerical simulations of the interaction between convection and magnetic fields using powerful supercomputers are providing deepened physical insight. Granulation, magnetic fields, and dynamo processes are being explored in the photospheres of other stars, guided by our improved understanding of the solar photosphere.

Not only are we beginning to understand the relation between the small-scale processes and the large-scale structures, but it is also becoming increasingly clear that the sun cannot be properly separated into a "quiet" and an "active" part. The sun (and other stars for that matter) should rather be looked upon as a complicated but indivisible organism, whose numerous and interrelated properties exhibit both cyclic and secular variations, on short as well as long time scales. The symposium has considered the photosphere of the sun within this broader context, as an integral part of a larger system.

The symposium took place in Kiev, USSR, May 15-20, 1989, and was attended by more than 200 participants from 24 countries. The presentations included 20 invited review papers, 37 orally contributed papers, and more than 100 poster papers. The meeting was sponsored by IAU Commission 12 and cosponsored by Commissions 10 and 36. We are grateful for the travel support provided by the IAU and by the Academy of Sciences of the Ukrainian SSR.

The Scientific Organizing Committee consisted of V. Bumba, C.J. Durrant, D.F. Gray, E.A. Gurtovenko, R. Howard, V.N. Karpinsky, R. Muller, Å. Nordlund, R.W. Noyes, R.J. Rutten, H.C. Spruit, J.O. Stenflo (Chairman), H. Yoshimura, and C. Zwaan. I am grateful to the SOC members for their support. In particular I am indebted to Rob Rutten, who was one of the originators of the idea of having this type of a symposium in Kiev, and who has provided an important link between SOC and the organizers in Kiev.

The Local Organizing Committee had the following members: V.V. Botvinova, E.A. Gurtovenko (Chairman), E.V. Kononovich, R.I. Kostik, V.N. Krivodubsky, V.N. Obridko, S.N. Osipov, N.G. Shchukina, V.A. Sheminova, Y.L. Spirin. The local organization was a great success, and the meeting was characterized by an open, cordial, and stimulating atmosphere, which allowed for the development of many new scientific contacts and friendships. It is a great pleasure to express our thanks to the LOC Chairman, Ernest Gurtovenko, to Roman Kostik and the other members of the Main Astronomical Observatory, and to the observatory director, Yaroslav Yatskiv, for all hospitality shown to us. The excellent conference facilities at the Trade House in Kiev, with simultaneous translation between English and Russian, further contributed to the success of the symposium.

During the meeting it was decided to speak out on one of the most important issues of our time — the militarization of space. A letter with an appeal to President G. Bush and President M. Gorbachev to take every pertinent measure to prevent that outer space will become a base for placing nuclear or other weapons was endorsed by acclamation by the symposium participants. The letter also expresses concern about atmospheric and space pollution with nuclear waste and radiation, which threatens life on our planet and will make it more and more impossible to explore the universe by astronomical observations.

The proceedings have been prepared from camera-ready manuscripts. In the case of a number of papers, mainly those by Soviet authors, it was necessary to make extensive language corrections and to retype the entire manuscripts, without changing anything of the scientific contents. I am responsible for any errors that may have been introduced in this process. It is finally a pleasure to express my sincere thanks to Mrs. S. Weber and Mrs. M. Szigeti of the Institute of Astronomy of the ETH Zurich for their very extensive secretarial assistance in the preparation of these proceedings.

Zurich, August 1989   Jan Olof Stenflo

I. GLOBAL PROPERTIES OF THE PHOTOSPHERE

MODELS OF THE SOLAR PHOTOSPHERE

EUGENE H. AVRETT
Harvard-Smithsonian Center for Astrophysics
60 Garden Street
Cambridge, Massachusetts 02138, USA

ABSTRACT. This review summarizes the main properties of the available theoretical models of the solar photosphere and the semiempirical models based primarily on observed spectra in different wavelength regions. LTE and non-LTE semiempirical models are compared with LTE and non-LTE theoretical (radiative equilibrium) models calculated with and without convective energy transport. Atomic and molecular lines throughout the spectrum affect the model calculations in essential ways. The temperature-minimum region, considered as the upper boundary of the photosphere, is discussed in detail, in view of the substantial non-radiative cooling in this region. Results are shown to illustrate the relationship between average one-dimensional models and Nordlund's three-dimensional convection models.

1. Introduction

One of the principal ways of learning how the solar atmosphere varies as a function of depth is to observe the Sun at different wavelengths, since variations of the opacity with wavelength correspond to a change of the depth at which the emerging radiation is formed. Opacity variations with wavelength occur between different regions of the spectrum, and within spectral lines.

The variation of atmospheric parameters also can be studied by means of center-to-limb observations, since the intensity at a given wavelength seen near the limb emerges from a higher level in the atmosphere than that seen near disk center. As discussed by Nelson (1978), other observational constraints on models of the atmosphere are the brightness variations and motions observed in solar granulation.

The present review will deal principally with models determined from observed spectra at different wavelengths, since the widest range of depths can be studied in this way.

2. The Solar Spectrum

Figure 1 is a sketch of the observed brightness temperature T_b of the Sun as a function of wavelength between 1 cm and 10 nm. The lower panel indicates the principal sources of opacity at the depths where the observed radiation is formed. Here $T_b (\lambda)$, is the temperature of an isothermal atmosphere in LTE without scattering that gives the same central intensity (or flux from the entire disk) as is observed.

The observed radiation is emitted from the greatest depth in the atmosphere at 1.6 μm where the opacity has a minimum. At this wavelength the central brightness temperature is about 6800 K. The brightness temperature of the disk is about 500 K lower because of limb darkening.

J. O. Stenflo (ed.), Solar Photosphere: Structure, Convection, and Magnetic Fields, 3–22.
© *1990 by the IAU.*

For $\lambda > 1.6$ μm the opacity becomes larger as λ increases, unit optical depth occurs higher in the atmosphere, and T_b decreases, reaching a minimum value of about 4500 K at 150 μm. Here there is little center-to-limb variation and the disk and central brightness temperatures are the same (apart from the presence of active regions at various locations on the disk that are brighter than the average quiet Sun). The increase of T_b for $\lambda > 150$ μm (where the opacity continues to increase) is due to the increase of temperature in the chromosphere.

For $\lambda < 1.6$ μm the opacity also becomes larger as λ decreases, and T_b decreases to a minimum value of about 4400 K near 160 nm. The subsequent increase of T_b in the region $\lambda < 160$ nm again is due to the chromosphere rise in temperature. We will discuss only the photosphere in this review, from the deepest observable layers to the temperature minimum region.

Figure 2 shows the observed flux from the entire disk between 2 μm and 250 nm, and the corresponding disk brightness temperature. The maximum in T_b at 1.6 μm corresponds to only a slight inflection in the flux distribution at this wavelength. These are broad-band flux measurements that do not distinguish between regions having a well defined continuum and regions dominated by absorption lines.

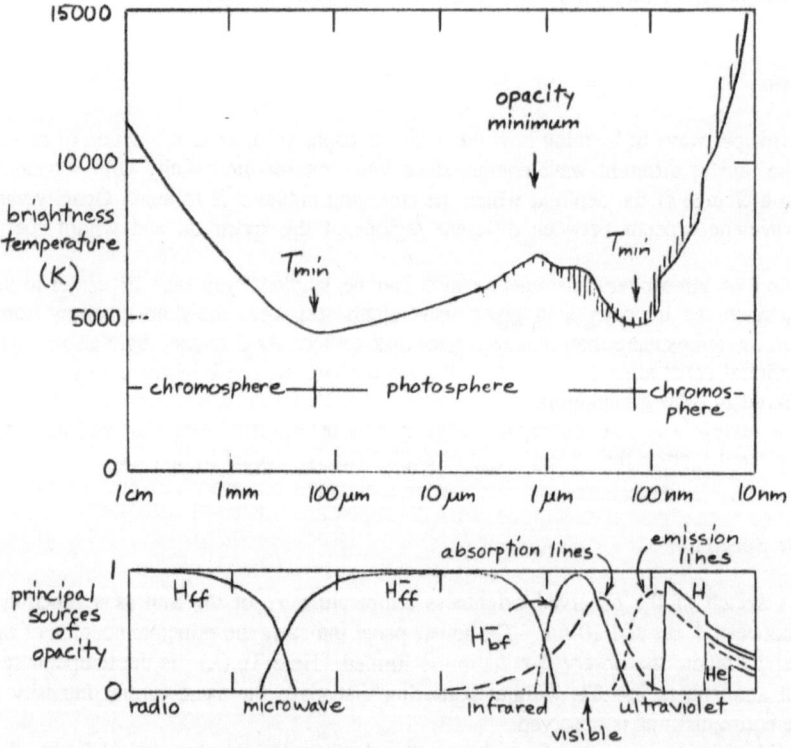

Figure 1. The approximate wavelength distribution of the solar brightness temperature from 1 cm to 10 nm, and the principal sources of opacity at the depths where the emitted radiation is formed.

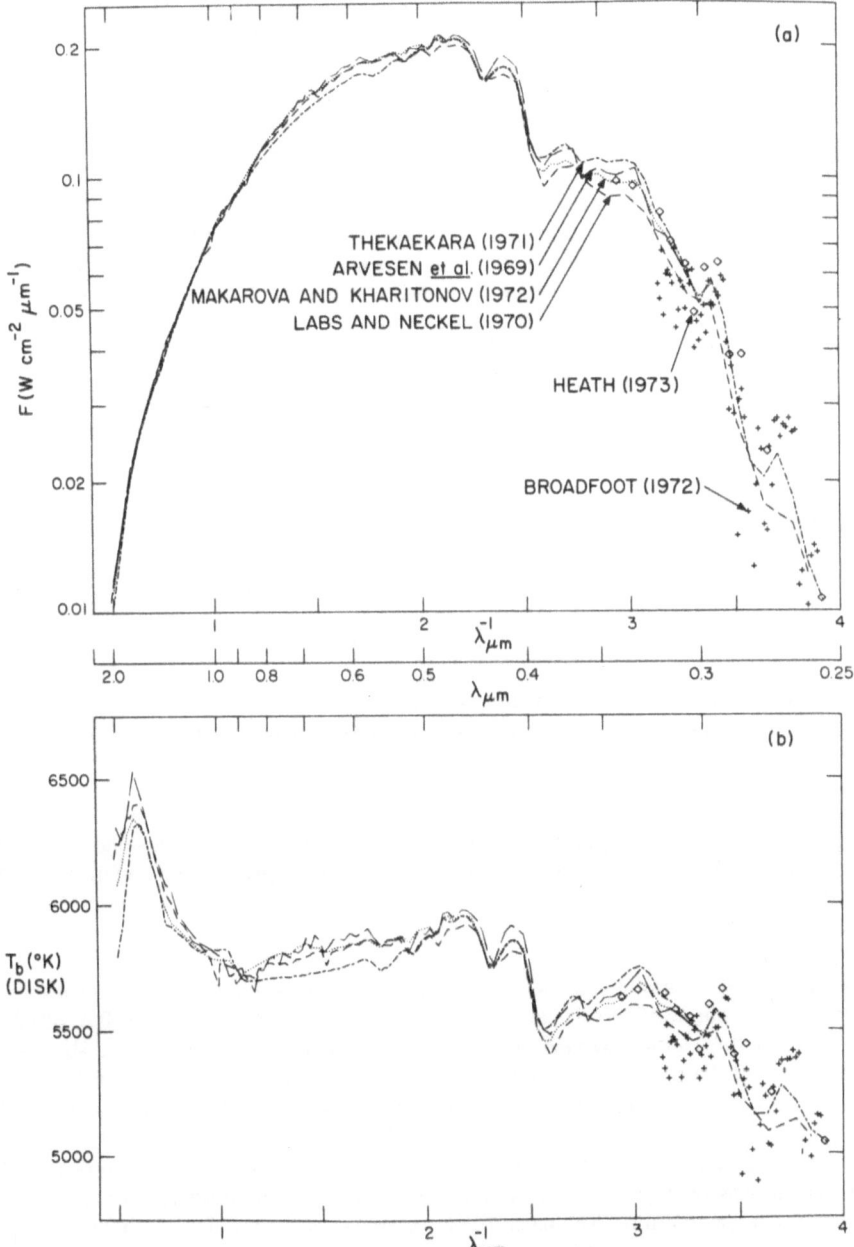

Figure 2. Observed flux in the wavelength range 250 nm - 2 μm (upper panel) and the corresponding disk brightness temperature (lower panel), from Vernazza, Avrett, and Loeser (1976).

Figure 3 shows two types of disk-center intensity distributions in the wavelength range from 1 μm to 200 nm: the mean broad-band distribution and the maximum intensities that occur between absorption lines. The corresponding disk-center brightness temperatures are shown in the lower panel.

The substantial differences between the mean and maximum values are due to the many lines in the spectrum. The observed spectrum in the range 380-420 nm appears in Figure 4. The broad features centered at 393.3 and 396.8 nm in Figure 4 are the K and H resonance lines of Ca II. For $\lambda > 400$ nm the maximum values follow the continuum distribution that would occur with zero line opacities, but for shorter wavelengths such a continuum is not as clearly defined from the observations. When we integrate over wavelength at each depth in the atmosphere to calculate photoionization rates and radiative cooling rates, it is important to accurately account for the many lines in the spectrum, such as those in Figure 4.

Figure 5 shows the continuation of Figure 3 from 200 to 125 nm. The observed central brightness temperature between spectral lines has a minimum of about 4400 K between 160 and 170 nm. The lines in the spectrum change from absorption lines in the range $\lambda > 200$ nm to emission lines in the range $\lambda < 170$ nm.

3. Atmospheric Models

In order to interpret the observed spectrum, we need a model that describes the temperature, density, and other atmospheric parameters as functions of depth. Also, we need to be able to calculate the spectrum from a given set of atmospheric parameters. If we can compute a spectrum that is in good agreement with observations, then the model should provide a description of the actual conditions in the atmosphere. The number of model parameters to be determined is generally much smaller than the number of features to be matched in the observed spectrum (cf. Figure 4), so that our confidence in the model increases in proportion to the agreement between computed and observed spectra.

There are two main types of model solar atmospheres: theoretical models and semiempirical models, which differ according to the way the temperature distribution is obtained. In the first case the temperature distribution is calculated according to physical theory with only a few basic parameters chosen to match properties of the solar atmosphere. In the second case the temperature distribution is adjusted by trial and error until the computed spectrum best agrees with observations.

A theoretical one-dimensional model in radiative and hydrostatic equilibrium is characterized by three basic parameters: the solar effective temperature which gives the total flux of energy passing through the atmosphere, the surface gravity, and the relative elemental abundances. Standard values of the first two parameters (from Allen 1973) are $T_{eff} = 5770$ K and $g = 2.740 \times 10^4$ cm s^{-2}. The best solar abundances are those given recently by Anders and Grevesse (1989).

Given such parameters, the goal of the modeler is to develop a computational procedure that includes all the essential physical effects that give rise to the observed spectrum. This is a major ongoing research effort. The most recent theoretical models of the solar photosphere are those of Kurucz (1979) and Anderson (1989).

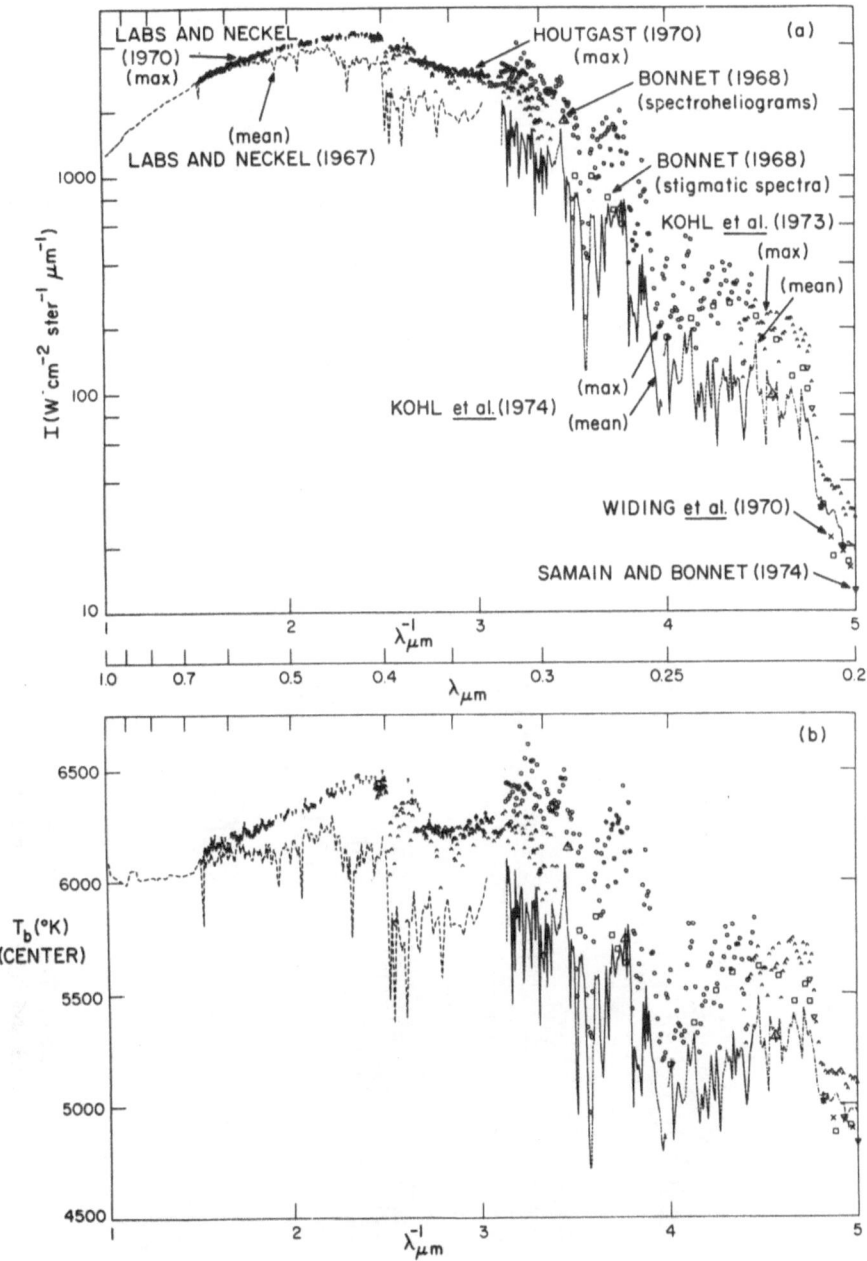

Figure 3. Central-intensity observations in the wavelength range 200 nm - 1 μ (upper panel) and the corresponding central brightness temperature values (lower panel), from Vernazza, Avrett, and Loeser (1976).

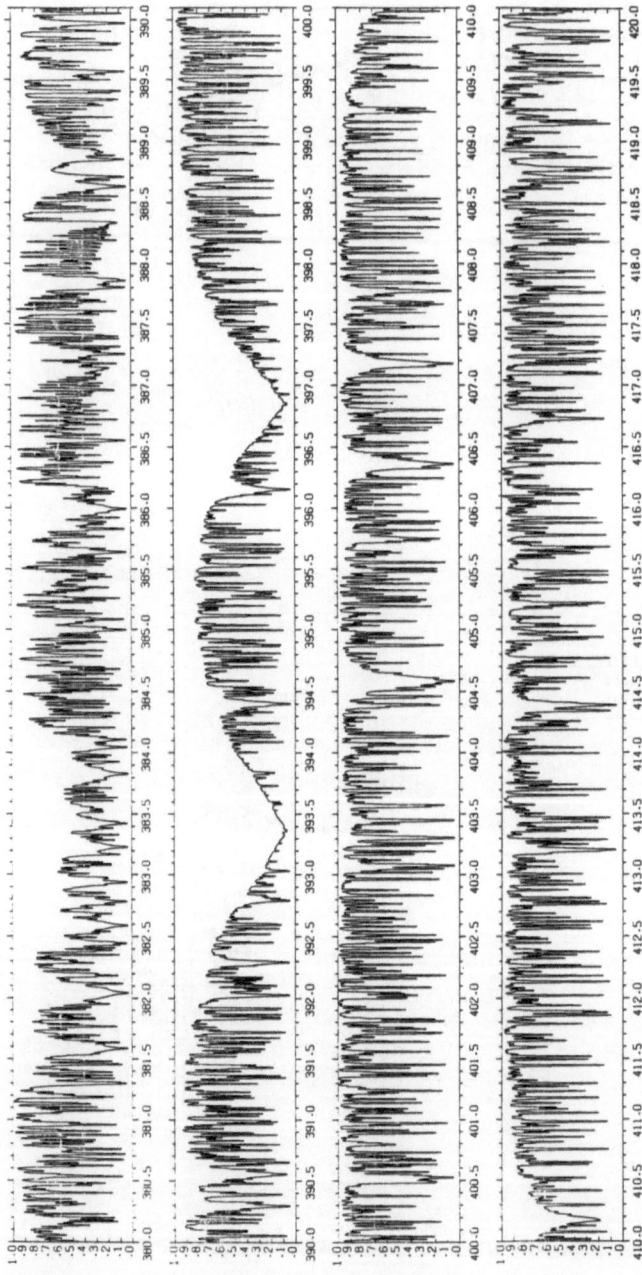

Figure 4. The solar flux between 380 and 420 nm, including the Ca II K and H profiles at 393.3 and 396.8 nm, from Kurucz, Furenlid, Brault, and Testerman (1984). The flux is plotted vs. wavelength in nm on a linear scale between zero and a pseudo-continuum level fitted to the observed flux maxima at 378 and 402 nm.

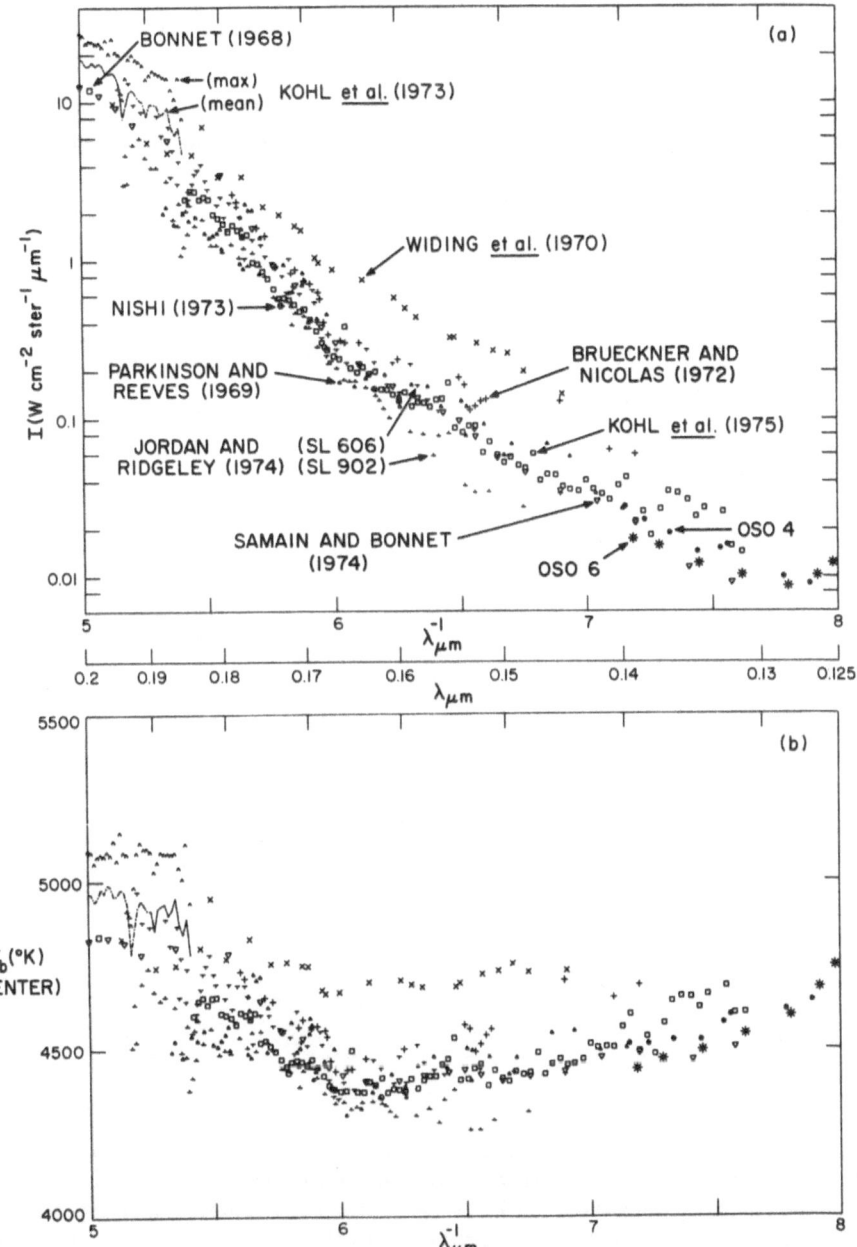

Figure 5. Central-intensity observations in the wavelength range 125-200 nm (upper panel) and the corresponding central brightness temperature values (lower panel), from Vernazza, Avrett, and Loeser (1976).

Both of these models attempt to include the effects caused by millions of atomic and molecular lines in the spectrum. Kurucz adopts the simplifying assumption of LTE (local thermodynamic equilibrium), while Anderson treats the lines throughout the spectrum by a non-LTE statistical method.

Kurucz provides a grid of LTE model atmospheres for effective temperatures between 5500 K and 50,000 K, for gravities from the main sequence down to the radiation pressure limit, and for solar, 1/10 solar, and 1/100 solar abundances. The models were computed with a statistical distribution-function representation of the opacity of almost 10^6 atomic and molecular lines. Kurucz includes convective energy transport using a mixing length theory (see Mihalas 1978, p. 187) in which the mixing length is assumed to be 2 times the pressure scale height.

Anderson also includes a very large number of line transitions, but rather than assuming LTE, he assigns non-LTE effects to lines throughout the spectrum based on a solution of the combined statistical equilibrium and radiative transfer equations for Fe II. The models given by Anderson do not include convective energy transport.

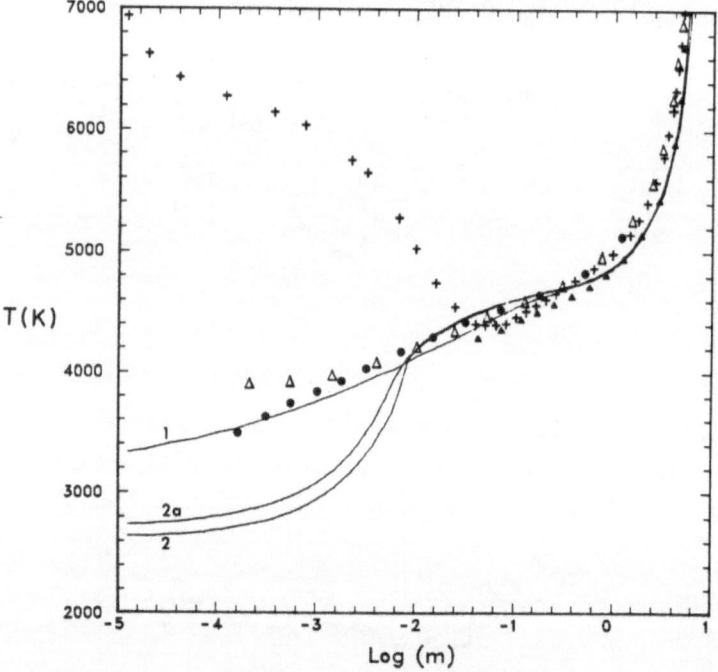

Figure 6. Temperature distributions for the theoretical models of Anderson (1989) and Kurucz (1979) compared with three semiempirical temperature distributions. *Curve 1*: Anderson's LTE theoretical (radiative equilibrium) model. *Curves 2 and 2a*: Anderson's non-LTE theoretical models. *Filled triangles*: Kurucz's LTE theoretical model. *Open triangles*: The LTE semiempirical model of Holweger and Müller (1974). *Plus signs*: The non-LTE semiempirical model of Maltby *et al.* (1986). *Filled circles*: The LTE semiempirical model of Ayres and Testerman (1981) based on carbon monoxide lines.

Figure 6 shows the temperature as a function of the column mass m(g cm^{-2}) for the theoretical models of Anderson and Kurucz compared with three semiempirical temperature distributions (discussed below).

The filled triangles indicate Kurucz's LTE solar model. The curve labeled 1 is Anderson's LTE model. Anderson and Kurucz use different approximations for the line opacities, and they adopt slightly different abundances, which may account for the small differences that occur near log m = -1.

The curve labeled 2 is Anderson's non-LTE model. (His model 2a is the same except that the metallic L → L transitions and the molecular transitions are in LTE.) We refer the reader to Anderson's paper for a detailed discussion of his non-LTE theoretical models.

We now consider the semiempirical models shown in Figure 6. The temperature distribution designated by plus signs in this figure is the model given by Maltby *et al.* (1986), which is an improved version of an earlier model by Vernazza, Avrett, and Loeser (1981). In this case the temperature distribution has not been determined theoretically from the constraint of radiative equilibrium, but has been adjusted to match observations. Otherwise, the semiempirical and theoretical models are calculated in similar ways.

Figure 1 showed that the observed brightness temperature increases for λ > 150 μm and for λ < 160 nm. The chromospheric temperature rise for log m < -1.4 in the Maltby *et al.* semiempirical model accounts for the brightness temperature rise in these high-opacity wavelength ranges. This model is based on detailed non-LTE calculations for the principal bround-free continua and for representative strong lines. Scattering is assumed for weak lines to simulate non-LTE effects in the chromospheric layers.

The Holweger and Müller (1974) model shown in Figure 6 (open triangles) is an updated version of the earlier LTE model of Holweger (1967), based on a fit to the observed equivalent widths of 900 selected lines, and on central-intensity and center-to-limb observations throughout the visible spectrum. This model shows no chromospheric temperature rise because the lines were assumed to be formed in LTE (LTE line opacities, and each line source function S assumed equal to the Planck function B), and because the analysis was based on absorption lines that did not show evidence of a chromospheric temperature rise. Thus the analysis did not include the strong resonance lines. However, lines were included such as Mg I λ5173 for which the central intensities are known to depart from LTE (*cf.* Mauas, Avrett, and Loeser 1988). Since this model is consistent with an extensive and reliable data set, the temperature decrease shown in Figure 6 for log m < -1.4 suggests that this is a decrease not of the kinetic temperature but of a line excitation temperature that is similar for all the lines used in the analysis. Thus, within the limits of uncertainty in central intensity measurements, one can match observations 1) without a chromospheric temperature increase, assuming "incorrectly" that S = B, or 2) with a temperature increase, assuming "correctly" that S decreases relative to B with decreasing gas density.

The semiempirical model of Ayres and Testerman (1981), shown in Figure 6 (filled circles), is based on observations of the fundamental and first overtone bands of carbon monoxide. They find that close to the limb (μ = 0.1) the strongest CO lines in the fundamental band (4.3 to 7.5 μm) have central brightness temperatures as low as 3700 K, and that if these lines are formed in LTE, the temperature decreases monotonically as indicated in Figure 6 with no chromospheric increase. Ayres and Wiedemann (1989) have carried out a detailed study of non-LTE effects in the CO fundamental lines. They find negligible departures from LTE for these lines and conclude that the low line-center brightness temperatures observed near the limb cannot be explained by non-LTE scattering effects, but must indicate the presence of kinetic temperatures as low as 3700 K. Since other observations clearly indicate that the temperature has a chromospheric rise, such as shown by the crosses in Figure 6, the chromospheric rise evidently does not occur everywhere on the solar surface. Sufficient cool material must be present to account for the low brightness temperatures seen in these CO lines close to the limb.

Above the temperature minimum region in Figure 6 there is a rough agreement between 1) Anderson's LTE theoretical model (curve 1), 2) the LTE semiempirical model based on lines that should be partly affected by non-LTE scattering (open triangles), and 3) the LTE semiempirical model based on the CO lines (filled circles). This agreement may be only fortuitous since a) the non-LTE theoretical model (curve 2) should be a better representation of the radiative-equilibrium atmosphere, without mechanical heating, than the LTE theoretical model, and b) above the temperature minimum region, the non-LTE semiempirical model should be a better representation of the observed atmosphere, with chromospheric heating, than the corresponding LTE model.

Further discussion of the chromosphere is inappropriate in this review of photospheric models. The reader should consult a recent paper by Anderson and Athay (1989) to see the results of adding mechanical heating in the non-LTE theoretical model calculations to produce a chromospheric temperature rise in approximate agreement with the non-LTE semiempirical model in Figure 6.

4. The Temperature Minimum

The Maltby *et al.* (1986) model in Figure 6 can be considered the most recent non-LTE semiempirical model in a sequence starting with the Utrecht Reference Model of the Photosphere and Low Chromosphere (Heintze, Hubenet, and de Jager 1964). The 1964-1986 lineage is indicated in Table 1. This table also gives the minimum temperature adopted for each model.

Table 1

Reference	T_{min} (K)
Heintze, Hubenet, and de Jager (1964)	4500
Gingerich and de Jager (1968)	4600
Gingerich, Noyes, Kalkofen, and Cuny (1971)	4170
Vernazza, Avrett, and Loeser (1973)	4100
Vernazza, Avrett, and Loeser (1976)	4150
Vernazza, Avrett, and Loeser (1981)	4170
Avrett, Kurucz, and Loeser (1984); Avrett (1985); Maltby, Avrett, Carlsson, Kjeldseth-Moe, Kurucz, and Loeser (1986)	4400

The Utrecht reference model was an improved version of an earlier photospheric model of Hubenet (1960). The temperature minimum value of 4500 K in this model was derived by de Jager (1963) from an investigation of the ultraviolet continuum and line spectrum. The model has a chromospheric temperature gradient obtained from eclipse observations by Heintze (1965).

In 1967 an international meeting was held at the Bilderberg Hotel near Arnhem, Netherlands, to establish a new reference model of the solar photosphere and low chromosphere. The papers from this meeting appear in *Solar Physics*, Vol. 3, No. 1, 1968, and are reprinted as a book (de Jager 1968). Agreement was reached on a model, called the Bilderberg Continuum Atmosphere (BCA), representing the continuum observations available at that time (Gingerich and de Jager 1968). The value of T_{min} = 4600K for this model was based on 1) the color temperature at 1600Å observed by Tousey (1963) and his colleagues, and 2) the analysis of the UV carbon monoxide spectrum by Rich (1966) who concluded that T_{min} = 4500 ± 100K.

At this meeting, Athay and Skumanich (1968) argued that $T_{min} \leq 4200$ K, based on studies of the Ca II H and K lines. Later work showed that their analysis underestimated T_{min} for two basic reasons. First, they used observed H and K profiles from Goldberg, Mohler, and Müller (1959) that indicated a minimum K_1 radiation temperature of 4200 K. (K_1 is the intensity minimum in the wing of the Ca II K line just outside the peak that occurs near line center. The K_1 brightness temperature is approximately the temperature minimum value. See Noyes and Avrett 1987.) More recent measurements have shown that this value is closer to 4400 K. Second, they assumed that the lines were formed according to the theory of complete frequency redistribution, which led them to choose $T_{min} \sim 4000$ K to obtain a minimum K_1 brightness temperature of 4200 K. As discussed below, it was realized five years later that the near wings of the H and K lines must be treated according to the theory of partial frequency redistribution, for which these two temperatures are roughly the same.

An improved version of the BCA model, called the Harvard-Smithsonian Reference Atmosphere (HSRA), was proposed by Gingerich et al. (1971) based on a variety of new photospheric and chromospheric observations. The minimum temperature was lowered from the BCA value of 4600 K to 4170 K for the following reasons. 1) New rocket observations of the 1650 Å region by Parkinson and Reeves (1969) indicated a minimum brightness temperature of 4400 K or smaller, compared with the 4600 K minimum value used for the BCA. 2) Cuny (1971) carried out a detailed study of how departures from LTE affect the continuum spectrum in the 1520-1680 Å region, and found that silicon and other neutral atoms are underpopulated relative to LTE in the temperature minimum region so that the minimum temperature should be lower, by perhaps 200 K, than the minimum observed brightness temperature. 3) Eddy, Lena, and MacQueen (1969) reported a value of about 4300 K for the brightness temperature at 300 μm. To obtain such a value from the model calculations, T_{min} needed to be in the range 4100-4200 K.

Vernazza, Avrett, and Loeser (1973, 1976, 1981) carried out more extensive non-LTE model calculations, incorporated further observational data, and extended the HSRA model into the chromosphere-coronal transition region. Only minor adjustments for T_{min} were made, since there were no changes in the ultraviolet continuum data used to define the minimum temperature in these models.

During the mid 1970s, however, it became clear that the Ca II and Mg II resonance lines imply higher values of T_{min}. As reviewed by Linsky (1985), the importance of partial frequency redistribution in the wings of the Ca II H and K lines was first recognized at IAU Colloquium No. 19 on Stellar Chromospheres (Jordan and Avrett 1973). Milkey and Mihalas

(1974) studied partial redistribution effects in the Mg II h and k wings and concluded that T_{min} ≥ 4400 K. Shine, Milkey, and Mihalas (1975) carried out a similar study of the Ca II H and K lines and found that the shape and center-to-limb variation of the computed profiles matched observations much better than the profiles obtained with complete redistribution, and that they could match the absolute K_1 intensity derived from the observations of White and Suemoto (1968) with a minimum temperature of about 4450 K. Ayres and Linsky (1976) carried out a partial redistribution analysis of the Ca II profiles observed by Brault and Testerman (1972) and the Mg II profiles of Kohl and Parkinson (1976). They found T_{min} = 4450 ± 130 K from the Ca II lines and T_{min} = 4500 (+80, -110) K from the lines of Mg II.

In 1978 the minimum brightness temperature in the far infrared was found to be higher than the value reported by Eddy, Lena, and MacQueen and others in 1969 (see the review by Mankin 1977). Rast, Kneubühl, and Müller (1978) determined that the minimum brightness temperature is 4530 (+100,-150) K near 130 μm.

As pointed out by Vernazza, Avrett, and Loeser (1981), comparison with the ultraviolet center-to-limb observations of Samain (1980) shows that at wavelengths where the continuum intensity originates on the photospheric side of the temperature minimum the average quiet-Sun model with T_{min} = 4170 K predicts less limb darkening than observed, while at wavelengths where the continuum intensity originates on the chromospheric side, less limb brightening than observed. Thus, the computed continuum source function is too flat on both sides of the temperature minimum. At these wavelengths the computed non-LTE continuum source function S is only loosely coupled to the Planck function B, and is much flatter than B in the minimum region. Figure 7 shows the variations of S and B with height h (km) at λ = 160.5 nm from model C of Vernazza, Avrett, and Loeser (1981). The computed intensities at disk center (μ = 1) and near the limb (μ = 0.3) are indicated in Figure 7 by I (1.0) and I (0.3), respectively.

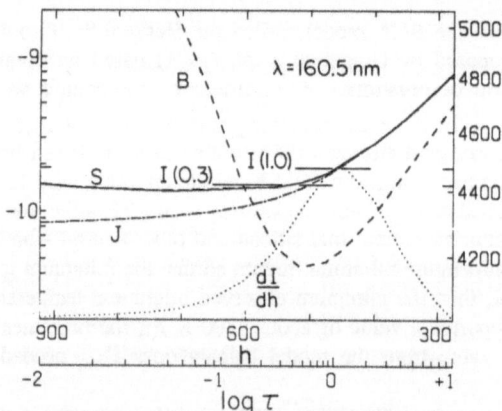

Figure 7. The Planck function B, the continuum source function S, the mean intensity J, and the contribution per unit height dI/dh to the emergent intensity at μ = 1, plotted as functions of height and the continuum optical depth, all at λ = 160.5 nm. The computed values of I(1.0) and I(0.3), the emergent intensities at μ = 1.0 and 0.3, are indicated. The units for B, S, J, and I on the left are ergs cm^{-2} s^{-1} sr^{-1} Hz^{-1}. The corresponding brightness temperature scale (in K) is given on the right. From Vernazza, Avrett, and Loeser (1981).

These computed intensities are in approximate agreement with observations. If S were closer to B near $\log \tau = 0$, we would have to increase T_{min} to keep the computed intensities the same. Neutral silicon and iron are the main contributors to the bound-free absorption and emission near 1600 Å. Thus it is important to examine the accuracy of the computed departures from LTE in Si I and Fe I.

A major uncertainty in the calculations is how to deal properly with the many lines throughout the spectrum. The lines are important 1) in calculating the photoionization-rate integrals that affect the ionization equilibrium of Si, Fe, Mg, C, Al, and other atoms, and 2) in calculating the spectrum for direct comparison with observations. The 1981 model calculations were repeated by Avrett, Kurucz, and Loeser (1984) using the line opacity tables available from R. Kurucz that included more than 1.7×10^7 atomic and molecular lines. The various photoionization rates were recomputed using the total line opacity (as a function of temperature and pressure) sampled at 7,000 wavelengths between 1490 and 6050 Å. See Avrett (1985) for further details. They found as a result that the ionization equilibrium of Si, Fe, and Mg in the temperature-minimum region is much closer to LTE than before, e.g., the level-1 Si I departure coefficient at h = 500 km changed from 0.3 to 0.8. In the new calculations that include the spectral lines more realistically, the silicon continuum source function at h = 500 km and $\lambda \sim$ 1600 Å is only about 30% larger than the Planck function, vs. 3 to 4 times larger in the Vernazza *et al.* calculations, and in the similar earlier calculations by Cuny (1971).

Consequently, T_{min} had to be raised from 4170 to 4400 K to maintain agreement between the computed and observed ultraviolet intensities. This modified version of the average quiet-Sun model of Vernazza, Avrett, and Loeser (1981) is tabulated in Appendix A of Maltby *et al.* (1986). (As discussed in the next section, the model given by Maltby *et al.* also has been adjusted in the deepest layers to match available observations as well as possible.)

The 4400 K minimum temperature is consistent 1) with the 130 μm observations of Rast, Kneubühl, and Müller (1978) and subsequent observations by Degiacomi, Kneubühl, and Huguenin (1985) at 50, 80, and 200 μ, 2) with observations in the ultraviolet minimum region (see Cook, Brueckner, and Bartoe 1983; Foing and Bonnet 1984), and 3) with Ca II and Mg II line observations (see Avrett 1985 for a comparison with the H line profiles observed with high spatial resolution by Cram and Dame 1983). As pointed out by Rutten (1988) in a recent review of the non-LTE formation of iron lines in the solar photosphere, the Maltby *et al.* non-LTE semiempirical model finally agreed with the Holweger and Muller (1974) LTE semiempirical model (except for the chromospheric temperature rise) after it was found that some of the earlier non-LTE effects arose artificially in the computer modeling rather than in the Sun.

The only observations that are clearly inconsistent with the Maltby *et al.* model, in the upper photosphere-low chromosphere region, are the central brightness temperature values of 3700 K deduced from the carbon monoxide lines observed near the limb by Ayres and Testerman (1981). These lines were observed earlier by Noyes and Hall (1972) who found a strong 5-minute oscillation of the CO central intensities near disk center which they suggested might be due to adiabatic cooling associated with the oscillation. The same effect could be due to expansion cooling of rising granules as suggested by Nordlund (1985). This work is discussed briefly in Section 7.

It should be noted that in Figure 6 the non-LTE semiempirical temperatures (plus signs) lie below the non-LTE radiative equilibrium temperatures (curve 2) in the minimum region -1.5 < log m < -0.5, and lie above the radiative equilibrium values deeper in the photosphere (log m > -0.5). This means that the flux derivative or net radiative cooling rate

$$\Phi = 4\pi \int \kappa_\nu \, (S_\nu - J_\nu) \, d\nu \quad (\text{ergs cm}^3 \text{ s}^{-1})$$

calculated for the semiempirical model is negative in the minimum region and positive deeper in the photosphere. ($\Phi = 0$ for the radiative equilibrium model.)

In the chromosphere Φ is also positive, and corresponds to the mechanical heating in ergs cm^{-3} s^{-1} responsible for the chromospheric temperature rise. Since the gas density is so much greater in the photosphere than in the chromosphere, small photospheric departures from radiative equilibrium can cause Φ to be numerically larger in the photosphere than in the chromosphere where radiative equilibrium no longer applies. Results given by Avrett (1985) show that the absolute value of $\int \Phi_{min}$, the integral of Φ over the minimum region, exceeds $\int \Phi_{higher}$, the integral over the chromosphere, transition region, and corona. Thus the amount of mechanical energy extracted from the temperature minimum region is larger than the amount of energy needed to heat the chromosphere and corona. The quantity $\int \Phi_{lower}$ defined as the integral over the region log m > -0.5 has not been evaluated for the Maltby *et al.* model, but it appears to be much larger than $|\int \Phi_{min}|$. These departures from radiative equilibrium may be the result of either oscillations or convective motions in the atmosphere.

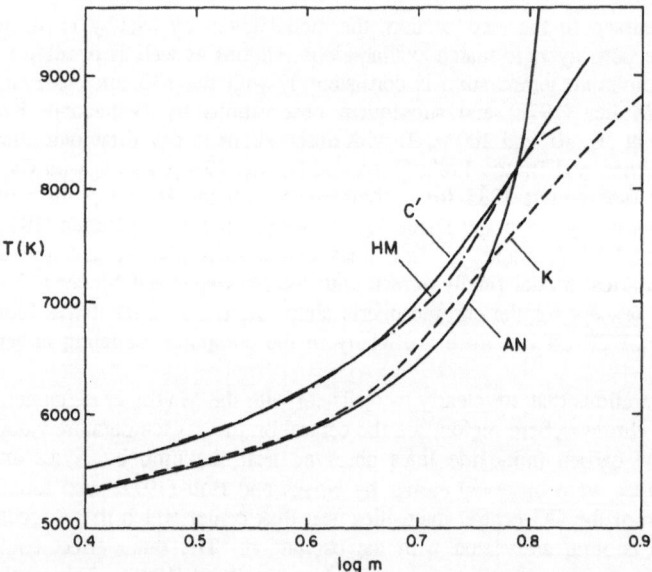

Figure 8. Temperature distributions in the deeper photospheric layers. HM: Holweger and Müller (1974). C′: Maltby *et al.* (1986). AN: Anderson (1989). K: Kurucz (1979).

5. The Deeper Layers

Figure 8 shows a comparison of four temperature models in the deeper part of the photosphere (log m > 0.4). HM is the LTE semiempirical model of Holweger and Muller (1974). C′ is the non-LTE semiempirical model of Maltby *et al.* (1986). At these depths LTE is a good approximation; these two models differ only as the result of a new attempt to calibrate the relative continuum observations of Pierce (1954) in the 1.3-2.5 μm region. See Maltby *et al.*, Appendix A, for details. (Note: this C′ model replaces model C of Vernazza, Avrett, and Loeser 1981.)

AN is the theoretical (i.e., radiative equilibrium) non-LTE model of Anderson (1989) which does not include convective energy transport. K is the theoretical LTE model of Kurucz (1979) which does include convective energy transport using a mixing length theory. Since LTE is a good approximation here, these two models are essentially the same except for the effects of convective transport.

We draw two conclusions from these results. 1) The mixing length approximation adopted by Kurucz (1979) gives a temperature gradient consistent with observations in the deepest layers, and 2) the semiempirical temperature distribution lies 200-500 K above the theoretical temperature distribution.

We stress that these are the properties obtained from average, one-component, static models. It may be that such results are as consistent as can be expected in view of the complex structures and motions that are observed.

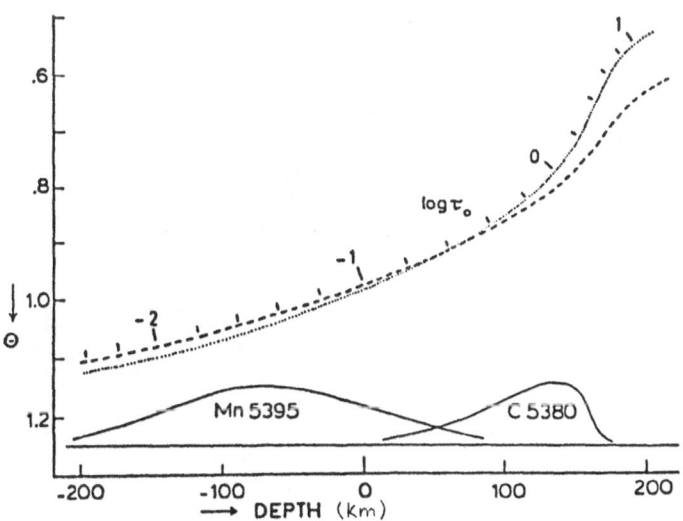

Figure 9. $\Theta = 5040/T$ vs. depth in km determined by Elste (1985) for regions with little or no magnetic field (dotted curve) and for plage regions with a magnetic field (dashed curve). τ_0 is the continuum optical depth at 5000 Å.

6. Two-Component Models

Nelson (1978) constructed a two-dimensional model of the photosphere including a physical model of granulation. The temperature distribution was adjusted so that the sum of radiative and convective fluxes remained constant with depth. He concluded that a one-dimensional model which reproduces the observed average intensities and center-to-limb behavior underestimates the temperature gradient in the deep photosphere and overestimates the amount of convective flux penetrating into the visible layers. His mean temperature distribution in the deep photosphere (T > 7000 K) has a steeper gradient than the semiempirical models in Figure 8.

Two recent studies provide insight into the relative photospheric temperature gradients in regions with and without magnetic fields. Elste (1985) derived the temperature gradients at two separate photospheric depths by observing the equivalent widths of the high excitation carbon line at 5380 Å and the ground-level manganese line at 5395 Å. This manganese line also serves as an indicator of longitudinal magnetic fields. Figure 9 indicates the depths at which these two lines are formed and the temperature distributions obtained from regions with little or no magnetic field (dotted curve) and from weak plage regions having a magnetic field (dashed curve). Here $\Theta = 5040/T$. It is well known that plage regions are brighter than quiet regions based on the radiation emitted from the outer photosphere and chromosphere. This work shows that plage regions have flatter temperature gradients in the deep photosphere.

Similar results were found by Foukal, Little, and Mooney (1989) from observations of the deepest observable layers at 1.63 μm. They found that facular flux tubes which are brighter than their surroundings in the visible continuum are darker than surrounding regions at 1.63 μm.

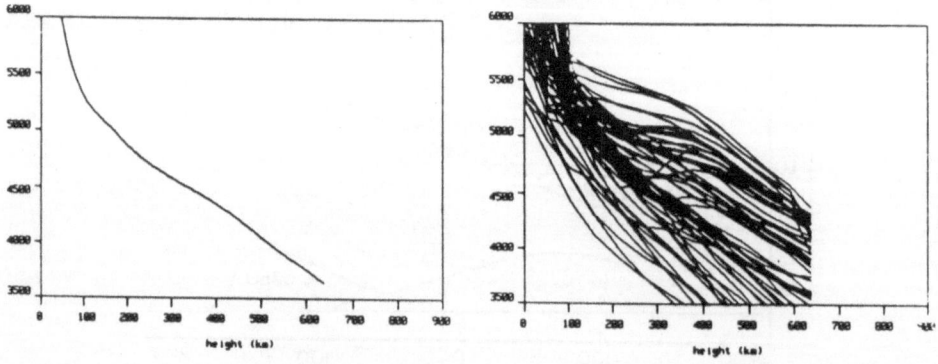

Figure 10. Results from the three-dimensional simulation of Nordlund (1985). *Left*: the horizontally averaged temperature distribution. *Right*: a selection of the individual distributions $T_{x,y}(h)$ that contribute to the average.

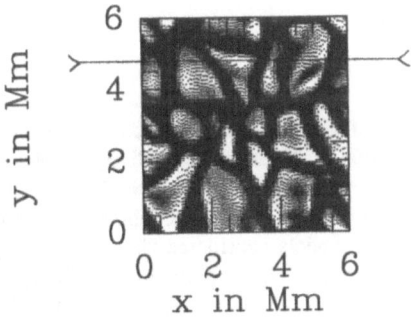

Figure 11. The calculated continuum intensity at 5000 Å in the normal direction as a function of position on the solar surface from one of Nordlund's convection simulations. From A. van Ballegooijen (private communication).

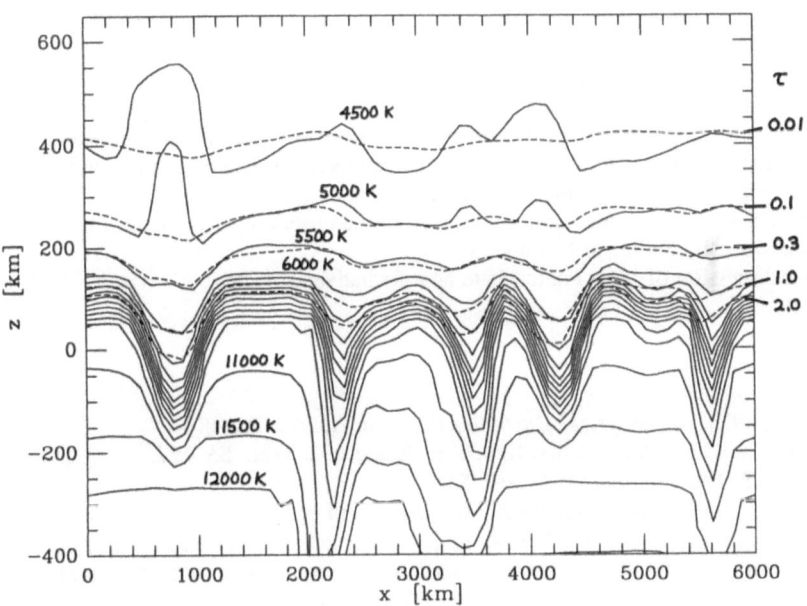

Figure 12. Temperature and optical depth contours in the vertical plane corresponding to the line indicated in Figure 11. From A. van Ballegooijen (private communication).

7. Three-Dimensional Simulations

A comprehensive review of the hydrodynamics of solar granulation was given by Nordlund (1985). (See also his chapter in this volume.) His three-dimensional simulations of granular convection give a very wide range of the calculated temperature as a function of height at different positions x,y on the disk. Figure 10 shows on the right a selection of his calculated $T_{x,y}(h)$ distributions, while the horizontally averaged temperature distribution is shown on the left. This range of values appears to be far beyond the limits of uncertainty indicated by the various semiempirical and theoretical models considered above. The range of temperature distributions appears exaggerated only because all are plotted on a common height scale. Because the H^- opacity is highly sensitive to temperature, the various curves in Figure 10 would appear much closer together if they were plotted on a common optical depth scale.

In the summer of 1988, Nordlund kindly gave this author a detailed tabulation of the results available at that time. Figure 11 and 12 were produced from those results by A. van Ballegooijen (private communicaton).

Figure 11 shows the calculated convective pattern in the visible continuum over a 6000 x 6000 km area near disk center. The horizontal line drawn across the upper part of this figure has been chosen to cross a number of cells and dark lanes. Consider a vertical cut into the atmosphere along this line. Figure 12 is a plot of the temperature contours in this vertical plane. Note that the horizontal scale extends over 6000 km as in Figure 11, but that the height scale ranges over only 1000 km. Temperature contours are given every 500 K between 4500 and 12,000 K. The dark lanes in Figure 11 are clearly apparent in Figure 12 as regions in which low temperatures extend deeply into the atmosphere. On an optical depth scale, however, these depressions are reduced. The dashed lines in Figure 12 are the optical depth contours for $\tau = 0.01$, 0.1, 0.3, 1, and 2 (in the continuum at 5000 Å). Were it not for the tendency for the τ contours to roughly follow the temperature contours, the computed contrast from such structures would be much greater than observed.

Further details of these simulations are given in the chapter by Nordlund in this volume and in a recent paper by Stein and Nordlund (1989).

Further discussion of "The Photosphere as a Radiative Boundary" is given by Anderson and Avrett (1989).

References

Allen, C.W. 1973, *Astrophysical Quantities* (London: Athlone Press).

Anders, E., and Grevesse, N. 1989, *Geochim. Cosmochim. Acta.*, **53**, 197.

Anderson, L.S. 1989, *Astrophys. J.*, **339**, 558.

Anderson, L.S., and Athay, R.G. 1989, *Astrophys. J.*, in press.

Anderson, L. S. and Avrett, E. H. 1989 in *The Solar Interior and Atmosphere*, ed. A.N. Cox, W.C. Livingston, and M. Matthews (Arizona: University of Arizona Press), in press.

Athay, R.G., and Skumanich, A. 1968, *Solar Phys.*, **3**, 181.

Avrett, E.H. 1985, in *Chromospheric Diagnostics and Modelling*, ed. B.W. Lites, (Sunspot, NM: National Solar Observatory), p. 67.

Avrett, E.H., Kurucz, R.L., and Loeser, R. 1984, *Bull. Amer. Astron. Soc.*, **16**, 450.

Ayres, T.R., and Linsky, J.L. 1976, *Astrophys. J.*, **205**, 874.

Ayres, T.R., and Testerman, L. 1981, *Astrophys. J.*, **245**, 1124.

Ayres, T.R., and Wiedemann, G.R. 1989, *Astrophys. J.*, **338**, 1033.

Brault, J., and Testerman, L. 1972, *Preliminary Edition of the Kitt Peak Solar Atlas* (Tucson: Kitt Peak National Observatory).

Cook, J.W., Brueckner, G.E., and Bartoe, J.-D.F. 1983, *Astrophys. J.*, **270**, L89.

Cram, L.E., and Dame, L. 1983, *Astrophys. J.*, **272**, 355.

Cuny, Y. 1971, *Solar Phys.*, **16**, 293.

Degiacomi, C.G., Kneubühl, F.K., and Huguenin, D. 1985, *Astrophys. J.*, **298**, 918.

deJager, C. 1963, *Bull. Astron. Inst. Neth.*, **17**, 209.

deJager, C. 1968, *The Structure of the Quiet Photosphere and the Low Chromosphere* (Dordrecht: Reidel).

Eddy, J.A., Lena, P.J., and MacQueen, R.M. 1969, *Solar Phys*, **10**, 330.

Elste, G. 1985, in *Theoretical Problems in High Resolution Solar Physics*, ed. H.U. Schmidt, Max Planck Institut für Astrophysik, MPA 212, p. 185.

Foing, B., and Bonnet, R.M. 1984, *Astrophys. J.*, **279**, 848.

Foukal, P., Little, R., and Mooney, J. 1989, *Astrophys. J.*, **336**, L33.

Gingerich, O., and deJager, C. 1968, *Solar Phys*, **3**, 5.

Gingerich, O., Noyes, R.W., Kalkofen, W., and Cuny, Y. 1971, *Solar Phys.*, **18**, 347.

Goldberg, L., Mohler, O.C., and Müller, E. 1959, *Astrophys. J.*, **129**, 119.

Heintze, J.R.W. 1965, *Rech. Obs. Astron. Utrecht*, **17** (2).

Heintze, J.R.W., Hubenet, H., and deJager, C. 1964, *Bull. Astron. Inst. Neth.*, **17**, 442.

Holweger, H. 1967, *Zeitz. f. Astrophysik*, **65**, 365.

Holweger, H., and Müller, E.A. 1974, *Solar Phys.*, **39**, 19.

Hubenet, H. 1960, *Rech. Astr. Obs. Utrecht*, **15** (1).

Jordan, S.D., and Avrett, E.H., eds. 1963, *Stellar Chromospheres*, NASA SP-317.

Kohl, J.L., and Parkinson, W.H. 1976, *Astrophys. J.*, **205**, 599.

Kurucz, R.L. 1979, *Astrophys. J. Suppl.*, **40**, 1.

Linsky, J.L. 1985, in *Progress in Stellar Spectral Line Formation Theory*, ed. J.E. Beckman and L. Crivellari (Dordrecht: Reidel), p. 1.

Maltby, P., Avrett, E.H., Carlsson, M., Kjeldseth-Moe, O., Kurucz, R.L., and Loeser, R. 1986, *Astrophys. J.*, **306**, 284.

Mankin, W.G. 1977, in *The Solar Output and Its Variation*, ed. O.R. White (Boulder: Colorado Associated University Press), p. 151.

Mauas, P.J., Avrett, E.H., and Loeser, R. 1988, *Astrophys. J.*, **330**, 1008.

Mihalas, D. 1978, *Stellar Atmospheres*, (San Francisco: Freeman).

Milkey, R.W. and Mihalas, D. 1974, *Astrophys. J.*, **192**, 769.

Nelson, G.D. 1978, *Solar Phys*, **60**, 5.

Nordlund, Å. 1985, in *Theoretical Problems in High Resolution Solar Physics*, ed. H.U. Schmidt, Max Planck Institut fur Astrophysik, MPA 212, p. 1.

Noyes, R.W., and Avrett, E.H. 1987, in *Spectroscopy of Astrophysical Plasmas*, ed. A. Dalgarno and D. Layzer (Cambridge: Cambridge University Press), p. 125.

Noyes, R.W., and Hall, D.N.B. 1972, *Astrophys. J.*, **176**, L89.

Parkinson, W.H., and Reeves, E.M. 1969, *Solar Phys.*, **10**, 342.

Pierce, A.K. 1954, *Astrophys. J.*, **120**, 221.

Rast, J., Kneubühl, F.K., and Müller, E.A. 1978, *Astron. Astrophys.*, **68**, 229.

Rich, J. 1966, *Silicon and Carbon Monoxide Absorption in the Solar Ultraviolet Spectrum*, Thesis, Harvard University, Cambridge, Mass.

Rutten, R.J. 1988, in *Physics of Formation of Fe II Lines Outside LTE*, ed. R. Viotti, A. Vittone, and M. Friedjung (Dordrecht: Reidel), p. 185.

Samain, D. 1980, *Astrophys. J. Suppl.*, **44**, 273.

Shine, R.A., Milkey, R.W., and Mihalas, D. 1975, *Astrophys. J.*, **199**, 724.

Stein, R.F., and Nordlund, Å. 1989, *Astrophys. J. (Lett.)*, in press.

Tousey, R. 1963, *Space Sci. Rev.*, **2**, 3.

Vernazza, J.E., Avrett, E.H., and Loeser, R. 1973, *Astrophys. J.*, **184**, 605.

Vernazza, J.E., Avrett, E.H., and Loeser, R. 1976, *Astrophys. J. Suppl.*, **30**, 1.

Vernazza, J.E., Avrett, E.H., and Loeser, R. 1981, *Astrophys. J. Suppl.*, **45**, 635.

White, O.R., and Suemoto, Z. 1968, *Solar Phys.*, **3**, 523.

THERMAL BIFURCATION OF THE OUTER PHOTOSPHERE

T. R. AYRES
Center for Astrophysics and Space Astronomy
University of Colorado
Boulder, Colorado, USA

ABSTRACT. The degree of thermal heterogeneity at the base of the solar chromosphere is substantially beyond that simulated in the best-available multi-component models; casting serious doubts on inferences drawn from them.

Everyone knows that the chromosphere of the Sun is hot. After all, spatially-averaged profiles of strong optical and ultraviolet resonance lines – Ca II H and K , Mg II h and k, and H I Lyα, for example – clearly show prominent emission reversals in their cores, indicating a temperature inversion at the top of the photosphere (e.g., Athay 1976). Virtually all empirical models of the solar outer atmosphere have incorporated such a temperature inversion. The most complete such model – that of Vernazza, Avrett, and Loeser (1981) – presents a range of thermal profiles to simulate spatial inhomogeneities. Nevertheless, each of the six distinct components has a temperature inversion at about the same altitude (≈ 500 km above $\tau_{5000} = 1$). The multi-component model successfully reproduces a wide range of spectral diagnostics – lines and continua – broadly covering the electromagnetic spectrum. The principal use of the model has been to calculate the *radiative cooling* as a function of altitude in the chromosphere, against which to compare the predictions of various *mechanical heating* scenarios.

Everyone knows that the chromosphere is hot, but somebody apparently neglected to tell the infrared bands of carbon monoxide! Initially, measurements of the strong $\Delta v = 1$ transitions (near 2150 cm^{-1} = 4.7μm) by Noyes and Hall (1972) revealed low core brightness temperatures ($T < 4200$ K; compared with $T_{min} \gtrsim 4400$ deduced from Ca II and Mg II) in the most opaque of the vibration-rotation lines; later, Ayres and Testerman (1981) confirmed the earlier results using the newly-commissioned Fourier transform spectrometer on the McMath telescope at Kitt Peak. At the extreme limb, where radition emerges from the highest accessible levels of the atmosphere in a given diagnostic, the strongest CO absorptions indicated the existence of very cool plasma ($T < 3800$ K) and no hint of any chromospheric ($T > 6000$ K) material, contrary to the expectations of the best-available homogeneously stratified models. Figure 1 illustrates the type of thermal profile derived from high temporal resolution and moderate spatial resolution recordings of the infrared CO bands in activity-free areas at disk center and near the extreme limb. The vertical

J. O. Stenflo (ed.), Solar Photosphere: Structure, Convection, and Magnetic Fields, 23–27.

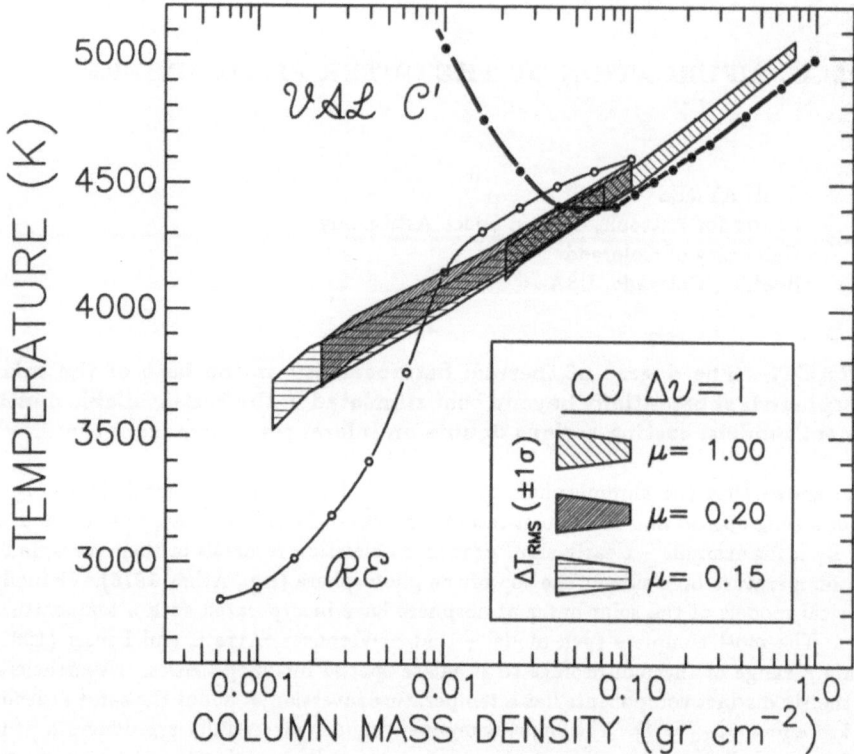

Figure 1: Measured CO brightness temperatures compared with an empirical chromospheric model (the VAL C' model of Maltby *et al.* [1986]), and a theoretical temperature profile calculated in radiative equilibrium including CO cooling (Anderson and Athay 1989).

extents of the shaded areas depict the maximum rms thermal fluctuations inferred to exist at the different levels of the atmosphere over horizontal size-scales comparable to a p-mode wavepacket (several Mm). The large response of the CO bands to p-mode excitations indicate a high altitude of formation.

The inescapable conclusion was that the "chromosphere" of the Sun must contain substantial amounts of relatively cool material in addition to the classical hot gas that one ordinarily associates with it. In the Ayres and Testerman model (later quantified by Ayres, Testerman, and Brault 1986) the weak central reversals in the cores of the Ca II lines, for example, might arise as a mixture of strong emission from small, intensely-heated regions diluted by pure absorption profiles from the more extensive cool component.

The juxtaposition of hot and cold gas in the chromosphere was attributed by Ayres (1981) to a thermal instability driven on the low-temperature side by the powerful surface cooling of the near-LTE CO $\Delta v = 1$ bands, and on the high-temperature side by strong radiative emission in the UV resonance lines of abundant species (e.g., Mg II, Ca II, and H I). The low-temperature cooling is enhanced for decreasing temperatures below about 4500 K by the exponentially increasing *formation* of the CO molecules, and stabilizes when the molecules have exhausted the available atomic carbon ($T \lesssim 4000$ K). The high-temperature cooling is promoted as T rises above about 5000 K by the exponential thermal sensitivity of the electron collisional excitation rates, as well as by the increasing population of electrons due to hydrogen ionization. However, in the intermediate range $4000 \lesssim T \lesssim 5000$ neither the CO infrared bands nor the atomic UV resonance lines are effective coolants. Thus, the plasma cooling function at chromospheric heights exhibits two distinct stable phases, with an unstable intermediate temperature range.

The bifurcation of the cooling function leads to a thermal instability when the chromospheric gas is subjected to a critical level of nonradiative heating: the gas remains cool – and near the radiative equilibrium stratification – when the mechanical heating falls below the critical value; but reverts to the hot, classical chromospheric phase when the nonradiative heating exceeds it. Ayres, Testerman, and Brault (1986) modelled simultaneous observations of the CO bands and Ca II K in quiet regions on the solar disk and in areas strongly disturbed by magnetic activity. The authors concluded that only a small fraction ($\lesssim 10\%$) of the solar chromosphere in quite regions is strongly-enough heated to produce hot plasma; and even in magnetic active regions only perhaps $\approx 50\%$ of the surface is truly chromospheric. Thermal inhomogeneities of that magnitude, particularly in the quiet Sun, are enough to invalidate most of the previous homogeneously-stratified models based on spatially-average solar spectra (or integrated starlight), and cause us to reexamine our understanding of chromospheric structure not only on the Sun, but also on the other stars of late spectral type.

Figure 2 illustrates more graphically the structural organization of the solar atmosphere. At the altitude of the "T_{min}", the chromosphere consists of two fundamental types of structures, both of small surface coverage: (1) long-lived *network bright points*, whose heating very likely is magnetic in origin; and (2) transient *cell-interior flashes*, whose heating might be electrodynamic as well, although the disturbances could possibly be purely acoustic. An excellent discussion of the two distinct types of bright points has been provided by Cram (1985). In the quiet Sun, the BPs cover perhaps $< 20\%$ of the surface (dominated by the cell-interior BPs in area, although the network points are brighter in Ca II); but in active regions the coverage might reach as much as 60% (dominated by the magnetic BPs). The (large) remaining volume at, and immediately above, the T_{min} (up to perhaps 700 km) is occupied by material that is substantially cooler than that in the heated structures. One might think of the cool gas as organized into *CO clouds*. The clouds continually form, perhaps through the adiabatic cooling of gas advected to high altitudes by convective overshoot, or by supergranular *plumes* (e.g., November 1989); and continually are disrupted by the sporatic disturbances responsible for the cell flashes.

Recent Non-LTE blanked models of the solar chromosphere with prescribed heating have pointed to the T_{min} region as the critical location of maximum nonradiative energy

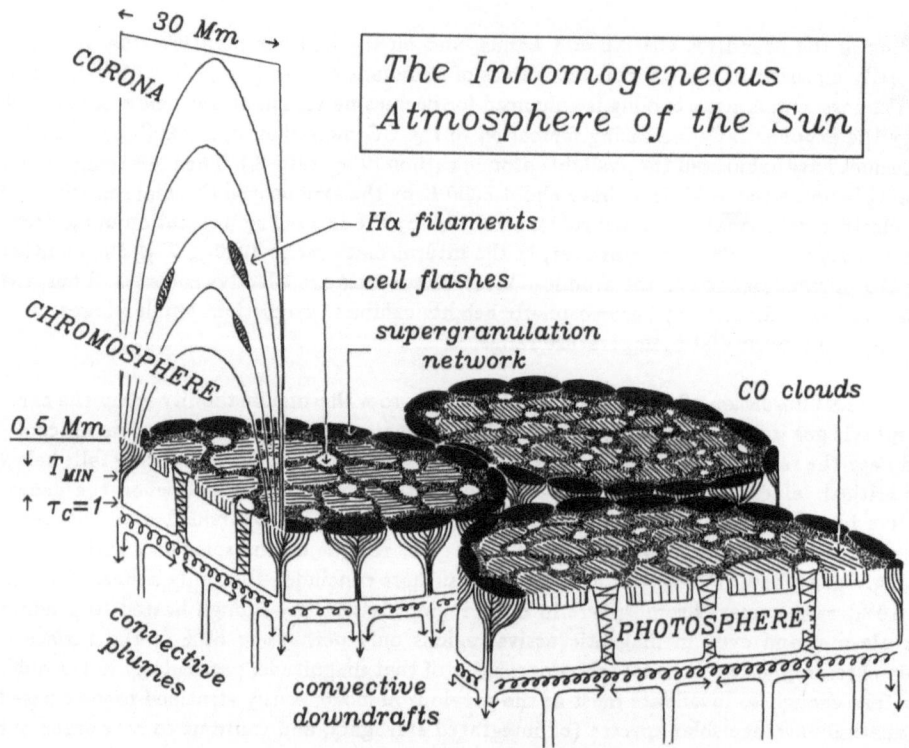

Figure 2: The inhomogeneous atmosphere of the Sun.

deposition (Anderson and Athay 1989). It is there that we have the best chance to isolate the elusive chromospheric heating mechanism. But, if the T_{min} in fact is dominated by cool structures, then the spatially-averaged thermal profiles derived by Anderson and Athay, and earlier by VAL, have little meaning. Instead, the energy-balance modelling should be focused on the specific structures – network and cell-interior BPs – where the strongly-heated gas at the base of the chromosphere truly resides. Of course, the issue is further complicated by the fact that the rapidly diverging fields of the network magnetic filaments must fill all of the available volume in the chromosphere above some critical level (the CO "clouds" have a small vertical scale height owing to their depressed temperatures). Thus, the middle chromosphere (at an altitude of about 1000 km) might be more thermally homogeneous than the underlying T_{min} region. Nevertheless, Ca II K$_2$ and Lyα filtergrams – characteristic of those high levels – exhibit considerable structure, and the network is clearly recognizable even at C IV temperatures (10^5 K).

Critically needed is a major effort to define a grid of thermally-distinct temperature profiles to more accurately describe the true physical conditions at the critical interface between photosphere and chromosphere. Such a grid of models can be used to explore the nature of the nonradiative heating mechanism in a more favorable light than present thermal profiles, which ignore the important role of cool gas in the low chromosphere.

ACKNOWLEDGEMENTS.   This work was supported by grants from the National Science Foundation and the National Aeronautics and Space Administration. The observations described in Fig. 1 were obtained at the Kitt Peak facility of the National Solar Observatories, operated by AURA under contract to the NSF.

REFERENCES

Anderson, L. S. 1989, *Ap. J.*, **339**, 558.

Anderson, L. S., and Athay, R. G. 1989, "Model Solar Chromosphere with Prescribed Heating", *Ap. J.*, (submitted).

Athay, R. G. 1976, *The Solar Chromosphere and Corona: Quiet Sun*, (Dordrecht: D. Reidel).

Ayres, T. R. 1981, *Ap. J.*, **244**, 1064.

Ayres, T. R., and Testerman, L. 1981, *Ap. J.*, **245**, 1124.

Ayres, T. R., Testerman, L., and Brault, J. W. 1986, *Ap. J.*, **304**, 542.

Cram, L. 1987, in *Cool Stars, Stellar Systems, and the Sun*, (eds.) J. L. Linsky and R. E. Stencel (New York: Springer-Verlag), p. 123.

Maltby, P., Avrett, E. H., Carlsson, M., Kjeldseth-Moe, O., Kurucz, R. L., and Loeser, R. 1986, *Ap. J.*, **306**, 284.

November, L. J. 1989, "The Vertical Component of the Supergranulation Convection", *Ap. J.*, (submitted) [*NOAO preprint No. 257*].

Noyes, R. W., and Hall, D. N. B. 1972, *Ap. J. (Letters)*, **176**, L89.

Vernazza, J. E., Avrett, E. H., and Loeser, R. 1981, *Ap. J. Suppl.*, **45**, 635.

TEMPERATURE DIAGNOSTICS OF THE UPPER PHOTOSPHERE

N.G. Shchukina[1], T.G. Shcherbina[1] and R.J. Rutten[2]

[1] Main Astronomical Observatory, Academy of Sciences of the Ukrainian SSR, 252127 Kiev, USSR
[2] Sterrekundig Instituut, Postbus 80000, 3508 TA Utrecht, The Netherlands

ABSTRACT. We use NLTE modelling of the alkali resonance lines and of C I and O I high-excitation multiplets to test their quality as temperature diagnostics of the upper photosphere, in the context of Avrett's NLTE recovery of the LTE Holweger-Müller model below the temperature minimum and Ayres' bifurcation into hot and cool components above the temperature minimum.

1. The two temperature issues

The two arrows in Figure 1 mark recent developments in spatially-averaged modelling of the upper solar photosphere. The arrow near $h = 500$ km marks an appreciable shift in the empirical NLTE continuum modelling by Avrett and coworkers, from the cool temperature minimum which characterised the earlier HSRA and VAL models (Vernazza et al. 1981) to the hotter minimum specified in the MACKKL model (Maltby et al. 1986). The primary diagnostics used are the infrared and ultraviolet continuum intensities; the change results from new, higher-intensity infrared data and from the outward shift in the computed height of formation of the near-ultraviolet continua due to the inclusion of more line blocking (Avrett, these proceedings).

The change brings the formation of the ultraviolet continua and of most optical metal lines close to LTE (Rutten 1988). The photospheric part of the new model is indeed much closer to the classical empirical LTE model of Holweger and Müller (1974), and also to theoretical radiative-equilibrium models (Bell et al. 1976, Anderson 1989). This similarity indicates that plane-parallel homogeneity, LTE and RE are reasonable assumptions for the upper photosphere, in conflict with the 3D granulation simulations of Nordlund (these proceedings) which predict spatial variations of large amplitude throughout the photosphere.

The arrow near $h = 900$ km marks the bifurcation into hot and cool components proposed by Ayres (these proceedings). His primary diagnostics are the Ca II K line for the hot component and the limb darkening of the infrared CO bands for the cool component, with CO molecules contributing strongly to the cooling wherever (or whenever) they associate—presumably in field-free medium between fluxtubes.

These two issues separated in height only recently. Originally Ayres (1981) started the bifurcation below the temperature minimum in order to reproduce the Ayres and

J. O. Stenflo (ed.), Solar Photosphere: Structure, Convection, and Magnetic Fields, 29–34.

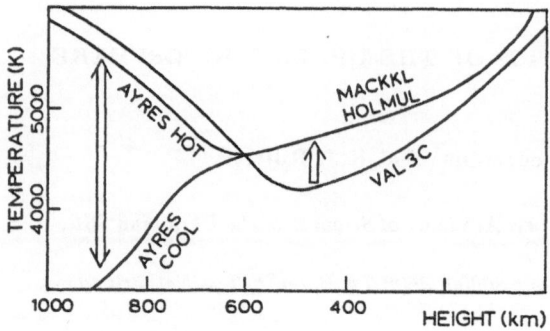

Figure 1. Two issues in current photospheric modelling. Schematic electron temperatures against height for various models of the solar photosphere. The righthand arrow marks the change below the temperature minimum in the modelling of Avrett *et al.*, from VAL3C to MACKKL. The lefthand arrow marks the bifurcation into hot and cool components proposed by Ayres.

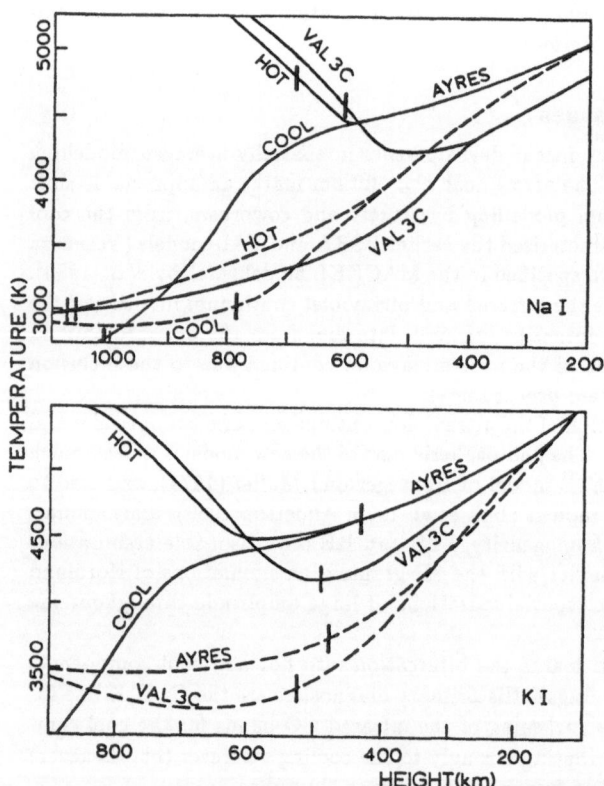

Figure 2. Formation of the Na D_1 line (*upper panel*) and the K I 770 nm line (*lower panel*).
Solid: electron temperature for the VAL3C model and for a model from Ayres which equals the hotter MACKKL model in the temperature minimum and splits into hot and cool components higher up.
Dashed: corresponding NLTE excitation temperatures.
Tick marks: $\tau = 1$ heights for viewing angle $\mu = 0.3$, respectively for LTE and NLTE line formation.

Linsky (1976) modelling of the Ca II line wings with the hot component. When the VAL to MACKKL change brought the continuum modelling into agreement with that, Ayres *et al.* (1986) shifted the split to higher layers. Avrett's criticism (these proceedings) that Ayres has overestimated the CO radiative cooling by using an inappropriate fixed radiation field is less pertinent there, but the criticism that the cool "CO clouds" are not seen in other data remains of interest.

Are there other observational diagnostics of these two temperature issues? Not from optical metal spectra such as Fe I and Fe II because their lines can be modelled as well with a hot as with a cool photosphere (Rutten and Kostik 1982). Which other lines might do? We have selected low-ionization and high-excitation lines as cool and hot diagnostics, respectively: the K I and Na I resonance lines which are from minority ionization stages with very low ionization energy and which should be enhanced in cool matter, and the C I and O I multiplets of about 8 eV excitation energy which should be enhanced in hot matter.

2. Low-ionization lines

Figure 2 shows results for the Na I and K I resonance lines; computational details will be given in Bruls *et al.* (in preparation). The line source functions (shown as excitation temperatures, dashed) decouple from the electron temperature (solid) in the upper and lower photosphere, respectively, and display typical scattering behaviour.

The computed heights of formation shift outward from LTE line formation ($\tau = 1$ ticks on solid curves) to NLTE line formation ($\tau = 1$ ticks on dashed curves). For K I this shift is due to photon losses in the resonance lines. These cause extra recombination from the ion reservoir which is not fully compensated by ultraviolet overionization in the temperature minimum region. For Na I the shift is very large for models with a chromospheric temperature rise because the rise is not followed by the ionising radiation originating from the photosphere. For Ayres' cool model, however, the ionising radiation is hotter than the electron temperature, resulting in radiative overionization and a reversed LTE–to–NLTE shift.

Figure 3 shows observed and computed cores for the Na D_1 line at two viewing angles. The differences between the various models are too small to be significant; the scattering behaviour makes the source function rather insensitive to the two temperature issues. The same holds for the K I resonance lines.

The inner wings of the Na I lines show better response. Figure 4 shows that these are sensitive to the temperature minimum region; the difference between the hot model (AYRES) and the cool model (VAL3C) is significant and exceeds the variations obtained by changing the microturbulent or collisional damping parameters within a reasonable range.

3. High-excitation lines

Figure 5 shows results for the infrared C I multiplet near 1070 nm; details are given elsewhere (Shchukina and Shcherbina 1989). The source function again displays photon-loss character, but with some sensitivity to the presence of a chromospheric temperature rise.

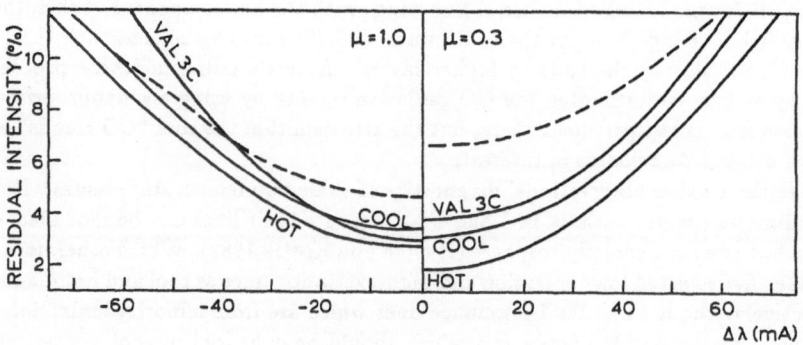

Figure 3. Computed (*solid*) and observed (*dashed*) line cores of the Na D_1 line. Viewing angles $\mu = 1.0$ (left) and $\mu = 0.3$ (right).

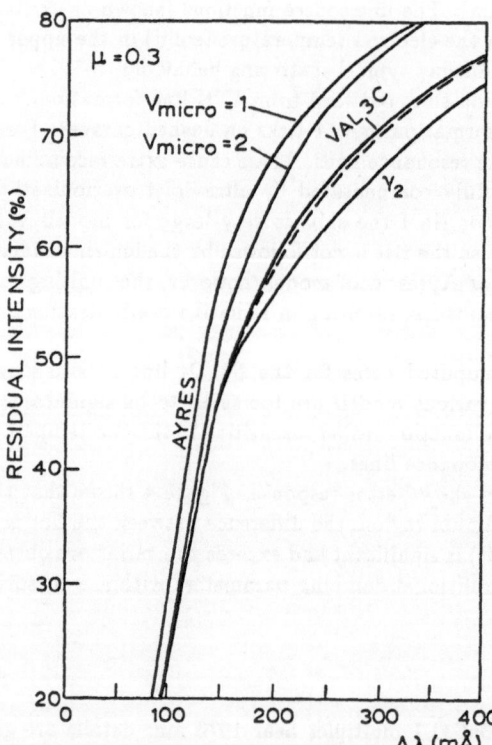

Figure 4. Computed (*solid*) and observed (*dashed*) inner wings of the Na D_1 line at viewing angle $\mu = 0.3$. The hot and cool components of model AYRES produce the same line wings. The AYRES model has been used with two microturbulence values (1 and 2 km/s). The VAL3C model has been used with two formalisms for collisional broadening. The effective damping parameter γ is larger by about 40% for the curve marked γ_2.

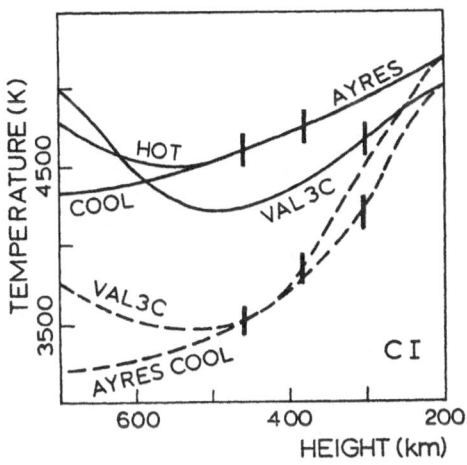

Figure 5. Formation of the C I triplet near 1070 nm.
Solid: model electron temperatures.
Dashed: C I excitation temperatures.
Ticks: $\tau = 1$ heights for $\mu = 0.3$, respectively for LTE line formation (solid curves) and NLTE line formation (dashed curves).

Figure 6. Computed (*solid*) and observed (*dashed*) line profiles (*left*) and equivalent widths (*right*) of the C I 1070 nm line.

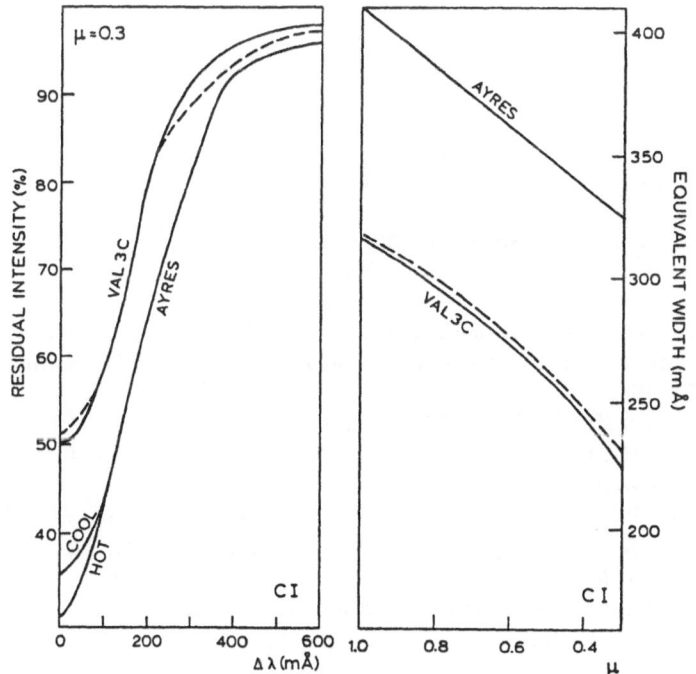

The line opacity scaling is about the same in LTE and NLTE. This equality results for models with a cool temperature minimum from the fortuitous cancellation between the overpopulation due to ultraviolet pumping in the resonance lines and the depopulation through pumped near-ultraviolet lines that connect the lower levels of the multiplet to levels close to the continuum.

However, the line opacity scaling does depend on the temperature structure. It causes the separation between the $\tau = 1$ ticks for the AYRES-hot (outermost tick), AYRES-cool (middle) and VAL3C (innermost) models and results in appreciable line profile sensitivity, as evident in Figure 6a in which the difference between a hot and a cool temperature minimum is again significant. There is even some sensitivity to Ayres' bifurcation in the line core, which is deeper for the hot model because that produces more opacity. This is surprising because one tends to think of such high-excitation lines as very deeply formed; the contribution function is doubly peaked, however, and feels the chromospheric temperature rise also.

Figure 6b shows that the difference between hot and cool minima is markedly present even in the integrated profiles, here shown center to limb. Similar results were obtained for the 777 nm triplet of O I (Shchukina 1987).

Conclusion

All lines discussed here require detailed NLTE modelling. None of the lines provides a good diagnostic of Ayres' bifurcation above $h = 700$ km, but the inner wings of the Na D lines and the C I 1070 nm and O I 777 nm multiplets are sensitive to the temperature around $h = 400$ km. The observations, both of the "cool" Na D wings and the "hot" C I multiplet, are better reproduced with the old VAL3C model than with the new MACKKL model. This contradicts the current trend towards a hot temperature minimum.

References

Anderson, L. S.: 1989, *Astrophys. J.* **339**, 558

Ayres, T. R.: 1981, *Astrophys. J.* **244**, 1064

Ayres, T. R. and Linsky, J. L.: 1976, *Astrophys. J.* **205**, 874

Ayres, T. R., Testerman, L., and Brault, J. W.: 1986, *Astrophys. J.* **304**, 542

Bell, R. A., Eriksson, K., Gustaffson, B., and Åke Nordlund: 1976, *Astron. Astrophys. Suppl.* **23**, 37

Holweger, H. and Müller, E. A.: 1974, *Solar Phys.* **39**, 19

Maltby, P., Avrett, E. H., Carlsson, M., Kjeldseth-Moe, O., Kurucz, R. L., and Loeser, R.: 1986, *Astrophys. J.* **306**, 284

Rutten, R. J.: 1988, in *Physics of Formation of FeII Lines Outside LTE*, Eds. Viotti, R., Vittone, A., and Friedjung, M., p. 185, IAU Colloquium 94, Reidel, Dordrecht

Rutten, R. J. and Kostik, R. I.: 1982, *Astron. Astrophys.* **115**, 104

Shchukina, N. G.: 1987, *Kinematika i Fizika Nebesnich Tel.* **3**, 36

Shchukina, N. G. and Shcherbina, T. G.: 1989, *Kinematika i Fizika Nebesnich Tel.* in press

Vernazza, J. E., Avrett, E. H., and Loeser, R.: 1981, *Astrophys. J. Suppl. Ser.* **45**, 635

SOLAR OSCILLATOR STRENGTHS AS A DIAGNOSTIC TOOL

E.A. Gurtovenko[1], R.I. Kostik[1] and R.J. Rutten[2]

[1] Main Astronomical Observatory, Academy of Sciences of the Ukrainian SSR, 252127 Kiev, USSR
[2] Sterrekundig Instituut, Postbus 80000, NL-3508 TA, Utrecht, The Netherlands

ABSTRACT. We briefly review the Kiev program for determining oscillator strengths of Fraunhofer lines from the optical solar spectrum, which has recently resulted in a compilation of solar gf-values for 1958 lines from 40 chemical elements (Gurtovenko and Kostik 1989). These gf-values were determined empirically by fitting solar lines using standard plane-parallel LTE modelling. Errors in this modelling propagate into the gf-values; reversely, the deviations in the latter may serve as diagnostics of the modelling and thus of spectral line formation in the solar photosphere. For a small subset, comparison with reliable laboratory data can be made; for other lines there is information in the differences between fits of the line area and of the line depth.

1. The Kiev oscillator strength program

The solar photosphere may be regarded as a natural furnace from which Fraunhofer lines originate in order to enable the measurement of their oscillator strengths. As such a furnace, the photosphere provides important advantages: (i)—the number of measurable lines is large; (ii)—the measurable lines are often precisely the ones needed in abundance determinations for other stars; (iii)—the furnace properties are rather well known. In contrast, laboratory measurements used to have very large errors until the precise Oxford measurements became available (Blackwell et al. 1982 and references therein), while the latter are mostly for lines that are less suitable for stellar abundance determinations (see Grotrian diagrams in Rutten and Kostik 1988).

In the early eighties a program was started at Kiev to determine empirical solar oscillator strengths, following the classical example set by Holweger (1967). Its first results were two extensive lists of Fe I gf-values (Gurtovenko and Kostik 1981a, 1981b). They were found to be of good quality, both in comparisons with laboratory data (Wiese 1983, Cowley and Corliss 1983) and in studies of their internal consistency (Rutten and Kostik 1982, Rutten and Zwaan 1983, Rutten and Van der Zalm 1984).

The sensitivity of such fits to the choice of atmospheric model and to NLTE effects was analysed by Rutten and Kostik (1982). They resolved the apparent conflict between the quality of these LTE fits and the large departures from LTE derived for Fe I lines by Lites (1972, Athay and Lites 1972). Rutten and Kostik found that the empirical LTE

J. O. Stenflo (ed.), Solar Photosphere: Structure, Convection, and Magnetic Fields, 35–40.
© 1990 by the IAU.

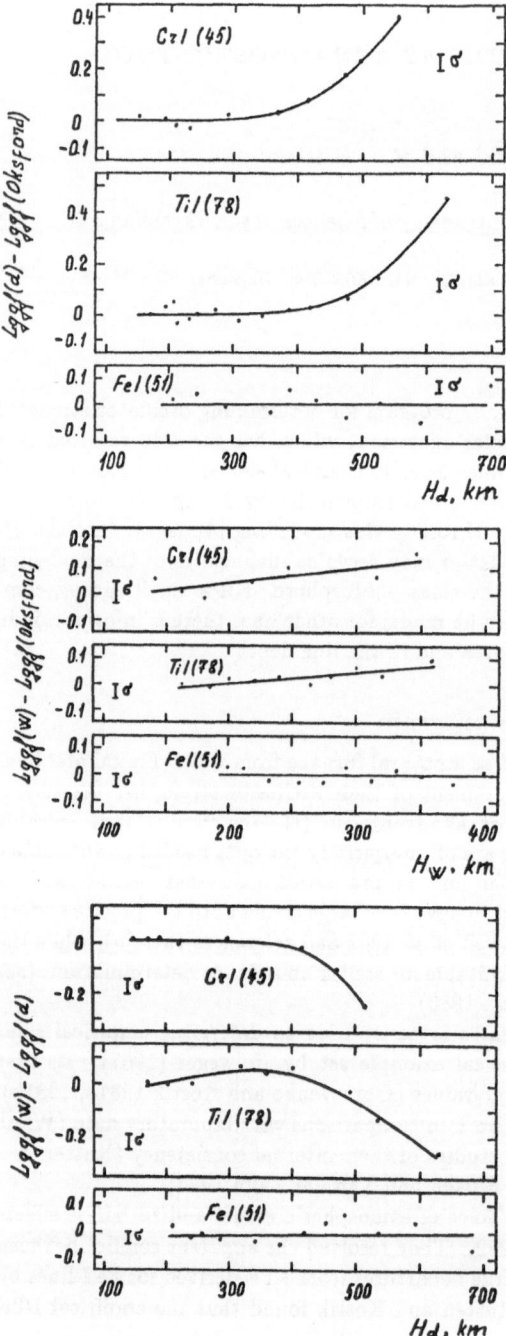

Figure 1. Logarithmic differences between *gf*-values determined by fitting line depths in the Jungfraujoch Atlas of the solar spectrum and laboratory measurements from Oxford.

Figure 2. Logarithmic differences between *gf*-values determined by fitting solar equivalent widths and laboratory measurements from Oxford.

Figure 3. Logarithmic differences between solar *gf*-values from line widths and line depths respectively, for the lines of overlap with Oxford measurements.

model photosphere of Holweger and Müller (1974) "masks" NLTE departures in iron lines even if the latter are large—as they should be if the solar atmosphere has as cool a temperature minimum as it has in the HSRA and VAL3C standard models (Gingerich et al. 1971, Vernazza et al. 1981). However, recent NLTE modelling with better ultraviolet line blanketing indicates that the upper photosphere is not so cool after all (Avrett 1985, Maltby et al. 1986, Anderson 1989), and that the LTE Holweger–Müller model has actually furnished a correct description of spatially-averaged photospheric line formation all along (Rutten 1988a). This issue is reviewed at length in Rutten (1988b); the point here is that the Kiev fits assuming LTE and the Holweger–Müller model provide accurate gf-values for most iron lines whether their actual departures from LTE are large or small. Thus, the solar furnace is an excellent one, without the sensitivity to deviations from LTE that beset earlier laboratory measurements of iron lines using free-burning arcs.

In addition, there are other error sources. The most important ones are unresolved blending and the use of empirical fitting parameters (micro– and macroturbulence, collisional damping enhancement) to correct for the effects of spatial and temporal averaging (Gurtovenko and Sheminova 1986). Their influence on the iron-line fits was tested by Rutten and Kostik (1988), who found that 0.1 dex precision is obtainable for most iron lines. Nordlund (1984) found larger differences when comparing such standard modelling with line profile fits based on his numerical simulation of the solar granulation. However, we expect that such differences will be smaller for his new elastic simulations (Rutten 1988b). In summary, the Kiev Fe I gf-values are generally reliable to within 0.1 dex (25%).

2. New results

The success described above has led the Kiev workers to continue their effort. A new compilation with solar gf-values for 1958 lines from 40 elements has just appeared (Gurtovenko and Kostik 1989). These are obviously useful for stellar abundance determinations, but they may also serve as diagnostics of photospheric line formation. We address the latter usage here. Since the fits are obtained from the comparison of theory and observations, they contain errors due to modelling assumptions of which the characteristics may vary with solar line strength and with atomic parameters. Such errors and their dependences can be employed to test the validity of the standard modelling and to study Fraunhofer line formation adding other spectra than Fe I.

The measured quantity per line is $\log L = \log(Agf)$ where $\log A = \log(N_{el}/N_H) + 12$. Thus, each measurement error $\delta \log L$ has $\delta \log L = \delta \log gf + \Delta \log A$ where $\Delta \log A$ is a correction to the adopted elemental abundance value which is the same for all lines from that element. The remaining error-per-line $\delta \log gf$ is the one which concerns us here. There are two sets of gf measurements, obtained from fitting equivalent widths and central depths respectively: gf_w values and gf_d values. There are two ways of studying errors in these: analysing the differences of each set with results from others, and analysing the differences between the two sets. We haven't performed such analysis yet in detail; here, we show first results that are indicative of the information content of the new measurements.

Figures 1 and 2 show the differences between the Kiev results and the Oxford measurements for the gf_d and gf_w values respectively, against the corresponding height of

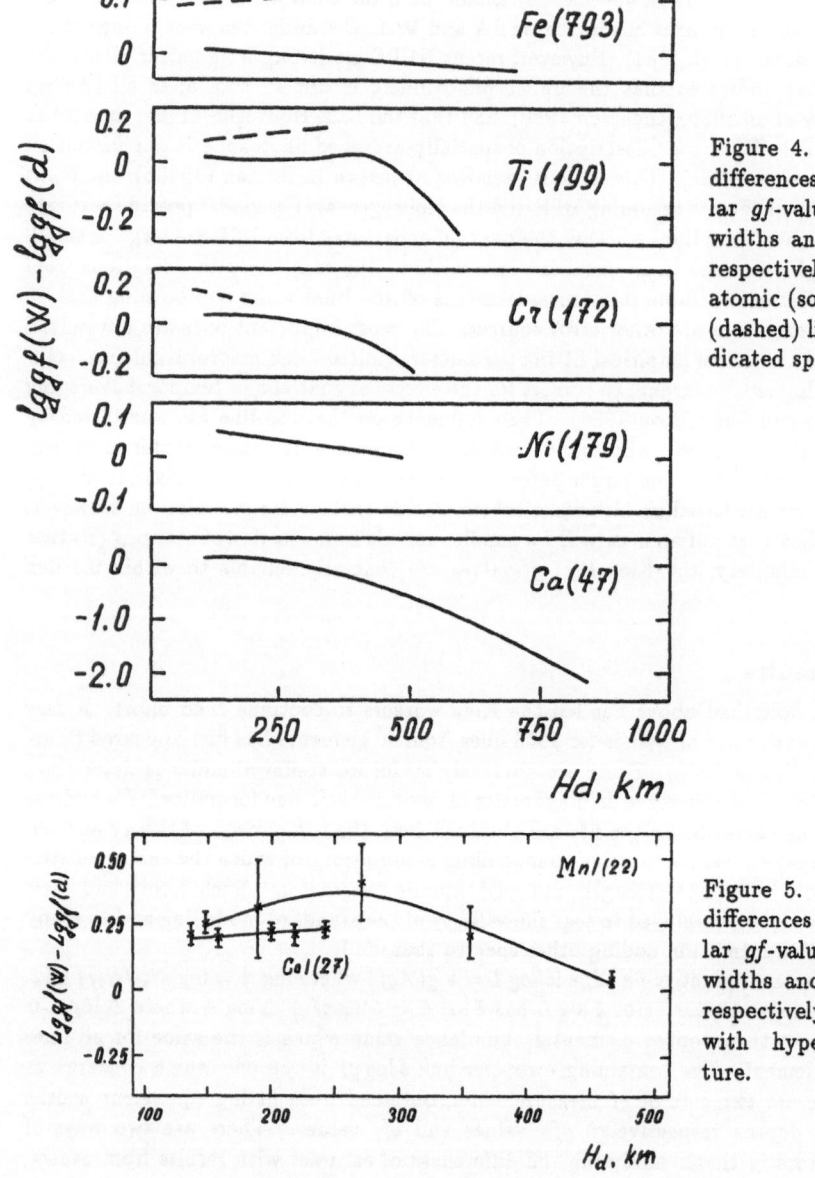

Figure 4. Logarithmic differences between solar *gf*-values from line widths and line depths respectively, for all atomic (solid) and ionic (dashed) lines of the indicated species.

Figure 5. Logarithmic differences between solar *gf*-values from line widths and line depths respectively, for lines with hyperfine structure.

formation in the photosphere. These diagrams contain averages for all lines for which Oxford results are available, from Fe I (51 lines), Ti I (78 lines) and Cr I (45 lines). The bars denote one-sigma standard deviations. The absence of a trend in the Fe I lines is to be expected because the nonthermal broadening parameters in the modelling were derived by fitting representative Fe I lines using the Oxford gf-values and the Holweger–Müller model itself was originally derived from iron lines. The differences are also small for the Ti I and Cr I lines, except for the strongest lines formed in the upper photosphere which have deeper cores in the observed spectrum than the modelling provides for if we assume that the Oxford values are correct.

Figure 3 displays the signature of these trends in a log gf_w – log gf_d plot against height for the same lines. Note the difference in the weak-line parts between the Cr I and the Ti I results.

Figure 4 shows the composite averages for all lines of iron (793 lines), titanium (199 lines), chromium (172 lines), nickel (179 lines) and calcium (47 lines), respectively from their neutral stages (full lines) and first ionization stages (dashed). Fe I now shows a slight trend, due to the addition of numerous other lines than the Oxford ones. A similar trend is shown by Ni I. The stronger lines of Cr I and Ti I display the downturns already evident in the smaller samples discussed above. Ca I shows similar behaviour but with a much larger amplitude. All ions show positive differences. Fe II displays a slight opposite trend, increasing with height.

It is already clear that these features have to do with solar line formation and not with inaccuracies in atomic parameters such as ionization potentials, partition functions etc. (see Grevesse 1984). Also, the trends cannot be removed within the constraints of the standard modelling, i.e. by changing the temperature structure or the turbulence parameters. Their explanation requires at least the introduction of departures from LTE in the excitation or ionization equilibria or both, and possibly of inhomogeneous structure not mimicked through nonthermal broadening.

A final point is that such difference plots may also serve to display hyperfine and isotope structure. These influence spectral line profiles just as microturbulence does; in general, their underestimation results in too large gf_w and too small gf_d values and in a specific pattern in the difference plots: first an increase and then a decrease (Gurtovenko and Sheminova 1988). Such behaviour is seen in Figure 5 which displays log gf_w – log gf_d differences for Mn I and Co I lines with hyperfine structure.

3. Conclusion

Our display of difference plots from the new Kiev compilation shows that there is interesting behaviour pointing to deficiencies in the standard modelling. Although the iron lines and probably also the nickel lines are represented quite well assuming LTE and the Holweger–Müller model, there are significant deviations for the strong lines from other spectra, requiring more detailed analysis. For Ti I and Cr I these deviations may simply display strong–line photon losses in excess of typical Fe I behaviour, but there is no immediate explanation for the very large deviations in Ca I and for the splits between atom and ion lines. We conclude that, again, empirical solar gf-values may be employed as a diagnostic of line formation in the solar atmosphere.

References

Anderson, L. S.: 1989, *Astrophys. J.* **339**, 558

Athay, R. G. and Lites, B. W.: 1972, *Astrophys. J.* **176**, 809

Avrett, E. H.: 1985, in Lites, B. W. (Ed.), *Chromospheric Diagnostics and Modeling*, p. 67, National Solar Observatory Summer Conference, Sacramento Peak Observatory, Sunspot, New Mexico

Blackwell, D. E., Petford, A. D., Shallis, M. J., and Simmons, G. J.: 1982, *Monthly Notices Roy. Astron. Soc.* **199**, 43

Cowley, C. R. and Corliss, C. H.: 1983, *Monthly Notices Roy. Astron. Soc.* **203**, 651

Gingerich, O., Noyes, R. W., Kalkofen, W., and Cuny, Y.: 1971, *Solar Phys.* **18**, 347

Grevesse, N.: 1984, *Physica Scripta* **T8**, 49

Gurtovenko, E. A. and Kostik, R. I.: 1981a, *Astron. Astrophys. Suppl.* **46**, 239

Gurtovenko, E. A. and Kostik, R. I.: 1981b, *Astron. Astrophys. Suppl.* **47**, 193

Gurtovenko, E. A. and Kostik, R. I.: 1989, *Fraunhofer Spectrum and the System of Solar Oscillator Strengths*, Naukova Dumka, Kiev

Gurtovenko, E. A. and Sheminova, V. A.: 1986, *Solar Phys.* **106**, 237

Gurtovenko, E. A. and Sheminova, V. A.: 1988, *Kinematika i Fizika Nebesnich Tel.* **4**, 18

Holweger, H.: 1967, *Zeitschr. f. Astrophysik* **65**, 365

Holweger, H. and Müller, E. A.: 1974, *Solar Phys.* **39**, 19

Lites, B. W.: 1972, *Observation and Analysis of the Solar Neutral Iron Spectrum*, NCAR Cooperative Thesis No. 28, High Altitude Observatory, Boulder

Maltby, P., Avrett, E. H., Carlsson, M., Kjeldseth-Moe, O., Kurucz, R. L., and Loeser, R.: 1986, *Astrophys. J.* **306**, 284

Nordlund, Å.: 1984, in Keil, S. L. (Ed.), *Small-Scale Dynamical Processes in Quiet Stellar Atmospheres*, p. 181, National Solar Observatory Summer Conference, Sacramento Peak Observatory, Sunspot, New Mexico

Rutten, R. J.: 1988a, in Cayrel de Strobel, G. and Spite, M. (Eds.), *The Impact of Very High S/N Spectroscopy on Stellar Physics*, p. 367, IAU Symposium 132, Reidel, Dordrecht

Rutten, R. J.: 1988b, in Viotti, R., Vittone, A., and Friedjung, M. (Eds.), *Physics of Formation of FeII Lines Outside LTE*, p. 185, IAU Colloquium 94, Reidel, Dordrecht

Rutten, R. J. and Kostik, R. I.: 1982, *Astron. Astrophys.* **115**, 104

Rutten, R. J. and Kostik, R. I.: 1988, in Viotti, R., Vittone, A., and Friedjung, M. (Eds.), *Physics of Formation of FeII Lines Outside LTE*, p. 83, IAU Colloquium 94, Reidel, Dordrecht

Rutten, R. J. and van der Zalm, E. B. J.: 1984, *Astron. Astrophys. Suppl.* **55**, 143

Rutten, R. J. and Zwaan, C.: 1983, *Astron. Astrophys.* **117**, 21

Vernazza, J. E., Avrett, E. H., and Loeser, R.: 1981, *Astrophys. J. Suppl. Ser.* **45**, 635

Wiese, W. L.: 1983, in West, R. M. (Ed), *Highlights of Astronomy*, **6**, 795

AVERAGE VARIATIONS OF PHOTOSPHERIC FEI AND FEII LINE PARAMETERS AS FUNCTION OF THE MAGNETIC FILLING FACTOR [*]

P.N. BRANDT[1] and M. STEINEGGER[2]

[1] *Kiepenheuer-Institut für Sonnenphysik*
Schöneckstr. 6
D-7800 FREIBURG, F.R.G.

[2] *Insitut für Astronomie*
Universitätsplatz 5
A-8010 GRAZ, Austria

ABSTRACT. A series of 17 Fourier transform spectra taken at the McMath telescope near disk center in regions of different magnetic field strengths were analyzed. Applying a multi-variate regression analysis magnetic filling factors $0 < \alpha \leq 0.11$ were determined. With α increasing from 0 to 0.11, line bisectors averaged over groups of lines of similar depth are found to show a blue shift decreasing from 0.35 km s^{-1} to nearly 0.1 km s^{-1}, when referred to the MgI line $\lambda5172.7$Å. The bisectors of FeII lines exhibit smaller blue shifts than FeI lines. The increase of bisector red shift near the continuum with increasing α, found earlier by Brandt and Solanki (1987), was confirmed and is tentatively interpreted as a manifestation of downdrafts in the vicinity of flux tubes (Deinzer et al., 1984).

A significant *increase of line width* (typically between 3 and 8%, depending on line strength) and a *decrease of line depth* is found with increasing filling factor. For strong lines the equivalent width W shows no variation or a slight increase, while for the weaker lines a reduction of W between a few % and > 10% is found.

1. Introduction

The asymmetries and wavelength shifts of solar line profiles have been known for over 30 years (e.g. Schröter, 1957; for a summary cf. Dravins, 1982). They are usually attributed to the brightness-velocity correlation of the granulation. The "C"-shaped line asymmetry can be observed in spectrograms of high spatial resolution (Keil and Yackovich, 1981; Kavetsky and O'Mara, 1984; Mattig et al., 1989) as well as — to a lesser degree — in spectra representing spatial averages (e.g. Adam et al., 1976; Dravins et al., 1981; Brandt and Schröter, 1982). The Sun observed "as a star" also exhibits this line asymmetry (Livingston, 1982) and thus provides a "Rosetta stone" for the indirect observation of surface convection on stars (cf. Gray, 1982; Dravins, 1987a, 1987b).

The influence of *magnetic fields* on solar and stellar convection is of basic astrophysical interest and has been studied theoretically (cf. reviews by e.g. Nordlund, 1986; Hughes and Proctor, 1988) and observationally (e.g. Mattig and Nesis, 1976; Livingston, 1982, 1983; Kaisig and Schröter, 1983; Brandt and Schröter, 1984; Miller et al., 1984; Title et al., 1986, 1989; Immerschitt and Schröter, 1987, 1989; Cavallini et al., 1985, 1988; Brandt and Solanki, 1989). Many of these investigations are based on the difference of the "C" shape and wavelength shift between quiet and active regions

---

[*] Mitteilungen aus dem Kiepenheuer-Institut No. 311

41

J. O. Stenflo (ed.), *Solar Photosphere: Structure, Convection, and Magnetic Fields, 41–46.*
© 1990 by the IAU.

on the Sun ('plages') and show as a result some straightening of the line bisectors and a slightly reduced blue shift in active regions which is interpreted as an indication of reduced dynamics in these areas. We present an analysis of Fourier transform spectrometer observations, which comprises a quantitative determination of the magnetic filling factors of the observed regions as well as a treatment of a large set of FeI and FeII lines.

2. Observations and Data Evaluation

For the analysis a set of 17 Fourier transform spectra were selected. They were observed in June 1984 at the McMath solar telescope in regions of varying magnetic activity and quiet Sun regions near disk center ($\cos \theta \geq 0.95$) with a slit of 5 by 25 arcsec^2. The observed wavelength range was $\lambda\lambda 5050 - 6650$ Å with a resolution of 180 000 and a S/N ratio of 2000 to 3000. An integration time of 13.7 min was used in order to average over several periods of the 5 min oscillation.

Continuum values were determined by parabolic fits through the highest intensity points in 75 Å sections of the spectra and for ease of treatment the data were Fourier interpolated to wavenumber steps of 0.00656 cm^{-1} (3 mÅ - 1.7 mÅ in wavelength). The magnetic filling factors α were estimated with the method developed by Stenflo and Lindegren (1977). This method is based on a multi-variate regression analysis of a line parameter (like line width or depth of the unpolarized profile) sensitive to Zeeman splitting. An average line weakening of 0.7, a continuum contrast of 1.4 and a magnetic field strength of 1500 G was assumed for the magnetic elements (for details cf. Stenflo and Lindegren, 1977; Brandt and Solanki, 1987, 1989). Using 182 unblended FeI lines in each spectrum, magnetic filling factors α ranging from 0.00 to 0.11 were obtained for the 17 spectra.

For further analysis 187 FeI lines and 23 FeII lines were selected from the list given by Solanki and Stenflo (1985). They fulfilled the conditions of having no blends and line depths $\geq$ 0.1. For each of these lines the position and residual intensity of the line minimum, the full width at half line depth 'FWHM', the equivalent width 'W' (by integration up to the nearest blend in the red and blue wing), and the bisector shape (between line minimum and $I/I_c = 0.92$) were determined.

3. Results

3.1 LINE BISECTORS: ASYMMETRIES AND SHIFTS

For the specification of the shape of the line bisectors in the lower and upper part the following definitions were used: the wavelength difference ($\lambda_{min} - \lambda_{0.7}$) between line minimum and $I/I_c = 0.7$ was denoted $\Delta\lambda_{\nabla}$, while correspondingly ($\lambda_{0.7} - \lambda_{0.9}$) was denoted $\Delta\lambda_{\Delta}$. Fig. 1 shows a plot of these wavelength differences averaged over the group of 38 of the strongest FeI lines (line depths between 0.7 and 0.9 I/I_c) as a function of the filling factor. The systematic *decrease* of $\Delta\lambda_{\nabla}$ in the lower part of the bisector and the *increase* from -2.2 mÅ to $\approx$ -5mÅ in the upper part can be seen clearly. The groups of the weaker lines show essentially the same behaviour somewhat less pronouncedly, due to the smaller range of heights of formation of these lines.

The wavelength positions of bisectors were determined with reference to the core of the MgI line $\lambda5172.7$ Å, which is supposed to show negligible granular blue shift (cf. Pierce and Breckinridge, 1973). The bisector positions were averaged in a two step process: 1st groups of lines of similar depths were averaged in each spectrum, and 2nd these averages were again averaged for spectra grouped according to increasing α. The results are shown in Fig. 2a-c for two groups of FeI lines and one group of FeII lines.

For the strongest FeI lines a consistent *decrease* of the blue shift of the bisectors with increasing α from about 0.35 km s^{-1}at half the line depth to nearly 0.1 km s^{-1} for α between 0.04 and 0.08 with

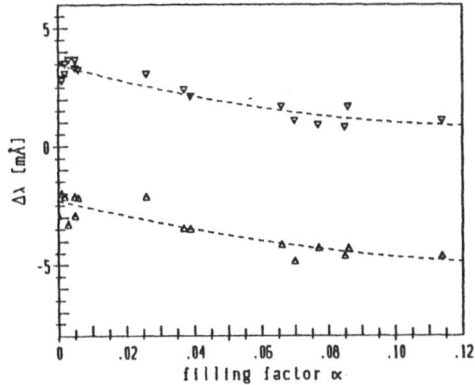

Fig.1: Wavelength differences in the lower part ($\nabla = \lambda_{min} - \lambda_{0.7}$) and upper part ($\Delta = \lambda_{0.7} - \lambda_{0.9}$) of bisectors averaged over 38 FeI lines of depths between 0.7 and 0.9 I/I_c.

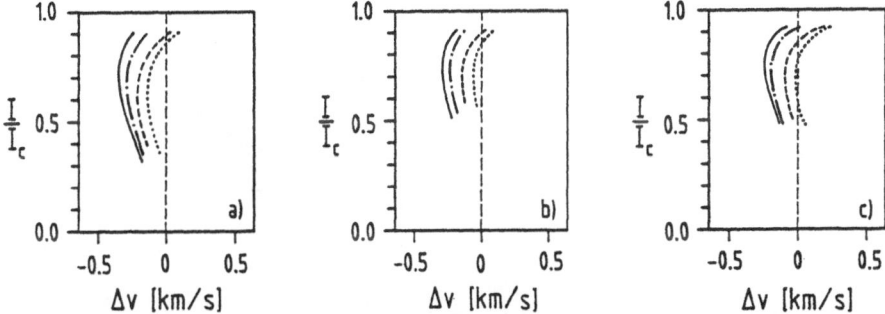

Fig. 2a-c: Average bisectors for groups of different line depths. a) 38 FeI lines of $0.7 < d \leq 0.9$; b) 73 FeI lines of $0.5 < d \leq 0.7$; c) 3 FeII lines of $0.5 < d \leq 0.7$. Additionally the bisectors are averaged over several spectra grouped with increasing α. Symbols: 8 spectra of $\alpha < 0.01$: ——— ; 3 spectra of $0.01 \leq \alpha < 0.04$: — · — · — · ; 3 spectra of $0.04 \leq \alpha < 0.08$: · · · · · ·; 3 spectra of $\alpha \geq 0.08$: – – – – .

a slight increase for higher values of α is found. The FeII bisectors show blueshifts that are smaller by about 0.15 km s^{-1}.

3.2 LINE HALFWIDTHS

In plots similar to Fig. 1 for each line second order polynomials were fitted into the variation of its FWHM with α, and the relative change between $\alpha = 0$ and $\alpha = 0.11$ was determined. For the group of the 38 strongest FeI lines ($0.7 < d < 0.9$) there is a consistent *increase* of the FWHM of 5 to 10%. Here the Zeeman broadening was not subtracted. But this will be done in a future analysis. This is not mainly an effect of Zeeman broadening, which is demonstrated by the fact that the 3 FeI lines with g=0 (λ 5576.1 Å, λ 5434.5 Å, λ 5123.7 Å) show a broadening of between 4 and 6%. The weaker lines increase their FWHM to a lesser extent: the group of $0.5 < d < 07$ between 3 and 8%, the group of $0.3 < d < 0.5$ between 0 and 6%, and the weakest lines between 0 and 5%. Lines with excitation potential $\chi_e > 3$ eV tend to exhibit smaller increases of the FWHM.

3.3 LINE DEPTHS

With α increasing from 0 to 0.11 all lines exhibit a significant *decrease of the line depth*, i.e. the lines become shallower. For the strongest lines ($0.7 < d < 0.9$) this decrease ranges between 5 and $> 10\%$, and reaches values between 7 and $\approx 17\%$ for the weaker lines. Here the lines of low excitation potential ($\chi_e \leq 3$ eV) show a significantly stronger decrease of d than those of high χ_e. The FeII lines consistently have a smaller decrease of line depth than the corresponding FeI lines of similar strength.

3.4 EQUIVALENT WIDTHS

Fig. 3a,b shows the variation of the equivalent width W with α increasing from 0 to 0.11 as a function of the equivalent width itself. While the strongest lines show no change at all or a slight increase of W, the medium-strong lines show a reduction between a few % and 10%. The strongest decrease of W ($\Delta W/W \approx 13\%$) is seen for the weakest lines, especially those of low excitation potential ($\chi_e < 3$ eV).

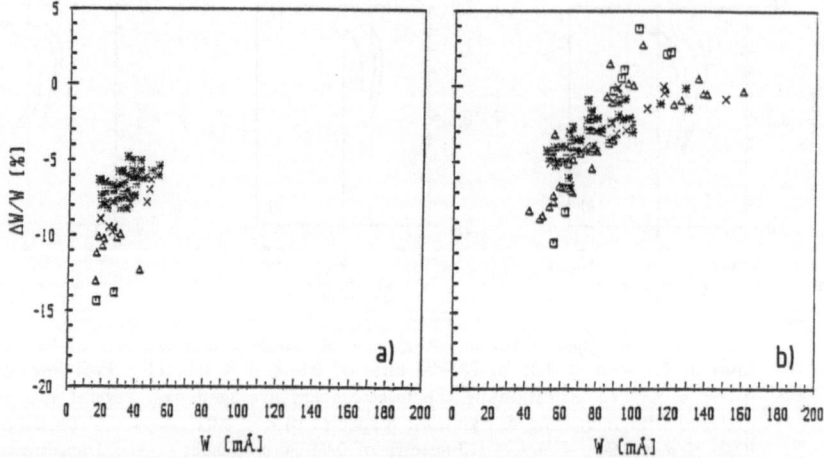

Fig. 3a,b: Changes of the equivalent widths between $\alpha = 0$ and $\alpha = 0.11$ as function of the equivalent width W at $\alpha = 0$. a) 76 lines of depth d< 0.5; b) 111 lines of depth d> 0.5. Symbols: □ $\chi_e \leq 2$ eV; △ $2 < \chi_e \leq 3$ eV; × $3 < \chi_e \leq 4$ eV; * $\chi_e > 4$ eV.

4. Discussion and Conclusion

The gradually reduced blue shift of the line bisectors seen in active regions of increasing filling factor α can be interpreted in terms of a *reduced* vertical velocity component of the granular velocity field, possibly coupled to a modified intensity contrast. This would agree well with spectroscopic observations of reduced rms velocities in active regions (Nesis et al., 1989), which are indirectly confirmed by the observation of Title et al. (1986, 1989) of increased granular lifetimes and smaller horizontal proper motion velocities of granules in active regions, both facts hinting at reduced dynamics. On the other hand, since no net flow velocities in fluxtubes could be detected hitherto (Solanki, 1986), the increase of the redshift of the uppermost bisector part with increasing α, already described by Brandt and Solanki (1987), may be due to *increased* downward velocities in the immediate vicinity of flux tubes, as postulated from model calculations by Deinzer et al. (1984). Also the broadening of most lines, especially of the g=0 lines not affected by Zeeman broadening, in active regions seems to indicate increased "macro-turbulence", rather than a reduced velocity field. Only detailed model calculations, taking into consideration a realistic fluxtube model and a modified structure of the convection outside the fluxtubes can help to understand the riddle. The observations presented here try to establish reliable reference data to which the results of such model calculations can be compared.

ACKNOWLEDGEMENTS

The observations were carried out at the McMath solar telescope (N.O.A.O., Tucson). The assistance of J. Brault, B. Graves, R. Hubbard and G. Ladd is gratefully acknowledged. The National Optical Astronomy Observatories are operated by the Association of Universities for Research in Astronomy, Inc., under contract with the National Science Foundation. We are also indebted to S. Solanki for carrying out the filling factor analysis. One of us (P.N.B.) gratefully acknowledges financial support from the Deutsche Forschungsgemeinschaft.

REFERENCES

Adam, M.G., Ibbetson, P.A., Petford, A.D.: 1976, *Mon. Not. Roy. Astron. Soc.* 177, 687

Brandt, P.N., Schröter, E.-H.: 1982, *Solar Phys.* 79, 3

Brandt, P.N., Schröter, E.-H.: 1984, *Small-Scale Dynamical Processes in Quiet Stellar Atmospheres*, ed. S. Keil, NSO Conf., 371

Brandt, P.N., Solanki, S.K.: 1987, *The Role of Fine-Scale Magnetic Fields on the Structure of the Solar Atmosphere*, eds. E.-H. Schröter, M. Vazquez, A.A. Wyller, Cambridge Univ. Press, 82

Brandt, P.N., Solanki, S.K.: 1989, in preparation

Cavallini, F., Ceppatelli, G., Righini, A.: 1985, *Astron. Astrophys.* 143, 116

Cavallini, F., Ceppatelli, G., Righini, A.: 1988, *Astron. Astrophys.* 205, 278

Deinzer, W., Hensler, G., Schüssler, M., Weisshaar, E.: 1984, *Astron. Astrophys.* 139, 435

Dravins, D.: 1982, *Ann. Rev. Astron. Astrophys.* 20, 61

Dravins, D.: 1987a, *Astron. Astrophys.* 172, 200

Dravins, D.: 1987b, *Astron. Astrophys.* 172, 211

Dravins, D., Lindegren, L., Nordlund, Å.: 1981, *Astron. Astrophys.* 96, 345

Gray, D.F.: 1982, *Astrophys. J.* 255, 200

Hughes, D.W., Proctor, M.R.E.: 1988, *Ann. Rev. Fluid Mech.* 20, 187

Immerschitt, S., Schröter, E.-H.: 1987, *The Role of Fine-Scale Magnetic Fields on the Structure of the Solar Atmosphere*, eds. E.-H. Schröter, M. Vazquez, A.A. Wyller, Cambridge Univ. Press, 53

Immerschitt, S., Schröter, E.-H.: 1989, *Astron. Astrophys.* 208, 307

Kaisig, M., Schröter, E.-H.: 1983, *Astron. Astrophys.* 117 305

Kavetsky, A., O'Mara, B.J.: 1984, *Solar Phys.* 92, 47

Keil, S.L., Yackovich, F.H.: 1981, *Solar Phys.* 69, 213

Livingston, W.C.: 1982, *Nature* 297, 208

Livingston, W.C.: 1983, *Solar and Stellar Magnetic Fields: Origins and Coronal Effects*, ed. J.O. Stenflo, IAU Symp. No. 102, 149

Mattig, W., Nesis, A.: 1976, *Solar Phys.* 50, 255

Mattig, W., Hanslmeier, A., Nesis, A.: 1989, *Solar and Stellar Granulation*, Proc. NATO Adv. Res. Workshop, Capri, June 21-25, eds. R. Rutten and G. Severino, Kluwer, Dordrecht, 187

Miller, P., Foukal, P., Keil, S.: 1984, *Solar Phys.* 92, 33

Nesis, A., Fleig, K.-H., Mattig, W.: 1989, *Solar and Stellar Granulation*, Proc. NATO Adv. Res. Workshop, Capri, June 21-25, eds. R.J. Rutten and G. Severino, Kluwer, Dordrecht, 289

Nordlund, Å.: 1986, *Small Magnetic Flux Concentrations in the Solar Photosphere*, eds. W. Deinzer, M. Knölker, H.H. Voigt, Vandenbeck & Ruprecht, Göttingen, 83

Pierce, A.K., Breckinridge, J.B.: 1973, Kitt Peak National Obs. Contr. No. 559

Schröter, E.-H.: 1957, *Z. Astrophysik* 41, 141

Solanki, S.K.: 1986, *Astron. Astrophys.* 168, 311

Solanki, S.K., Stenflo, J.O.: 1985, *Astron. Astrophys.* 148, 123

Stenflo, J.O., Lindegren, L.: 1977, *Astron. Astrophys.* 59, 367

Title, A.M., Tarbell, T.D., Simon, G.W., and the SOUP Team: 1986, *Adv. Space Res.* Vol. 6, No. 8, 253

Title, A.M., Tarbell, T.D., Topka, K.P., Ferguson, S.H., Shine, R.A., and the SOUP Team: 1989, *Astrophys. J.* 336, 475

II. PHOTOSPHERIC FINE STRUCTURE

High Resolution Observations of the Photosphere

A. M. Title, R. A. Shine, T. D. Tarbell, K. P. Topka
Lockheed Palo Alto Research Laboratory
3251 Hanover Street, Palo Alto, California 94304, USA

and G. B. Scharmer
Stockholm Observatory, S-133 00 Saltsjöbaden, Sweden

ABSTRACT. High resolution observations, theoretical models, and simulations are discovering many new and exciting phenomena in the solar atmosphere. In recent years, there have been a number of very high quality observations of the solar surface and lower photosphere made on the ground at Sacramento Peak Observatory, Pic du Midi, and at the Swedish Solar Observatory, La Palma. In space the Solar Optical Universal Polarimeter (SOUP) has made diffraction limited (30 cm aperture) time sequences completely free from atmospheric disturbances. The recognition that significant progress is possible in non-linear dynamics has encouraged a number of theoretical groups to attack the problem of convection in the solar atmosphere. Two, two and a half, and three dimensional simulations yield the geometry of the flow below the surface and a prediction of the response of the atmosphere above the surface. Models of magnetic flux tubes are now very sophisticated, and modern high resolution observations should be able to test these theories. The development of the technique of Local Correlation Tracking (LCT) has allowed the direct measurement of horizontal velocities in the atmosphere near disk center. The combination of Doppler and LCT measurements allows a direct measurement of the photospheric vector flow field. Measurements from SOUP, Sacramento Peak, Pic du Midi, and La Palma have shown that mesoscale flows cover the surface and that there exist still larger scale flows associated with emerging pores and active regions. Much of the recent experimental and theoretical progress in processing and understanding high resolution data has resulted from the availability of powerful scientific workstations for user interaction, large amounts of memory for image storage, and supercomputers for the massive fluid dynamics calculations. We are now in the very early stages of learning how to use these new computer tools to identify and follow processes in the solar atmosphere.

1. Introduction

High resolution observations of the Sun are now yielding important insights into

49

J. O. Stenflo (ed.), Solar Photosphere: Structure, Convection, and Magnetic Fields, 49–66.

the basic physical processes of convection, magnetic field structure, and the interactions between magnetic and velocity fields. The development and operation of new telescopes, observing techniques, camera systems, and computer systems with mass storage and interactive image displays are responsible for this progress.

The past few years has seen the completion of a number of new telescopes in the Canary Islands, Spain. Although only in their first years of operation, results of exceptional quality have already been produced. Figures 1 (a), (b), and (c) show images of quiet sun (a) and penumbra (b) taken at the Swedish Solar Observatory (SSO) and a spectrum (c) from the Vacuum Tower Telescope of the Federal Republic of Germany. Further development of these observatories and their instrumentation will occur as they come into full operation. In addition more telescopes for the Canary Islands are either under construction or in the planning stage. These include the French Themis and the Utrecht Open Tower Telescope. In the more distant future the 2.5 meter Large Earth based Solar Telescope (LEST), whose site has not as yet been selected, will greatly increase the potential for obtaining high resolution data from the ground.

On August 5 and 6, 1985 the Solar Optical Universal Polarimeter (SOUP) mounted on the European Space Agency's (ESA) Instrument Pointing System (IPS) in the United States Space Shuttle collected about 6000 diffraction limited (30 cm aperture) images of the solar surface. Because of the short orbital observing day, the time sequences are limited to 30 – 45 minutes. These sequences are pointed stably to 0.003 arc second root mean square (RMS) and are completely free from atmospheric blurring and distortion. Besides being a valuable scientific asset for studies of granulation, sunspots, and pores, they provide a baseline for evaluation of new data analysis techniques for ground based data.

Similar results can also be obtained by high altitude balloon flights. However, even at 35 to 40 km there is sufficient atmosphere to degrade images, if the telescope is not well designed thermally. SOUP will be reflown on balloons during the Max '91 period. The later half of the next decade should see the flight of the Orbiting Solar Laboratory (OSL), which includes 1 meter optical, 30 cm UV, and 30 cm XUV telescopes which feed filtergraphs, spectroheliographs, and spectrographs. OSL is a free flying polar - orbiting satellite which will have uninterrupted solar observations for approximately 240 days per year. The design life of OSL is three years and it should operate at least partially for a decade.

In the instrument area narrowband filter systems have been developed that are tunable over a broad spectral range (3500–6800 Å) and have bandpasses as narrow as 20 milliangstroms. New spectrographs with spectral resolutions of 500,000 and more are demonstrating spatial resolution of 0.3 arc seconds. Initial experiments with IR arrays show great promise for exploring the deep photosphere and the temperature minimum region and for directly measuring magnetic field strengths.

Agile mirrors (tilt correction) are in regular use in several observatories. When the seeing is good, an agile mirror can remove a sizable fraction of the wavefront error. Of course, they also remove virtually all drive errors and shake introduced

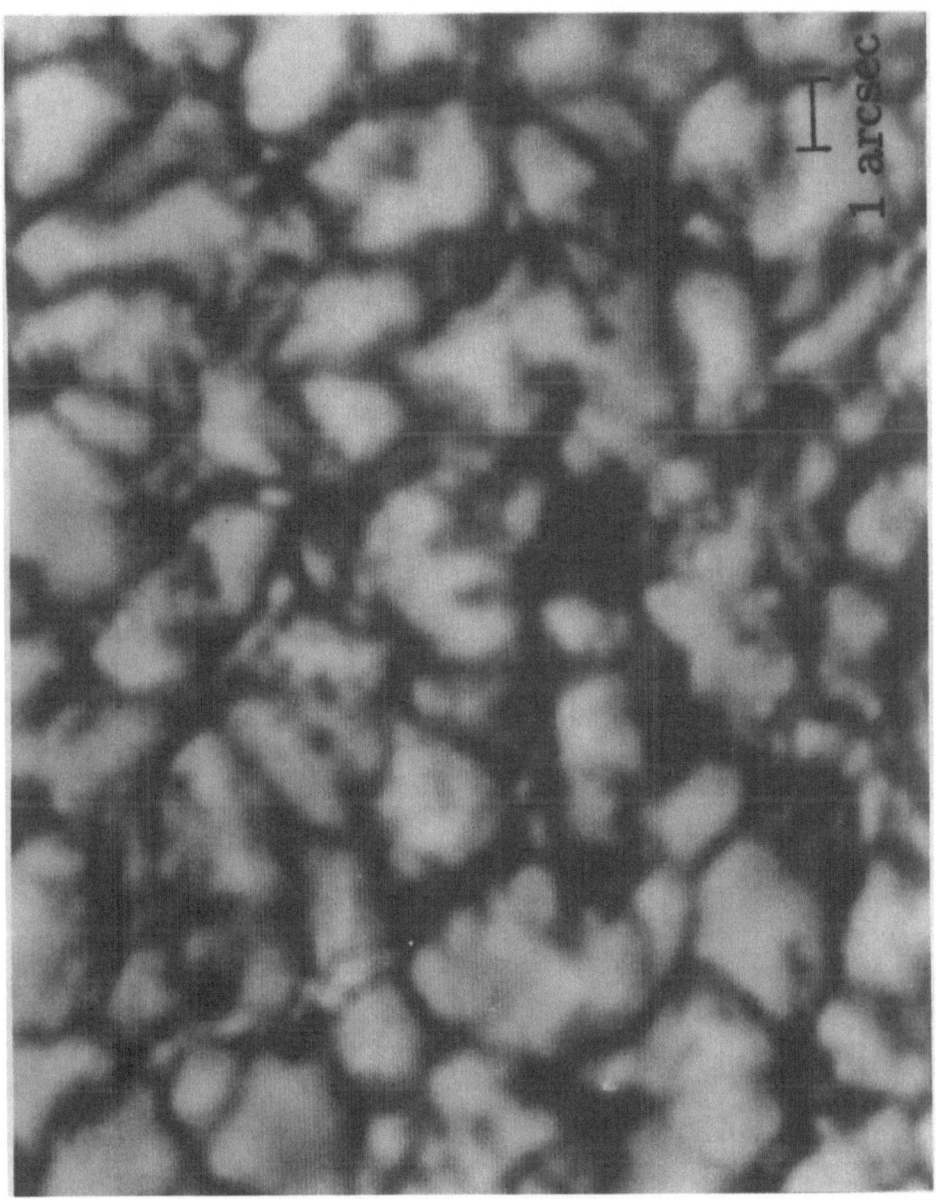

Fig. 1 (a)   Image of quiet sun taken at Swedish Solar Observatory, La Palma.

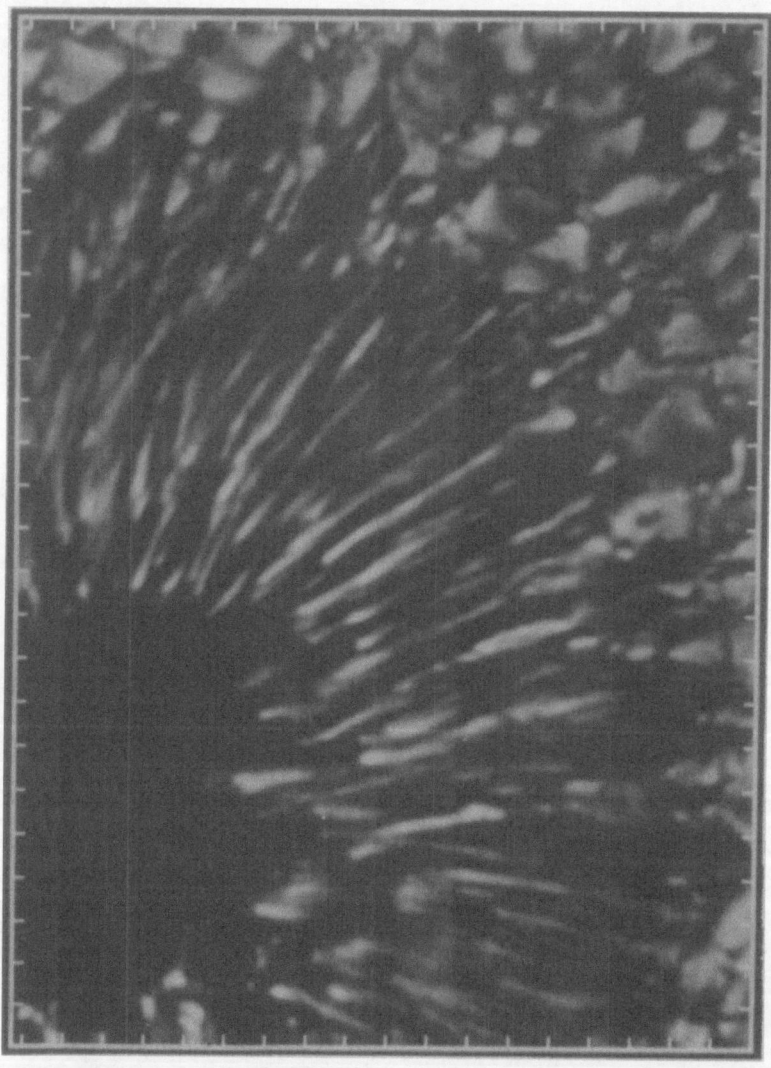

Fig. 1 (b) Image of penumbra taken at the Swedish Solar Observatory. Tick marks are at 1 arc second intervals.

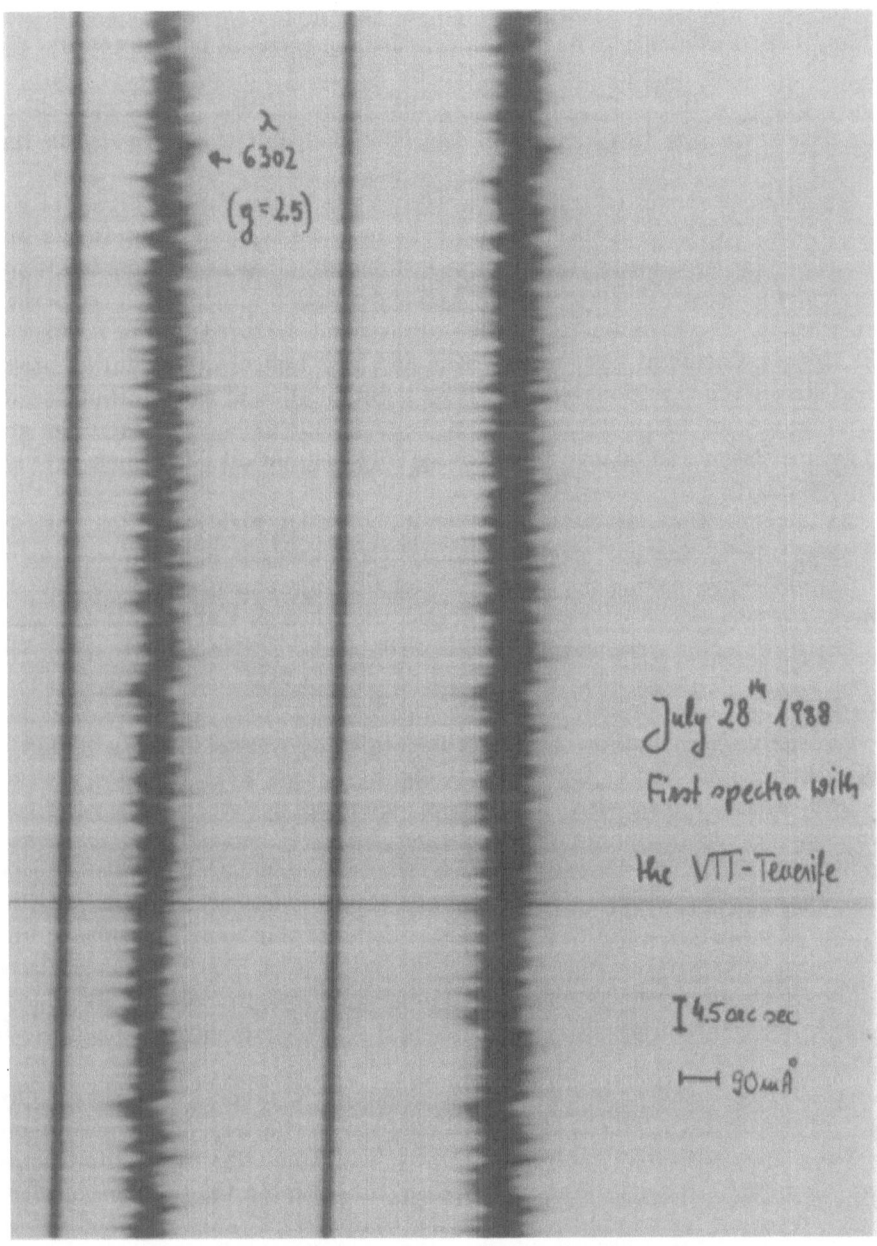

Fig. 1 (c)   Images of Solar Spectrum near Iron 6302 taken at the Vacuum Tower Telescope of the Federal Republic of Germany on Tenerife.

by the telescope itself. Measurements have shown that there is sufficient power in atmospheric turbulence to degrade images at frequencies as high as 100 Hz. Control signals for agile mirrors are generated by quadrant photodetectors operating on spots, pores, or the vertices in intergranular lanes. Several groups have developed correlation trackers to generate the control signal, but these are still in the test phase.

An adaptive mirror (wavefront correction) was successfully operated at the National Solar Observatory/Sacramento Peak by the Lockheed group in the summer of 1988. This technology should allow future diffraction limited observations over regions of at least a few arc seconds and significant improvement over perhaps 10 arc seconds. Figure 2 illustrates the improvement achieved by the Lockheed mirror. Measurements at Sacramento Peak and SSO indicate that, during periods of good seeing, the variations are caused by a combination of weak turbulence at high altitudes (about 10 km) and rather stronger turbulence much nearer the ground. In such situations an adaptive mirror may yield significant corrections over regions of arc minutes.

At the same time, as the capabilities for collecting high resolution data are expanding rapidly, the potential for reduction and analysis is also increasing. Solar observatories are making the transition from photographic film to solid state arrays. Detectors with 512×512 pixels are commonly available and at least one manufacturer sells detector systems with 1024×1024 pixels. It is reasonable to expect that 2048×2048 and even 4096×4096 camera systems will come in to use in the next few years.

Currently a combination of data transfer rate and storage capacity of mass memory systems limit some applications of digital camera systems. Each time a 1024×1024 camera is read with 12 bit per pixel accuracy, 12×10^6 bits must be stored. One read per second implies a system transfer rate of 1.5 megabytes per second and a storage capacity of 5.4 gigabytes per hour. This combination of high transfer rate and high capacity is currently beyond the state of the art in commercial devices which are relatively inexpensive. But it is now possible to collect 1024×1024 arrays at an average rate of one every 6 seconds and store 1500 such images or nearly 2.5 hours of observations on a single 8mm video cassette using a recorder which costs about $5000. Laser disk video recorders allow rapid and very agile viewing of the data.

When data are taken in digital form, computers can be used to extract physical parameters directly. This is in marked contrast to the very recent past where the first step in quantitative analysis was microdensitometering many film frames. (In fact the new CCD's can be used to reduce considerably the problem of digitizing film.) Once an image sequence is in the computer, a range of processes can be applied. These include removal of atmospheric distortions, image differencing and ratioing, local correlation tracking to measure horizontal velocities, 3D Fourier filtering to eliminate or isolate particular classes of waves, overlays of simulations or predictions on movies, and overlays of movies on movies to name but a few. These

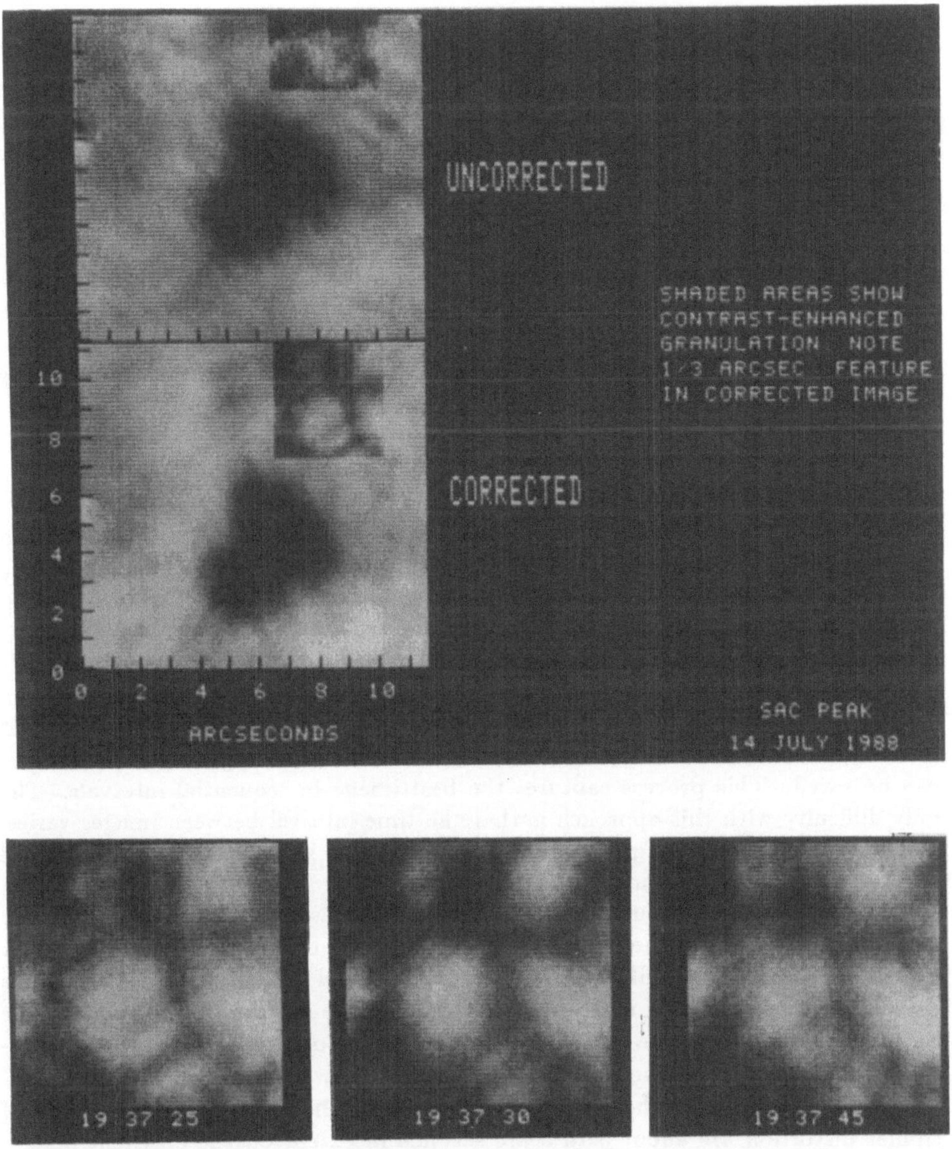

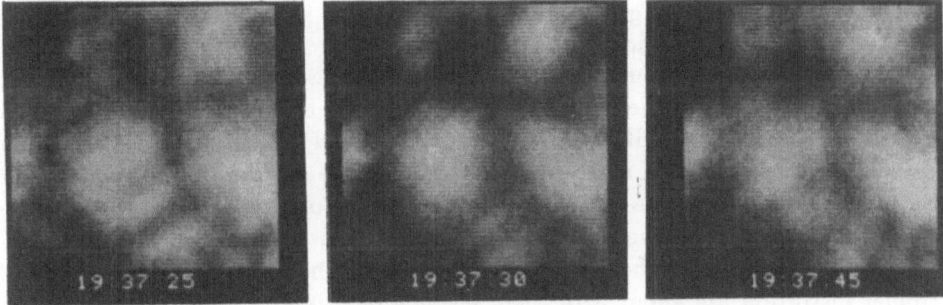

Fig. 2   Corrected and uncorrected granulation images which illustrate the improvement achieved by the Lockheed adaptive mirror at Sacramento Peak Observatory. Shown in the bottom of the figure are images from the corrected field of view. The region is 4.5 arcseconds square.

data manipulation techniques are now possible on workstation class computer systems with high resolution data displays and several gigabytes of mass storage.

In parallel with the new high resolution observations there have been significant advances in theory and simulation of solar phenomena. There are now detailed models of small flux tubes which can be tested against observations of magnetic fields, velocity fields, and line and continuum intensities. Simulations of granulation are now making predictions of phenomena in and above the surface. Predictions are also made about flow below the surface. These may be tested by using a combination of helioseismological techniques and measurements of surface flows. Progress in nonlinear dynamics, Chaos theory, is yielding new insights into convection, turbulence, and dynamo generation of large and small scale magnetic fields.

In the sections below we will discuss some of the more recent and interesting discoveries using high resolution observations.

2. Continuum Observations

Over the past several decades the solar observatory on Pic du Midi has produced the very highest quality images of the solar surface. The basic technique for data collection has been to take periodic bursts of images on film. The best images in these bursts are then manually selected for study. Recently at the SSO, this process has been automated and implemented in real time. A video camera records an image every 20 milliseconds and a seeing monitor evaluates the image quality on every video field. The image and image quality value are stored in fast computer memory, so that the "best" image in a chosen time interval, typically 5–10 seconds, can be saved. This process captures the best image in sequential intervals. The only difficulty with this approach is that the time interval between images varies. But the immense advantages are that a quality movie is available in real time and all the poorer images do not have to be saved.

For studying time evolution of solar phenomena, as many artifacts as possible must be removed from the images. A time sequence of images from an excellent groundbased site still exhibits serious artifacts caused by atmospheric distortions even after image selection and removal of gross image motion. Unfortunately, the better the images the more visually disturbing the distortions are to movie observation. Differential image motion across the field of view, "rubber sheet distortion," remains even in images whose sharpness approaches the diffraction limit. Areas of similar distortion are about 3 to 5 arc seconds in extent on the excellent images from the Canary Islands and Pic du Midi. (The small size of the distortion patches is most likely due to weak turbulence at 10 km, and probably should be expected at any excellent ground based site.) Because of the sharpness of the best images, distortions are not clear from single images and are hard to recognize on sequences of prints. However, the difference between movies with and without distortions is remarkable. Fortunately, distortions can be removed from rapid time sequences by procedures which assume that the rapid displacements of distortion patches are atmospheric rather than solar. Of course, such procedures must be used with some caution because of the possibility of introducing artificial flows into the data in the process of removing distortions. Distortion removal codes are now relatively costly in computer time. "Destretching" a set of 500 512 × 512 16 bit images takes approximately 30 hours on a Vaxstation 3200.

High resolution observations of the solar surface show a complex array of interacting phenomena – the global oscillations, granulation, mesogranulation, supergranulation, thermal and internal gravity waves, and quasi-stationary flow fields with various scales. Furthermore, all of these properties change with the magnetic field filling factor. To understand the physics of the solar surface, it is essential first to separate these processes.

The data from SOUP showed the importance of the global oscillations in the intensity fluctuations observed in the surface, and how these obscured the evolution of the solar granulation. Figure 3 shows the temporal autocorrelation function (ACF) of the intensity field from groundbased data and SOUP. The agreement is remarkable, especially considering the fact that the spatial resolution varies from about 0.3 arc second in the La Palma data to at least one arc second in some of the other data. Figure 4 further illustrates the robustness of the ACF to resolution by the comparison of the ACF of original (a) and data smeared to one arc second (c). Curves (a) and (c) do not differ significantly. The ACF's and the movies suggest that the global oscillations play a dominant role in the temporal autocorrelation function.

The spatio-temporal $(k - \omega)$ dispersion relation for the global oscillations is well known. It was therefore possible to remove global oscillations from the SOUP data by three dimensional (3D) Fourier transforming from the $x - y - t$ space of the original data to $k_x - k_y - \omega$ space, applying a filter in this space to delete the region of the oscillations, and then inverse transforming back to $x - y - t$ space. Figure 4 (b) and (d) show the ACF functions for the original (a) and degraded (b) image set, respectively, after filtering. Figure 4 demonstrates that the five minute oscillation does dominate the ACF's and that after the oscillations are removed the difference between original and degraded data is distinct.

Removal of the global oscillations allows the difference between quiet and magnetic Sun to be measured. Figure 5 shows the ACF's for regions of quiet and magnetic sun (average magnetogram signal greater than 75 gauss) before (a) and (b) and after (c) and (d) Fourier filtering. Before filtering there was some difference between quiet and magnetic ACF's, but after filtering the $1/e$ decay time differs by nearly a factor of three.

Because of the absence of seeing it was possible to use local correlation tracking (LCT) to follow local displacements of the intensity field in pairs of images and thus infer the horizontal velocities. The spatial resolution of the horizontal velocity measurements is determined by the field-of-view, "mask size," used for correlation tracking. Since the success of LCT on SOUP data, the technique has been used with great success on ground based data. Using the SOUP horizontal flow measure-

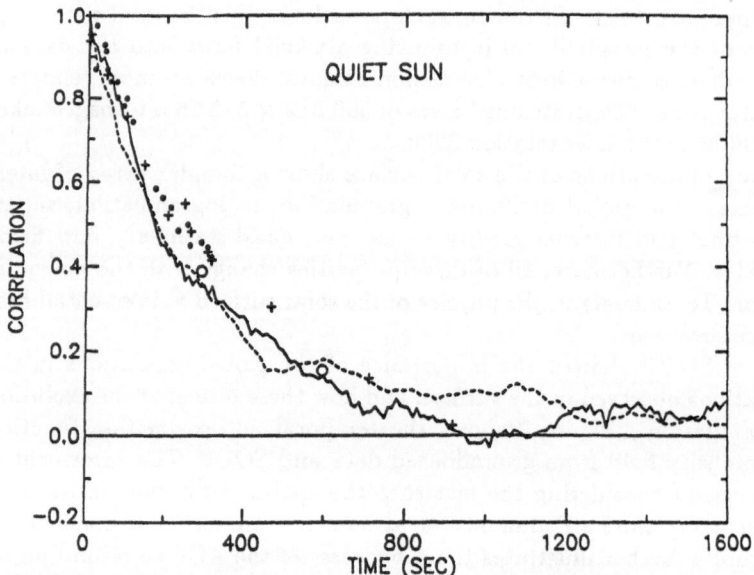

Fig. 3 The temporal autocorrelation function of the intensity field from two different quiet sun regions from SOUP (solid and dashed) and several different groundbased measurements (dots, crosses, and circles).

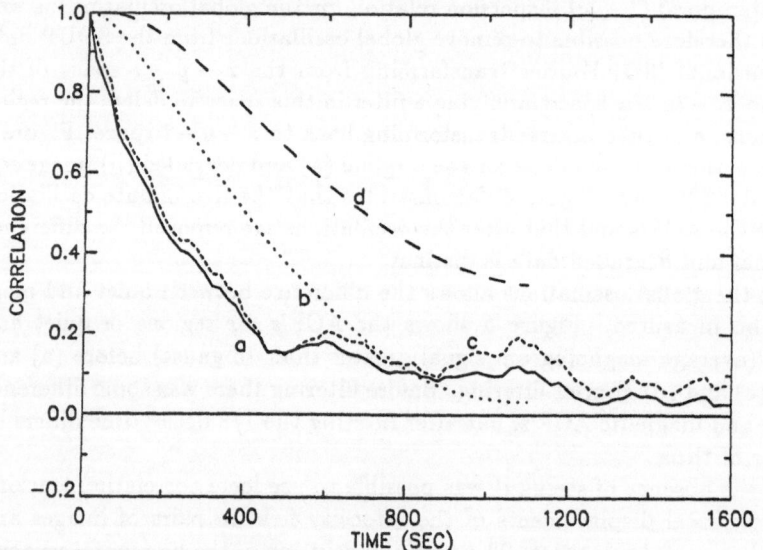

Fig. 4 Autocorrelation functions from 0.134 arc second/pixel SOUP quiet sun for original data (a), the same data Fourier filtered to remove oscillations (b), the same data as (a) smeared to 1 arc second resolution before autocorrelating (c), the same data as (b) smeared to 1 arc second resolution before correlating.

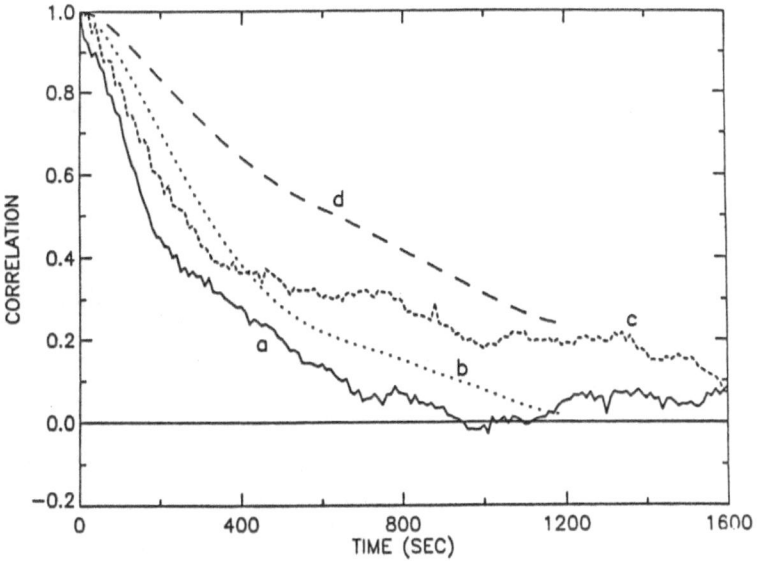

Fig. 5 Autocorrelation functions for quiet sun before (a) and after Fourier filtering (b) and for magnetic sun before (c) and (d) after Fourier filtering.

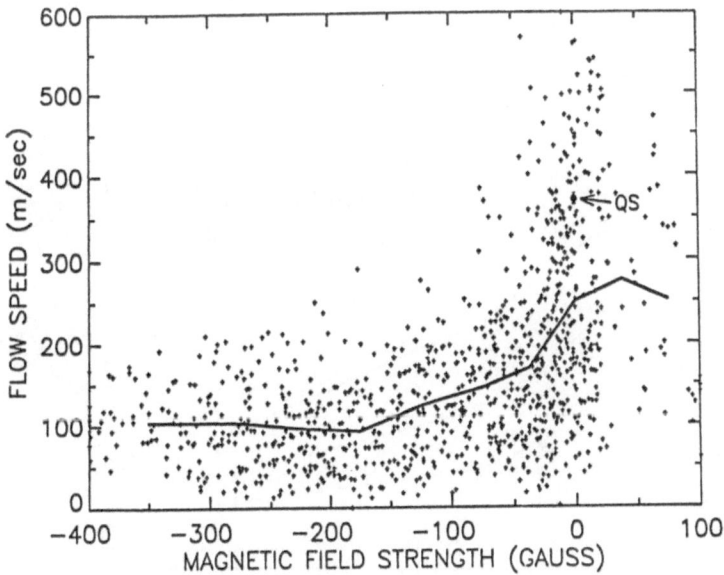

Fig. 6 Scatter plot of horizontal flow speed versus magnetogram signal. The mean flow speed versus magnetogram signal is shown solid. The dot labeled QS is the mean flow speed in zero gauss quiet sun.

ments, most of the decay of the temporal ACF could be explained by the spatial decorrelation of the granulation pattern because of local flows. Figure 6 shows a scatter plot of 25 minute (one orbit) averaged horizontal velocity as measured with a 3 arc second gaussian mask, versus magnetogram signal. In regions of magnetic field, the granulation pattern is advected by flows at a rate a third that in quiet Sun.

The instantaneous horizontal velocities measured by LCT can be used to calculate the RMS velocity fluctuations. Figure 7 shows the RMS velocity versus aperture mask size in regions with different amounts of magnetic flux. In quiet sun, apertures of the size of granulation yield RMS velocities of 1.6 km/s. This is sufficient to account for the width of solar lines. It can be seen from the figure that the vigor of the granulation is suppressed in magnetic regions, but not in regions between field structures.

Once the global oscillations were removed, it was evident from the SOUP movies that exploding granules were a very common phenomenon rather than a reasonably rare occurrence as previously reported. The birth rate and total area covered in the expansion phase is sufficient to cover the solar surface in about 20 minutes. Maps of the divergence of the horizontal flow fields reveal a structure with a mean diameter of about 7 arc seconds. This scale is clearly larger than granulation and smaller than supergranulation and can be fairly represented as mesogranulation. Exploding granules are almost always associated with areas of positive divergence which define the mesogranulation pattern. Thus the two processes are intimately

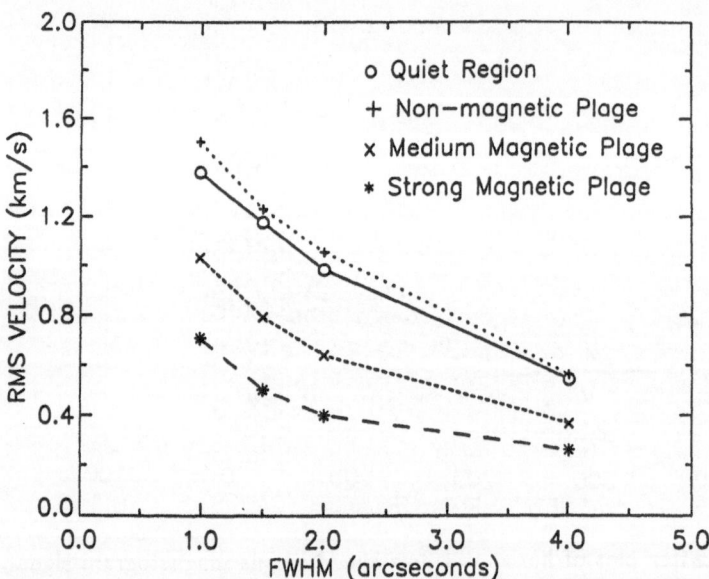

Fig. 7   RMS velocity versus width of the LCT gaussian mask for several different solar regions.

associated.

Measurements based on SOUP, La Palma, and Sacramento Peak data have yielded somewhat different results on mesogranulation inferred from LCT. Direct Doppler measurements from Sacramento Peak and Big Bear observations yield data on the mesogranulation which differ among themselves and with the LCT measurements. The LCT measurements are sensitive to the quality of the original data and the size of the sampling aperture used for the tracking. The Doppler measurements are sensitive to the image quality and time average required to remove 5 minute oscillations. Doppler measurements are also made at a very different height in the atmosphere. We believe that LCT has shown that there is an important flow scale in the photosphere that is larger than granulation with a lifetime on the order of an hour, that enough good data has not yet been reduced to define the flow sufficiently, and that the different interpretations are due to differences in temporal and spatial resolution and analysis procedures.

The existence of a scale larger than granulation in the surface flow is best exhibited by "corks" which are free floating test particles moved by the measured horizontal flow field. Corks are observed to collect in linear structures with a size of 6 to 7 arc seconds in about 30 minutes. Figure 8 shows a set of consecutive 30 minute evolutions of corks that cover 75 minutes. From the figure it is clear that the mesogranular structure evolves on the time scale of an hour.

3. Filtergram Observations

The techniques developed for obtaining and analyzing continuum images have also been applied to narrow band filter images. Because of the longer exposure times and more elaborate optical systems required, the image quality achieved is not as good as that of continuum images but the relative improvement over unstablized images is great. Figure 9 (a),(b),(c), and (d) show a continuum image, a line center image, a Dopplergram, and a magnetogram taken through a 80 milliangstrom tunable filter. All of the images were taken with exposure times of about 0.3 second using a 1024 × 1024 Texas Instrument virtual phase CCD.

The images shown in figure 9 are part of a time sequence of three hours duration. Every 50 seconds images were obtained in the blue wing of Fe 6302 Å (-60 mÅ) in right and left circularly polarized light, in four wavelengths through Ni 6768 (-90, -30, 30, 90 mÅ), and in the continuum near Ni 6768 (1500 mÅ). From these data movies have been created of all the wavelength steps, magnetograms, Dopplergrams, and the line center corrected for Doppler shift. All of the images have been corrected for atmospheric distortion and aligned with respect to each other to a fraction of a pixel (one pixel = 0.17 arc second.)

Comparison of the magnetogram, Dopplergram, and line center images in figure 9 shows that the vertical component of the velocity, the size of the of granules, the bright line center structures ("filigree") and the magnetic field are interrelated. In areas where the vertical component of velocity has average amplitude and the granules have their normal 2 arc second spatial structure, the line center bright

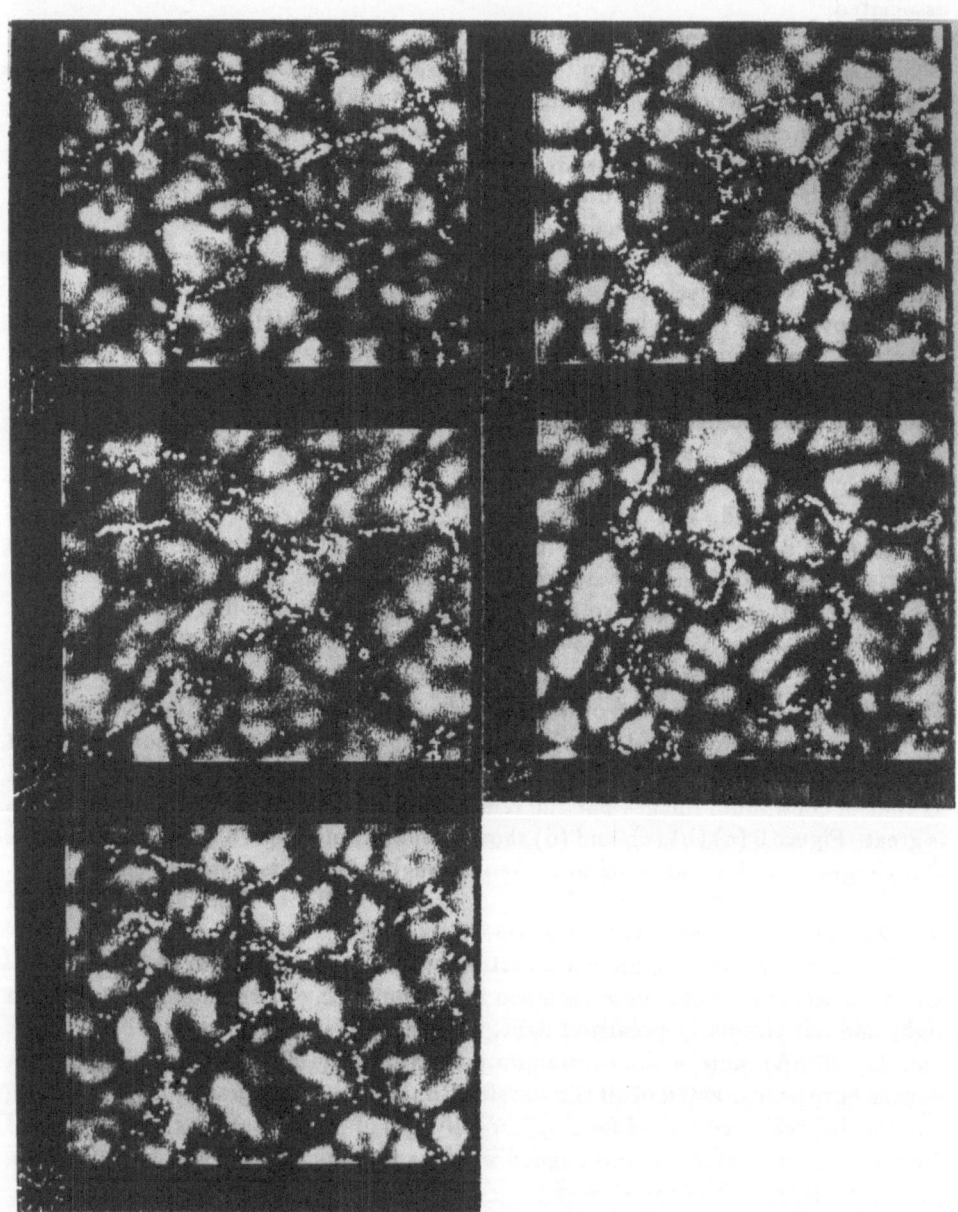

Fig. 8   A set of consecutive 30 minute cork evolutions covering 75 minutes.

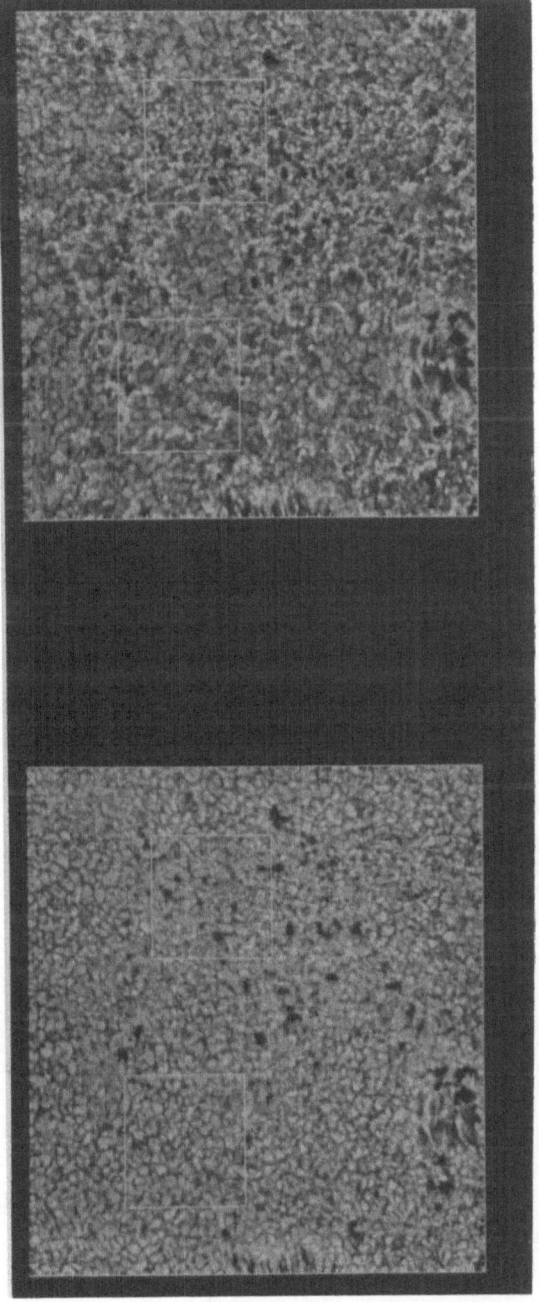

Fig. 9 (a), (b) Continuum image (a) and line center image (b) taken within 50 seconds at SSO through a 80 mÅ tunable filter. The white boxes overlayed on the images are shown expanded in figure 10. Tick marks are at 2 arc second intervals the field of view is 85 × 85 arc seconds

64

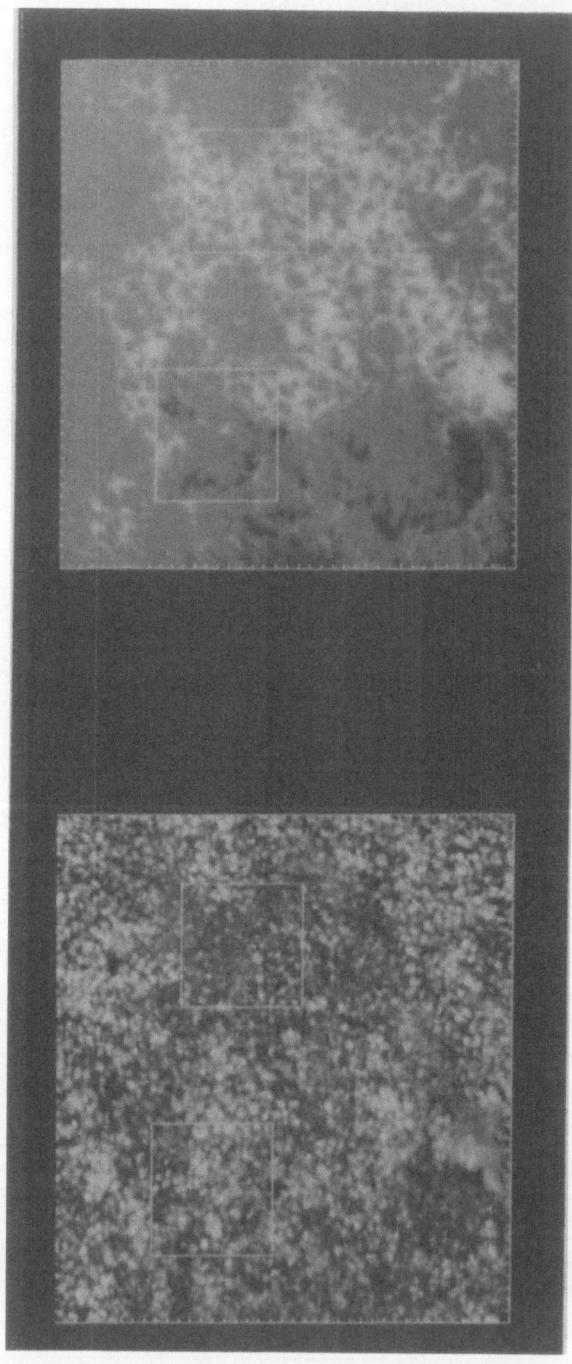

Fig. 9 (c), (d) Dopplergram (c) and a magnetogram (d) taken within 50 seconds at SSO through a 80 mÅ tunable filter. The white boxes overlayed on the images are shown expanded in figure 10. Tick marks are at 2 arc second intervals the field of view is 85 × 85 arc seconds

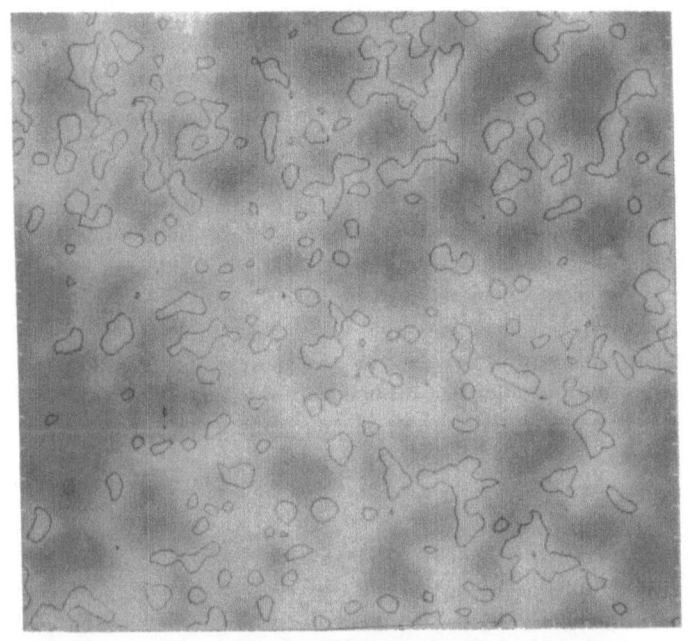

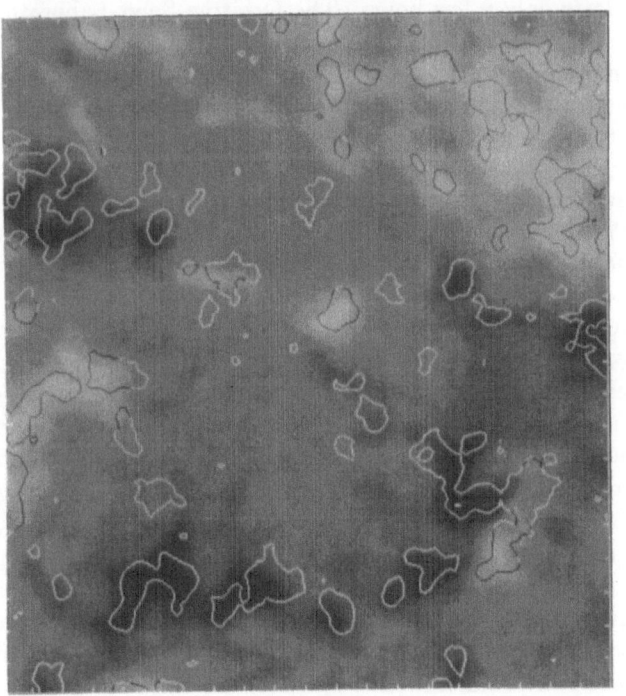

Fig. 10 Magnetograms from regions where the Doppler signal is average (a) and lower than average (b). Overlayed on the magnetograms are contours of the line center bright points. The locations of the magnetograms are marked on the images in figure 9. The tick marks are at 1 arc second intervals.

structures and magnetic field are very nearly co-spatial. This is illustrated in figure 10 (a) which shows the location of bright structures overlayed on the corresponding magnetogram. But in regions where both the vertical velocity is lower than average and the granules are an arc second or less, the brightness in line center breaks into many small points and the magnetic field is spread out compared to brightness structure. This is illustrated in figure 10 (b).

Individual magnetograms show that the plage field is distributed around 2 to 5 arc second cellular structures. The plage magnetic field looks like "swiss cheese." The magnetogram movie shows that in less than a hour the area inside the outer boundary of the plage is swept by the moving interior swiss cheese cell boundaries. In regions of flux emergence near the edge of a spot flux elements are observed to move with velocities of about 3 km/second. In the coming few years it will be a major task to understand the detailed interrelations between flows, convection, and magnetic field structures in plage and quiet sun.

4. For the Future

While ground based data is invaluable, more than 95% of the effort in reducing the data is in the alignment of the image set and the removal of atmospheric distortions to the extent possible. Even under the best of circumstances outstanding seeing – 1/3 arc second – lasts only a few hours. The Orbiting Solar Laboratory (OSL), which will be placed in a free flying polar orbit, will be able to observe the sun continuously for hundreds of days at a time. Its 1 meter aperture will allow 1/10 arc second resolution in the mid-visible and has the possiblity of resolving nearly 1/20 arc second in the near ultraviolet.

LEST using adaptive optics should allow collection of data in a limited region of the sun, 2 to 5 arc seconds in diameter, for limited periods of time, hours, with a resolution of 1/25 arc second in the mid-visible. LEST is being designed to be nearly free of induced polarization which should allow it to measure the vector magnetic fields of very small flux tubes.

A major task for solar physics in the next few years will be to extract physical processes from movies of the atmosphere obtained at several heights. Compared with the development of 3-D Fourier filtering and local correlation tracking of granulation movies, this will be a very difficult task requiring innovations in analysis procedures, manipulation of video displays, and image presentation techniques.

As techniques are developed for extracting the essential features of convection and magnetic fine structures, it will be required to develop, in parallel, detailed theoretical models and carry out both analysis and simulations to compare with the measurements. At the present state of the art in convection modeling and three dimensional radiative transfer this will require significant developments in the very largest computers in existence. Hopefully, the next few years will see new ideas emerge which permit realistic simulations to be done with the supercomptuers then available.

PROPERTIES OF THE SOLAR GRANULATION

V.N. KARPINSKY
Central Astronomical Observatory
of the USSR Academy of Sciences
196140, Leningrad, USSR

To the memory of the late Prof. V.A. Krat

Until recently the theory of the photospheric structure has had a modest role. This situation has qualitatively changed with the appearance of the numerical simulations of Nordlund [1], Uus [2], Gadun [3], and others.

The possibilities of 3-D physical interpretation of the observations are limited, remaining qualitative and semiempirical in general. Thus simulations are needed.

The foundation for the fine structure investigations is observation with high spatial resolution, including direct photographs and spectrograms. We already have a set of such observational data now.

A number of facts can be established with confidence, but others are on the verge of observational possibilities at present. The last word still lies in the future. We will examine the facts, analyse them, compare them with corresponding conclusions from modelling, and try to outline the general picture of the photospheric fine structure, as seen by an observer.

We start with the high resolution direct photographs of the solar surface. The two-dimensional brightness field reflects the horizontal structure of temperature, and perhaps also density, in the low photosphere. We are confident now that the brightness inhomogeneities in the granulation are large. At $\lambda = 5000$ Å the RMS $\Delta I_5 = 22$ % . This value was obtained from observations by the Soviet Stratospheric Solar Observatory (SSSO) [4], and significantly exceeds previous estimates, e.g. [5]. At first it was received with some mistrust, but then it was confirmed by independent observations [6], and is in good agreement with predictions of numerical simulations [1, 3]. The mean difference in brightness temperature between granule – porule peaks is 700 K. The largest difference exceeds 1000 K, which is 1/6 of the mean temperature of the photosphere. Thus it is impossible to consider the granulation as a minor fluctuation around the mean level.

Fig. 1 gives an estimate of the true two-dimensional spatial spectrum of brightness deviations at solar disk center, corrected for image MTF of blurring and noise. Here $A(s)$ is the total power density, and

$$s = \sqrt{s_x^2 + s_y^2} = \frac{1}{\Lambda}$$

is the radial spatial frequency. An optical image of the spatial spectra of the granulation was formed from direct photographs of the SSSO, using coherent optical methods [4]. The

67

J. O. Stenflo (ed.), Solar Photosphere: Structure, Convection, and Magnetic Fields, 67–79.
© *1990 by the IAU.*

obtained spectrum has a simple form, and can be well represented by two power functions with ascending and descending branches. The spatial period $\Lambda = 960$ km corresponds to the maximum, and is smaller than the main period of the granulation ($\Lambda = 1300$ km) [7]. If we consider the brightness inhomogeneities as temperature fluctuations caused by turbulent convection [9], it is possible to make the conclusion that a considerable part of the inertial and all of the dissipative domains have been in the range of high spatial frequencies ($\Lambda < 300$ km), beyond the resolution that can be attained so far.

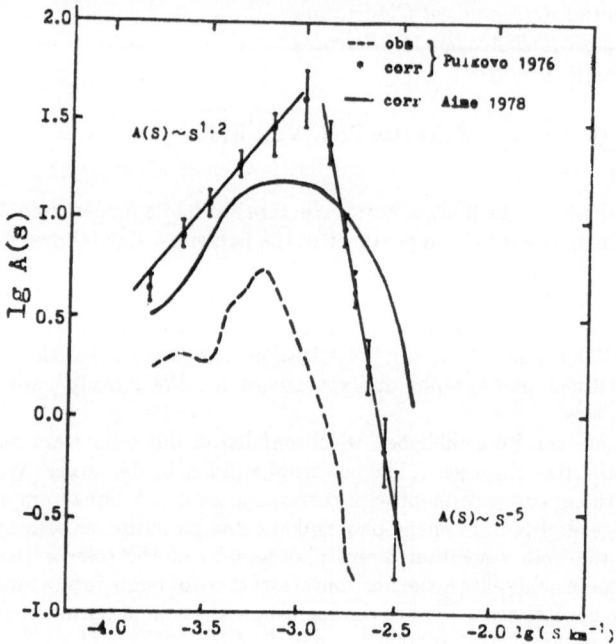

Figure 1. The two-dimensional spatial power spectrum $A(s) = 2\pi\, P(s)$, corrected for MTF, as a function of spatial frequence s. $A(s)$ is the power in frequency intervals $\Delta s = 1$ km^{-1}. The power of the constant brightness component is assumed to be equal to unity. The scales are logarithmic. The error bars indicate the uncertainty of the MTF estimate. $\lambda_{eff} = 5000$ Å.

It has been shown that RMS $\Delta I\,(\theta)$ decreases slowly and monotonically from 22 % at disk center to 4 – 7 % at the limb [9]. This result differed qualitatively from previous data, but is in satisfactory agreement with latest results [10]. This reliably determined key fact contradicted the concepts of Wilson [11] and formed the basis for the next empirical models (Altrock, Musman, Nelson, and others [12]). The height dependence of RMS ΔT in such models (taking into account the above-mentioned large value of the brightness inhomogeneities) are given in Fig. 2 (curve 1), together with data from the numerical model of Gadun [3] (curve 2), while curve 3 refers to Wilson's model [11]. It follows from this that the temperature deviations strongly decrease with height from 1000 K at a depth of 25 km to 300 K at a height of 30 km. The vertical temperature gradient in the fine structures may exceed 20 K km^{-1}, and a temperature inversion in the photosphere is possible. The temperature inhomogeneities are absent in the layers higher than 150 km according to all the empirical models [14].

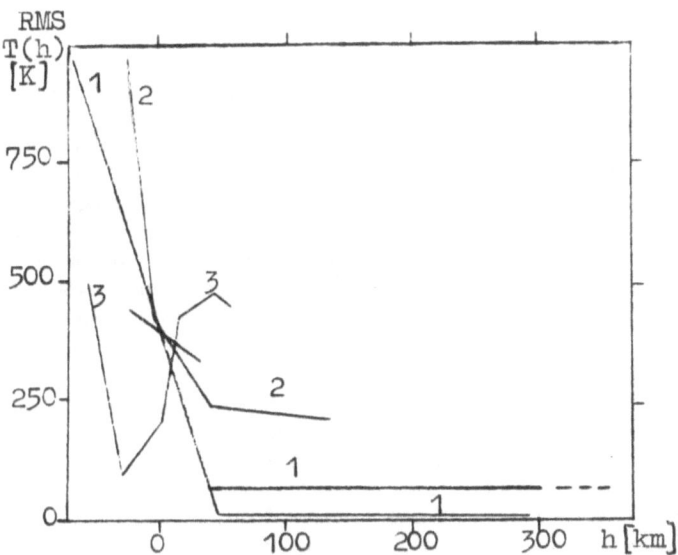

Figure 2. The height dependence of the temperature inhomogeneities. 1: empirical models. 2: numerical model of Gadun. 3: Wilson model. The short line corresponds to the vertical gradient of the mean temperature at $h = 0$ in HSRA.

For the first time fine structures in the brightness were discovered in our stratospheric frames at the extreme limb where the curvature of the limb profile has changed sign, and even 'behind' the limb [4]. They can be clearly seen in the photometric tracings parallel to the limb at the extreme limb $\alpha = 0$, and at a distance of $\alpha = 2.4$ arcsec from it. In the latter case the radial brightness gradient is reduced by a factor of 10. Therefore limb displacements with scales smaller than 3000–5000 km do not exceed 15–20 km, to result in brightness inhomogeneities with RMS $\Delta I_5 = 4 - 7 \%$. These fine structures were unexpected and interesting. According to generally accepted concepts the length $\Delta l = 1000$ km along the line of sight corresponds to $\Delta \tau_5 = 0.05$ at the effective height in the photosphere by the limb ($\tau_5 = 0.003$). In this case $\Delta T/T \approx 0.2$ at the limb. We expect that 20–40 such elements with opposite sign of the fluctuation should occur along the line of sight. In this case, however, the contrast in the image must be strongly reduced. The detailed analysis of Lites [15] gives no explanation of this phenomenon. The structure at the limb can give evidence for the existence of a considerable variance in the temperature and, possibly density at heights of about 350 km. There are reasons to look for strong inhomogeneities in the region of the temperature minimum too ($h = 500$ km).

The existence of the structures at the extreme limb seems to be of fundamental importance for our knowledge of the photosphere. No doubt independent verification of it is necessary.

We may consider the photospheric brightness field in terms of random fluctuations. But it is not 'purely random'. It is a non-Gaussian random field, different from a Gaussian one through the existence of substantial inner bonds and good organization. This is the basis for the morphological and morphodynamical approach to such fields, with spatial-temporal predictions that can be more effective than Wiener stochastic extrapolation.

For an estimate of finer peculiarities of the brightness field an excursion analysis of a random field has been made [16]. If the brightness is higher than a given level we have an upwards excursion, if it is lower, a downwards one. The selected brightness level plays a special structural role in our model. There are only upward excursions above this level and downward ones below it in the field. We call it the starting level or starting plane, because both kinds of excursions start from it. It is essential that this starting level is significantly lower than the mean level of the photospheric brightness (by 4.5 –10 %).

The photospheric brightness field may be well represented as an ensemble of two-dimensional, separately connected brightness pulses of both signs, upwards and downwards from the starting level [17]. This field differs significantly from a gaussian one. We will identify bright pulses from the starting level with granules, dark pulses with porules. The term 'porule' was introduced by Rösch (1959). One granule may contain several maxima of higher brightness forming a 'subline' structure. The characteristics of granule and porule ensembles are different, but there is no essential topological distinction between them. Only few granules have darkness in their centers, as mentioned by Kitai and Kawaguchi [19]. This tendency probably becomes stronger after correction for blurring, but its real existence is questionable. The precise, conceptual definition of these two kinds of elements allows us to distinguish single granules as structural elements on the isophotal maps and photographs, and to estimate granular areas with high accuracy. Previous identifications of granules with maximum brightness have a considerable uncertainty. Our model of the brightness structures is illustrated in Fig. 3.

The proposed model is in good agreement with the real brightness field and its dynamics. It would not be successful to represent it as a gaussian field, or as an ensemble of some kind of local bright formations, separated by a multiconnected infinite dark grid of intergranular lanes.

The mean size of the granules is 700 – 800 km, but the largest ones exceed 2000 km.

The perimeter – area connection for granules has been investigated. Mandelbrot has pointed out the importance of the dependence between $\lg(s)$ and $\lg(p)$ as a characteristic of an ensemble of structures [20]. In the linear case $\lg(p) = \lg(a) + (D/2)\lg(s)$, where D is the fractal dimension. $D = 1$, if the shape of the elements is the same for different sizes.

This dependence for granules was first determined by Roudier and Muller [21] and is given in Fig. 4 (dots and thicker line) together with our data (crosses and thinner line). Both determinations are in good qualitative agreement with each other. There are two straight and distinct branches with a sharp break between them around a granule size of 1000 – 1400 km. The fractal dimension is constant for each branch. For $s < 10^6$ km^2, $D = 1.2$. The indentations of the granular boundary lines grow strongly for $s > 10^6$ km^2 ($D = 2.2 - 2.5$). Such a peculiarity needs an explanation. We have however not discovered any morphodynamical peculiarities connected with this break.

The main elementary evolutionary events which cause qualitative changes of a granule are fragmentation into two or several granules, and merging of two or several granules. They are sophisticated processes. It is impossible to consider that large granules are only fragmented, while the small ones are only merging. The birth of new granules and their fading away are five times less frequent than the other events. During the intervals between such events the granules remain qualitatively unchanged. At this stage of qualitative invariability they exhibit a tendency of decreasing their area with time instead of increasing it as supposed

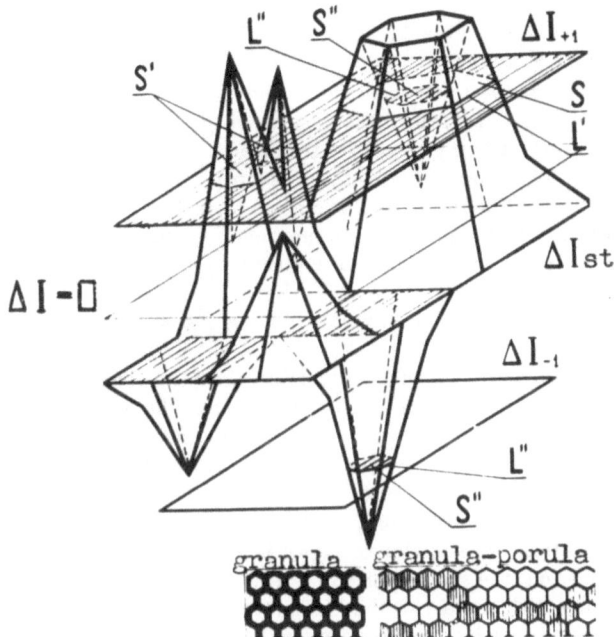

Figure 3. Sketch of the morphological model of the granular brightness field.

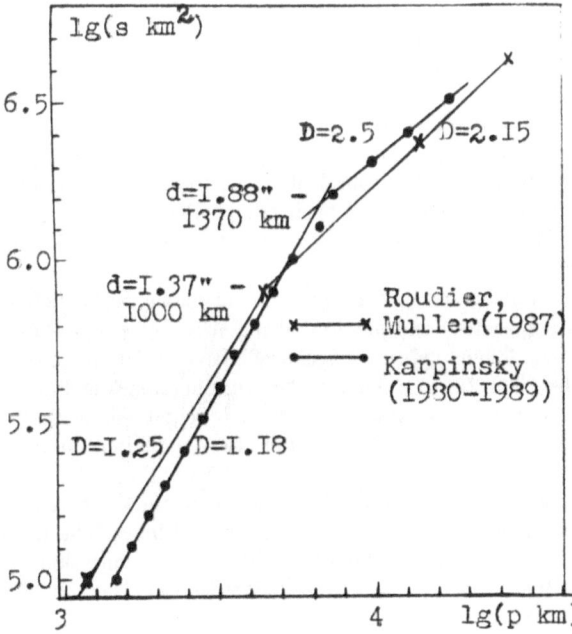

Figure 4. The relation between the perimeters of granules and their areas (in logarithmic scale).

earlier [22]. The corresponding lifetime of a granule as an individual structure is 5 – 6 minutes, in agreement with the estimate in [23].

The evolution of the granules differs qualitatively from the scheme of birth, growth, and fragmentation into several independent granules, which then are expelled and fade away.

There are reasons to distinguish a subset 'dot granules' with diameters smaller than 200 km (0.3″) and high brightness peaks. Their number is 30 % of the total number of all the granules, but they occupy less than 2 % of the total area. The mean distance between them is about 4″, although their distribution over the surface of the Sun is quite irregular [18, 4]. The relation between these objects and small magnetic fluxtubes is a very interesting subject.

Our data do not clearly reveal the predominating role of exploding granules, mentioned by a number of authors [24, 25].

The brightness inhomogeneities in the continuum and spectral lines are quite different from each other, as noted by Evans [26], Krat [27], Evans and Catalano [28]. The dependence of the correlation coefficient $r(\bar{J}_i)$ between the brightness in the continuum and at various levels of intensity inside the spectral lines is given in Fig. 5. It is assumed that the mean continuum intensity is unity. No significant differences in this dependence between various spectral lines were discovered, including lines with different sensitivities to the magnetic field.

The map of the brightness in a vertical cross-section is shown in Fig. 6. The unshaded areas correspond to an excess of emission, the shaded ones to a deficit. Because the height scale has been magnified twenty times with respect to the horizontal scale, the inclinations of the boundaries between the bright and dark structures are distorted. In reality they are inclined at large angles to the vertical (more than 80°). The brightness inhomogeneities at various levels of intensity in the Fe I 4690.1 Å line are illustrated. For this we only use inhomogeneities of the source function, assume LTE and the Eddington-Barbier approximation, and that the level with $\tau = 1$ has a constant geometrical depth. We are of course speaking about a basically qualitative picture.

It is evident that the brightness structures are fundamentally three-dimensional and nonuniform over height scales of less than 70 km. They differ from earlier suggested patterns of thin layers with slow decline towards the horizontal plane, or from mushroom-like, pancake-like structures, by the variety of their shapes and their complexity.

A second alternative model has also been considered. It is assumed that only deviations of the line absorption coefficient change with height, while the source function inhomogeneities are instead constant over the region of line formation for a given point on the solar surface.

These two opposite models show temperature and possibly also density inhomogeneities, which manifest themselves in brightness. Their typical vertical size is an order of magnitude smaller than the horizontal size, i.e., less than 100 km. The width of the emission contribution function is about 200 km. It seems that such small-scale changes with height should not be observable so clearly. Like the inhomogeneities at the extreme limb, they should not be seen. This is the second 'hot topic'.

Now let us turn to spatially resolved motions along the line of sight in the photosphere. The velocities manifest themselves as displacements or wiggles of spectral lines. The RMS $V(h)$ is given in Fig. 7. The data of Pravdjuk [29] without correction for blurring (line 1), the estimates of Durrant et al. [30] (line 2), as well as those of Bässgen and Deubner [31] (lines 3, 4) for various versions of the correction are presented here.

The general conclusion is that considerable velocities are typical over the whole range of heights from 50 km to 450 km, with small variations. The typical sizes of the velocity and

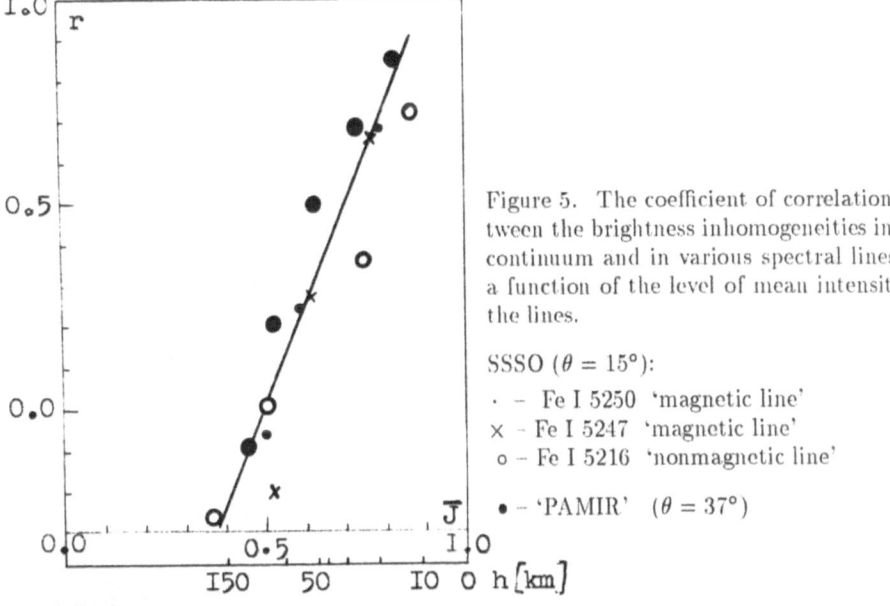

Figure 5. The coefficient of correlation between the brightness inhomogeneities in the continuum and in various spectral lines, as a function of the level of mean intensity in the lines.

SSSO ($\theta = 15°$):

· – Fe I 5250 'magnetic line'
× – Fe I 5247 'magnetic line'
o – Fe I 5216 'nonmagnetic line'

● – 'PAMIR' ($\theta = 37°$)

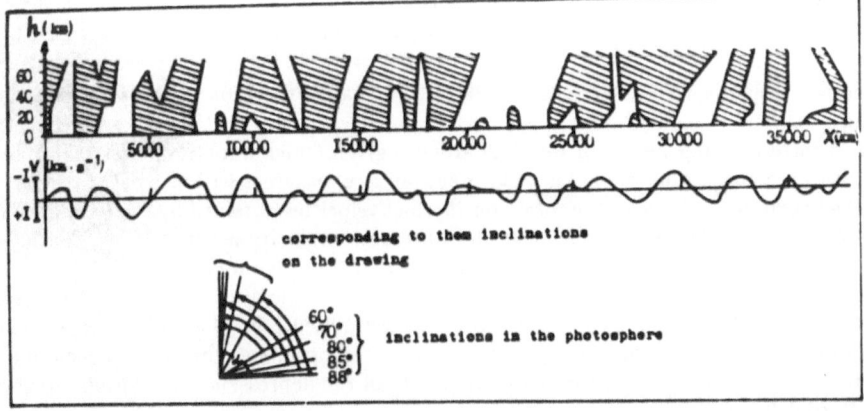

Figure 6. Map of the brightness (emissivity) for a vertical section of the photosphere, and the corresponding vertical velocities. The transformation of real angles in the photosphere to those in the figure is shown in the bottom of the diagram.

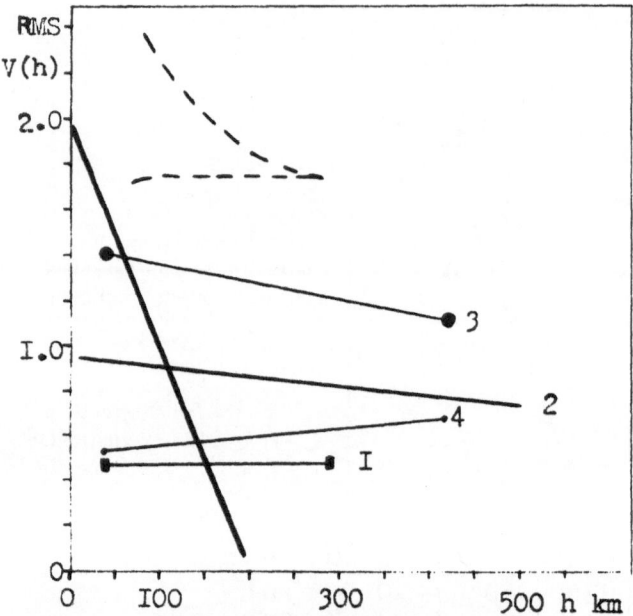

Figure 7. Height dependence of the RMS velocities. 1: Pravdjuk (1982). 2: Deubner, Mattig, Nesis (1979). 3, 4: Bässgen, Deubner (1982). Thick line: Keil (1980). Dashed line: Total non-thermal velocity (Kostik, Kondrasheva, Sheminova, 1981-85).

brightness inhomogeneities are similar [29]. An upward expansion of the structures is not found.

The corresponding results of Keil [32] are also given (thick line), indicating that vertical velocities are absent above 200 km. This qualitatively disagrees with the other results.

The height dependence of the total non-thermal velocities obtained by the Kiev group [33] is also shown (dashed line). The spatially unresolved velocity accounts for a considerable part of the total velocity. This follows from the spatial spectrum (Fig. 1) as well.

The coefficients of correlation between the Doppler velocities in the low photosphere (height 100 km) and in higher layers have been estimated [29]. The height dependence is given in Fig. 8. We use the effective heights for the Doppler measurements, which in a first approximation correspond to the centre of gravity of the depression contribution function of the Unsöld-Pecker form.

It is possible to speak of perfect correlations between the velocities for heights up to 240 km. This fact was discovered by Pravdjuk [29] and Durrant and Nesis [34], and has been confirmed later [35]. Hence the velocity structure has the shape of a set of vertical, parallel columns in this part of the photosphere [34, 7]. On the other hand the velocities at these heights should vary greatly according to the 'oscillatory' or 'convective' concepts of the photospheric velocity field [34].

This correlation, however, drops rapidly above 240 km. At a height of 275 km it is only $r = 0.5$. This was found from the behaviour of the displacement of the CN 3882.6 Å line [29, 36] at various levels of the intensity inside the line, and was confirmed using the Na

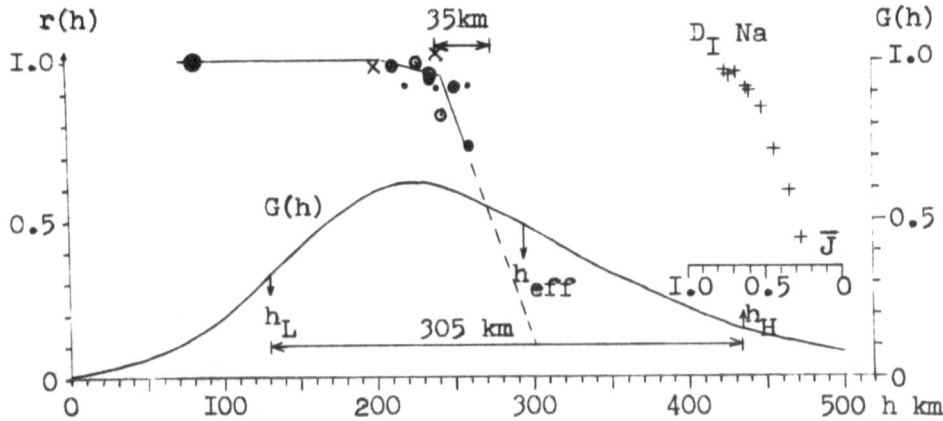

Figure 8. The dependence on height of the correlation coefficient between the Doppler velocities in the low photosphere ($h_{eff} = 100$ km) and at different height levels h.

D_1 line. The height scale for this drop of the correlation is much smaller than the width (≈ 300 km) of the CN contribution function (illustrated at the bottom of Fig. 8). In this case the calculations show that the correlation coefficient should not be smaller than 0.96. The disagreement with the observation is obvious. The given arguments are of a qualitative nature, but it seems to me that this is another 'hot topic' in our concepts of the photospheric structure.

The mentioned qualitative difference between the structures of vertical velocities and brightness should lead to a reduced correlation between them [37, 38]. The coefficient of correlation between the total velocity at a height of about 100 km and the brightness in the continuum is $r = 0.54$. For this value the completely uncorrelated additive component is 1.6 times larger than the correlated one [39]. The differences are due to fine structures to a considerable degree [53]. For fine structures only with $\Lambda_x < 2000$ km, $r = -0.66$ [39]. The 5-minute oscillation does not contribute here. The coefficient of correlation between velocity and brightness in spectral lines quickly goes to zero when moving towards the line centre, and then changes sign to positive. Note that minus corresponds to an upward motion of the bright elements. The coherence spectrum (linear coherence, similar to the coefficient of correlation) drops considerably (to 0.5 and less) for $\Lambda_x < 1750$ km [7], and in particular for $\Lambda_x < 700 - 1000$ km [40, 41, 42], although for 2000 km $< \Lambda_x < 4000$ km the coherence is high.

It is possible that a correlation between the vertical velocities and the brightness exists in the lower layers of the photosphere ($\tau \geq 1$) [42], but that it is disturbed considerably at heights above 50 km, for the photosphere as a whole, in agreement with the results of numerical simulations. Decorrelation may be the result of turbulence, internal gravity waves, or inhomogeneities of non-convective nature. It is difficult to separate and estimate the relative roles of these effects now [42].

Considerable disagreements between the results of various investigators often occur in current granulation research. In a number of cases these discrepancies may be the result of errors or observational difficulties. In some cases, however (and this is very important), they may reflect real variations in the photospheric structure. Abnormal granulation and filigree were discovered in the beginning of the 1970s [43].

Changes of the granulation with the solar cycle have been established [44], including their connection with the radio emission [45]. Distinctions between the fine structures inside supergranules and at its boundaries have been determined [24, 46], and mesogranulation has been discussed [47, 48]. Fast, large-scale changes of the granulation ('perestroika' of its structure) have been discovered on the SSSO photographs. The number of granules change by a factor of 1.5 and their mean area by a factor of 2 during 5 minutes. This change is coherent over an area of at least 5×10^8 km^2. The general restructuring of the ensemble as a whole happened within the lifetime of its granule elements [49, 50]. This non-stationarity demands a stricter approach in comparison with the individual definition of the granular characteristics.

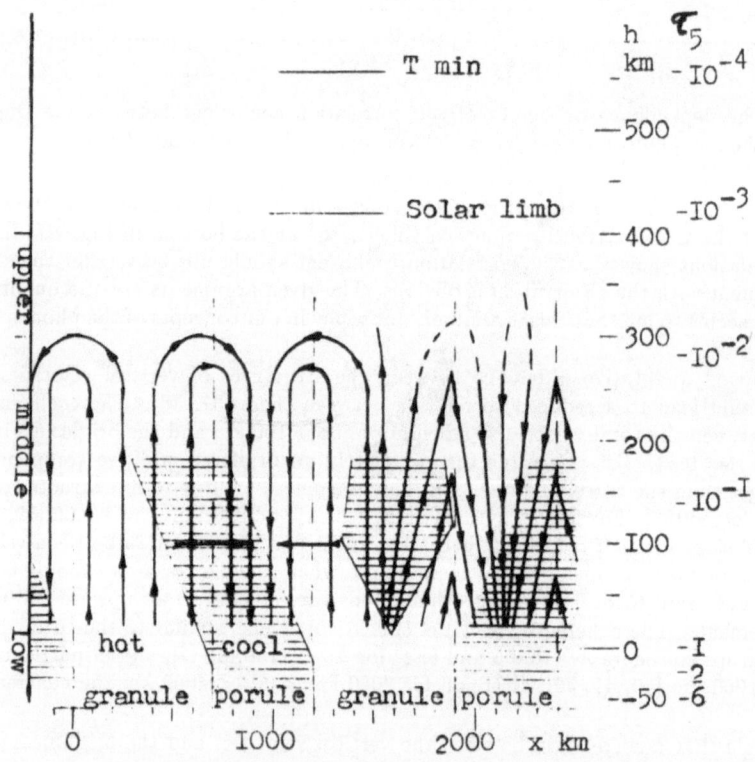

Figure 9. Illustration of the structures and the movements in the photosphere.

We present in Fig. 9 such a scheme of structures and movements in the photosphere. We may distiguish four layers: the low photosphere up to a height of 50 km, the middle ($h = 50 - 240$ km), the upper photosphere above 240 km, and the temperature minimum region.

In the low photosphere the vertical temperature gradient is high, and the density is almost constant. The velocity and temperature structures may spatially coincide here. The concept of 'granule' is reasonable here for the designation of some indivisible, three-dimensional structure. Outflow of matter from a granule through its motionless boundary takes place. It is accompanied by a cooling of $5 - 10$ K km^{-1}. This feature is well reproduced by the numerical models. The observed structure is closely described by laminar convection here.

This unity is violated in the middle photosphere. It is impossible to speak of unique cellular patterns there. Vertical, cylindrical velocity structures are typical, and the horizontal outflow does not change this. The brightness structures, however, do not coincide with the velocity structures. There is no correlation between them.

The transition to the upper photosphere occurs where the column velocity structure is breaking up. Here and in higher layers there may be very significant inhomogeneities of non-convective origin, variations of the absorption coefficient, density, and magnetic field. This is in accordance with the picture proposed by Nesis [51]. So considerable small-scale inhomogeneities are characteristic of the whole photosphere, not only of its lower part, directly adjacent to the convection zone. The sophisticated structure of the middle photosphere arises due to the convection, but is only indirectly connected with it.

The nature of the inhomogeneities in the high photosphere and the temperature minimum region is probably not convective [51]. The non-thermal energy, which produces non-equilibrium conditions in the chromosphere and corona, may be generated here [52].

The density drop between heights $h = 100$ km and $h = 240$ km is about a factor of 5, according to HSRA. In this case the cylindrical velocity structures without upwards expansion must smooth out density fluctuations in the middle photosphere. The density fluctuations at height 250 km should be greater. If this is so, we have no constant downward outflow (effect of 'the pipe being full of holes'). This downward outflow without expansion has not been observed directly yet.

The hydrostatic equilibrium may be violated under these circumstances. Hydrostatic equilibrium is postulated for all photospheric models, but as far as I know, we have no direct observational proof that this assumption is valid.

It is possible that we may be able to explain through such effects the observability of the fine structure at the extreme limb, as well as the strong reduction in the correlation between the brightness in the continuum and the spectral lines and the velocities over small height differences.

Let us in conclusion formulate some general problems:
1. The nature, meaning, and significance of the inhomogeneities at heights above 200 km.
2. The role of magnetic fields for the mere existence of the photospheric fine structure.
3. The ultra small-scale, subgranular structures, and their role for the generation of turbulent viscosity and magnetic diffusion.
4. It seems to me to be important to learn to predict the granulation pattern over 15 minutes in time, for 10 000 km in space. Some possibilities for this exist.
5. The change in the nature of the granulation structure in space and time. The connection with the large-scale and global structures and characteristics.
6. It is necessary to learn to diagnose better from the observations the inhomogeneities of the temperature, density, macro- and microturbulent velocities, and, of course, the magnetic field in the three-dimensional photosphere.

The granulation is a large, natural, organized system. It is not so hopelessly sophisticated as active solar features. There is hope that we may be able to understand it in general as well as in detail, and make contributions not only to science about the Sun, but also to the understanding of the nature of dissipative structures — one of the most important problems of our time.

The author considers it his pleasant duty to express his sincere gratitude to Dr. L.M. Pravdjuk, Prof. G.B. Gelfreich, and Mrs. N.S. Petrova for their assistance in the preparation of the present paper.

References

1. Nordlund, Å. (1985) *Solar Phys.* **100**. 209-235.
2. Uus, U. (1986) *Tartu Astroph. Observ. Publ.* **51**, 20-34.
3. Gadun, A.S. (1986) Preprint Inst. Theoret. Phys., Kiev, ITF-80-106, 24.
4. Karpinsky, V.N., Mekhanikov, V.V. (1977) *Solar Phys.* **54**, 25-30.
5. Edmonds, F.N. (1962) *Astrophys. J. Suppl. Ser.* **6**, 357-406.
6. Nordlund, Å. (1984) in 'Small-Scale Dynamical Processes in Quiet Stellar Atmospheres' (ed. S. Keil), 135-137.
7. Karpinsky, V.N. (1985) *Lect. Notes in Physics* **233**, 259-260.
8. Ledoux, P. et al. (1961) *Astrophys. J.* **133**, 184-197.
9. Pravdjuk, L.M. et al. (1974) *Solnechnye Dannye Bull.* No. 2, 70-86.
10. Keil, S.L. (1977) *Solar Phys.* **53**, 359-368.
11. Wilson, P.R. (1969) *Solar Phys.* **9**, 303-314.
12. Bray, R.J., Loughhead, R.E., and Durrant, C.J. (1984) *The Solar Granulation* (Cambridge University Press), p. 256.
13. Muller, R. (1985) *Solar Phys.* **100**, 237-255.
14. Kneer, F. (1984) in 'Small-Scale Dynamical Processes in Quiet Stellar Atmospheres' (ed. S. Keil), 110-129.
15. Lites, B.W. (1983) *Solar Phys.* **85**, 193-214.
16. Sveshnikov, A. (1968) *Prikladnye Metody Teorii Sluchainych Funktsyi*, Nauka, Moscow, p. 464.
17. Karpinsky, V.N.(1980) *Solnechnye Dannye Bull.* No. 2, 91-102.
18. Karpinsky, V.N.(1980) *Solnechnye Dannye Bull.* No. 7, 94-103.
19. Kitai, R., Kawaguchi, I. (1979) *Solar Phys.* **64**, 3-12.
20. Mandelbrot, B. (1982) *The Fractal Geometry of Nature*, Freeman and Co., New-York, p. 480.
21. Roudier, Th., Muller, R. (1987) *Solar Phys.* **107**, 11-26.
22. Mehltretter, J.P. (1978) *Astron. Astrophys.* **62**, 311-316.
23. Kawaguchi, I. (1980) *Solar Phys.* **65**, 207-220.
24. Carlier, A. et al. (1984) *C.R. Acad. Sci. Paris* **226**, 199-201.
25. Title, A.M. et al. (1988) 'Statistical Properties of Solar Granulation Derived from the Soup Instrument on Spacelab 2', LPARL, p. 51.
26. Evans, J.W. (1964) *Astroph. Norveg.* **9**, 33-54.
27. Krat, V.A. (1973) *Solar Phys.* **32**, 307-310.
28. Evans, J.W., Catalano, C.P. (1972) *Solar Phys.* **27**, 290-300.
29. Pravdjuk, L.M. (1982) *Solnechnye Dannye Bull.* No. 2, 103-112; No. 5, 110.
30. Durrant, C.J. et al. (1979) *Solar Phys.* **61**, 251-270.
31. Bässgen, M., Deubner, F.L. (1982) *Astron. Astrophys.* **11**, L1-L3.
32. Keil, S. (1980) *Astrophys. J.* **237**, 1024-1034.
33. Sheminova, V.A. (1985) *Kinematika i Fisika Nebesnikh Tel.* **1**, No. 2, 50-52.
34. Durrant, C.J., Nesis, A. (1982) *Astron. Astrophys.* **11**, 272-278.

35. Kushnir, M.V. (1982) *Solnechnye Dannye Bull.* No. 10, 80-87.
36. Belenko, V.I. et al. (1983) *Solnechnye Dannye Bull.* No. 12, 61-69.
37. Karpinsky, V.N. et al. (1977) *Solnechnye Dannye Bull.* No. 12, 79-83.
38. Karpinsky, V.N. (1979) *Pisma v Astronomicheskij Journal* 5 , 552-556.
39. Karpinsky, V.N. et al. (1987), in 'Solar Physics Conference', Tesisis, Alma-Ata.
40. Aime, C. et al. (1985) *Lect. Notes in Physics* **233**, 103-107.
41. Wiehr, E., Kneer, E. (1988) *Astron. Astrophys.* **195**, 310-314.
42. Deubner, F.-L. (1988) *Astron. Astrophys.* **204**, 301-305
43. Dunn, R.B., Zirker, J.B. (1973) *Solar Phys.* **33**, 281-304.
44. Alissandrakis, C.E. et al. (1982) *Solar Phys.* **76**, 129-136.
45. Macris, K.J. (1988) *C.R. Acad. Sci. Paris*, in press.
46. Simon, C.W. et al. (1988) *Astrophys. J.* **327**, 964-967.
47. November, L.J. et al. (1981) *Astrophys. J.* **245**, L123-L126.
48. Oda, A. (1984) *Solar Phys.* **93**, 243.
49. Karpinsky, V.N. (1988) *USSR Astron. Circ.* **1525**, 19-20.
50. Karpinsky, V.N. (1989) *Kinematika i Fisika Nebesnykh Tel* **5**, No. 3, 22-35.
51. Nesis, A. (1985) *Lect. Notes in Physics* **233**, 248-253.
52. Bialko, A.V., Avrett, E.H. (1985) 'A Mechanism for Chromospheric Heating by Fast Electrons Generated in the Temperature Minimum Region' (Preprint), The Academy of Sciences of the USSR, L.D. London Institute for Theoretical Physics, Chernogolovka.
53. Parfinenko, L.D. (1985) *Solnechnye Dannye Bull.* No. 8, 68-73.

ANALYSIS OF THE SOLAR GRANULATION IN THE OPACITY MINIMUM REGION

Serge Koutchmy

Institut d'Astrophysique CNRS, 98bis Bd Arago, F-75014 Paris, France

and

NSO–Sacramento Peak, Sunspot, 88349 NM, USA

Abstract : An observing program in the spectral region of the opacity minimum region near 1650 nm was started at NSO/SP–VTT many years ago and few results were presented. Thanks to an image processing unit connected to a rather conventional TV camera, IR images taken at 1650 nm can be displayed on the TV monitor and images stored on a VCR for subsequent comparison with corresponding images in the visible spectrum. This rather qualitative analysis is substantiated thanks to results coming from one–dimensional time series obtained with an IR pin–hole spectrophotometer. Power spectra of the o.m.r. granulation as well as its temporal behaviour are compared with results obtained using the same method in the optical region of the solar spectrum. Noticeable differences are emerging, including the reduced effect of 5-min oscillations.

1. Introduction

The solar atmosphere radiative opacity reaches its absolute minimum in the 1650 nm spectral region of the continuum, see Vernazza et al. 1981, which is also a region of excellent Earth atmosphere spectral transmission, making ground-based observations at this particular wavelengths of high quality. Unfortunately, the specificity of infrared technology and method, see Turon and Léna, 1973, has limited the progress which has been anticipated. Indeed the access of this solar opacity minimum region (o.m.r.) offers a great potential for the analysis of the structures of the solar atmosphere, starting with granulation. Here, we will concentrate on the "quiet" granulation (outside of active region) analysed during years of the solar minimum (1975–76 and 1987–88), see also Turon, 1975.

We first notice that the contribution function deduced for the homogeneous standard solar atmosphere of Vernazza et al., 1981, shows that in the o.m.r. at 1650 nm the radiation comes from layers deeper by approximately 40 km than the radiation of the optical continuum in the 500–600 nm region. However, although this value is rather small compared to the typical "scale" of granules, see Bray et al., 1984, it is interesting to notice also the extension of the contribution function : in the o.m.r. the radiation is coming from a rather limited vertical extension layer, contrary to the optical case where a substantial part of the radiation is still produced in the upper part of the "continuum" photosphere where 5-min oscillations start to become of considerable amplitude, see Keil, 1980. Taking into account the results of Keil extrapolated to the deepest layers, 5-min oscillations should become there evanescently small and, accordingly, convective motions should be dominant (especially in vertical direction). Concerning the analysis of the optical granulation, 5-min oscillations have been shown to give a substantial contribution, see Koutchmy and Lebecq, 1986, and Title et al., 1989, which corresponds

J. O. Stenflo (ed.), Solar Photosphere: Structure, Convection, and Magnetic Fields, 81–84.

to temperature fluctuations not to be confounded with the convective of origin temperature fluctuations of the granulation. Finally, let us notice that intensity modulations (the "contrast") corresponding to these temperature fluctuations are considerably smaller in o.m.r. which corresponds to the infrared (IR) part of the solar spectrum; a 100 K temperature fluctuation correspond to a 8.5 % intensity variation at 500 nm and only 3.5 % at 1690 nm.

There exists no obvious imager in the IR The linear array built by T & L, 73 to perform there o.m.r. observations at the Mac Math telescope failed to produce useful 2D–images; their best results have been deduced using a single detector and telescope scanning. Furthermore, a telescope of large enough aperture should be used in order to limit the influence of the diffraction, the wavelength of the o.m.r. being of order of 3 times longer than the optical one. Fortunately, an other limitation due to the Earth Atmosphere smearing and image motion is improved in the IR.

2. Imaging of granules in the o.m.r. and brightness distribution

Following the experience of T & L, 73, I started to image the IR granulation at the VTT of NSO/S.P. in 1975, during the sunspot minimum. Typical matrices of 32 x 32 were recorded in 43 sec of time by scanning the center of the Sun using a computer controlled motion of one of the primary plan mirror of the heliostat and a specially designed pin-hole photometer (unsuccessful tests were performed in 1972 at the Pic du Midi Observatory and in 1974, at the 1 m aperture Crimean Astrophysical Observatory) working in the o.m.r. with different passbands (from 100 nm to 3 nm), see Koutchmy et al., 1977. Time series of several tens of images were obtained showing convincingly the improvement of the seeing in IR. However, images are now obtained more easily with an IR vidicon equiped with a video-digitizer. The first movies showing the IR granulation have been obtained during the summer of 1988, with a conventional IR tube and TV camera. Time sequences of IR nearly diffraction-limited granulation images were obtained at a slow rate of 2 images of 512 x 512 px^2 per sec of time, see Figure 1, through a 110 nm interference filter centered at 1650 nm; the signal level is adjusted, as well as the signal-noise ratio, with both the target voltage and the video amplifiers; unfortunately, the sensitivity of the tube is inhomogeneous by up to 40 % over the field and the precision of digitization limited by the 8 bits A/D converter.

Before a time series is started, a gain table is built using a flat field and stored in

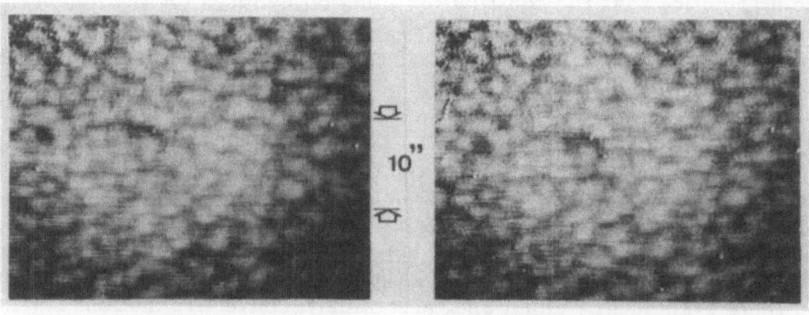

t2 2mn 02sec   t1 1mn 31sec

Figure 1. Direct images of the 1650 nm o.m.r. granulation taken from a slow rate video-movie. Each image is processed in real time to remove the px to px variation of sensitivity.

the SUN computer which operates a VICOM multi-processor; accordingly, a pixel to pixel scaling is performed after each digitization, i.e. 2 times per sec, thanks to a loop giving each time the result on a central monitor; processed images are stored on a VCR. More sophisticated algorithm has been also used to look at umbral dots, etc. In this experiment, a second channel was simultaneously recorded in real time, to show the optical granulation with a conventional video-CCD camera fed by a splitter. Although both movies are obviously showing the same bright and dark features, the behaviour of the images is quite different, partly because the effect of the Earth atmospheric induced seeing effects are different for each channel and partly because a real difference exists in brightness distribution.

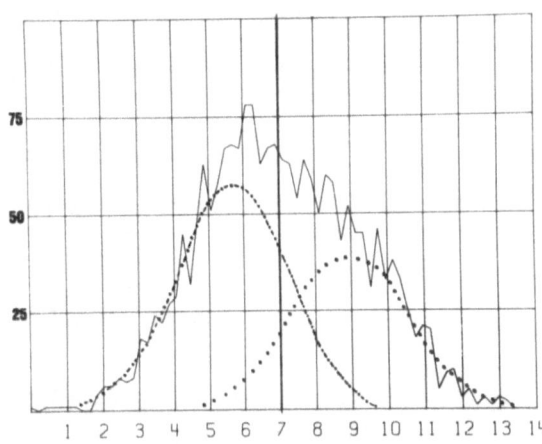

Figure 2. Histogram analysis of the brightness distribution of the the o.m.r. granulation. One unit on the abscissa corresponds to a 1 % intensity variation. The bin width is 10^{-3} of the average intensity (with abscissa value 7).

In the visible region, it is well known that even with a good image quality, when the uncorrected RMS reaches 5 to 7 %, the histogram of brightness fluctuations is nearly gaussian, see Pravdjuk et al. 1974. We performed the same analysis with data taken in the o.m.r. at 1750 nm (see further) and, surprisingly, the histogram shows clearly, see Figure 2, a skewed bi-distribution which can be perceived in the optical region only when a very high quality picture (RMS $\geq$ 9%) is used.

3. Results of the statistical analysis of granules in o.m.r.

Using computer controlled scans of the same region at different $\mu = cos\ \theta$ values, a great amount of data representing the brightness distribution of the quiet Sun o.m.r. granulation has been collected and analysed; a similar study has also been performed at 600 nm, see Koutchmy and Lebecq, 1986. The typical digitization rate was 100 px-sec^{-1}, with a 14 bits.px^{-1} precision; linear synchronous detection of the modulated at 1100 Hz signal was achieved, resulting in a S/N ratio of 2000 in the o.m.r. (at the center of the Sun with a broad filter) and of 175 at 600 nm. A round pin-hole of .75 arcsec effective diameter and a .19 arcsec sampling interval was systematically used; the corresponding smearing function and MTF have been evaluated both theoretically and with numerous scans of the extreme-limb, see Koutchmy et al. 1977 and Koutchmy, 1977. We first computed the RMS (after subtraction of the best 2^d order polynomial from each scan of at least 200 arcsec length), the auto-correlation function, power-spectra, etc.. Scans were also cross-correlated, to evaluate the "life-time" of granules and even, using the drift of the center of gravity of the cross-correlation function, the solar rotation. Data collected at the VTT of NSO/S.P. are significantly better than the best results coming from the Mac Math telescope : higher RMS, good reproducibility of the scans, significant temporal cross-correlation up to 16 min, etc.. The typical values of the corrected for the foreshortening ($\sim \mu^{-1/2}$) RMS of observed intensity fluctuations in the o.m.r. at 1750 nm are : ± 2.04 at $\mu = 1$; ± 2.02 at $\mu = .8$ and ± 1.96 at $\mu = .7$. Only a slight center-limb effect is perceived on this range of values of μ. The 2-D power spectrum has also been computed and even corrected for the smearing; compared to the spectrum obtained at 600 nm, <u>more power appears in the high spatial frequencies region</u> (although results are very sensitive to the values of the correction) and, more significantly, far less power appears in the low frequencies region, see Figure 3.

Finally, the cross-correlation analysis permits the determination of the "lifetime" of granules. Exactly the same procedure was used at 600 nm and in the o.m.r. giving there a definitly smaller value (3.25 min instead of 6 min) of the decrement of the best exponential function which fits the data in each case. This strongly suggests a smaller characteristic time of evolution of the o.m.r. granulation, consistent with an increase of the convective velocities with the depth, see Keil, 1980. However, it is not clear how the 5-min oscillations clearly affecting the intensity fluctuations at 600 nm, see Koutchmy and Lebecq, 1986, bias the results. Title et al. 1989 suggested a method of correction which seems to work for comparing the "quiet" and the "active" region granulation. I performed a spectral analysis of the o.m.r. time series using different methods of averaging : the 5-min oscillations are clearly of smaller amplitude in the o.m.r. and even at large spatial scales they barely emerge from the noise, see Figure 4.

References

Bray, R.J., Loughhead, R.E. and Durrant, C.J. : 1984, *"The Solar Granulation"*, Cambridge Univ. Press.

Keil, S.L. : 1980, Astron. Astrophys. **82**, 144

Koutchmy, S., Koutchmy, O. and Kotov, V. : 1977, Astron. Astrophys. **59**, 189

Koutchmy, S. and Lebecq, C. : 1986, Astron. Astrophys. **169**, 323

Title, A.M., Tarbell, T.D., Topka, K.P., Ferguson, S.H. and Shine, R.A. and the SOUP team : 1989, Ap.J. in press

Turon, P. : 1975, Solar Phys. **41**, 271

Turon, P.J. and Léna, P. : 1973, Solar Phys. **30**, 3

Vernazza, J.E., Avrett, E.H. and Loeser, R. : 1981, Ap.J. Suppl. **45**, 635

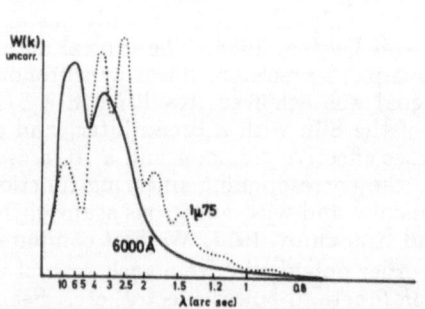

Figure 3. 2-D power spectra per unit surface, computed for the o.m.r. granulation (dotted line) and the 600 nm granulation (full line), uncorrected and unsmoothed. The small-scale features are due to statistical fluctuations, but the general behaviour is significant.

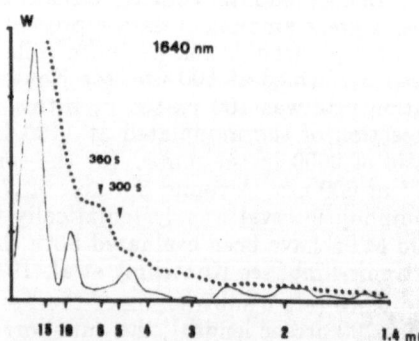

Figure 4. Power spectra of the temporal brightness variation of the o.m.r. granulation. The dotted line is the average of 1100 spectra computed for different points on the Sun from a 46 min time sequence, while the full line has been computed using the brightness values averaged over a 206 arcsec length. The repetition time of each scan is 42.8 s.

FINE STRUCTURE OF PHOTOSPHERIC FACULAE

R. MULLER
Pic Du Midi Observatory
65200 Bagnères de Bigorre

ABSTRACT

Properties of the photospheric bright points associated with magnetic flux tubes are reviewed both in faculae (facular points) and in the photospheric network (network bright points – NBPs) out of active regions. A special attention is given to their size distribution, to their location relative to the granular, mesogranular and supergranular patterns, and to their relation with the small scale magnetic features, both in active and quiet regions. In particular a new granulation movie reveals that NBPs form in large intergranular spaces, compressed by the surrounding granules.

At the center of the solar disk, bright points are much brighter than the mean photosphere ; their contrast increases toward the limb up to μ = 0.3 – 0.2 and then decreases to the limb, as it is now widely accepted. But, all the published contrasts are of little significance because of center-to-limb selection effects. New center-to-limb contrast variations of individual network bright points are presented, which take into account the selection effects.

1. Introduction

According to their historical definition, faculae are extended bright areas visible in active regions when they are close enough to the solar limb. High resolution observations show that faculae are formed with many small features, most of them being smaller than 1 " (Figure 1, from Muller, 1975), called facular points or facular granules (Bray and Loughead, 1964, Muller, 1975, 1977). Facular points are difficult to study near the limb because of foreshortening. It isn't very easy to identify facular features in white light near the disk center because their contrast does not exceed the contrast of granules. Nevertheless, in the core of photospheric lines or in the wings of strong chromospheric lines like Hα or CaIIH and K, facular features can be observed very well at the disk center, as their contrast increases with the height in the photosphere. In addition these sub-arcsecond features are not affected by the foreshortening when they are far away from the limb. Actually, instead of using narrow filters in Fraunhofer lines, I use a quite wide 10A bandpass interferential filter centered on the CN band at 4308 A ; it allows me to observe with exposure times which are short enough to resolve successfully sub-arcsecond features down to the diffraction limit of the telescope (O".2 : Figures 2 and 4). Facular points are produced by magnetic flux tubes. Consequently they can be used to locate the magnetic field in the photosphere, which is very difficult on magnetograms because of insufficient resolution ; this allows us to obtain indirect important information about the interaction between flux tubes and convective patterns at granular, mesogranular and supergranular scales, and to derive the physical structure of flux tubes from their contrast.

Bright photospheric features are not only observed in active regions but also in the quiet atmosphere of the sun, where they form the photospheric network, which is cospatial with supergranule boundaries. Faculae are also visible at the solar poles, especially around minima of activity when the polar magnetic field is relatively strong ; they are known as polar faculae.

85

J. O. Stenflo (ed.), Solar Photosphere: Structure, Convection, and Magnetic Fields, 85–96.
© 1990 by the IAU.

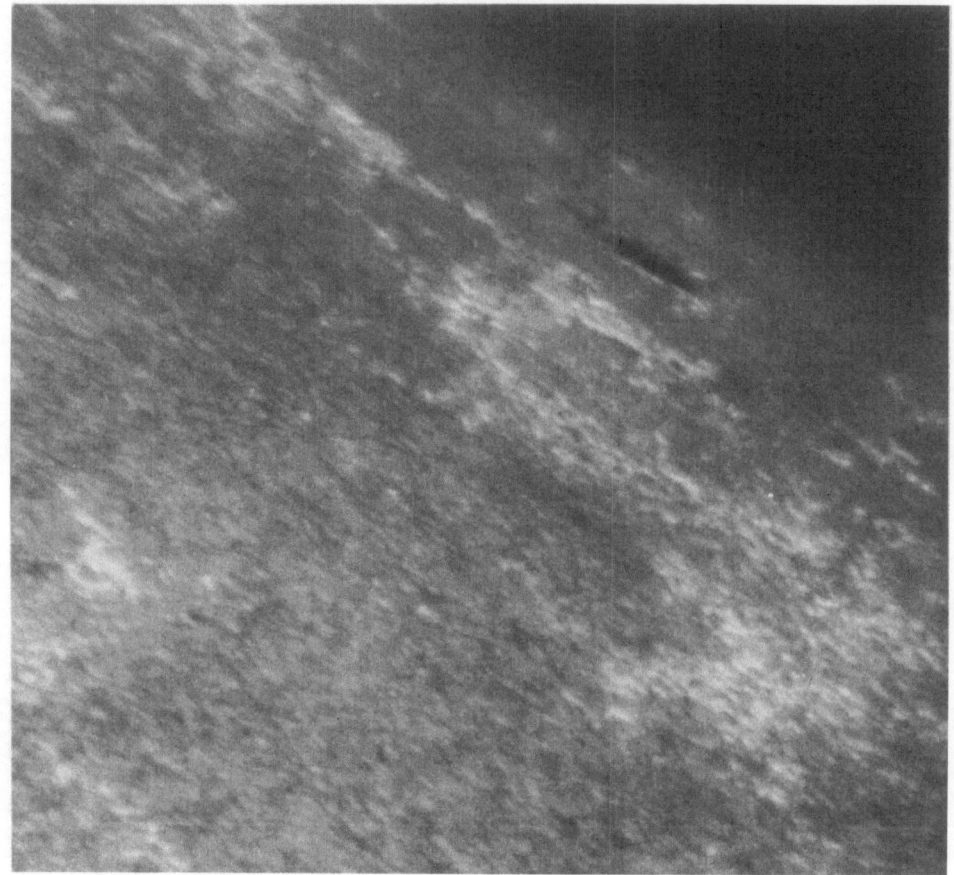

Figure 1. White light faculae near the solar limb.

Both the faculae in active regions and the photospheric network in the quiet sun are formed of bright points, most of them being smaller than 0".5. In quiet and active regions, their properties (size, brightness, lifetime) are very similar and they are believed to have the same magnetic origin (Chapman and Sheeley, 1968). In the photosphere, network bright points often appear isolated or surrounded by only a few neighbours. In active regions the situation is more complicated because the density of facular points is very high, forming a compact pattern of adjacent points.

Consequently the basic properties of facular points will be much easier to study in the quiet sun ; they will be described in the next section. Then the properties of facular points in active regions will be compared to those of network bright points.

The facular contrast and its center-to-limb variation deserves a particular attention because they are used to derive empirical facular and magnetic flux tube models, as well as to test theoretical models. It will be shown that the center-to-limb variations published so far are of little significance and cannot be used to derive models ; more significant and realistic new measurements will be presented.

2. Properties Of Network Bright Points (NBPs) In The Quiet Photosphere

At the photospheric level, the size of NBPs rarely exceeds 0".5 (Figure 2) ; their characteristic size has been found to be close to 0"2 - 150 km - (Muller and Keil, 1983; Mehltretter, 1974 ; Dunn and Zirker, 1973) , it may even be smaller if the size of many NBPs is beyond the resolution of existing telescopes. They are located in the intergranular lanes (Mehlhetter, 1974 ; Muller, 1983, Figure 2) and at the supergranular boundaries (Muller, 1983) ; no clear relation with the mesogranular pattern is known, except that they avoid diverging regions in the granule horizontal proper motions (Figure 3).
A detailed description of the behavior of NBPs has been given by Muller (1983) : they appear at the supergranular boundaries, seldom inside the cells ; on a smaller scale they are formed in large spaces at the junction of several granules, never inside a granule nor in a common space between two granules : they appear preferentially at the boundaries of convective cells. The lifetime ranges from 5 to 50 min, with an average value of 18 min; their size rarely exceeds 0".5; they remain in intergranular spaces throughout their lifetime ; they tend to form very close to an already existing bright point ; 15 % of them seem to split into two facular points ; they disappear simply by fading away in an intergranular space ; merging with another bright point has not been clearly observed. It is also worth mentioning that granules surrounding NBPs are disturbed during their formation phase : they become elongated in the direction of the NBP, pointing to it (Muller, Roudier, Hulot, 1989).

3. Properties of facular Points (FPs) In Active Regions

3.1 MORPHOLOGY

In active regions, faculae extend over large areas and are mainly formed of chains and of compact clusters of tiny bright points, when observed at very high resolution like in Figure 4 (which was taken in the CN band at 4308 A) ; the basic feature appears to be a very small bright point smaller than 0".5, even than 0".3 for many of them. These facular points (FPs) are very similar to the NBPs described above, in size, brightness and lifetime (Muller and Mena, 1986). They can be aligned forming chains, crinkles (Dunn and Zirker, 1973) and clusters of adjacent points a few arcseconds in size. Their relationship with granules (which are significantly smaller than in the quiet sun) is not very clear, although many of them are located in intergranular lanes. In the continuum, the granulation appears smeared if the seeing is not perfect (abnormal granulation, Dunn and Zirker, 1973). Facular points form a pattern of cells, 2" to 5" in size, including several granules (Title, Tarbell, Topka, 1987) ; why such a cellular pattern shows up is still unknown. Supergranular size cells can also be delineated. Unlike in the quiet sun, the size of FPs can significantly exceed 0".5 or even 1"0 in a few cases (Figure 4). According to Spruit and Zwaan (1981) and Schüssler and Solanki (1983), facular points of size in the range 0".5- 1".5 should not be bright ; this point should be clarified. Near the limb probably only large and bright FPs and unresolved clusters of small FPs are visible (see section 5).

3.2 RELATION TO THE PHOTOSPHERIC MAGNETIC FIELD

At moderate resolution, down to about 1", both the photospheric network and faculae coincide well with the photospheric magnetic field (Chapman and Sheeley, 1968). At higher resolution the detailed correspondence is less clear ; there is however   strong

88

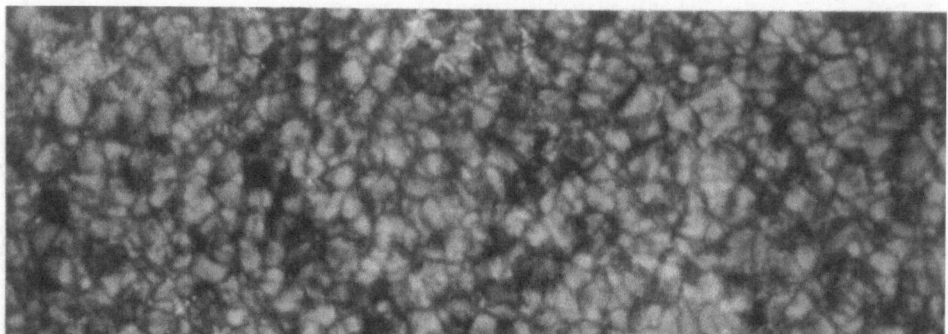

Figure 2. Network Bright Points (NBPs) imbedded in the granular pattern. Filtergram taken at 4308 Å in the CN band.

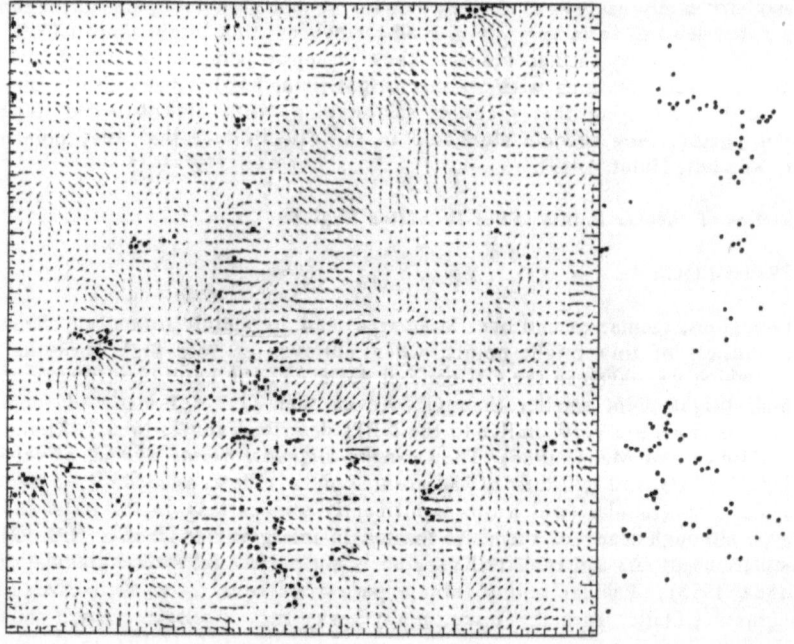

Figure 3. Location of NBPs in the granular flow pattern in a quiet area (55" x 53") at the disk center. The flow pattern was derived from an outstanding granulation movie obtained at the Pic du Midi Observatory and processed with the Title technique at Lokheed Palo Alto Research Laboratory (Title et al. 1989). NBPs concentrate at the boundary of a large supergranule which is also delineated by large outward flows ; they also concentrate in several converging flow regions. The large concentration of NBPs in the lower half of the area corresponds to an abnormal granulation area. Many diverging flow regions can be recognized easily in the processed area ; they could correspond to mesogranules ; NBPs seem to avoid these diverging regions. There are very few NBPs identified inside the large supergranule.

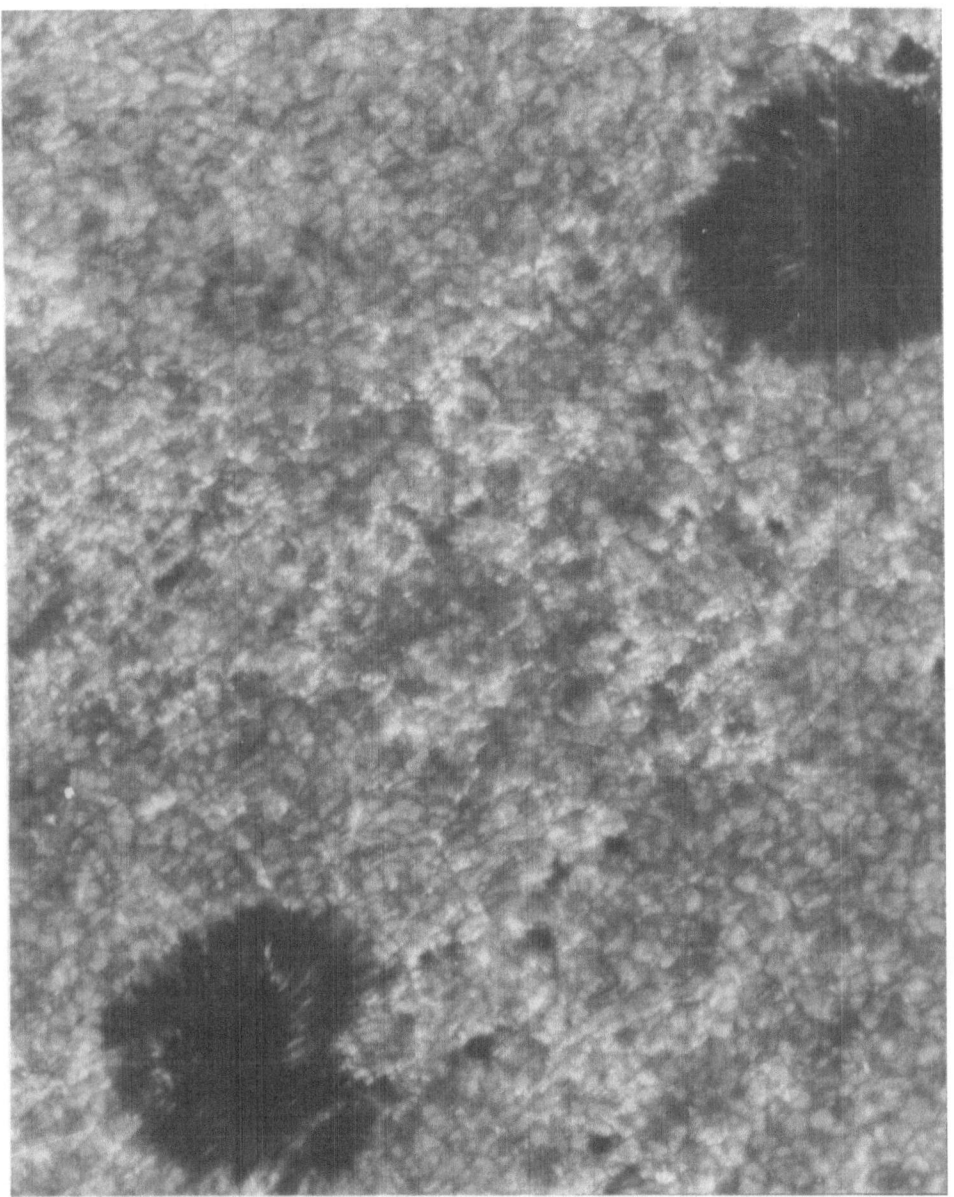

Figure 4. Photospheric faculae observed near the disk center, with the 50 cm refractor at the Pic du Midi Observatory, in the CN band at 4308 A (10 A bandbass filter). Most facular points are smaller than 0".5, many of them approaching the diffraction limit of the telescope (0".2) ; on the other hand larger bright features, sometimes exceeding 1".0 are not rare.

evidence that NBPs and FPs are closely related to kG magnetic fields (Stenflo, 1973, Title et AL.1985). The close relation is confirmed by new high resolution magnetograms obtained by Title and co-workers at La Palma (private communication) ; however the magnetic field cross-section seems to be a little larger than that of the corresponding bright points. In fact, a close inspection of Figure 3 shows that adjacent bright points are very close to each other ; this means that if every bright point is associated to a flux tube, its cross-section cannot be much larger than the size of the point.

In the quiet sun, very few NBPs can be identified inside supergranule cells (Muller, 1983) Figure 3 in this paper. Nevertheless intranetwork magnetic field can be detected with sensitive magnetographs (Harvey, 1977, Martin, 1988). Why ? The magnetic flux associated with an NBP is 4×10^{17} Mx if we assume a field strength of 2000 G and a diameter of 150 km. The magnetic flux measured in the intranetwork field is about 10^{16} Mx, which means that either the size or the field strength is much smaller than for the magnetic features visible at the supergranule boundaries. This may explain why flux tubes are not visible in filtergrams as bright points : either their size is too small to be resolved with the best 50 cm class solar telescopes, or the magnetic field is too weak for a significant brightening to be produced.

4. Contrast At The Disk Center

Faculae are visible as bright features in white light only near the limb. Near the center of the disk, their observed brightness is about the same as that of granules, which makes them uneasy to identify. Muller and Keil (1983) have measured an average brightness of 1.08 times the mean brightness of the photosphere. However, owing to their very small size, their real brightness should be much higher ; actually, Muller and Keil (1983) derived an average value corrected for blurring, in the range 1.3 - 1.5 ; it can even be much higher for the brightest NBPs. It seems that the background intensity of the photosphere in facular areas is slightly lower than in the quiet sun (Hirayama, Hamano and Mizugaki, 1985).

In photospheric lines or in the wings of chromospheric lines, faculae are visible as bright features even at the disk center.

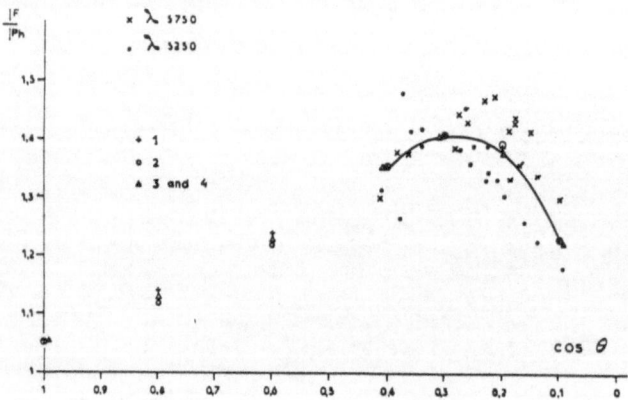

Figure 5. Center-to-Limb Variation of the intensity of facular points (relative to the sunsunding quiet photosphere), corrected for blurring.
Observation : Pic du Midi 50 cm refrator ; resolution : 0".3 (Muller, 1975, Figure 3).

5. Center–To–Limb Variation of The Contrast

The center–to–limb variation of the contrast is used to derive empirical facular and flux tube models or to test theoretical models. This explains why it is an important parameter to be measured accurately.
Most of the published Center-to-Limb Variations (CLV) of the contrast of faculae were obtained from low resolution measurements. That kind of variations are useful for solar oblateness determinations and for evaluating the contribution of faculae to the energy budget of active regions (variation of solar irradiance and luminosity problem). The only high resolution CLV published so far is the one by Muller (1975), who claimed a resolution of O".3. He found a maximum of contrast of 1.4 (corrected for blurring) near $\mu = 0.3$, then a decrease toward the limb (Figure 5). The extrapolated value at the disk center is strongly underestimated, owing to the measurements of Muller and Keil (1983). The Muller's CLV is confirmed by Hirayama (1978) who analysed O".8 resolution observations.

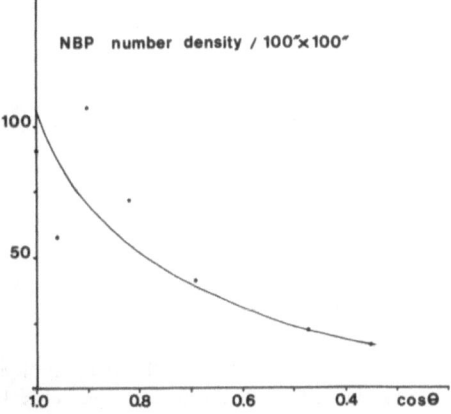

Figure 6. Center-to-Limb Variation of the number density of NBPs number per surface unit of 100" x 100").

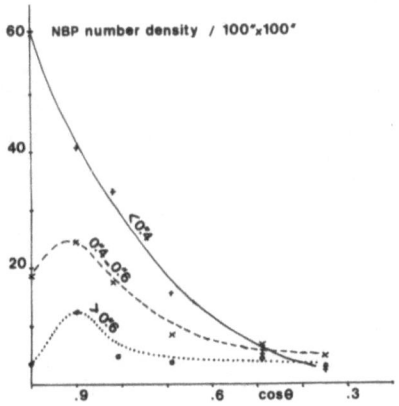

Figure 7. Center-to-Limb Variation fo the number of NBPs of various classes of sizes.

92

CLV of facular contrast have to be used with caution when deriving models, for the following reason pointed out by Muller and Roudier (1984) : the number of NBPs decreases between the center of the disk and the limb by a factor as large as 90 or 95 % (Figure 6 ; Figure 5a in Muller and Roudier, 1984). This means that most NBPs are not visible at the limb, because of the combined effect of foreshortening and resolution, or because of transparency if faculae are elevated hot clouds as suggested by cloud models (Hirayama and Moriyama, 1979 ; Stenflo and Solanki, 1980). This also means that near the limb the visible facular features are the largest ones (selection effect), or are clusters of adjacent unresolved NBPs, especially in active regions where all published measurements were made. Thus even Muller's high resolution CLV measurements do not correspond to isolated characteristic NBPs and are then also irrelevant for deriving and testing facular and flux tube models.

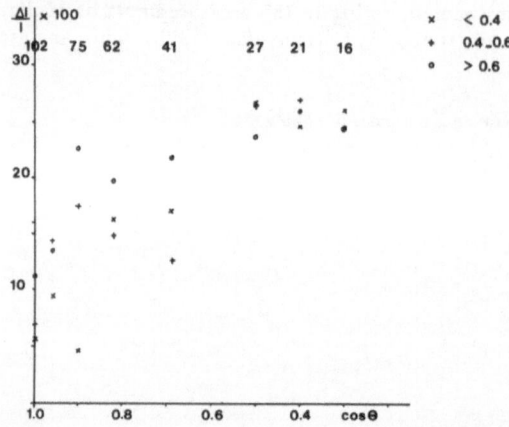

Figure 8. Center–to–Limb Variation of the contrast, all NBPs included. The number density of identified NBPs at various positions on the disk is indicated below the curve.

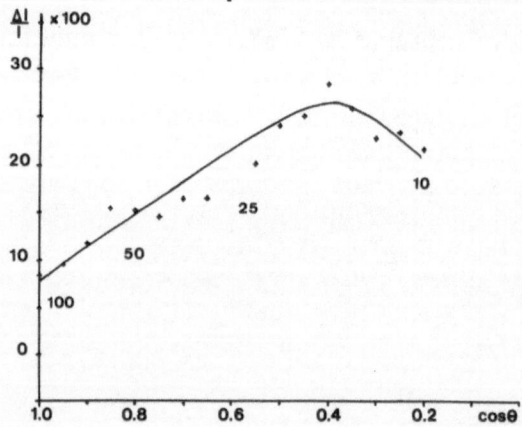

Figure 9. Center–to–Limb Variation of the contrast, for NBPs of various classes of size. The number density of identified at various positions on the disk is indicated above the curve.

6. Center–to–Limb Variation of Isolated NPBs

Figure 7 shows the varation of the number of NBPs of different sizes : as expected NBPs smaller than 0".4 desappear much faster than the larger ones. The increase of the number of points larger than 0".4 around μ = 0.9 is still unexplained. Near the center of the disk, NBPs smaller than 0".4 are dominant ; near the limb there are less. The variation of the average brightness (i.e. all NBPs, large, small, bright and weak included, Figure 8) is very similar to that I obtained for facular elements in active regions. A part of the contrast increase is due to the selection effect mentionned above : at the center of the disk about 100 NBPs per surface unit of 100" x 100" are identified ; as we move toward the limb, this number decreases to a value of about 10. The variation is almost size independant (Figure 9), although near the disk center large NBPs seem to be brighter in average than the small ones ; there is almost no brightness difference between large and small NBPs near the limb ; however the selection is still effective, especially for the points smaller than 0".4 (see Figure 7). Therefore it is hard to know the CLV of a typical bright point of a given size, except for the brightest ones. Because near the limb only 5 NBPs are identified in each class of size, we may plot the contrast variation of the five brightest NBPs present at each observed heliographic position (Figure 10). Again the CLV is almost size independant (however the CLV for each class of size is very unaccurate because of the low number of bright NBPs visible near the limb). The contrast increase from the center of the disk to μ = 0.3 is relatively flatter than for the case when all NBPs are included (Figure 8), because of the higher brightness at the disk center. Toward the limb the contrast decreases. Very probably the contrast should also increase rather smoothly for weaker NBPs. Models should take this smooth increase into account, resulting probably in a smaller vertical temperature gradient inside flux tubes.

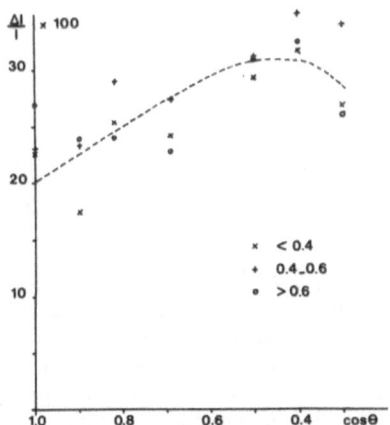

Figure 10. Center–to–Limb Variation of the contrast of the five brightest NBPs of each class of size and at every observed position on the solar disk.

7. Addendum : formation of NBPs by granule compression

The movie from which we have derived the granulation flow pattern shown in Figure 3 has also been used to study the dynamics of the formation of NBPs. As no simultaneous CaIIK3933 or CN4308 filtergrams have been performed, NBPs have been directly

94

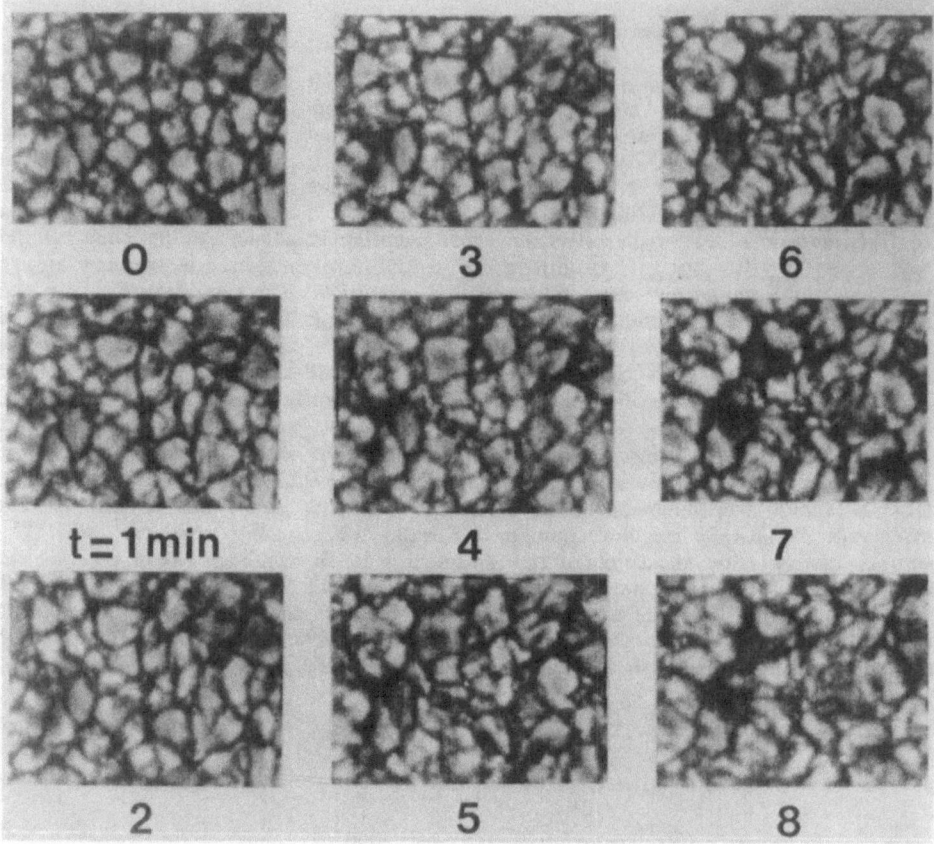

Figure 11. Formation of an NBP in the center of a large intergranular space by the compression of the converging surrounding granules (in the center of the field of view). Its maximum of visibility accurs after 5 minutes.

identified from the white light movie as small (smaller than 0".5), relatively bright and sharp features, of lifetimes longer than 10 min (small granules with which they can be mistaken have much shorter lifetimes, less than 3 min). NBPs appear in large intergranular spaces at the junction of several granular when they converge towards the center of the space (Figure 11) ; their brightness increases during the process and reaches a maximum when they are compressed by the surrounding granules, which form at this stage a "daisy-like pattern" (Figure 11, at t=1,2,3,4,5 min ; several conspicuous examples are also shown in Figure 2 in the paper by Muller, Roudier and Hulot, 1989). When the surrounding granules evolve, the daisy-like pattern is destroyed ; the NBPs remain visible in the intergranular lanes for another 15 min, on an average, although they are less bright and less sharp; they undergo a chaotic random walk controlled by evolving granules, especially the expanding ones. In general, when a large space appears in the intergranular pattern, for example when a large granule fades away, the surrounding granules quickly fill this "empty" space ; usually nothing special occurs, but if some magnetic flux is present there (probably not concentrated enough to be visible as an

NBP), it can be compressed and concentrated by the converging granules, producing an NBP. Thus, NBPs probably correspond to a concentration phase of the magnetic flux. If most of the magnetic field in the photosphere has a strength of 1 to 2 kG, as inferred from the line ratio technique (Stenflo,1973), then the strength may even reach higher values during the concentration phase ; another possibility which cannot be excluded is that of a normally rather weak magnetic field, being concentrated to kilogauss strengths only for a while during the phase of granule compression. In order to understand the formation of flux tubes properly, it will be necessary to mesure the evolution of the magnetic field and the flow during the granule compression process with a spatial resolution well below 0".5.

8. Conclusion

The importance of the photospheric faculae lies in the fact that they are produced by magnetic fields. They allow us to learn about the structure and the behavior of the magnetic flux tubes. However some basic properties like the size, the brightness, the relation to the magnetic field, are not yet known with sufficient accuracy and detail ; the center-to-limb variation of the contrast of a typical NBP or FP is not even known at all. The spatial resolution of the observations have to be improved for filtergrams as well as for magnetograms ; this requires : a) telescopes larger than the 50 cm ones currently in operation ; b) shorter exposure times which can now be obtained with CCD cameras ; c) image stabilization devices or telescopes in the space. From a theoretical point of view, faculae and flux tube models should be revised so that they produce a flatter center-to-limb variation of the contrast between $\mu = 1.0$ and $\mu = 0.3$.

Acknowledgements.

H. Auffret has participated in the analysis of the center-to-limb variation of the contrast of NBPs. The granulation movie has been digitized and processed at Lockheed Palo Alto Research Laboratory in collaboration with Th. Roudier, J. Vigneau, Z. Frank, R. Shine, T. Tarbell and A. Title and at the Solar National Observatory at Sunspot in collaboration with G. Simon. They are all aknowledged for their kind hospitality in the United States. A financial support was obtained from the US Air Force, through a Window-On-Science trip WOS-89-0047.

References

Bray, R.J. and Loughhead, R.E. (1964), Sunspots, Chapman and Hall, Ltd, London.

Chapman, G.A. and Sheeley, N.R. (1968) 'The photospheric network', Solar Phys. 5, 442-

Dunn, R.B. and Zirker, J.B. (1973) 'The solar filigree', Solar Phys. 33, 281-304.

Harvey, J.W. (1977) 'Photospheric magnetic and velocity fields in active regions', in E.A. Müller (ed.), Highlights of Astronomy, 4, 223 .

Hirayama, T. (1978) Publ. Astron. Soc. Japan 30,337.

Hirayama, T., Hamano, S. and Mizugaki, K. (1985) 'Precise wideband photometry of photospheric faculae with an emphases on the disk center', Solar Phys. 99, 43.

Hirayama, T. and Moriyama, F. (1979) 'Center-to-limb variation of the intensity of the photospheric faculae', Solar Phys. 63, 251.

Martin, S.F. (1988) 'The identification and interaction of network, intranetwork and ephemeral-region magnetic fields', Solar Phys. 117, 243–259.

Mehltretter, J.P. (1974) 'Observations of photospheric faculae at the center of the solar disk', Solar Phys. 38, 43–57.

Muller, R. (1975) 'A model of photospheric faculae deduced from white light high resolution pictures', Solar Phys. 45, 105–114.

Muller, R. (1977) 'Morphological properties and origin of the photospheric facular granules', Solar Phys. 52, 249–262.

Muller, R. (1983) 'The dynamical behavior of facular points in the quiet photosphere', Solar Phys. 85, 113–121.

Muller, R. and Keil, S.L. (1983) 'The characteristic size and brightness of facular points in the quiet photosphere', Solar Phys. 87, 243–250.

Muller, R. and Ména B. (1987) 'Motions around a decaying sunspot', Solar Phys. 112, 295–303.

Muller, R. and Roudier, Th. (1984) 'Variability of the quiet photospheric network', Solar Phys. 94, 33–47.

Muller, R., Roudier, Th. and Hulot, J.C. (1989) 'Perturbation of the granular pattern by the presence of magnetic flux tubes', Solar Phys. 119, 229–243.

Schüssler, M. and Solanki, S.K. (1988) 'Continuum intensity of magnetic flux concentrations : are magnetic elements bright points?', Astron. Astrophys. 192, 338–342.

Spruit, H.C. and Zwaan, C. (1981) 'The size dependance of contrasts and numbers of small magnetic flux tubes in an active region', Solar Phys. 70, 207–228.

Stenflo, J.O. (1973) 'Magnetic-field structure of the photospheric network', Solar Phys. 32, 41–63.

Title, A.M., Tarbell, T.O. and Topka, K.P. (1987) 'On the relation between magnetic field structures and granulation', Astrophys. J. 317, 892–899.

BRIGHT FEATURES IN THE INTERGRANULAR REGION

Z. SUEMOTO and E. HIEI
National Astronomical Observatory
Mitaka, Tokyo 181
Japan

ABSTRACT. There are two kinds of bright threads in spectrograms of the Ca II H and K lines: the ones in the outer line wing are due to enhancements of the continuum inside granules, while the others in the inner wing are due to bright features in the intergranular regions.

1. Introduction

It has generally been believed that granulation is the only inhomogeneity of the normal photosphere inside the supergranulation cells. In a paper by Suemoto et al. (1987) we have suggested the possible existence of another inhomogeneity (bright features) in the upper photosphere, which shows up as a bright continuum (provisionally called K_0-continuum) in the inner wing ($3 \text{ Å} > \Delta\lambda > 0.5 \text{ Å}$) of the Ca II K line, and appears in the intergranular lanes. The granular continuum and the K_0-continuum are confirmed to be located more or less complementarily. Further the absorption lines are found to be weakened at the location of the K_0-continuum.

2. Material and Reduction

Spectrograms with a wavelength range from 3885 Å to 4015 Å were taken in one exposure in quiet regions at disk center with a linear dispersion of 0.56 Å mm^{-1}. Two spectrograms were selected to be raster scanned with an aperture of 50 μm × 50 μm (0.32 arcsec × 28 mÅ) and a step size of 20 μm × 20 μm (0.13 arcsec × 11 mÅ) from 3918 Å to 3985 Å. The total area of 24 mm×120 mm of the spectrogram gives 1200×6000 data points.

The contrast $C(\lambda_j, k)$ at a particular position k along the slit at wavelength λ_j is defined as the ratio between the intensity at that position and the mean intensity averaged over all positions along the slit at that wavelength. We selected 13 'windows' in the wings of the Ca II H and K lines, and derived 13 contrast curves along the slit, one for each window.

J. O. Stenflo (ed.), *Solar Photosphere: Structure, Convection, and Magnetic Fields, 97–100.*

3. Results

3.1. SIZE AND DISTRIBUTION

The widths of the granular continuum and the K_0-continuum are inferred from their auto-correlation curves. The full width at half maximum of the granular continuum was estimated from the first minimum, and is about 1000 km. It is to be noted here that the size of the K_0-continuum is also about the same as that of granules.

Correlation coefficients of all the combinations of the 13 windows were calculated and are shown in Figure 1. The ordinates and the abscissae are the correlation coefficients from -1.1 to 1.0, and the effective wavelength distance from the Ca II K line center, $\Delta\lambda_k^*$, respectively. The anti-correlation between the granular continuum and the K_0-continuum is clearly seen.

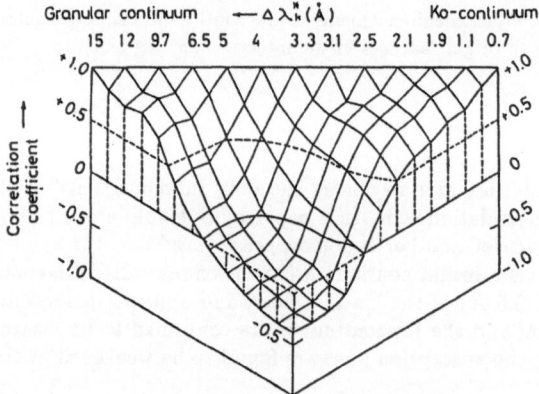

Figure 1. Correlation coefficient diagram for the brightness fluctuations. The correlation coefficients of all the combinations of $\Delta\lambda_k^*$'s are plotted, where $\Delta\lambda_k^*$ is the effective wavelength distance from the Ca II K line center.

It has long been believed that the visibility of the granulation decreases towards the limb but persists far beyond $\cos\theta = 0.3$ until the pattern finally becomes invisible. Edmonds (1962) measured the relative RMS of the brightness fluctuation of the granulation across the solar disk, which is characterized by a slow increase from disk center ($\theta = 0°$) to $\theta \approx 50°$, and a sharp decrease to $\theta \approx 70°$, and then an increase again to $\theta \approx 74°$ (θ is the heliocentric angle). Keil (1977) also measured the center-to-limb variation of the RMS granular contrast which decreases monotonically between $\theta = 0°$ and $\approx 55°$ and then increases slightly at $\theta \approx 70°$. Beyond $\theta \approx 70°$ there are various estimates of the distance from the limb at which the granulation disappears: $15 - 10$ arcsec (Edmonds, 1962), $10 - 4$ arcsec (Rösch, 1957; Loughead and Bray, 1960), and less than 2 arcsec (Pravdjuk et al., 1974; Muller, 1977). Although the distance of the disappearance is different among the authors, all the photographs show the persistence of the so-called 'granular pattern' towards the extreme limb.

Since the granular continuum does not extend much beyond the 'transition wavelength' towards the K line center, many granular regions could not extend much higher than $\tau \approx 0.3$.

On the other hand, by similar reasoning, intergranular hot features would be located much higher than $\tau \approx 0.1$ in the photosphere. It is very likely, then, that what appears to be the granular pattern seen near the limb, say, around a heliocentric angle of $\theta \approx 70°$, may be a mixture of normal granules and the bright features in the intergranular regions. The intensity at $\cos\theta \approx 0.2$ is about 0.3 times that at the disk center, which is equal to the residual intensity at the 'transition wavelength' mentioned above. Beyond $\theta \approx 70°$ the bright features in the intergranular regions may become more and more important. Thus a higher value of the relative RMS of the brightness fluctuations beyond $\theta \approx 74°$ derived by Edmonds (1962) might be due to the increasing predominance of the K_0-continuum.

3.2. CENTRAL INTENSITIES OF ABSORPTION LINES

Since the K_0-continuum is located much higher than, say, $\tau \approx 0.1$ in the photosphere, the central intensities of strong absorption lines may also be affected by the higher temperature responsible for the bright features of the K_0-continuum.

The central intensities of strong lines clearly have a positive correlation with the K_0-continuum, while the correlation is only slightly positive for intermediately strong lines. This is in good agreement with the suggestion mentioned above: central intensities of strong absorption lines, which are formed in the layer of the upper photosphere where the K_0-continuum also originates, would become bright, but medium strong lines are formed at a somewhat lower height than that of the K_0-continuum and the correlation becomes slightly worse. These correlations suggest that the source function of both the lines and the K_0-continuum should have a hump in the upper layer of the photosphere in intergranular regions.

The correlations with the granular continuum are random for strong lines, while being positive for intermediately strong lines. This indicates that the strong lines have almost no relation with granules, while the central intensities of the intermediately strong lines change with the continuous background. This suggests that the source function of the lines and the continuum above granules should decrease monotonically with height, and that the high temperature zone may not extend to the upper layer of the photosphere.

Keil and Canfield (1978) studied the correlation between the central intensities of absorption lines and the granular continuum: the correlation is good for weaker lines, but becomes worse and even negative for the strongest lines, in good agreement with our results.

3.3 RADIAL VELOCITY

The line shift is defined as the central wavelength of the half-level chords of the absorption lines. The radial velocity is derived from the difference between the line shift at each position along the slit and the value averaged over the slit length.

There are good correlations between the line shifts and both continua, which means that upward velocities are predominant above the granules and downward velocities below the layer where the K_0 bright features exist. This result is consistent with previous investigations (Krat and Shpitalnaya, 1974; Bray et al., 1984; Muller, 1985).

References

Bray, R.J., Loughead, R.E., and Durrant, C.J. (1984), *The Solar Granulation*, 2nd edition, Cambridge Univ. Press.

Edmonds, F.N. Jr. (1962) *Astrophys. J. Suppl.* **6**, 357.

Keil, S.L. (1977) *Solar Phys.* **53**, 359.

Keil, S.L. and Canfield, R.C. (1978) *Astron. Astrophys.* **70**, 169.

Krat, V.A. and Shpitalnaya, A.A. (1974) *Solnechnye Dannye* No.2, 63.

Muller, R. (1977) *Solar Phys.* **52**, 249.

Muller, R. (1985) *Solar Phys.* **100**, 237.

Pravdjuk, L.M., Karpinsky, V.N., and Andreiko, A.V. (1974) *Solnechnye Dannye* No.2, 70.

Rösch, J. (1957) *l'Astronomie* **71**, 129.

Suemoto, Z., Hiei, E., and Nakagomi, Y. (1987) *Solar Phys.* **112**, 59.

III. SMALL-SCALE MAGNETIC FIELDS

EMPIRICAL MODELS OF PHOTOSPHERIC FLUX TUBES

S.K. SOLANKI
Department of Mathematical Sciences,
Unversity of St Andrews,
St Andrews, KY16 9SS,
Scotland

ABSTRACT. The empirically derived properties of magnetic flux tubes at both ends of the size spectrum, i.e. magnetic elements and sunspots, are reviewed. Emphasis is placed on quantitative results. The following parameters are discussed in greater detail: The strength and structure of the magnetic field, the temperature stratification and the structure of the velocity field.

1. Introduction

The concept of flux tubes has been successfully used to describe solar photospheric magnetic structures as diverse as sunspots (penumbral diameters generally larger than 5000 km) and magnetic elements (diameters less than approximately 300 km). A photospheric flux tube can be roughly described as a bundle of field lines passing with a more or less vertical axis through the non-magnetic solar photosphere, although highly inclined flux tubes can also be envisaged, for example during flux emergence and submergence.

This review covers some of the *quantitative* observational results obtained on these structures. Although I have tried to treat both sunspots and magnetic elements in equal detail, an imbalance is unavoidable, due to my restricted experience with sunspots. A lack of space forces the exclusion of flux tubes of intermediate size, like pores or magnetic knots. Also, at present it is still unclear whether the latter are really single flux tubes, or whether they are conglomerates of smaller flux tubes (cf. Knölker and Schüssler, 1988). The present review also does not cover possible "weak" mixed polarity fields, since their nature has not yet been resolved and it is not known whether the flux tube picture applies to them. For more details on such features see Stenflo (1988, 1989) and Martin (1989 and these proceedings).

Even within the domain of small scale concentrated fields, there are a number of important topics not covered by this review. For example, it does not deal with those aspects of small scale magnetic fields reviewed elsewhere in these proceedings. Thus I shall leave theoretical aspects and a description of our basic physical understanding of these structures almost entirely to Schüssler and Ryutova, the discussion of their evolution, as derived from magnetograms, to Martin, and questions of morphology to Muller and Title. Before starting on the three main topics, strength and structure of the magnetic field, thermal stratification and velocity structure, let me make a few remarks on the modelling of solar magnetic features in general.

2. General Remarks on the Empirical Modelling of Magnetic Features

There are three main approaches to determining the atmospheric structure of solar magnetic features.

a) **Purely empirical:** As many quantities as possible are individually and independently derived from the observations. For those quantities for which no direct observational information is available, "reasonable" values are assumed.

b) **Semi-empirical:** Key quantities are determined from observations, but the remaining parameters are

103

J. O. Stenflo (ed.), Solar Photosphere: Structure, Convection, and Magnetic Fields, 103–120.
© *1990 by the IAU.*

derived self-consistently from these using simple physics (e.g., hydrostatic equilibrium and thin tube approximation). Semi-empirical models are generally derived iteratively through a succession of steps. Synthetic spectra are compared with observed spectra and the model is changed until the discrepancies become sufficiently small.

c) **Ab initio:** All magnetic and hydrodynamic quantities are derived from basic physical laws, with a minimum of initial observational input. Clues to missing physics or inapropriate assumptions can be obtained by comparing with semi-empirical models, or directly with observations.

Each of the three approaches has a place within the study of solar magnetic features. However, the emphasis has been shifting from purely empirical to the semi empirical and the purely theoretical approaches, as the data and our idea of the underlying physics continue to improve.

All empirical models are no better than the data they are based on and, even more restrictively, the exact diagnostic used. It is, therefore, crucial to develop good diagnostics which are, ideally, very sensitive to a given model parameter and insensitive to all other model parameters. Unfortunately, ideal or nearly ideal diagnostics are rare, so that generally many different diagnostics must be applied simultaneously to complement each other and constrain the structure of the magnetic feature sufficiently. In practice, this often implies the use of more spectral lines. However, only in very few cases can a model be completely prescribed by the observations and additional physical input is generally required. For this reason the semi-empirical approach is generally to be preferred to the purely empirical approach.

There are broad parallels between the construction of empirical models for magnetic elements and sunspots, particularly the need for multi-component, or 2-D models.

Unpolarized low spatial resolution data from network elements and active region plages or faculae can be reproduced by single component models. However, high resolution and polarimetric data show that the field is concentrated into small elements surrounded by a relatively field free atmosphere, calling for a 2-component model: A magnetic component covering a fraction α of the surface within the spatial resolution element of the telescope, and a non-magnetic component covering a fraction $1 - \alpha$, where α is the magnetic filling factor. In such a model the observed Stokes parameters, averaged over both components ($\langle I \rangle$ = unpolarized light, $\langle Q \rangle$, $\langle U \rangle$ = difference between two orothoganal linear polarizations, $\langle V \rangle$ = difference between the two senses of circular polarization) can be written as:

$$\langle I \rangle = \alpha I_m + (1 - \alpha) I_{nm}, \qquad \langle Q \rangle = \alpha Q_m, \qquad \langle U \rangle = \alpha U_m, \qquad \langle V \rangle = \alpha V_m.$$

I_m, Q_m, U_m and V_m are the Stokes profiles emanating from the magnetic feature, while I_{nm} is the only Stokes parameter with a contribution from the non-magnetic region in the resolution element. Stokes V, Q and U are totally independent of the non-magnetic component. Relatively unequivocal information on the magnetic features can be obtained if at least one of these three parameters is observed. Due to the often large linear polarization produced by most solar telescopes, Stokes V has traditionally been used the most, although all four Stokes profiles are required to determine the full magnetic field vector. As we shall see, even at disk centre some observations require proper 2-D models of magnetic elements due to their rapid expansion with height. Closer to the limb 2-D becomes an absolute must (Van Ballegooijen, 1985a, b), since one of the basic assumptions of empirical two- or multi-component models breaks down. Due to the small size of magnetic elements and their expanding canopy a single ray may pass through more than one component, so that the various components are no longer diagnostically decoupled. By this I mean that spectra cannot be calculated individually in each component and the averaged spectrum is no longer simply a sum of the suitably weighted spectra from the individual components.

Although single component models of sunspot umbrae have been quite successful in reproducing low spatial resolution continuum data, the presence of spatially unresolved small-scale structure (umbral dots) also requires the construction of 2-component models of the umbra if higher resolution observations are to be reproduced, with one component each for the dark and the bright features. Due to the small size of umbral dots, models based on data obtained away from disk centre should be two-dimensionsal, since some rays pass through both atmospheric components and the assumption of decoupled atmospheres breaks down. The highly structured penumbra, obviously, also requires at least multi-component models.

Note that since all components of sunspots are permeated by a magnetic field, all four observed Stokes parameters now obtain contributions from both the bright and dark features. If we assume these to cover a fraction α and $(1 - \alpha)$ of the resolution element, respectively, then the observed Stokes parameters can be written as

$$\langle I \rangle = \alpha I_b + (1 - \alpha) I_d, \quad \langle Q \rangle = \alpha Q_b + (1 - \alpha) Q_d, \quad \langle U \rangle = \alpha U_b + (1 - \alpha) U_d, \quad \langle V \rangle = \alpha V_b + (1 - \alpha) V_d.$$

Subscripts 'b' and 'd' denote the bright and dark components respectively. Some of the main advantages of observing Stokes Q, U, or V are therefore lost when studying sunspot fine structures (although the problem of stray light is substantially reduced). This makes the empirical determination of sunspot fine structure properties an extremely difficult task.

3. Magnetic Field

3.1. Sunspots

The resolvable, large scale structure of the sunspot magnetic field strength is relatively straightforward to measure (e.g. from the wavelength difference between the two σ-components of a completely Zeeman split line, although more refined techniques have also been applied) and has been determined by a number of authors with relatively high accuracy. A comparison between the results of different authors can best be made for large and relatively symmetrical sunspots. Mattig (1961), Stepanov (1965), Ioshpa and Obridko (1965), Beckers and Schröter (1969), Wittmann (1974), Gurman and House (1981), Kawakami (1983) and Lites and Skumanich (1989), have, amongst others, derived the field strength as a function of distance from the centre of such sunspots. A selection of $B(r)$ dependences are plotted in Fig. 1.

The field strength near the sunspot centre can vary from sunspot to sunspot, but generally lies in the range 2000–3000 G. In particular, Brants and Zwaan (1982) have shown using purely umbral lines (i.e. lines which are much stronger in the umbra than in the photosphere) that no sunspot umbra, not even the smallest, has a central field strength lower than 2000 G which is considerably higher than values derived from lines also present in the penumbra and the quiet photosphere. Due to the presence of fine scale structure in both the umbral (light bridges, umbral dots) and the penumbral photospheres (penumbral filaments, penumbral grains), the field strengths derived above are averages over the field strengths in the bright and dark components.

The final aim is to obtain the complete magnetic field vector at all points in space above a sunspot. Important contributions have been made by, among others, Beckers and Schröter (1969) (only from Stokes I and V), Deubner and Göhring (1970), Wittmann (1974), Gurman and House (1981), Hagyard et al. (1983), Kawakami (1983) and Lites and Skumanich (1989). Ioshpa and Obridko (1965), Hagyard et al. (1983), and Lites and Skumanich (1989) find that the observed field is close to potential, although Wittmann (1974) and Lites and Skumanich (1989) do find evidence for a twist around the vertical axis in some symmetric sunspots, even after taking magnetooptical effects into account. Fig. 2 shows the inclination angle of the field with respect to the vertical as a function of the normalised sunspot radius, as derived by various authors. One interesting result of the most recent determinations is that the field in the penumbra is not completely horizontal, not even at the outermost edge. However, seeing fluctuations and the unresolved penumbral fine structure may affect this result.

The determination of the field strength variation over the fine scale structure in the sunspot umbra and penumbra is a difficult undertaking and widely different results have been published. Empirical estimates of field strengths in umbral dots range from: 10% of the field strength in the dark umbral core (with opposite polarity) by Beckers and Schröter (1969) (A misinterpretation of magnetooptical effects? See Wittmann, 1971), 50% by Obridko (1968) and Kneer (1973), 90% by Buurman (1973), Adjabshirzadeh and Koutchmy (1983) and Pahlke (1988) and, finally, Zwaan et al. (1985) and Lites and Scharmer (1989), who find no evidence of a weaker field in umbral dots. Although the newer observations favour higher B values, the problem is not yet completely resolved, mainly due to its intractability (see end of Sect. 2).

In the penumbra Beckers and Schröter (1968) and Abdussamatov (1976) found 100–400 G higher B values in dark penumbral filaments than in bright ones from direct high spatial resolution observations. We are also fortunate that the $12\,\mu$ lines are observed in emission in the penumbra. The discovery of the great Zeeman sensitivity of these lines (they are approximately 7–8 times more sensitive than Fe I 5250.2 Å) by

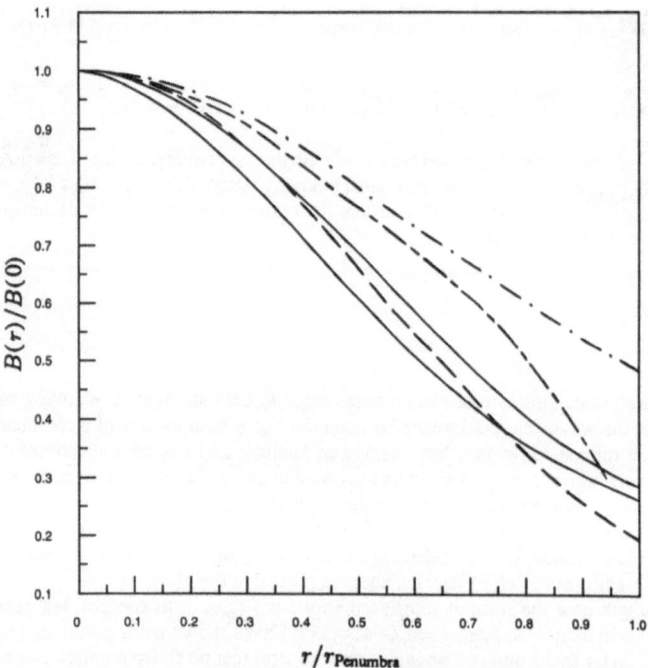

Fig. 1: Magnetic field strength normalised to the central field strength $B(r)/B(0)$, of large symmetrical sunspots vs. radial distance from sunspot centre normalised to the outer penumbral radius, r/r_{Penumbra}. Dot-dashed curve: Beckers and Schröter (1969), long and short dashes: Wittmann (1974); dashed curve: Kawakami (1983); Solid curves: two sunspots studied by Lites and Skumanich (1989). For the spots observed by Lites and Skumanich the umbral radius corresponds to $r/r_{\text{Penumbra}} = 0.4$.

Brault and Noyes (1983) and their identification by Chang and Noyes (1983) has provided us with an exciting new diagnostic of B near the height of the photospheric temperature minimum. Deming et al. (1988) have compared the broadening of the σ-components of a $12\,\mu$ Mg I line with the width of its π-component. Since the σ-components are considerably broader and the vertical B-gradient is expected to be relatively small, they conclude that there is a distribution of field strengths with a full width at half maximum of approximately 450 G. Note, however, that due to uncertainties in the formation mechanism of the line, this value is not a stringent limit. If this line is temperature sensitive, then it may be formed preferentially in one of the penumbral components, so that the field strength may differ by more than 450 G. On the other hand, if part of the broadening is due to inhomogenieties in the Evershed flow (cf. Sect. 5.1), then the field strength variation may be smaller than 450 G.

At least a part and perhaps all of the horizontal variation of the field strength between bright and dark features in umbrae and penumbrae may be caused by different heights of formation of the spectral lines in these features. It is premature to talk of inhomogenieties in the actual magnetic field before this point has been settled.

3.2. Magnetic Elements

Since magnetic elements have generally not been spatially resolved and their field strength is insufficient to completely split spectral lines in the visible, indirect techniques have had to be developed to measure the true

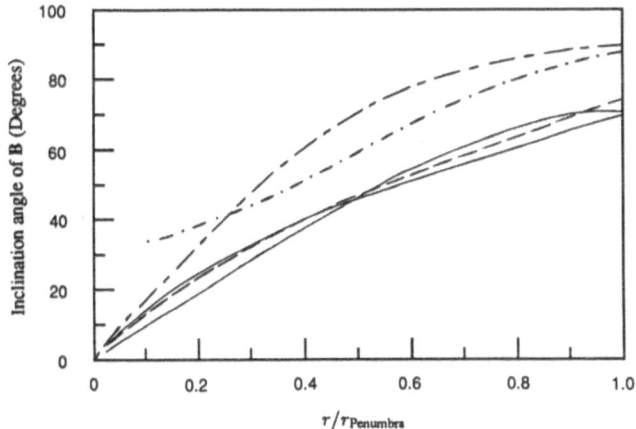

Fig. 2: Inclination angle to the vertical of the magnetic field vector vs. r/r_{Penumbra}. The various curves correspond to the same measurements as in Fig. 1.

field strength in such structures. "Field strengths" derived without such techniques reflect not only the intrinsic field strength, but also depend on the magnetic filling factor (which is influenced by the spatial resolution achieved by the observations), on the assumptions made for the atmospheric structure of the magnetic elements and, for Stokes I observations, their surroundings. I begin by reviewing the field strength measured at a single height in the photosphere.

The first relatively model independent diagnostic, the so-called line-ratio technique, was introduced by Stenflo (1973). It is based on the comparison of the Stokes V profile of a line having a large Landé factor to that of a line with a smaller Landé factor. Relative model independence is achieved if both lines have similar line strengths, excitation potentials and wavelengths. For sufficiently weak fields the ratio of the amplitudes of the two Stokes V profiles is simply proportional to the ratio of their Landé factors, while for sufficiently strong fields the ratio is unity (when both lines are completely split). However, for fields of intermediate strength (0.5–2.0 kG) the ratio of the two Stokes V profiles varies as a function of field strength (Zeeman saturation effect). The measurement of this line-ratio is, therefore, a straightforward diagnostic of field strength, although the exact calibration generally requires some model calculations.

This technique has been extended and applied by Frazier and Stenflo (1978), Wiehr (1978), Stenflo and Harvey (1985), Rachkovsky and Tsap (1985), Solanki et al. (1987), Sánchez Almeida et al. (1988b) and Lozitskij and Tsap (these proceedings). Although the results of Wiehr (1978) and Lozitskij and Tsap differ slightly from the others, the basic result is that at the heights at which the Stokes V profiles of lines in the visible are formed (i.e. the middle photosphere) the field strength averaged over the flux tube cross-section is approximately 1000–1200 G.

In the mean time a host of other techniques has also been developed, most of which, unfortunately, are of lower accuracy. Indirect support for kG fields has come from empirical flux tube models. Thus, Chapman (1974) found that a field strength of approximately 1800 G is required to obtain consistency within his empirical flux tube model. Similarly Koutchmy and Stellmacher (1978) derived field strengths of approximately 1500 G by comparing observations of $I + V$ and $I - V$ with model calculations. Tarbell and Title (1977) applied a technique based on the Fourier transform of the Stokes V profile (under some assumptions on the profile shape) to solar data and derived field strengths between 1000 G and 1800 G. Robinson et al. (1980) and later Sun et al. (1986) applied a technique developed by Robinson (1980) which is based on the Fourier transform of the Stokes I profile and which has had its main application to the detection of fields on late type stars (cf. Saar, these proceedings). They also generally found kG fields with a couple of exceptions. However

Saar (1988) and Hartmann (1987) have shown that in the simple form used by the above authors this techniqe can give wrong values of B. Solanki and Stenflo (1984) derived a kG intrinsic field strength from a statistical analysis of a large number of Stokes V profiles of mostly weak Fe I lines. Recently Del Toro Iniesta et al. (1989) have extended the centre of gravity technique (e.g. Rees and Semel, 1979). When applying it to data of approximately 1″ resolution they find field strengths between 550 G and 1700 G, with a preponderance of lower field strengths. They suggest that the difference between their results and those obtained from other sources is due to differences in spatial resolution. On the other hand, Keller et al. (1989) have applied the line ratio technique to data with a spatial resolution of 0.5″–1″ and find that the observations are compatible to a kG field strength in all the features. They show that the scatter in the observed line ratio is explained in a natural manner by noise in the data. It may be concluded from the above that, given the fact that not all techniques are equally reliable, the results are relatively consistent: The field strength in the middle photosphere is between 1000 and 1500 G.

The main advances regarding our knowledge of the magnetic field structure have, in recent years, come from observations in the infrared. The 12 μ lines have been exploited by Brault and Noyes (1983), Deming et al. (1988) and Zirin and Popp (1989). In active regions (but outside sunspots) $B \approx 250$–550 G is found near the temperature minimum, although the exact height of formation of these lines within magnetic elements is all but clear and even their formation mechanism is not beyond doubt (cf. Lemke and Holweger, 1987). However, they are probably formed in either the upper photosphere or lower chromosphere and the smaller field strengths observed in them do imply a general decrease of field strength with height, as expected from pressure balance.

At present it is simpler to map the derived magnetic field to a certain height by using lines formed at shorter wavelengths in the infrared, for example, the Fe I $g = 3$ line at $1.56485\,\mu$. Harvey and Hall (1975) first noted that this line is three times more Zeeman sensitive than, e.g., the $g = 3$ Fe I 5250.2Å line and derived kG field strengths. Stenflo et al. (1987b), Solanki et al. (1989) and Zayer et al. (1989) confirmed and refined the findings of Harvey and Hall (1975), showing that at the height of formation of this line (lower photosphere) the field strength is approximately 1500–1600 G. This very sensitive and model independent diagnostic, therefore, fully confirms the predominance of kG fields in the lower and middle photosphere. From the profile of this line, and by comparing it with a low g line in the infrared as well as with lines in the visible formed higher in the atmosphere, Solanki et al. (1989) and Zayer et al. (1989) obtained a good idea of the vertical and horizontal structure of the magnetic field strength in the lower and middle photosphere. A 2-D model containing a field with $B = 2000$ G at $\tau = 1$ which is stratified according to exact pressure balance, so that it drops approximately exponentially with height, reproduces the observational constraints on the field strength in the lower and middle photosphere to a high degree. Their analysis also implies that in the lower photosphere, in contrast to sunspots, the field strength is horizontally constant within the tube with a relatively sharp boundary to the non-magnetic atmosphere. The observations, therefore, favour flux tube models of magnetic elements having a boundary current sheet.

The determination of the full magnetic vector is still at an early stage and the reliability of the measurements carried out so far and their interpretation may, perhaps, be questioned. Due to their limited spatial resolution, observations can so far only give an idea of the inclination of whole groups of magnetic elements. Deubner (1975) mentions that he finds evidence of an almost random distribution of field line inclinations. Solanki et al. (1987) derived limits to the inclination from Stokes I, Q and V line profiles and found that the magnetic elements averaged over several thousand km^2 on the solar surface can be inclined in a preferred direction by more than 10° to the vertical. Lites and Skumanich (1989) support these findings.

4. Thermal Structure

4.1. Sunspots

The empirical modelling of the thermodynamical structure of sunspot umbral photospheres is a mature subject, with a large body of work. I shall, therefore, concentrate mainly on more recent investigations.

One problem facing sunspot modellers is that properties of sunspots may depend on their size, shape, age and even on the phase of the solar cycle. If so, then no unique model may be able to describe all sunspots or even all stages in the life of a single sunspot. Maltby et al. (1986) argue that there is no evidence of variations in dark sunspot umbral cores from one sunspot to another, except with the solar cycle (Albregtsen and Maltby, 1978). Also, Sobotka (1988) finds no dependence of the temperature on age of small sunspots, but, on the other hand, he does see the need of different umbral dot filling factors in different umbrae. Finally, Grossmann-Doerth et al. (1986) find differences between the properties of umbral dots in different parts of the umbra.

As a way out of this dilemma modellers have either restricted the applicability of their models to, e.g., large sunspots (Maltby et al., 1986; Obridko and Staude, 1988) or have presented different models, e.g., for different sunspot sizes and different umbral dot fillings (Sobotka, 1988), or for different phases of the solar cycle (Maltby et al., 1986).

If inhomogenieties like umbral dots are ignored, or only that part of the umbra which appears to be uniform is considered, then single component models can be constructed. Such models exist in profusion and now rival the quiet photosphere models with regard to the breadth of data which they are based on. Recent models of the darkest part of the umbral photosphere have been constructed by Kollatschny et al. (1980) and Van Ballegooijen (1984) which are based on line profiles, and by Maltby et al. (1986), based on purely continuum data. These models are relatively similar and satisfy the observations almost equally well. Although the latter authors do not see any signs for the need of two-components, Van Ballegooijen finds such a need, possibly because he is fitting line profiles which can be more sensitive to inhomogenieties.

A number of authors have taken the alternative approach and have constructed 2-component models of umbrae, with a cold component describing the umbral core, and a warmer component describing umbral dots. Perhaps a third component describing light bridges in old or complex sunspots should also be introduced. Specific models for light bridges have been made by Sobotka (1989).

Adjabshirzadeh and Koutchmy (1983) and Obridko and Staude (1988) have published 2-component models. The most comprehensive is that of the latter authors which satisfies a wide variety of observational data on large sunspots. Among other things its cold component reproduces the continuum observations of Albregtsen et al. (1984), on which the deeper layers of the Maltby et al. model are based, while a 50% mixture of the hot component and the cold component can reproduce the umbral dot brightness observations of Wiehr and Stellmacher (1985). For smaller sunspots Sobotka (1988) has proposed 2-component models based on a few spectral lines. These are simpler than the models described above, being "down-scaled" photospheric models. He finds large variations in the filling factors of the bright component of small umbrae. Whether this is due to the rapidly increasing problem of stray light with decreasing umbral diameter is unclear. The temperature stratification of a few recent umbral models is illustrated in Fig. 3.

The temperature of umbral dots is still a debated subject, with two schools of thought. The first claims that dots have photospheric brightness and hence temperature (e.g. Beckers and Schröter, 1968; Koutchmy and Adjabshirzadeh, 1981; Pahlke, 1988), while the other suggests that they are a few 100 to 1000 K cooler than the quiet photosphere, depending partly on their position in the umbra (e.g. Grossmann-Doerth et al., 1986; Obridko and Staude, 1988; Sobotka, 1988). Since umbral dots are generally spatially unresolved and both umbral components have a magnetic field, the derivation of true dot temperatures is an unresolved issue. Indirect techniques, such as using Stokes V profiles of lines with different excitation potentials (Pahlke, 1988) can only reduce the problem, but not resolve it. In many ways the problem is similar to that of deriving the temperature structure of magnetic elements from Stokes I profiles (see Sect. 4.2). The disadvantages mentioned in that connection largely also apply here.

The penumbra is possibly the most visible example of a solar photospheric structure requiring multi-component models. In white light the pattern of bands of varying shades of brightness is quite striking. Krat et al. (1972) and Muller (1973) applied the 2-component approach to the brightness distribution of the penumbra and sorted the penumbral filaments into bright and dark bins with approximately 0.7–1 and 0.3–0.7 times the photospheric brightness, respectively. Kjeldseth-Moe and Maltby (1974) used these continuum data to derive 2-component models of the penumbra which, when appropriately combined, also reproduce low resolution multi-wavelength continuum observations obtained by Maltby (1972). Their model has been included in Fig. 3. Although a simple 2-component scheme has been called into question by Grossmann-Doerth and Schmidt (1981), who point out that the histogram of the continuum intensities of penumbral filaments is single peaked (with brightness ranging between approximately 0.5 and 1.0 times the photospheric value), a 2-component

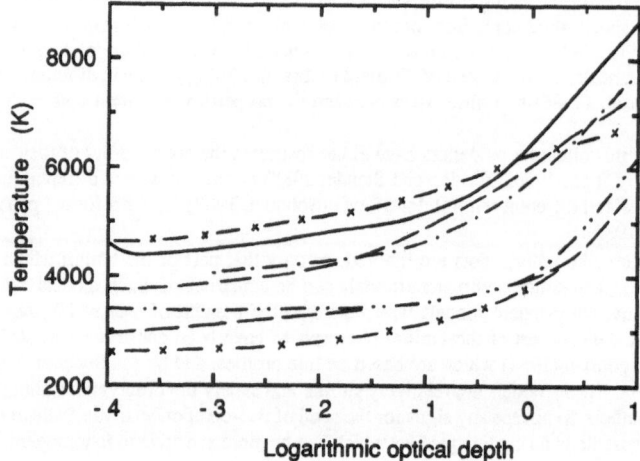

Fig. 3: Temperature vs. logarithmic optical depth of a selection of models of various sunspot components and of the quiet sun (solid line). Maltby et al. (1986) umbral core model: —— · · ——; Obridko and Staude (1988) umbral core and umbral dot models: —— ——; Adjabshirzadeh and Koutchmy (1983) umbral core and dot models: —— × ——; Kjeldseth-Moe and Maltby (1974) dark penumbral filament model: —— · ——; (their bright penumbral filament model is similar to the quiet photosphere model and has not been plotted separately).

approach is a first step on the way towards reliable empirical penumbral models. However, the caveat of Grossmann-doerth and Schmidt (1981) demonstrates tha compared to umbrae, penumbral modelling is still at an early stage and high quality, high resolution data are very desirable.

4.2. Magnetic Elements

Empirical models of magnetic features outside of sunspots also have a long and varied history. A plethora of 1-component, later 2-component and recently 2-D models have been constructed with the help of data ranging from the centre to limb variation (CLV) of continuum contrast measurements at one wavelength to Stokes V profiles of many lines.

The oldest types of models are one-component models based on low spatial resolution data. Such models describe some ill-defined average over the magnetic elements and their surroundings. They generally have a temperature very similar to the quiet sun at equal optical depth near $\tau = 1$ at 5000 Å and a gently increasing temperature *difference* to the quiet sun with height throughout the photosphere (i.e. decreasing optical depth, cf. Model by Shine and Linsky, 1974, in Fig. 4). Such models have been constructed by, e.g., Schmahl (1967), Stellmacher and Wiehr (1973), Shine and Linsky (1974) and Vernazza et al. (1981, model F). The general $T(\tau)$ shape of such models may easily be guessed by simply looking at an image of the solar disk in white light: Faculae are not seen near the centre of the disk, but are clearly visible as bright features near the limb, where the continuum is formed higher in the atmosphere. In summary, as long as the spatial resolution is low, single component models are relatively well defined, although they do not describe any part of the real solar atmosphere in active regions.

In contrast to this, the results of 2-component modelling depend very strongly on the data and, in some cases, on additional assumptions. We can divide such models into basically two groups, those based on Stokes I and those based on Stokes V (and, in future, possibly Stokes Q and U as well, see Solanki et al.,

1987). Both groups share one major uncertainty, namely the value of the continuum intensity near disk centre, which determines the lower photospheric layers of the models. However, whereas models can be derived from Stokes V with a minimum of additional assumptions (if the correct combination of lines is used), the approach based on Stokes I requires another two major assumptions: a) the filling factor of the magnetic features in the resolution element, b) the temperature of the atmosphere surrounding the magnetic elements. Recall that Stokes I obtains contributions from both magnetic and non-magnetic components.

Before discussing the various types of models, let me briefly discuss the continuum contrast observations, i.e., the ratio of the continuum intensity in the magnetic elements to that in the quiet sun. The directly observed continuum contrast is one of the most uncertain parameters of magnetic elements. Values range from 1 to 2 and depend on the spatial resolution of the observations. Higher resolution observations give larger contrasts (compare Koutchmy, 1977; Muller and Keil, 1983; Foukal and Fowler, 1984; Hirayama et al., 1985, etc.). Schüssler and Solanki (1988) introduced another technique, based on the line ratio of two Stokes I profiles, which specifically takes the unknown temperature of the surroundings into account. After assuming an upper limit of 25% for the filling factor of their observed region they were able to set a lower limit of 1.4 on the continuum contrast. Particularly exciting are speckle observations which allow very high spatial resolution to be achieved and which show points, associated with Ca II brightenings and therefore probably with magnetic fields, that are around 1.5 times brighter than the quiet sun (Von der Lühe, 1989). The difference between low and high spatial resolution results can best be reconciled in active regions if the continuum forming layers of the non-magnetic atmosphere between magnetic elements is cooler than the quiet sun by approximately 100 K (Schüssler and Solanki, 1988). This lower temperature has been confirmed using other data and another technique by Brandt and Solanki (1989).

One single component and a selection of 2-component models derived from Stokes I measurements are plotted in Fig. 4. They show tremendous differences from one to another. I shall now try to roughly explain the main features of these models.

All the plotted 2-component models have an appreciably higher temperature in the middle and upper photosphere than the quiet sun and the 1-component model. The reason is that now the observed line weakening (or, in some models, the continuum contrast enhancement near the limb) is assumed to be produced entirely within the small fraction of the surface covered by the magnetic elements. Note that the modellers using Stokes I data have assumed the magnetic elements to be surrounded by the quiet sun. Around the height where the lines are formed ($-2.5 \lesssim \log \tau \lesssim -1.5$) the 2-component models are reasonably similar, given the uncertainty in filling factor. The difference between the dashed (Chapman) and the double dashed (Walton) curves in Fig. 4 illustrates the effect on the derived stratification of choosing different filling factors, since Walton (1987) also obtained a model very similar to Chapman's simply by increasing the assumed value of α. Some of the differences between models in the upper layers may be solar (e.g. Hirayama, 1978, derived two different models from his continuum contrast data, of which only the hotter is shown, see also the discussion of the Stokes V based models, below). However, the models differ substantially in the deeper layers where they rely on continuum contrast near solar disk centre. Models based on low spatial resolution, i.e., low continuum contrast data (e.g. Chapman, 1970, 1977, 1979; Walton, 1987) usually have a temperature close to the quiet sun value in the deeper layers, while models based on high spatial resolution, i.e., high continuum contrast data (e.g. Koutchmy and Stellmacher, 1978; Stellmacher and Wiehr, 1979), have a high temperature in their lower layers.

The Stokes I models therefore reflect mainly 1) the data they are based on, 2) the choice of the modeller regarding filling factor and external atmosphere and 3) partly real differences between solar magnetic features. The general features of the Stokes I models can, therefore, be understood if only a few particulars are kept in mind.

There are fewer models based on Stokes V. The first Stokes V based model was that of Stenflo (1975) derived from a few lines and assuming a continuum contrast close to unity which was an accepted value at that time. Consequently it differs considerably in its lower layers from the more recent models of Solanki (1986) and Keller (these proceedings). However, at the level of line formation it is quite similar (cf. Fig. 5). The models of Solanki satisfy selected line parameters of Stokes V profiles of hundreds of simultaneously observed unblended Fe I and II lines. However, they are not based on a least squares fit (inversion technique). The models derived by Keller (these proceedings) have been calculated both in 1-D and 2-D (which produce

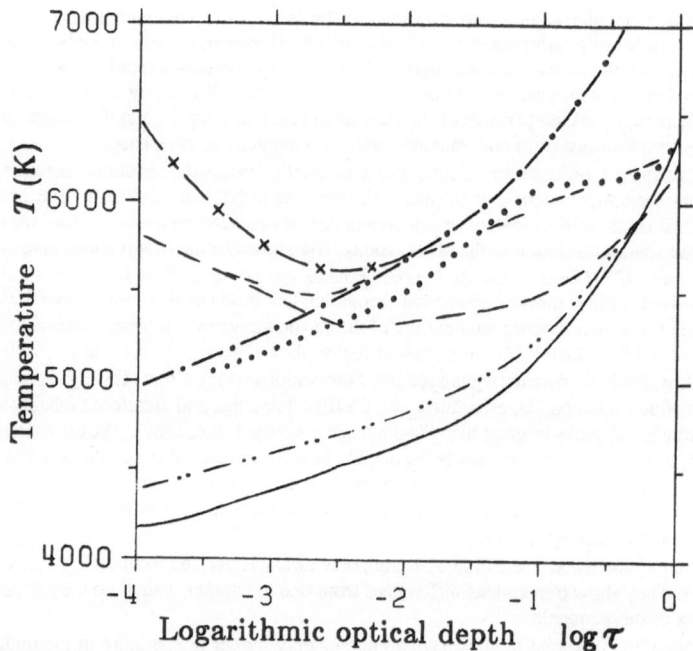

Fig. 4: Temperature of facular models based on Stokes I vs. logarithmic optical depth. Quiet sun (HSRA, Gingerich et al., 1971): solid line. 1-component model of Shine and Linsky: ———··———; 2-component models of Chapman (1977): ——— ———; Walton (1987): —— ——; Hirayama (1978): ·····; Koutchmy and Stellmacher (1978): ——— × ———; Stellmacher and Wiehr (1979): ———·———.

almost identical temperature stratifications). They are based on a smaller data set (10 lines), but, on the other hand, they have been derived using an "inversion" of these data, i.e. these models give the best least squares fit to the 10 Stokes V profiles from the starting values chosen.

Despite the large data sets underlying them, these models are only reliably determined from Stokes V between approximately, $-3 \lesssim \log \tau \lesssim -1$. In particular, below $\log \tau \approx -1$ the models are mainly determined by the continuum intensity. This has been taken to lie around 1.3–1.4 by Solanki (1986) and at 1.8 by Keller. Both Solanki (1986) and Keller (these proceedings) derive different models for active region plages and the network with flux tubes being warmer when the filling factor is lower. Although this result was originally based on observations of only a few regions, the work of Pantellini et al. (1988) and more recently Zayer et al. (to be published) has confirmed this trend for many regions covering a large range of filling factors.

The models of Keller (these proceedings) are reasonably similar to the models of Solanki (1986) in the height range over which they are reliable, which is not all too surprising since they are based on partially the same data. The models by the above-mentioned authors based on data from the enhanced network differ from each other more than those based on the plage data due to the lower S/N in the network data.

In summary, the models based on Stokes V overcome some of the main disadvantages faced by purely Stokes I based models, and also appear to be reasonably reliable, within their limitation. The three main limitations are: 1) The current models are based on relatively low resolution data, so that a single model may be trying to describe magnetic elements having different properties. 2) All current models rely on continuum contrast measurements for the determination of the temperature in the deeper photospheric layers. 3) LTE

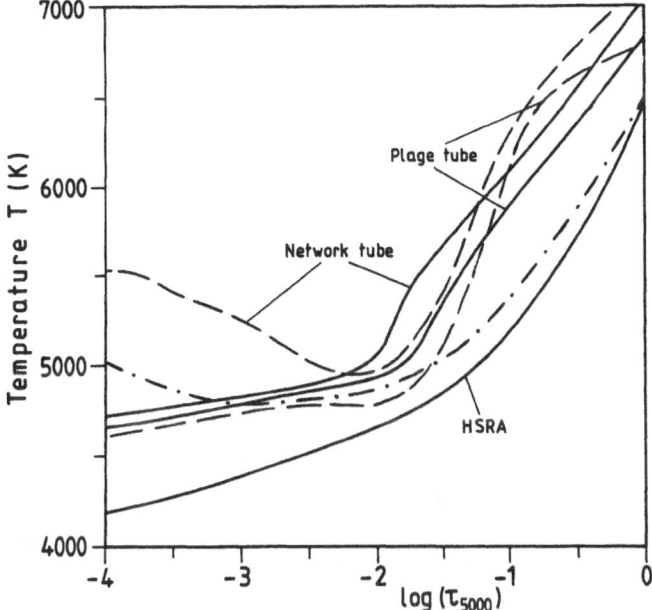

Fig. 5: Temperature of magnetic element models based on Stokes V vs. logarithmic optical depth, compared to the model of the quiet sun, the HSRA (marked). Stenflo (1975): dot-dashed curve; Solanki (1986): solid curves; Keller (these proceedings): dashed curves. Where separate models for the network and for plages exist these have been marked.

is assumed. The importance of NLTE effects in a 2-D flux tube with hot walls has been demonstrated by Stenholm and Stenflo (1978) using a 2-level atom. Solanki and Steenbock (1988), on the other hand, used only a 1-D flux tube model, but a very comprehensive iron atomic model and an empirical temperature stratification to show that departures from LTE are larger in magnetic elements than in the quiet sun. One important consequence of neglecting NLTE is that the present models cannot reproduce the onset of the chromospheric temperature rise. Since the source functions in the cores of the strongest low excitation lines, which are formed highest in the atmosphere, begin to decouple from the Planck function, in LTE calculations the temperature will generally have to keep dropping with height to mimic the source function.

5. Velocities: Flows, Oscillations and Waves

Mass-motions in all solar photospheric features are studied with basically similar techniques. Some of the generally used diagnostics in the quiet sun and in sunspots are: the wavelength of the Stokes I line core, the line width, the line asymmetry and the time dependence of these parameters. In complete analogy the diagnostics of mass-motions in and around magnetic elements are: The zero-crossing wavelength, line width and asymmetry of Stokes V and the temporal variation of these quantities. * The Stokes V diagnostics can, naturally, also be used in sunspots. Conversely the Stokes I velocity sensitive parameters can also provide

---

* The variation of the Stokes V amplitude or wing area as a function of time has also been interpreted in terms of mass-motions which is analogous to looking at variations in the line depth of Stokes I.

an idea of the combined velocity field of the magnetic elements and of their surroundings. The evolution of features on filtergrams or magnetograms may be used to trace horizontal motions, but these shall not be discussed here (cf. Martin, Title, these proceedings).

5.1. Velocities in Sunspots

The oldest known dynamic effect in magnetic features is the Evershed flow, an outflow of matter at photospheric heights in the sunspot penumbra which becomes an inflow in the penumbral chromosphere (inverse Evershed effect), while lines formed near the temperature minimum (e.g. $12\,\mu$ emission lines, cf. Deming et al., 1988) do not show any Evershed effect.

Measurements in the wings of photospheric lines give sizeable velocities of 1–4 km s^{-1} (Beckers, 1968; Abdussamatov and Krat, 1970; Ichimoto, 1987). However, line cores show shifts of typically only 0–1.5 km s^{-1} depending on position in the penumbra and on the spectral line (e.g. Wiehr et al., 1986; Ichimoto, 1987). A combination of the two measurements implies an asymmetry in the Stokes I line profiles which can only be explained by a vertical and/or horizontal velocity gradient. Evidence that the Evershed flow is not homogeneous but is concentrated in the darker filaments has been found by, e.g., Abdussamatov and Krat (1970) and Mamadazimov (1972). Since this represents a gradient across the field lines (in a horizontal direction) it is at least partly responsible for the asymmetry in Stokes I. However, it is probably not the only source of Stokes I asymmetry, since the change in sign of the Evershed effect with height also suggests the presence of a vertical velocity gradient. Finally, the observations of broad-band circular polarization (see Sect. 5.3) imply that a gradient along the (inclined) field lines is also present. It is still not clear whether the Evershed flow disappears within 2–3″ of the outer boundary of the penumbra (e.g., Wiehr et al., 1986) or whether it continues to be seen further out as well (e.g., Küveler and Wiehr, 1985).

Umbral dots show sizeable blueshifts of 2–3 km s^{-1} compared to the dark umbral core (Kneer, 1973; Pahlke, 1988). Umbral Stokes V profiles are asymmetric which Pahlke (1988) interprets as being due to different flow velocities in umbral dots and the umbral core.

Both the umbra and the penumbra are rich in oscillations and waves. Their amplitudes, periods and, in some cases, propagation speeds can be directly determined. Umbral Oscillations at the photospheric level generally have amplitudes substantially smaller than 0.5 km s^{-1} and periods close to 5 minutes (e.g. Beckers and Schultz, 1972; Soltau et al., 1976; Kneer et al., 1981; Lites and Thomas, 1985; Abdelatif et al., 1986). Lines formed close to the temperature minimum still see power in the 5 minutes band, but also see oscillations with a period of 3 minutes, which is the main period observed in the chromosphere (Lites and Thomas, 1985). The oscillations can have slightly different periods in different parts of the umbra (Lites, 1986), suggesting the presence of more than one independently oscillating cell within an umbra. The observed oscillations are probably not the only non-stationary mass-motions in umbrae. Line broadenings give rms velocities of 1–2 km s^{-1} (both vertical and horizontal, Beckers, 1976), suggesting that there may be considerable power at very small scales (perhaps in connection with umbral dots). Penumbral oscillations have also been observed (Beckers and Schultz, 1972; Musman et al., 1976; Balthasar et al., 1987; Lites, 1988) and Musman et al. (1976) have detected signs of outwards propagation (running penumbral waves). Balthasar et al. (1987) and Lites (1988) find evidence that the penumbral oscillations are aligned along the magnetic field. The power in both the umbral and the penumbral 5-minute oscillations is considerably lower than in the quiet photosphere.

5.2. Velocities in Magnetic Elements

Stationary flows within magnetic elements are diagnosed from time-averaged Stokes V zero-crossing measurements. The watershed in such measurements came towards the middle of the present decade. Earlier measurements all showed downflows between 0.5 and 2 km s^{-1} (e.g. Giovanelli and Ramsay, 1971; Giovanelli and Brown, 1977; Harvey, 1977; Giovanelli and Slaughter, 1978; Wiehr, 1985; Scholier and Wiehr, 1985), while later observations do not show any significant downflows (Stenflo and Harvey, 1985; Solanki, 1986; Stenflo et al., 1987a, Wiehr, 1987; Solanki and Pahlke, 1988; Muglach, private communication). In particular Solanki (1986), from the analysis of the zero-crossings of a few hundred unblended lines, could set an upper limit of ±250 m s^{-1} on the stationary flow velocity in magnetic elements. The most recent confirmation of an absence of downflows comes from measurements in the infrared. In a study currently underway

Karin Muglach has looked at all unblended Fe I lines between 1.5 and 1.7 μ and finds that the data are consistent, within the scatter, with an absence of a stationary flow. This removes the remaining uncertainty (due to the measurements of a single line, Fe I 15648.5 Å, by Harvey, 1977, and Stenflo et al., 1987b) surrounding the unshifted zero-crossing of spatially and temporally averaged Stokes V profiles. Of course, local flows of short duration in either direction cannot be excluded, but their confirmation awaits good S/N high spatial resolution observations.

The discrepancy between the older and the newer observations can be explained either by the improvement in instrumentation (see Harvey, private communication, with regard to Harvey, 1977), selection effects (Solanki and Pahlke, 1988, with regard to Scholier and Wiehr, 1985) and differences in spectral resolution (Solanki and Stenflo, 1986, with regard to the rest of the measurements showing downflows). The observations showing downflows in Stokes V generally have a lower *spectral* resolution than those which do not. The presence of an asymmetry in Stokes V, with the blue wing of V being stronger than its red wing for nearly all lines near disk centre (cf. Sect. 5.3) means that after the spectral smearing caused by the spectrograph such a profile is always redshifted. Since the Stokes V asymmetry was unknown before the middle of the current decade, this instrumentally produced redshift was (quite naturally) falsely interpreted as due to a downflow within the magnetic elements.

The variation in the zero-crossing wavelength has been followed as a function of time by Giovanelli et al. (1978) and Wiehr (1985). They observed waves with periods close to 300 s and the former authors also found evidence for upward propagation. However, the amplitudes suggested by these studies are only $\approx 0.2-0.3$ km s^{-1} in the upper photosphere, i.e. similar to those of photospheric umbral oscillations. If taken at face value, these numbers suggest that flux tube waves are insignificant for the energetics of the outer solar atmosphere. However, as we shall see below, this conclusion is probably false.

Another source of information on non-stationary mass motions within magnetic elements are the line widths of Stokes V profiles. Solanki (1986) has studied line widths of Stokes V profiles formed in magnetic elements at solar disk centre, while Pantellini et al. (1988) have analysed them at various positions on the disk. The results clearly show the presence of non-stationary motions in small magnetic elements with amplitudes larger than found in the quiet sun, both along the field lines and across them. Although the exact amplitudes may be somewhat affected by uncertainties in the detailed temperature structure, this main result appears well established (cf. Keller, these proceedings). The line broadenings cannot be caused by velocities outside the magnetic elements (Solanki, 1989).

The difference between the velocity amplitudes derived from the zero-crossing time series and from line broadening measurements is easily explained if the small size of magnetic elements is taken into account. If numerous magnetic elements are present in the resolution element and the waves or oscillations in them are not exactly in phase then the Stokes V zero-crossing signal gets strongly smeared and becomes difficult to measure. Also, if the wavelength of the wave is less than the width of the contribution function of the Stokes V profile (defined by Van Ballegooijen, 1985a, and Grossmann-Doerth et al., 1988a), then the wave amplitude seen in the Stokes V zero-crossing is smaller than the true amplitude (Solanki and Roberts, 1989).

5.3. Stokes V Asymmetry

The final velocity diagnostic, the asymmetry of the Stokes V profile, is discussed separately since its observation and interpretation in sunspots and in magnetic elements are closely related to each other and its causes in magnetic elements have only recently been clarified.

The first observational suggestion of an asymmetry in Stokes V came from the discovery of broad band circular polarization in sunspots (Illing et al., 1974a), later confirmed by Illing et al. (1974b, 1975), Kemp and Henson (1983), Henson and Kemp (1984) and Makita (1986). These observations measure the net circular polarization averaged over many spectral lines. Recently broad band circular polarization has also been measured outside sunspots (Kemp et al., 1987). Spectrally resolved observations of the asymmetry of individual Stokes V profiles formed in magnetic elements have been published by, e.g., Stenflo et al. (1984), Solanki and Stenflo (1984, 1985), Wiehr (1985) and Pantellini et al. (1988). Near solar disk centre the area and amplitude of the blue Stokes V wing of most lines is larger than the area, respectively amplitude of the red wing. When moving towards the limb the sign of the area asymmetry changes, first for the stronger lines, later for the weaker ones. For magnetic elements (and probably also for sunspots) the source of the broad band circular polarisation is dominantly the area asymmetry of the Stokes V profiles of individual spectral lines as shown by Mürset et al. (1988).

Auer and Heasley (1978) demonstrated that the Stokes V profiles of lines formed in LTE are antisymmetric in the absence of velocities. The mechanisms proposed to explain the area asymmetry of Stokes V are based on the violation of one or the other of the above two assumptions. An approach based on departures from LTE has been taken by Kemp et al. (1984) and Landi Degl'Innocenti (1985). However, any quantitative application to a solar situation is formidably complicated and has so far not been attempted. Also, the rather high densities in the photosphere make it unlikely that the extremely strong domination by the radiation field (compared to collisions), required by this "atomic orientation" model, is present. The fact that the Stokes V profiles of Fe II lines, formed very close to LTE, are just as asymmetric as Fe I lines also speaks strongly against this mechanism. The use of velocities appears more promising. Illing et al. (1975) first showed that a velocity gradient along the line of sight, combined with a gradient in the magnetic field strength, produces an asymmetry (cf. Grigorjev and Katz, 1975). Auer and Heasley (1978) later pointed out that the magnetic field gradient is not required if the magnetic vector is not parallel to the line of sight.

Further work on sunspots has concentrated mainly on approximately reproducing the spacial distribution, the absolute level and the CLV of the measured broad band circular polarization (e.g. Landmann and Finn, 1979; Makita, 1986; Skumanich and Lites, 1987). Qualitative agreement with the observations of the spatial dependence of the broad band circular polarization of Henson and Kemp (1984), which show a maximum of the asymmetry near the inner border of the penumbra, can be reached within the confines of a self-similar sunspot model (Schlüter-Temesvary theory) with a considerable Evershed flow. The Evershed flow is therefore currently expected to be the main contributer to the V asymmetry. The spectral dependence of the broad band circular polarization is mainly a result of the spectral density of absorption by lines (Mürset et al., 1988). A more thorough analysis awaits better data.

In magnetic elements the interpretations and models have tried to reproduce the observed Stokes V line profiles. Obtaining the asymmetry without producing a zero-crossing shift in Stokes V has been one of the main problems. Sánchez Almeida et al. (1988a, 1989) have found that they can reproduce the Stokes V asymmetry and zero-crossing of a number of lines with a flow within the magnetic elements if they assume a field whose strength increases with height. However, such a model contradicts directly measured field strengths (Sect. 3.2). On the other hand, Solanki and Pahlke (1988) have shown that an internal stationary downflow in a physically reasonable flux tube model, which satisfies observations of the magnetic field structure, produces synthetic spectra that conflict with the observations in various ways.

Van Ballegooijen (1985c) suggested that downflows outside the magnetic elements should produce the proper sign of the asymmetry. Grossmann-Doerth et al. (1988b, 1989) proved that in this case, i.e. when velocity and B do not overlap, the Stokes V profile is asymmetric but its zero-crossing wavelength corresponds exactly to the rest wavelength (cf. Fig. 6). One implication of this result is that 2-D models incorporating the expansion of the magnetic elements with height are necessary to reproduce the Stokes V area asymmetry. Another implication is that the Stokes V area asymmetry provides an excellent diagnostic of the velocity in the immediate surroundings of magnetic elements. Solanki (1989) has studied these diagnostic capabilities at disk centre in greater detail. He finds that for a 2-D flux tube model fulfilling the observational constraints on field strength and temperature, the area asymmetry of four lines with widely different properties can be satisfied by an external downflow of approximately 1 km s^{-1}, if the immediate surroundings of the flux tube are 250–350 K cooler than the quiet sun. This conforms very well to the picture, derived from white light images and magnetograms, of magnetic elements being concentrated in cool, downflowing intergranular lanes (e.g. Dunn and Zirker, 1974, Mehltretter, 1974; Muller, 1983; Title et al., 1987).

However, Solanki (1989) also found that a velocity outside the flux tube alone cannot reproduce the shape and width of Stokes V completely. In order to reproduce both these quantities, without affecting the zero-crossing wavelength too strongly, longitudinal waves or oscillations inside the magnetic elements, with a slight difference between the upflowing and downflowing phase had to be introduced. The data require wave amplitudes between 1 and 3.0 km s^{-1}. Further information on the nature of the wavelike motion or more precise values of the amplitudes cannot be derived from the very simple model used (two velocity components only).

Fig. 6: Schematic illustration using a two layered atmosphere of how an asymmetric Stokes V profile may be produced without any shift of its zero-crossing wavelength. Only the special case of a magnetic field aligned along the line of sight is considered here. Bottom frame: Profiles of $\eta_{\pm}$ (absorption coefficients for right and left circularly polarized light, respectively) in the lower atmospheric layer, where there is no field but a downflow (the two profiles are identical and redshifted). Second lowest frame: $\eta_{\pm}$ in the upper layer where there is a magnetic field but no velocity. Third lowest frame: $I_{\pm}$, the emergent intensity profiles for the two polarizations. Due to saturation effects I_{+} has a larger equivalent width (η_{-} in both atmospheric layers lies at almost the same wavelength, while η_{+} is at widely different wavelengths). Topmost frame: $V = I_{+} - I_{-}$, with a larger area of the blue wing than of the red wing. Note that since in both the upper and lower atmosperic layers $\eta_{+} = \eta_{-}$ at λ_0, $V(\lambda_0) = 0$, i.e. V is asymmetric, but unshifted.

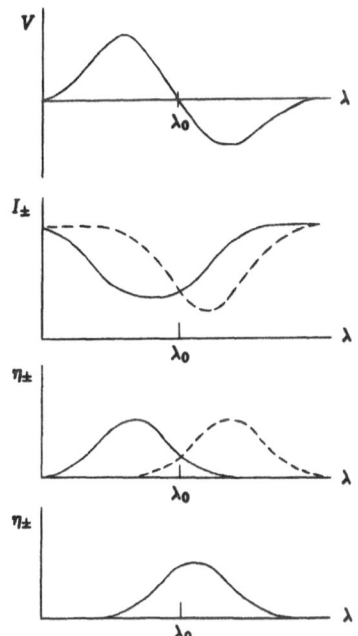

In summary, the Stokes V profiles observed in magnetic elements at solar disk centre can be reproduced by a flux tube model with the field fanning out with height, having no stationary internal flow, but waves or oscillations with different up- and downflow phases. The flux tube is surrounded by a ring of cool downflowing gas. The change in sign of the asymmetry closer to the limb is probably due to the surrounding gas flowing horizontally towards the flux tubes (horizontal component of granular motion).

Acknowledgements: It is a pleasure to thank Dr. Gurtovenko and the other members of the Local Organising Committee for organising and financing my travel and stay in the USSR. Hania Allen kindly helped with the drawing of some of the figures, while David Evans and Alan Miles helped me to find the relevant literature on umbral and penumbral oscillations and waves.

References

Abdelatif, T.E., Lites, B.W., Thomas, J.H.: 1986, *Astrophys. J.* **311**, 1015
Abdussamatov, H.I.: 1976, *Solar Phys.* **48**, 117
Abdussamatov, H.I., Krat, V.A.: 1970, *Solar Phys.* **14**, 132
Adjabshirzadeh, H., Koutchmy, S.: 1983, *Astron. Astrophys.* **122**, 1
Albregtsen, F., Jorås, P.B., Maltby, P.: 1984, *Solar Phys.* **90**, 17
Albregtsen, F., Maltby, P.: 1978, *Nature* **274**, 41
Auer, L.H., Heasley, J.N.: 1978, *Astron. Astrophys.* **64**, 67
Balthasar, H., Küveler, G., Wiehr, E.: 1987, *Solar Phys.* **112**, 37
Beckers, J.M.: 1968, *Solar Phys.* **3**, 258
Beckers, J.M.: 1976, *Astrophys. J.* **203**, 739–752.
Beckers, J.M., Schröter, E.H.: 1968, *Solar Phys.* **4**, 303
Beckers, J.M., Schröter, E.H.: 1969, *Solar Phys.* **10**, 384

Beckers, J.M., Schultz, R.B.: 1972, *Solar Phys.* **27**, 61

Brandt, P.N., Solanki, S.K.: 1989, *Astron. Astrophys.* submitted

Brants, J.J., Zwaan, C.: 1982, *Solar Phys.* **80**, 251

Brault, J.W., Noyes, R.W.: 1983, *Astrophys. J.* **269**, L61

Buurman, J.: 1973, *Astron. Astrophys.* **29**, 329

Chang, E.S., Noyes, R.W.: 1983, *Astrophys. J.* **275**, L11

Chapman, G.A.: 1970, *Solar Phys.* **14**, 315

Chapman, G.A.: 1974, *Astrophys. J.* **191**, 255

Chapman, G.A.: 1977, *Astrophys. J. Suppl. Ser.* **33**, 35

Chapman, G.A.: 1979, *Astrophys. J.* **232**, 923

Del Toro Iniesta, J.C., Semel, M., Collados, M., Sánchez Almeida, J.: 1989, *Astron. Astrophys.* in press.

Deming, D., Boyle, R.J., Jennings, D.E., Wiedemann, G.: 1988, *Astrophys. J.* **333**, 978

Deubner, F.L.: 1975, *Osserv. Mem. Oss. Astrofis. Arcetri* **105**, 39

Deubner, F.L., Göhring, R.: 1970, *Solar Phys.* **13**, 118

Dunn, R.B., Zirker, J.B.: 1973, *Solar Phys.* **33**, 281

Foukal, P., Fowler, L.: 1984, *Astrophys. J.* **281**, 442

Frazier, E.N., Stenflo, J.O.: 1978, *Astron. Astrophys.* **70**, 789

Gingerich, O., Noyes, R.W., Kalkofen, W., Cuny, Y.: 1971, *Solar Phys.* **18**, 347

Giovanelli, R.G., Brown, N.: 1977, *Solar Phys.* **52**, 27

Giovanelli, R.G., Livingston, W.C., Harvey, J.W.: 1978, *Solar Phys.* **59**, 49

Giovanelli, R.G., Ramsay, J.V.: 1971, in *Solar Magnetic Fields*, R. Howard (Ed.), *IAU Symp.* **43**, 293

Giovanelli, R.G., Slaughter, C.: 1978, *Solar Phys.* **57**, 255

Grigorjev, V.M., Katz, J.M.: 1975, *Solar Phys.* **42**, 21

Grossmann-Doerth, U., Larsson, B., Solanki, S.K.: 1988a, *Astron. Astrophys.* **204**, 266

Grossmann-Doerth, U., Schmidt, W.: 1981, *Astron. Astrophys.* **95**, 366

Grossmann-Doerth, U., Schmidt, W., Schröter, E.H.: 1986, *Astron. Astrophys.* **156**, 476

Grossmann-Doerth, U., Schüssler, M., Solanki, S.K.: 1988b, *Astron. Astrophys.* **206**, L37

Grossmann-Doerth, U., Schüssler, M., Solanki, S.K.: 1989, *Astron. Astrophys.* in press

Gurman, J.B., House, L.L.: 1981, *Solar Phys.* **71**, 5

Hagyard, M.J., Teuber, D., West, E.A., Tandberg-Hanssen, E., Henze, W., Beckers, J.M., Bruner, M., Hyder, C.L., Woodgate, B.E.: 1983, *Solar Phys.* **84**, 13

Hartmann, L.: 1987, in *Cool Stars, Stellar Systems, and the Sun*, V., J.L. Linsky, R.E. Stencel (Eds.), Lecture Notes in Physics Vol. 291, Springer-Verlag, Berlin, p. 1

Harvey, J.W: 1977, in *Highlights of Astronomy*, E.A. Müller (Ed.), Vol. 4, Part II, p. 223

Harvey, J.W., Hall, D.: 1975, *Bull. Amer. Astron. Soc.* **7**, 459.

Henson, G.D., Kemp, J.C.: 1984, *Solar Phys.* **93**, 289

Hirayama, T.: 1978, *Publ. Astron. Soc. Japan* **30**, 337

Hirayama, T., Hamana, S., Mizugaki, K.: 1985, *Solar Phys.* **99**, 43

Ichimoto, K.: 1987, *Publ. Astron. Soc. Japan* **39**, 329

Illing, R.M.E., Landman, D.A., Mickey, D.L.: 1974a, *Astron. Astrophys.* **35**, 327

Illing, R.M.E., Landman, D.A., Mickey, D.L.: 1974b, *Astron. Astrophys.* **37**, 97

Illing, R.M.E., Landman, D.A., Mickey, D.L.: 1975, *Astron. Astrophys.* **41**, 183

Ioshpa, B.A., Obridko, V.N.: 1965, *Soln. Dannye* **3**, 54

Kawakami, H.: 1983, , *Publ. Astron. Soc. Japan* **35**, 459

Keller, C.U.: 1988, *Diplomarbeit*, E.T.H. Zürich.

Keller, C.U., Tarbell, T.D., Title, A.M., Solanki, S.K., Stenflo J.O.: 1989, *Astron. Astrophys.* submitted

Kemp, J.C., Henson, G.D.: 1983, *Astrophys. J.* **266**, L69

Kemp, J.C., Henson, G.D., Steiner, C.T., Powell, E.R.: 1987a, *Nature* **326**, 270

Kemp, J.C., Macek, J.H., Nehring, F.W.: 1984, *Astrophys. J.* **278**, 863

Kjeldseth-Moe, O., Maltby, P.: 1974, *Solar Phys.* **35**, 101

Kneer, F.: 1973, *Solar Phys.* **28**, 361

Kneer, F., Mattig, W., Von Uexküll, M.: 1981, *Astron. Astrophys.* **102**, 147

Knölker, M., Schüssler, M.: 1988, *Astron. Astrophys.* **202**, 275

Kollatschny, W., Stellmacher, G., Wiehr, E., Falipou, M.A.: 1980, *Astron. Astrophys.* **86**, 245

Koutchmy, S.: 1977, *Astron. Astrophys.* **61**, 397

Koutchmy, S., Adjabshirzadeh, A.: 1981, *Astron. Astrophys.* **99**, 111

Koutchmy, S., Stellmacher, G.: 1978, *Astron. Astrophys.* **67**, 93

Krat, V.A., Karpinsky, V.N., Pravidjuk, L.M.: 1972, *Solar Phys.* **26**, 305

Küveler, G., Wiehr, E.: 1985, *Astron. Astrophys.* **142**, 205

Landi Degl'Innocenti E.: 1985, in *Theoretical Problems in High Resolution Solar Physics*, H.U. Schmidt (Ed.), Max Planck Inst. f. Astrophys., Munich, p. 162

Landman, D.A., Finn, G.D.: 1979, *Solar Phys.* **63**, 221

Lemke, M. Holweger, H.: 1987, *Astron. Astrophys.* **173**, 375

Lites, B.W.: 1986, *Astrophys. J.* **301**, 992

Lites, B.W.: 1988, *Astrophys. J.* **334**, 1054

Lites, B.W., Scharmer, G.: 1989, in *High Spatial Resolution Solar Observations*, O. Von der Lühe (Ed.), Sacramento Peak, NM, in press.

Lites, B.W., Skumanich, A.: 1989, *Astrophys. J.* in press

Lites, B.W., Thomas, J.H.: 1985, *Astrophys. J.* **294**, 682

Makita, M.: 1986, *Solar Phys.* **106**, 269

Maltby, P.: 1972, *Solar Phys.* **26**, 76

Maltby, P., Avrett, E.H., Carlsson, M., Kjeldseth-Moe, O., Kurucz, R.L., Loeser, R.: 1986, *Astrophys. J.* **306**, 284

Mamadazimov, M.: 1972, *Solar Phys.* **22**, 129

Martin, S.F.: 1988, *Solar Phys.* **117**, 243

Mattig, W.: 1961, *Mitt. Astron. Gesell. Hamburg* **14**, 47

Mehltretter, J.P.: 1974, *Solar Phys.* **38**, 43

Muller, R.: 1973, *Solar Phys.* **32**, 409

Muller, R.: 1983, *Solar Phys.* **85**, 113

Muller, R., Keil, S.L.: 1983, *Solar Phys.* **87**, 243

Mürset, U., Solanki, S.K., Stenflo, J.O.: 1988, *Astron. Astrophys.* **204**, 279

Musman, S., Nye, A.H., Thomas, J.H.: 1976, *Astrophys. J.* **206**, L175

Obridko, V.N.: 1968, *Bull. Astron. Inst. Chechoslovakia* **19**, 183

Obridko V.N., Staude, J.: 1988, *Astron. Astrophys.* **189**, 232

Pahlke, K.-D.: 1988, Ph.D. Thesis, University of Göttingen

Pantellini, F.G.E., Solanki, S.K., Stenflo, J.O.: 1988, *Astron. Astrophys.* **189**, 263

Rachkovsky, D.N., Tsap, T.T.: 1985, *Izv. Krymskoj Astrofiz. Obs.* **71**, 79

Rees, D.E., Semel, M.D.: 1979, *Astron. Astrophys.* **74**, 1

Robinson, R.D.: 1980, *Astrophys. J.* **239**, 961

Robinson, R.D., Worden, S.P., Harvey, J.W.: 1980, *Astrophys. J.* **236**, L155

Saar, S.H.: 1988, *Astrophys. J.* **324**, 441

Sánchez Almeida, J., Collados, M., Del Toro Iniesta, J.C.: 1988a, *Astron. Astrophys.* **201**, L37

Sánchez Almeida, J., Collados, M., Del Toro Iniesta, J.C.: 1989, *Astron. Astrophys.* in press

Sánchez Almeida, J., Solanki, S.K., Collados, M., del Toro Iniesta, J.C.: 1988b, *Astron. Astrophys.* **196**, 266

Scholiers, W., Wiehr, E.: 1985, *Solar Phys.* **99**, 349

Schüssler, M.: 1986, in *Small Scale Magnetic Flux Concentrations in the Solar Photosphere*, W. Deinzer, M. Knölker, H.H. Voigt (Eds.), Vandenhoeck & Ruprecht, Göttingen, p. 103

Schüssler, M., Solanki, S.K.: 1988, *Astron. Astrophys.* **192**, 338

Shine, R.A., Linsky, J.L.: 1974, *Solar Phys.* **37**, 145

Schmahl, G.: 1967, *Z. Astrophys.* **66**, 81

Skumanich, A., Lites, B.W.: 1987, *Astrophys. J.* **322**, 483

Sobotka, M.: 1988, *Bull. Astron. Inst. Czechoslovakia* **39**, 236

Sobotka, M.: 1989, *Solar Phys.* in press

Solanki, S.K.: 1986, *Astron. Astrophys.* **168**, 311

120

Solanki, S.K.: 1989, *Astron. Astrophys.* in press
Solanki, S.K., Keller, C., Stenflo, J.O.: 1987, *Astron. Astrophys.* **188**, 183
Solanki, S.K., Pahlke, K.D.: 1988, *Astron. Astrophys.* **201**, 143
Solanki, S.K., Roberts, B.: 1989, in Proc. Chapman Conference on Magnetic Flux Ropes, C.T. Russell (Ed.), in press
Solanki, S.K., Steenbock, W.: 1988, *Astron. Astrophys.* **189**, 243
Solanki, S.K., Stenflo, J.O.: 1984, *Astron. Astrophys.* **140**, 185
Solanki, S.K., Stenflo, J.O.: 1985, *Astron. Astrophys.* **148**, 123
Solanki, S.K., Stenflo, J.O.: 1986, *Astron. Astrophys.* **170**, 120
Solanki, S.K., Zayer, I., Stenflo, J.O.: 1989, in *High Spatial Resolution Solar Observations*, O. Von der Lühe (Ed.), Sacramento Peak, NM, in press.
Soltau, D., Schröter, E.H., Wöhl, H.: 1976, *Astron. Astrophys.* **50**, 367
Stellmacher, G., Wiehr, E.: 1973, *Astron. Astrophys.* **29**, 13
Stellmacher, G., Wiehr, E.: 1979, *Astron. Astrophys.* **75**, 263
Stenflo, J.O.: 1973, *Solar Phys.* **32**, 41
Stenflo, J.O.: 1975, *Solar Phys.* **42**, 79
Stenflo, J.O.: 1988, *Solar Phys.* **114**, 1
Stenflo, J.O.: 1989, *Astron. Astrophys. Rev.* 1, 3.
Stenflo, J.O., Harvey, J.W.: 1985, *Solar Phys.* **95**, 99
Stenflo, J.O., Harvey, J.W., Brault, J.W., Solanki, S.K.: 1984, *Astron. Astrophys.* **131**, 333
Stenflo, J.O., Solanki, S.K., Harvey, J.W.: 1987a, *Astron. Astrophys.* **171**, 305
Stenflo, J.O., Solanki, S.K., Harvey, J.W.: 1987b, *Astron. Astrophys.* **173**, 167
Stenholm, L.G., Stenflo, J.O.: 1978, *Astron. Astrophys.* **67**, 33
Stepanov V.E.: 1965, *IAU Symp.* **22**, 267
Sun, W.-H., Giampapa, M.S., Worden, S.P.: 1987, *Astrophys. J.* **312**, 930
Tarbell, T.D., Title, A.M.: 1977, *Solar Phys.* **52**, 13
Title, A.M., Tarbell, T.D., Topka, K.P.: 1987, *Astrophys. J.* **317**, 892
Van Ballegooijen, A.A.: 1984, *Solar Phys.* **91**, 195
Van Ballegooijen, A.A.: 1985a, in *Measurements of Solar Vector Magnetic Fields*, M.J. Hagyard (Ed.), NASA Conf. Publ. 2374, p. 322
Van Ballegooijen, A.A.: 1985b, in *Theoretical Problems in High Resolution Solar Physics*, H.U. Schmidt (Ed.), Max Planck Inst. f. Astrophys., Munich, p. 167
Van Ballegooijen, A.A.: 1985c, in *Theoretical Problems in High Resolution Solar Physics*, H.U. Schmidt (Ed.), Max Planck Inst. f. Astrophys., Munich, p. 177.
Vernazza, J.E., Avrett, E.H., Loeser, R.: 1981, *Astrophys. J. Suppl. Ser.* **45**, 635
Von der Lühe, O.: 1989, in *High Spatial Resolution Solar Observations*, O. Von der Lühe (Ed.), Sacramento Peak, NM, in press.
Walton, S.R.: 1987, *Astrophys. J.* **312**, 909
Wiehr, E.: 1978, *Astron. Astrophys.* **69**, 279
Wiehr, E.: 1985, *Astron. Astrophys.* **149**, 217
Wiehr, E.: 1987, in *Cool Stars, Stellar Systems, and the Sun*, V., J.L. Linsky, R.E. Stencel (Eds.), Lecture Notes in Physics Vol. 291, Springer-Verlag, Berlin, p. 54
Wiehr, E., Stellmacher, G.: 1985, in *High Resolution in Solar Physics, Lect. Notes in Phys.* **233**, 254
Wiehr, E., Stellmacher, G., Knölker, M., Grosser, H.: 1986, *Astron. Astrophys.* **155**, 402
Wittmann, A.D.: 1971, *Solar Phys.* **20**, 365
Wittmann, A.D.: 1974, *Solar Phys.* **36**, 29
Zayer, I., Solanki, S.K., Stenflo, J.O.: 1988, *Astron. Astrophys.* in press
Zirin, H., Popp, B.: 1989, *Astrophys. J.* in press
Zwaan, C., Brants, J.J., Cram, L.E.: 1985, *Solar Phys.* **95**, 3

EMPIRICAL PHOTOSPHERIC FLUXTUBE MODELS FROM INVERSION OF STOKES V DATA

C. U. KELLER
Institute of Astronomy
ETH-Zentrum
CH-8092 Zürich
Switzerland

ABSTRACT. We present results of an inversion procedure that derives the turbulent velocity, the magnetic field strength, and the temperature stratification of the photospheric layers of solar magnetic fluxtubes from 10 FeI and FeII Stokes V line profiles around 5250 Å and from the continuum contrast. The free parameters of two-dimensional magnetohydrostatic fluxtube models are determined by minimizing the difference between observed and calculated Stokes V parameters in an iterative manner. Results of this inversion procedure applied to observations of a plage and a network region at disk center indicate a temperature deficit (at equal geometrical height) of the fluxtubes at the level of continuum formation and a temperature excess at the highest levels of line formation in general agreement with the latest theoretical fluxtube models.

1. Introduction

Most existing empirical models of solar magnetic fluxtubes have been obtained by fitting synthetic spectra from simple fluxtube models to observed Stokes I profiles or synthetic continuum intensities to the observed center-to-limb variation of the facular continuum contrast. A few one-dimensional models have been derived from Stokes V observations (see the references in Solanki, this volume). Only in the latter case the analysis can be performed independently of the spatial resolution and the filling factor. In this work we present empirical, two-dimensional, self-consistent fluxtube models obtained by an inversion of Stokes V line profiles. These models take into account the spreading of the fluxtube with increasing height, the current sheet, and tension forces, in contrast to earlier models. A more precise description of the inversion procedure and its application to observations can be found in Keller et al. (1989).

2. Inversion of Stokes V Profiles

The inversion of Stokes V profiles is based on the determination of a few model fluxtube parameters by the iterative least-squares fitting algorithm of Marquardt (1963). The free parameters of the axisymmetric, magnetohydrostatic fluxtube models of Steiner et al. (1986) are the magnetic field strength at optical depth unity inside the fluxtube $B(\tau_i = 1)$, the macroturbulence velocity as a function of the strength and the excitation potential of

J. O. Stenflo (ed.), Solar Photosphere: Structure, Convection, and Magnetic Fields, 121–124.
© 1990 by the IAU.

the spectral lines, and the temperature difference between the fluxtube interior and the quiet photosphere at equal geometrical height at five grid points. The zero level of the geometrical height scale is defined at optical depth unity of the quiet photosphere at 5000 Å ($\tau_e = 1$). The horizontal temperature distribution inside the fluxtube is homogenous, the microturbulence velocity has a value of $0.6\,\mathrm{km\,s^{-1}}$ independent of the height, and the radius of the fluxtube at $\tau_e = 1$ is 100 km. These models are used to calculate synthetic Stokes V profiles which are parameterized such that the chosen parameters vary dominantly with one model fluxtube parameter. The minimization of the difference between observed and synthetic Stokes V parameters leads to a determination of the free model fluxtube parameters.

We have selected 8 FeI and 2 FeII lines around 5250 Å for this inversion procedure. The observed Stokes V profiles have been symmetrized around their zero-crossings to avoid complicate fluxtube models that can explain the observed Stokes V asymmetry (Grossmann-Doerth et al., 1988b). The following Stokes V parameters have been used: The magnetic line ratio (Stenflo, 1973), formed between the FeI 5247.1 Å and the FeI 5250.2 Å Stokes V amplitudes, is insensitive to all fluxtube parameters except the magnetic field strength and the macroturbulence velocity (Solanki et al., 1987). A 'thermal' line ratio is formed between the FeI 5247.1 Å and FeI 5250.6 Å lines (Stenflo et al., 1987). The difference in the excitation potential of these two lines is the reason for the sensitivity of this ratio to the temperature. The ratios of the areas of the Stokes V wings between the FeI lines and the FeII 5197 Å line depend strongly on the temperature because FeII lines are rather insensitive to the temperature compared to FeI. These FeI to FeII Stokes V ratios with lines of different strength and excitation potential are the main diagnostics for the temperature stratification. Note that the Stokes V signal is proportional to the filling factor. Thus, ratios of Stokes V amplitudes or areas are independent of the filling factor. When deriving the temperature structure of fluxtubes it is essential to include the broadening of spectral lines by turbulent velocities (Solanki, 1986). We, therefore, fit the FWHM of the Stokes V wings and the distance between the two Stokes V extrema. It was necessary to include an estimated continuum contrast of magnetic fluxtubes (1.8; Koutchmy, 1977 even states a lower limit of 2) to stabilize the inversion code. The influence of the estimated continuum contrast on the resulting model, however, is negligible at the levels of line formation.

3. Results

In this chapter we present two-dimensional fluxtube models obtained by applying the inversion procedure to high spectral resolution Fourier Transform Spectrometer observations of a plage and a network region at disk center (Stenflo et al., 1984). When starting the inversion procedure from different initial values we obtain nearly the same models; this shows that the inversion applied to this specific data set gives unique solutions. Figure 1 shows the temperature on the axis of the fluxtube model as a function of the optical depth and as a function of the geometrical height. The magnetic field strength at $\tau_i = 1$ is 2400 G for the plage and 2160 G for the network fluxtubes. The macroturbulence velocities of weak lines are comparable to the values measured from Stokes I profiles in the quiet photosphere; however, strong lines show macroturbulence velocities which exceed the values found in the quiet photosphere by roughly $2\,\mathrm{km\,s^{-1}}$. This is in agreement with earlier models derived from the same data with a different technique (Solanki, 1986). The estimated accuracy of the results is 100 K for the temperature, $0.2\,\mathrm{km\,s^{-1}}$ for the turbulent velocity, and 100 G for the magnetic field strength.

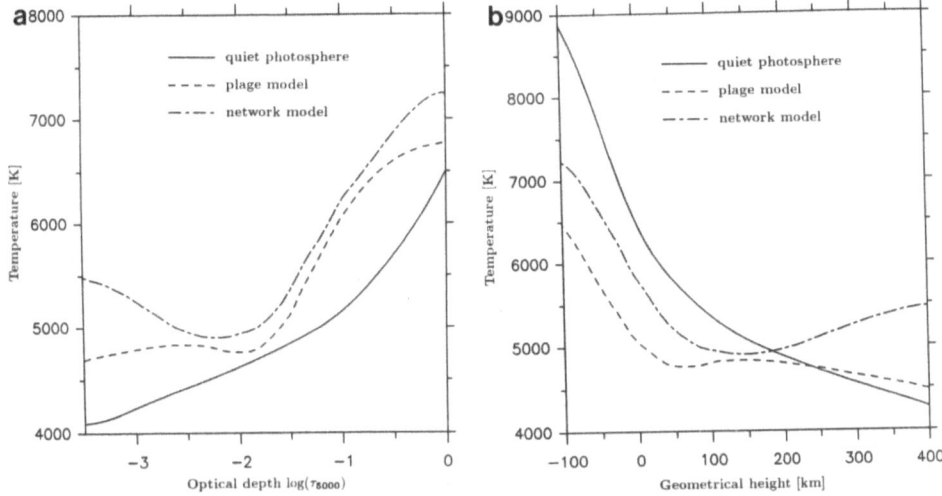

Figure 1. The plage and network models compared with the quiet solar photosphere at (a) equal optical depth and (b) equal geometrical height.

The influence of different microturbulence velocities, radii of the fluxtubes, errors in the determination of the oscillator strengths of the spectral lines, and different grid point locations on the resulting temperature stratification have been investigated and found to have no significance. Due to NLTE effects the fluxtube temperature in the upper photospheric layers is only a lower limit.

4. Discussion

Most empirical temperature stratifications of fluxtubes have been obtained as a function of the optical depth; no geometrical height scale is associated with these models. However, the temperature stratifications of the models presented in this work have been obtained as a function of the geometrical height scale. This is a large advantage for the interpretation of the results and the comparison with theoretical models. The optical depth scale can easily be computed from the temperature and the pressure startifications. Our fluxtube models show a temperature deficit compared to the quiet photosphere below $z = 0$ km. The partial inhibition of convective motion inside magnetic elements seems to be the source of this temperature deficit. Deinzer et al. (1984) even found a temperature deficit of up to 3000 K at $z = -125$ km in their theoretical models. The temperature excess in the higher layers have recently been explained by radiative transfer effects (see Fig. 2b). The hot bottom illuminates and heats the higher levels of the fluxtubes (Kalkofen et al., 1988; Grossmann-Doerth et al., 1988a). Our temperature startifications are in good agreement with earlier models (see Fig. 2a) derived from the same Stokes V data (Solanki, 1986). We confirm that the fluxtubes in the network region are hotter than those in the plage region at equal optical depth and find the same behavior at equal geometrical height. Although there is now a general agreement between theoretical and empirical models we want to emphasize that existing theoretical models cannot explain the low temperature occuring around $\tau_i = -2$.

124

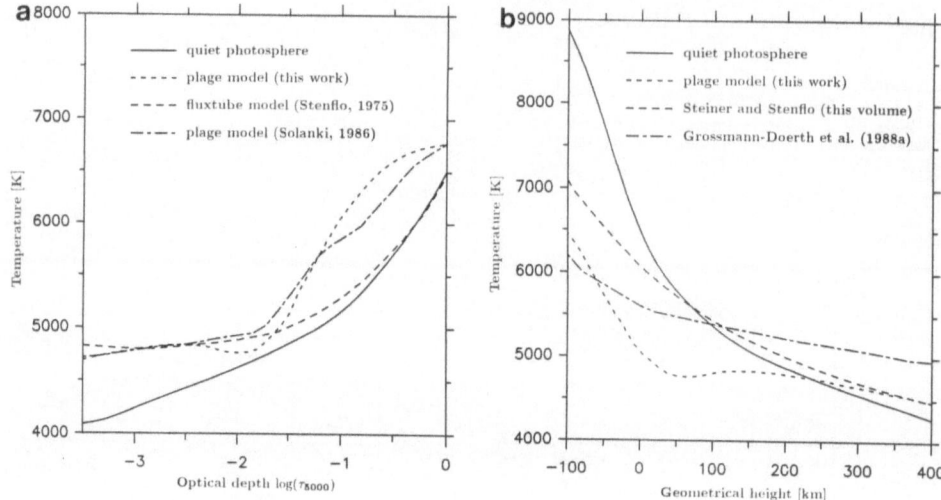

Figure 2. The plage model compared with empirical (a) and theoretical models (b) from the literature.

ACKNOWLEDGEMENTS. It is a great pleasure for me to thank S.K. Solanki, O. Steiner, and J.O. Stenflo for many fruitful discussions. The Swiss Society for Astrophysics and Astronomy provided a travel grant which is thankfully acknowledged. Part of this work was supported by the Swiss National Science Foundation under grant No. 2000-5.229.

References

Deinzer, W., Hensler, G., Schüssler, M., Weisshaar, E.: 1984, *Astron. Astrophys.* **139**, 435
Grossmann-Doerth, U., Knölker, M., Schüssler, M., Weisshaar, E.: 1988a, in *Solar and Stellar Granulation*, NATO Advanced Research Workshop, R.J.Rutten, G.Severino (Eds.), in press
Grossmann-Doerth, U., Schüssler, M., Solanki, S.K.: 1988b, *Astron. Astrophys.* **206**, L37
Kalkofen, W., Bodo, G., Massaglia, S., Rossi, P.: 1988, in *Solar and Stellar Granulation*, NATO Advanced Research Workshop, R.J.Rutten, G.Severino (Eds.), in press
Keller, C.U., Solanki, S.K., Steiner, O., Stenflo, J.O.: 1989, in preparation
Koutchmy, S.: 1977, *Astron. Astrophys.* **61**, 397
Marquardt, D.W.: 1963, *J. Soc. Ind. Appl. Math.* **11**, 431
Solanki, S.K.: 1986, *Astron. Astrophys.* **168**, 311
Solanki, S.K., Keller, C., Stenflo, J.O.: 1987, *Astron. Astrophys.* **188**, 183
Steiner, O., Pneumann, G.W., Stenflo, J.O.: 1986, *Astron. Astrophys.* **170**, 126
Stenflo, J.O.: 1973, *Solar Phys.* **32**, 41
Stenflo, J.O.: 1975, *Solar Phys.* **42**, 79
Stenflo, J.O., Harvey, J.W., Brault, J.W., Solanki, S.K.: 1984, *Astron. Astrophys.* **131**, 33
Stenflo, J.O., Solanki, S.K., Harvey, J.W.: 1987, *Astron. Astrophys.* **171**, 305

PROPERTIES OF PHOTOSPHERIC FLUXTUBES DERIVED FROM MAGNETOGRAPH OBSERVATIONS

V.G. LOZITSKIJ[1] and T.T. TSAP[2]
[1] *Kiev Shevchenko University Astronomical Observatory*
  252053, Kiev, USSR
[2] *Crimean Astrophysical Observatory*
  334413, p/o Nauchny, Crimea, USSR

ABSTRACT. Data from magnetograph observations, obtained in six Fe I and Mg I lines with the Crimean magnetograph with a spatial resolution of 1×1 and 1×2 arcsec have been interpreted using the line-ratio method and a two-component model with spatially unresolved fluxtubes and a background field. Magnetograph data obtained by other authors have also been taken into account. The magnetic field strength, its variation with distance from the fluxtube axis, the characteristics of the spectral line profiles, as well as other properties of the fluxtubes have been determined.

1. Introduction

The first important results in the study of the fine structure of solar magnetic fields were obtained by Severny (1957). Further investigations led to the picture of 'subtelescopic fluxtubes with kilogauss field strength', which are present over practically the whole surface of the Sun (e.g. Stenflo, 1973; Solanki and Stenflo, 1984; Rachkovskij and Tsap, 1985; Lozitskij and Tsap, 1989). Though these structures have been studied by many authors, the obtained results are still tentative. The main reason for this is the complexity of the indirect methods that have to be used because of the subtelescopic dimensions of the fluxtubes.

In the present paper we are making an analysis of our own magnetograph data, including the most important data of other authors, which are informative from the point of view of predicting the properties of small-scale fields.

2. Observations

The observations were made with the double-channel Crimean magnetograph at the center of the solar disk. The longitudinal component $H_\parallel$ of the magnetic field was recorded simultaneously in two of the following lines: Fe I 4808, 5233, 5250, 5253, 6302 Å, and Mg I 5184 Å. As a rule we used the Fe I 5253 line in one of the channels, and one of the other above-mentioned lines in the other. The exit slits were approximately the same, sampling a window of about 40–100 mÅ from the line centre (Lozitskij and Tsap, 1989). Besides,

J. O. Stenflo (ed.), Solar Photosphere: Structure, Convection, and Magnetic Fields, 125–128.

simultaneous observations of $H_\parallel$ were made in the Fe I 5250 line with different exit slits, at 10–34 and 68–94 mÅ from the line centre.

For the photospheric lines the measured strength H appeared to be increasing with equivalent width D (see Fig.1), which is in good agreement with the measurements of Gopasyuk (1985). With the Mg I line we have measured two ratios in two different regions near the central portion of the disk: $H_\parallel(5184)/H_\parallel(5253) = 1.50$ and 1.15. In the 5250 line we have found $H_\parallel(10-37)/H_\parallel(68-94) = 0.86$. We have analysed these data together with the results of Stenflo (1973) and Rachkovskij and Tsap (1985), obtained in the Fe I 5247 and 5250 lines with three different exit slits.

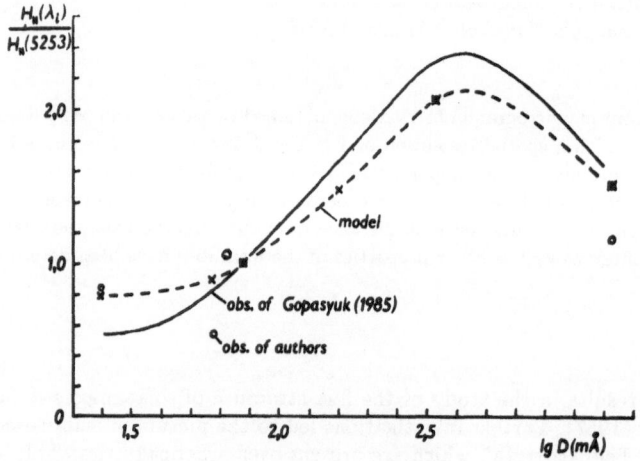

Figure 1. The field strength ratio $H_\parallel(\lambda_i)/H_\parallel(5253)$ vs. equivalent width D according to observations and model calculations.

3. Model

We have assumed that the magnetic field has two components: the fluxtubes, each of which having the same characteristics, and the background field. We also assume that the spectral line profiles of the background field are equal to those of the undisturbed atmosphere, while the line profiles in the fluxtubes might differ, both by width and shape. The Doppler shifts inside and outside the fluxtubes have been assumed to be the same. This last assumption has been well confirmed for the spatially resolved concentrations of magnetic flux (Fig.2), as well as for the subtelescopic fluxtubes (Stenflo et al., 1987).

4. Results of the computations

Good agreement between theory and observations was obtained when the magnetic field strength H_0 at the axis of the fluxtubes is assumed to be 2.2 kG at the level of formation of the Fe I 5250 line. The line profiles in the fluxtubes need to be 30–40 % narrower

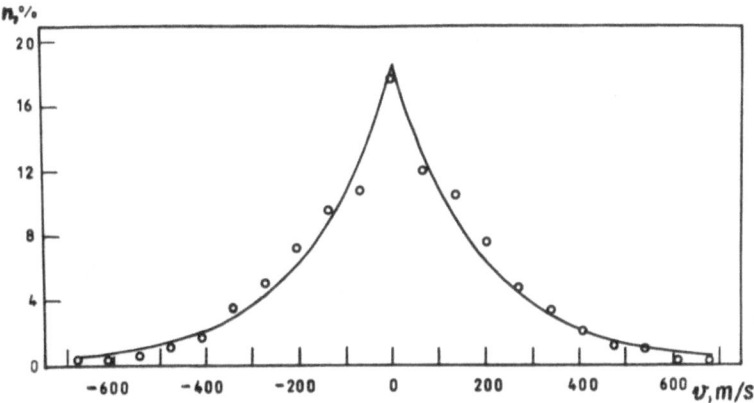

Figure 2. The line-of-sight velocity distribution in magnetic field concentrations, providing evidence that the convective motions in the magnetic elements are very small ($\lesssim 20$ m s^{-1}). The open circles represent the observations with a resolution of 1 × 1 arcsec, the solid line is an exponential curve fitting the observations (Tsap, 1989).

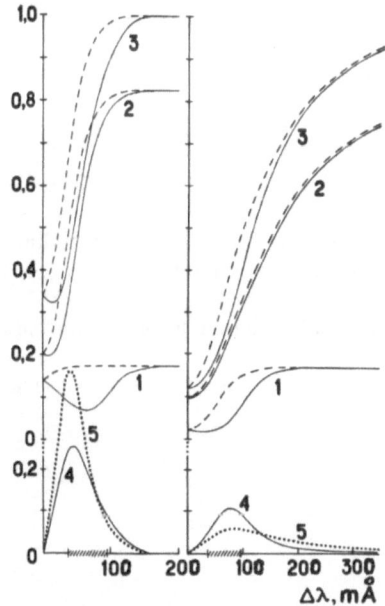

Figure 3. Variation with $\Delta\lambda$, the distance from line centre, of the model Stokes $I \pm V$ and V profiles for the line Fe I 5250 (left) and Fe I 5233 (right), assuming a filling factor of 17.5 % . 1: $I \pm V$ profiles from the fluxtubes; 2: analogous profiles from the background field; 3: total $I \pm V$ profiles (sum of components 1 and 2), which are directly accessible to observations; 4: Corresponding V profiles of the model; 5: V profiles in the case of a homogeneous field of 422 G, when the magnetic flux through the aperture is the same as in the two-component model. The shaded parts on the $\Delta\lambda$ axis show the positions of the exit slits of the magnetograph.

as compared with the undisturbed profiles and have weakened wings (Fig.3), which might indicate a decrease of the gas pressure. The magnetic profile of the fluxtubes, $H(x)$, has been represented by the expression $H(x) = H_0(1 - x^4)$, where x is the distance from the axis, in units of the fluxtube radius. This profile has practically the same shape as that found in pores (Steshenko, 1967). The total magnetic flux of the fluxtubes is 1.6 times larger than the flux of the background field.

The two empirical values $H_\|(5184)/H_\|(5253) = 1.50$ and 1.15 refer to two different situations at the level of the temperature minimum. The first value is expected for fluxtubes with small height divergence, whereas the second value (1.15) is very close to what would be expected for a homogeneous magnetic field (1.17). This may mean that in some cases the height divergence of the field lines is so rapid that the field becomes nearly homogeneous already at the level of the temperature minimum.

References

Gopasyk, S.I. (1985) 'Measurements of solar magnetic fields outside the spots using different strength lines', *Izv. Krymsk. Astrofiz. Obs.* **72**, 159-171.

Lozitskij, V.G and Tsap, T.T. (1989) 'An empirical model of the small-scale magnetic element in quiet region of the Sun', *Kinem. i fizika nebesnich tel* **5**, 50-58.

Rachkovskij, D.N. and Tsap, T.T. (1985) 'The magnetic field investigation by the line-ratio method', *Izv. Krymsk. Astrofiz. Obs.* **71**, 79-87.

Severny, A.B. (1957) 'Some results of investigation of nonstationary processes on the Sun', *Astron. Zh.* **34**, 684-693.

Solanki, S.K. and Stenflo, J.O. (1984) 'Properties of solar magnetic fluxtubes as revealed by Fe I lines', *Astron. Astrophys.* **140**, 185-198.

Stenflo, J.O. (1973) 'Magnetic field structure of the photospheric network', *Solar Phys.* **32**, 41-63.

Stenflo, J.O., Solanki, S.K., and Harvey, J.W. (1987) 'Center-to-limb variation of Stokes profiles and the diagnostics of solar magnetic fluxtubes', *Astron. Astrophys.* **171**, 305-316.

Steshenko, N.V. (1967) 'The magnetic field of the small solar spots and pores', *Izv. Krymsk. Astrofiz. Obs.* **37**, 21-28.

Tsap, T.T. (1989) 'The convective motions in the magnetic field elements', *Izv. Krymsk. Astrofiz. Obs.*, in press.

SMALL-SCALE MAGNETIC FEATURES OBSERVED IN THE PHOTOSPHERE

Sara F. Martin
Big Bear Solar Observatory
Solar Astronomy 264-33
California Institute of Technology
Pasadena, CA 91214, U.S.A.

ABSTRACT. Small-scale solar features identifiable on the quiet sun in magnetograms of the line-of-sight component consist of network, intranetwork, ephemeral region magnetic fields, and the elementary bipoles of ephemeral active regions. Network fields are frequently observed to split into smaller fragments and equally often, small fragments are observed to merge or coalesce into larger clumps; this splitting and merging is generally confined to the borders and vertices of the convection cells known as supergranules. Intranetwork magnetic fields originate near the centers of the supergranule convection cells and appear to increase in magnetic flux as they flow in approximate radial patterns towards the boundaries of the cells.

Large ephemeral active regions which develop magnetic flux in excess of 5×10^{19} Mx often exhibit a secondary substructure of 'elementary bipoles' identical with the substructure that larger active regions exhibit during their first hours or day of development; the elementary bipoles often appear to berandomly oriented with respect to the axis of the initial bipole and these elementary poles often cancel with the initial poles or other elementary poles of opposite polarity.

Network, intranetwork and ephemeral region magnetic fields all encounter and interact with one another. Encounters of the same polarity result in the merger and adding of the magnetic flux from different features. Encounters of opposite polarity usually result in cancellation - the mutual disappearance of magnetic flux of opposite polarity at their common boundary. It is deduced that the mixed-polarity network originates primarily from the separated poles of ephemeral regions and secondarily from merged clusters of intranetwork fields.

1. Introduction

The small-scale magnetic features discussed in this paper are small relative to supergranules but are much larger than the smallest known magnetic elements in the the solar atmosphere. They consist of intranetwork magnetic fields, fragments of network, ephemeral active regions and the elementary bipoles of ephemeral active regions and active regions. Moving magnetic features around sunspots also are an important subject in the domain of small-scale magnetic features but their inclusion is beyond the scope of this paper. This paper will emphasize all other directly observable magnetic field structures in the range of 2-20 arc seconds in diameter. Structures of this scale are observable on an every-day basis. An advantage of studying structures in this range of scales is our current ability to observe their entire apparent lifetimes. A disadvantage is that the clarity of the observations is often affected by atmospheric turbulence.

Recent excellent reviews have thoroughly discussed the existence of much smaller magnetic

129

J. O. Stenflo (ed.), Solar Photosphere: Structure, Convection, and Magnetic Fields, 129–146.
© 1990 by the IAU.

structures than presented here. Both theory and techniques for determining the true dimensions of such small-scale magnetic elements have already been presented (Stenflo, 1989 and reviews cited therein). Thus, the features discussed here consist of clusters of smaller magnetic field elements and this knowledge is useful in the interpreting the observable evolution of features seen on time-series of magnetograms taken over intervals of several hours.

A full disk magnetogram from the USA National Solar Observtory site at Kitt Peak is shown in Figure 1. The date is 9 October 1988. On this magnetogram is a rectangle that corresponds to the field of view observed at the Big Bear Solar Observatory on the same day. The first objective is to examine the small-scale magnetic features within this limited field near the center of the solar disk and to describe how they change in a typical observing day in videomagnetograms from the Big Bear Solar Observatory.

2. Videomagnetogram Images and Data

A typical set of images taken at Big Bear Solar Observatory on 9 Oct 1988 in the 6103 A line of CaII is shown in Figure 2. Each set consists of a magnetogram at 512, 1024, 2048, and 4096 integrations. An integration is the smallest complete spatial and temporal unit of a videomagnetogram; it is pair of video images, one in each polarization, that have been electronically added to display the positive polarity as white and the negative polarity as black. This minimum magnetogram unit containing magnetic fields of both polarities can be displayed in 1/15 second, twice the standard video frame rate for a single frame. However, in order to see the magnetic fields, many such successive videomagnetogram units must be integated (added). Our local convention is to stop the integration of successive images in integral powers of 2. With the present system, it is necessary to take at least 2^{10} (1024) integrations to observe the weak magnetic fields on the quiet sun. However, the weak fields are more completely detected at 2048 and 4096 integrations.

On 9 Oct. 1988, a set of 4 such integrated images, as shown in Figure 2, was taken at intervals of 5 minutes throughout the 7.5 hour observing day. A five minute interval was chosen to minimize the effects of the solar 5 minute oscillations when one views the images of a selected number of integrations as a time-lapse movie. In the remaining illustrations in this paper, we use only the magnetograms with the highest number of integrations recorded on this day (4096).

The polarity in these videomagnetograms is determined by the color immediately outside of the contours. Positive polarity is white and negative polarity is black. Weaker fields are seen as shades of light gray for positive polarity and shades of dark gray for negative polarity. The areas within the contours represent magnetic fields that are stronger than can be displayed by the gray-scale range of our current image display system. For these images on 9 Oct. 1988, the lowest contour at 2048 integrations corresponds to fields of approximately 50 Gauss. The contours are introduced as a technique to show a more extended range of magnetic field strengths than can be readily distinguished in the gray-scale range alone. However, there also is a practical limit to the number of contours that can be displayed in small magnetic features on the sun at a given spatial resolution. If too many contours are introduced, the contours cannot be decoded for the purpose of calculating magnetic flux. For this reason, the magnetograms are taken in sets. The magnetograms with the higher numbers of integrations allow us to see and calculate the weak magnetic fields but the information in the strongest fields is lost. The magnetograms with lower integrations and

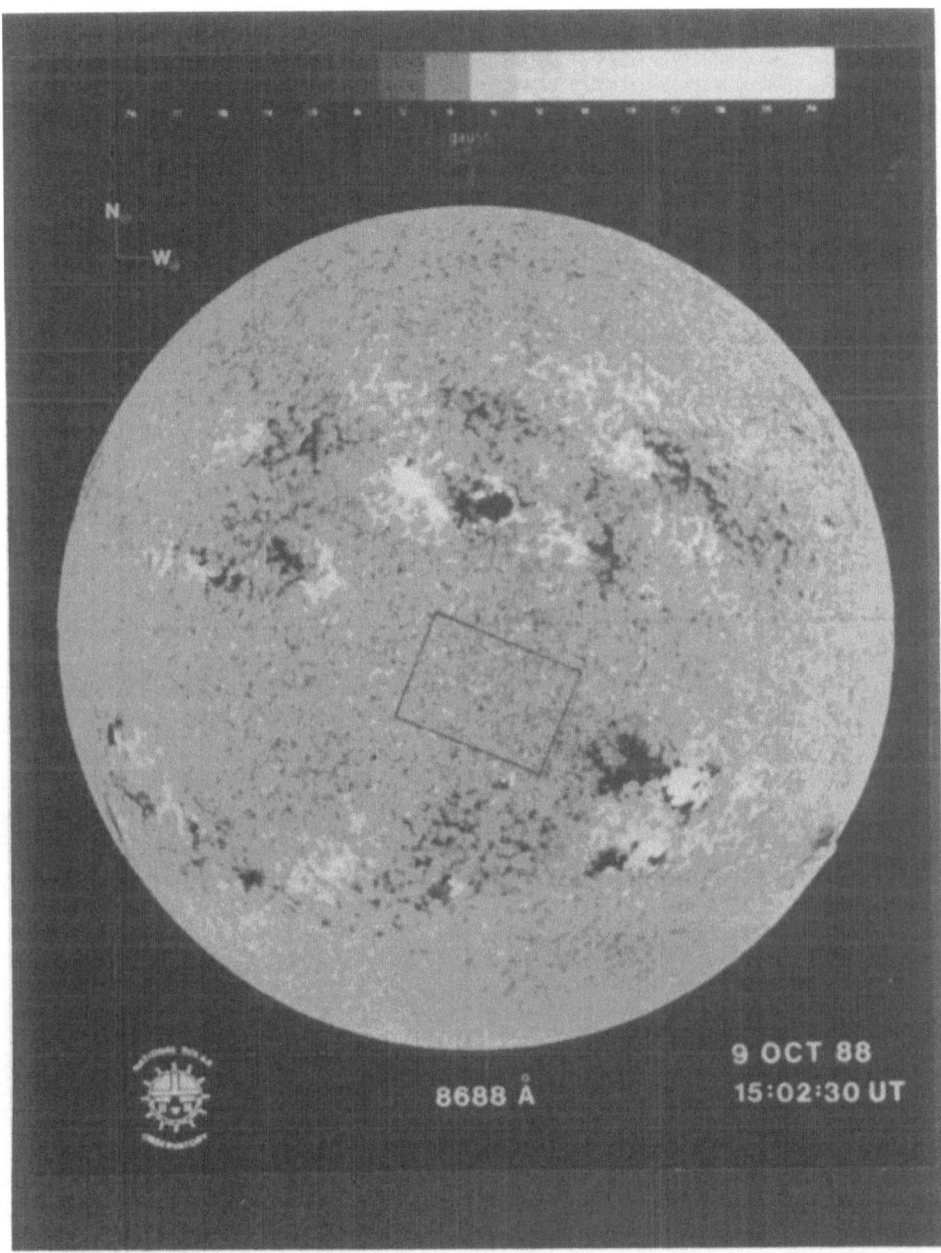

Figure 1. Magnetogram from USA National Solar Observatory taken on 9 Oct. 1988. The rectangle outlines the field of view containing weak-field, mixed-polarity network observed at Big Bear Solar Observatory on the same day.

Figure 2. A set of consecutive magnetograms taken at the Big Bear Solar Observatory on 9 Oct. 1988 in order of increasing numbers of integrated video frame pairs: upperleft, 512; lower left, 1024; upper right, 2048; lower right, 4096. North is at the bottom and west at the left. The images must be rotated 180 degrees plus the position angle to be in the same orientation

only a few contours are retained for combining images and for calculating the magnetic flux of the strong fields.

The magnetograms are recorded both in digital format on magnetic tape and also are displayed and photographed on a television monitor. Because the data is on magnetic tape, images can be selected and redisplayed in a variety of ways and at different stages of data processing. In our analyses we use both original and processed images. Decoding the contours and redisplaying the images in shades of gray makes the images easier to look at. Casual observers often would rather view the decoded images. However, during our analyses, we often prefer to use the original images with the contours because they better show a wider range of magnetic flux. For live presentations, different colors can be introduced in the contours of each polarity to make the visual interpretation easier.

3. The Identification of Magnetic Features

Reliable identifications of intranetwork, network and ephemeral region magnetic fields only can be done by studying time series of magnetograms. Even when one is very familiar with this type of data, it is not possible to correctly identify every feature on a single magnetogram. That is because the features evolve so rapidly that one often cannot distinguish between single poles of ephemeral regions and network fragments or between intranetwork patches and very small network fragments. These ambiguities are usually solved by taking time series of magnetograms. Then one can easily distinguish between ephemeral regions, network and intranetwork fields because each evolves and changes in characteristic ways.

If all features are studied together as a statistical sample, as done by Wang et al. (1989), no characteristic flux is identifiable. They found that the number of features always increases with decreasing feature flux above the threshhold of sensitivity. This means that network, ephemeral regions and intranetwork fields all have overlapping distributions of total magnetic flux per feature.

Figure 3 shows a magnetogram early in the observing day and another magnetogram late in the observing day. On the second magnetogram the network magnetic fields are enclosed in white polygons; new ephemeral regions are enclosed in solid-line black ovals; old ephemeral regions are enclosd in dashed black ovals. The remaining features, all notably small features, are intranetwork magnetic fields. The total flux in this magnetogram is 3.3×10^{21} Mx. Of this flux, 42% is in network magnetic fields (including the old ephemeral regions), 12% in new ephemeral regions and 46% in the intranetwork magnetic fields.

Because of the good image quality in these sets of observations, there is the opportunity to study the following:

1. the scale and dynamics of the intranetwork magnetic fields
2. characteristics of the mixed polarity network and its origin
3. the evolution and size distribution of ephemeral regions
4. the relationship of all of the above with respect to the evolution of supergranules

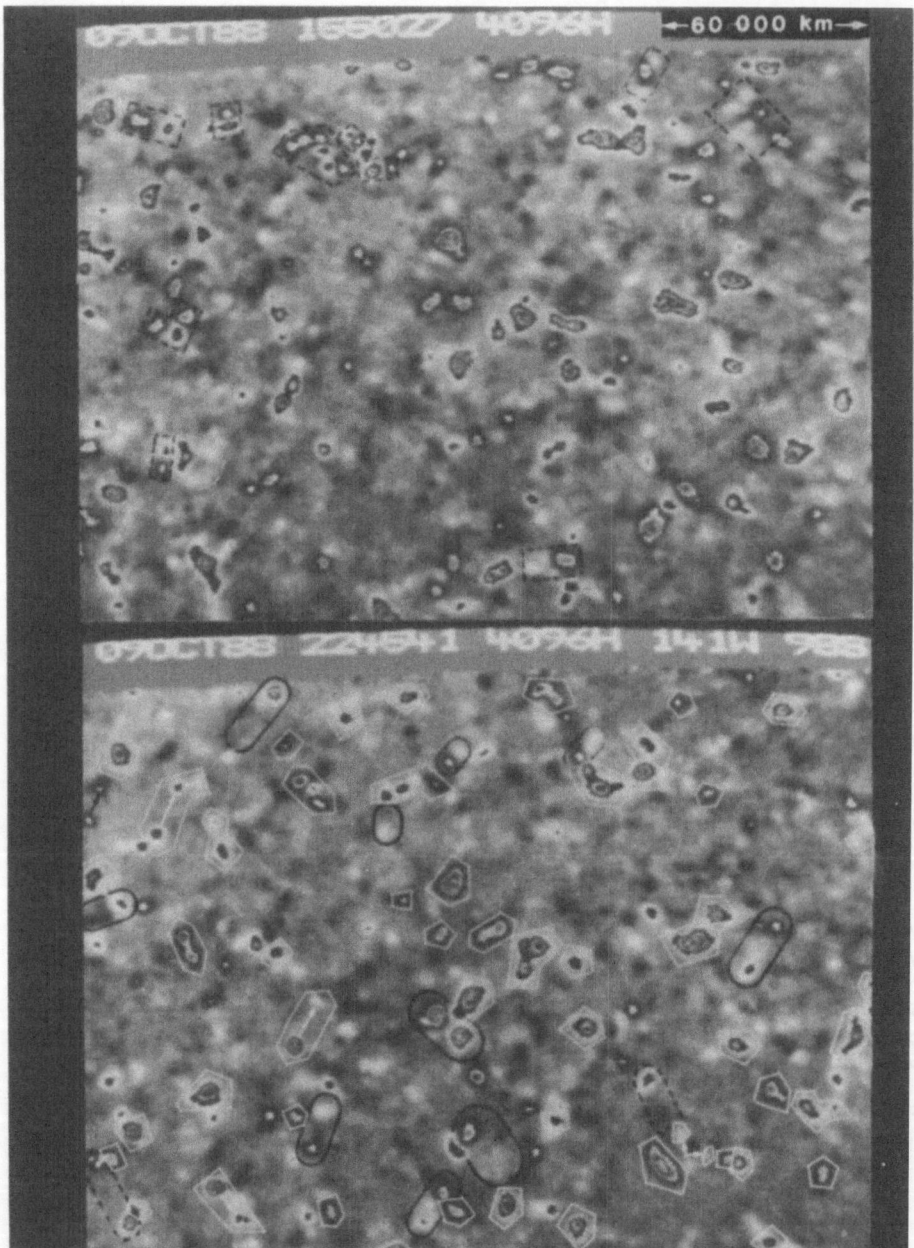

Figure 3. Magnetograms taken early and late on 9 Oct 1988.   In the early magnetogram,
prominent cancelling magnetic fields are enclosed in black, dashed rectangles or polygons.
The dashed white polygon encloses intranetwork patches which coaslesce around the larger area
of positive polarity network. In the later magnetogram new ephemeral regions observed from
birth are enclosed in solid-line, black ovals and older ephemeral regions in dashed ovals.
Network fragments that were observed throughout the day are enclosed in solid-line white
polygons. The remaining small patches of field in between the ephemeral regions and network
are the intranetwork magnetic fields.

4. Network Magnetic Fields

Network magnetic fields are identified within the polygons in the lower half of Figure 3. These network features are all of the ones that could be traced as a persistent features since the beginning of the observing day. Persistent does not imply a lack of change - only that part of a network cluster is traceable for many hours. As seen in Figure 3, the majority of the network features have changed substantially from the beginning until the end of the day. The types of changes that can be seen in the network in the time-series of magnetograms are:

(1) splitting into smaller fragments,
(2) gain in flux due to merging with neighboring network, intranetwork, or ephemeral region fields of the same polarity,
(3) decrease in flux from encountering and polarity, or
(4) combinations of (1)-(3).

Examples of the merging and splitting of network fragments are illustrated by Martin (1988). The network includes most of the largest areas of magnetic flux in the field-of-view but also includes many small fragments which are comparable to the scale of the small intranetwork fields. The smallest areas of network magnetic flux shown here are ones that originated from the fragmentation (splitting) of larger network elements. Such small network elements do not usually survive as independent entities for more than a few hours without merging with other network or intranetwork patches or cancelling with any opposite polarity fields that are encountered. The arrows show the direction and distance of travel for some of the fragmented areas of network.

5. Ephemeral Active Regions

An ephemeral active region is a small bipole in which its opposite polarities:
(1) appear adjacent to each other at nearly the same time
(2) increase in magnetic flux and
(3) move in opposite directions as the region increases in magnetic flux, at least until one of the poles encounters external opposite-polarity field comparable to its magnitude at the time of encounter.

Every ephemeral regions in Figure 3 developed at least one saturation contour at sometime during the observing day. Our threshhold for identification is therefore a peak flux of at least 50 Gauss in at least one pole. However, no ephemeral regions were observed on this day that did not fit this criteria. The solid ovals include the 10 of the 11 ephemeral regions whose birth was observed during the 7.5 hour observing day. The one not shown was on the extreme right edge of the field of view and observations of it were incomplete. The two dashed ovals are ephemeral regions not observed from birth but are identifiable from their characteristic pattern of growth and motion. Old ephemeral regions (ones that have reached or passed their peak magnetic flux) are usually indistinguishable from the mixed polarity network in single images. The two old ephemeral regions in Figure 3 were identified while viewing the data as a time-lapse movie; the one in the lower right quadrant of Figure 3, at the time shown, lost its negative pole from cancelling with opposite polarity network. Both poles of the one in the lower left corner have merged with network of the same polarity and similar magnitude. Both of the old ephemeral regions were included in the 42% of the fields designated as network magnetic fields.

It is seen also in Figure 3 that many (at least 6) of the 11 new ephemeral regions have also encountered and either merged or cancelled with network fields. Thus it is apparent that ephemeral regions are one of the possible sources of the mixed polarity network. The ephemeral regions contribute to both cancellation and resupplying of the mixed polarity network. This topic is further discussed in Section 7.

In addition to interacting with the network magnetic fields, ephemeral regions also interact with intranetwork magnetic fields. The fine structure of ephemeral regions is also at least as small as the observable intranetwork fields. Figure 4 illustrates the development of an ephemeral region and typical interactions with intranetwork fields. It is the same ephemeral region with several contours seen in the middle of the second image in Figure 3. This ephemeral region, like many this size and larger, is first seen as more than just a single pole of each polarity. At 1714 UT, the positive pole is initially seen as three small, separate dots. These merge by 1748 to form the primary positive pole. The negative pole of the ephemeral region grows at a site where a small, negative polarity (black) intranetwork patch already existed at 1628. The intranetwork patch and the negative ephemeral region pole are indistinguishable from each other at 1702. An adjacent positive-polarity (white) intranetwork patch also partially cancels with the negative ephemeral region pole but then migrates toward the positive pole and merges with it. An new elementary bipole is seen at 1834 within the oval between the original poles of the ephemeral region. The negative (black) pole is visible as early as 1748, also between the primary poles of the ephemeral region.

To the right of the positive pole, another positive (white) intranetwork magnetic patch concentrates and develops a single contour. By 1903 it merges with the positive pole of the ephemeral region to form a single magnetic feature.

We tend to think of ephemeral regions just as growing bipoles because that is an essential part of our definition of an ephemeral region. However, this example illustrates how parts of an ephemeral region can merge with adjacent features of the same polarity and cancel with adjacent magnetic features of opposite polarity. Cancellation is defined as the disappearance of both polarities at a common boundary in line-of-sight magnetograms (Martin, Livi and Wang, 1985; Livi, Wang and Martin, 1985). Cancellation is not a rare, isolated phenomenon but rather a process that is typical when opposite polarity features encounter one another. The interpretation of cancellation is briefly discussed in Section 9.

The cancellation and merging of ephemeral regions with adjacent network and intranetwork fields is the primary reason that the magnetic flux of ephemeral regions is rarely balanced between the two poles (Livi, Wang and Martin, 1985). The distribution of flux in the poles of ephemeral regions and between their individual poles is shown in Figure 5 at the time of the second image in Figure 3. The magnetic flux in the ephemeral regions at 2257 UT (excluding one which is partly out of the field of view) is 2.0 X 10 Mx. The individual ephemeral regions range in total flux from 4.8 to 47.7 X 10 Mx. There is no known lower limit to the size of an ephemeral region. The upper limit is arbitrary because there are no distinctive physical properties of individual ephemeral active regions that separate them from small active regions (Harvey, Harvey and Martin 1975). The statistical distributions of ephemeral regions by latitude, magnitude, and inclination of their magnetic axes have all been shown to blend smoothly into the statistical distributions of active region (Harvey, Harvey and Martin 1975; Martin and Harvey 1979, Harvey 1988) with the exception of one study by Tang, Howard and Adkins (1984). This single discrepancy is resolved by the study of Harvey (1989) which shows the spectrum of active regions, small active regions and ephemeral

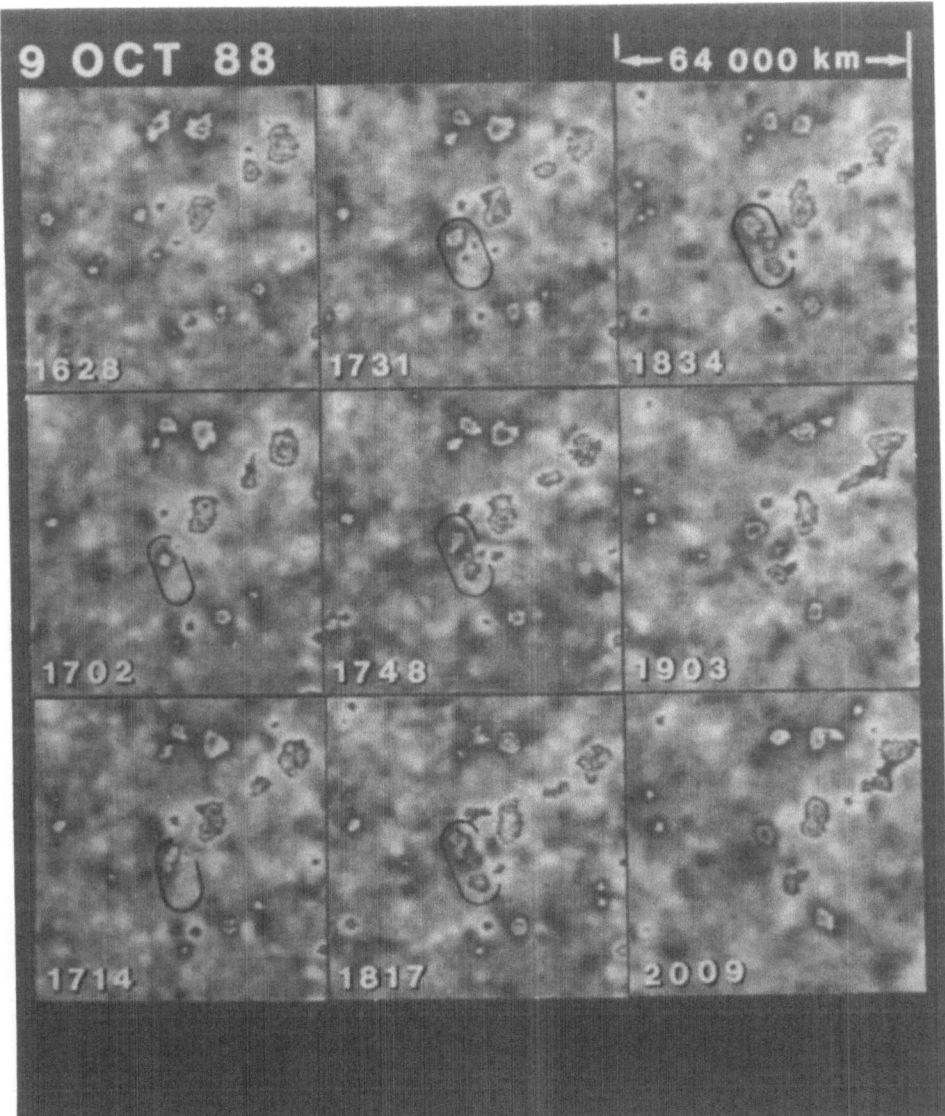

Figure 4. The new magnetic flux of an ephemeral region is first recognizable at 1702. The new negative pole coincides with a pre-existing negative polarity intranetwork patch. At 1714 the new positive pole is initially seen as 3 separate positive patches which then merge together by 1748. A secondary elementary bipole is visible between the initial poles at 1834. The negative elementary pole became visible at 1748 before the appearance of the positive pole.

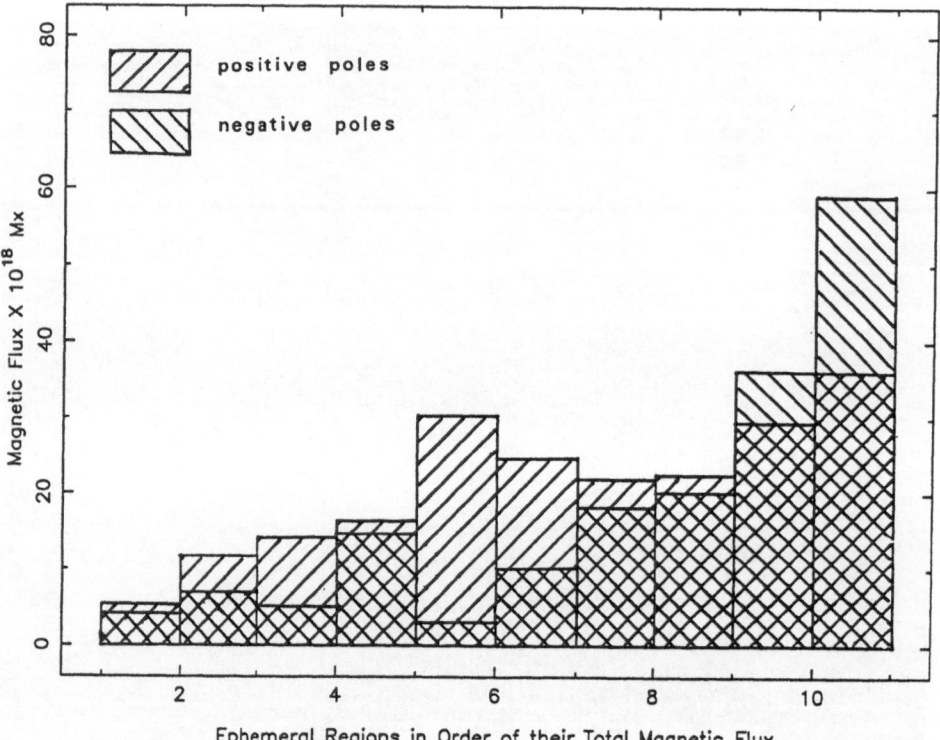

MAGNETIC FLUX OF EPHEMERAL REGIONS OBSERVED ON 9 OCT 88, 2257 UT

Figure 5. The magnetic flux in the poles of the ephemeral regions ranged from 3 to 36 X 10^{18} Mx at 2257 UT on 9 Oct. 1988. The imbalance between the poles of the ephemeral regions was found to be due to the coalescence and cancellation of the ephemeral region magnetic flux with adjacent network and intranetwork magnetic features.

active regions to be continuous rather than discontinuous. Our arbitrary convention is to call new bipolar regions ephemeral regions if they develop insufficient concentrated magnetic flux to form sunspots. Sunspot pores form whenever concentrations of flux exceed approximately 10^{20} Mx (Harvey and Martin, 1973; Zwaan 1978, 1985; Chou and Wang, 1987).

6. Intranetwork Magnetic Fields

Intranetwork patches rarely survive more than a few hours. They originate in the centers of supergranules and typically flow at a rate of approximately 0.35 km s^{-1} (Zirin, 1987) until they encounter either network, ephemeral regions or other intranetwork fields from adjacent supergranule cells. If the encountered fields are the same polarity, the intranetwork fields merge with them. If the encountered fields are of opposite polarity, the intranetwork fields cancel with the encountered fields.

As the intranetwork fields flow from the centers to the boundaries of the supergranule cells, they increase in magnetic flux and area. The increase in flux might be due to the merging of adjacent fields or to a change in the direction of the magnetic field such that a larger fraction of the line-of-sight component is detected. The merging of adjacent intranetwork fields is observable both before the intranetwork patches reach the network boundaries as well as at the boundaries. A cluster of merging intranetwork patches are enclosed in the dashed white polygon in the lower left quadrant of the upper frame in Figure 3. In the lower frame in Figure 3, it is seen that the merger of these intranetwork patches has resulted in about a 50% increase in the flux of the network fragment around which the intranetwork fields coalesce.

In magnetograms composed of successively lower integrations (2048, 1024 and 512), the intranetwork field patches appear to originate within increasingly larger annuli with respect to the centers of the cells. This effect is due to both the lower magnetic sensitivity in magnetograms with lesser numbers of integrations and the apparent increase in flux as the intranetwork fields flow approximately radially toward the cell boundaries.

Figure 6 shows the migration of intranetwork patches toward a isolated concentration of negative polarity network magnetic field. The network fragment is isolated and relatively stationary with only small variations in flux throughout the day. It is located at the vertices of several supergranule cells. It is continuously interacting with approaching intranetwork fields which converge towards it from the adjacent supergranule cells. The positve intranetwork fields cancel with (diminish) the network fragment while a comparable number of negative polarity intranetwork fields merge with (add to) the network fragment. Since each cancellation and merger occurs slowly and involves small quantities of flux (on the order of 10^{18} Maxwells per hour), the total flux of the network fragment changes very little during the course of the day.

7. The Elementary Bipoles of Ephemeral Active Regions and Active Regions

The appearance of new elementary bipoles are characteristic small-scale features seen in ephemeral active regions whose total flux is of the order 10^{19} Mx and in larger, new active regions. Only the ephemeral region with the greatest magnetic flux, observed on 9 Oct. 1988 (Figure 4), developed a secondary elementary pair of poles after the development of the initial pair of poles. However, most similar high integration, high quality observations of

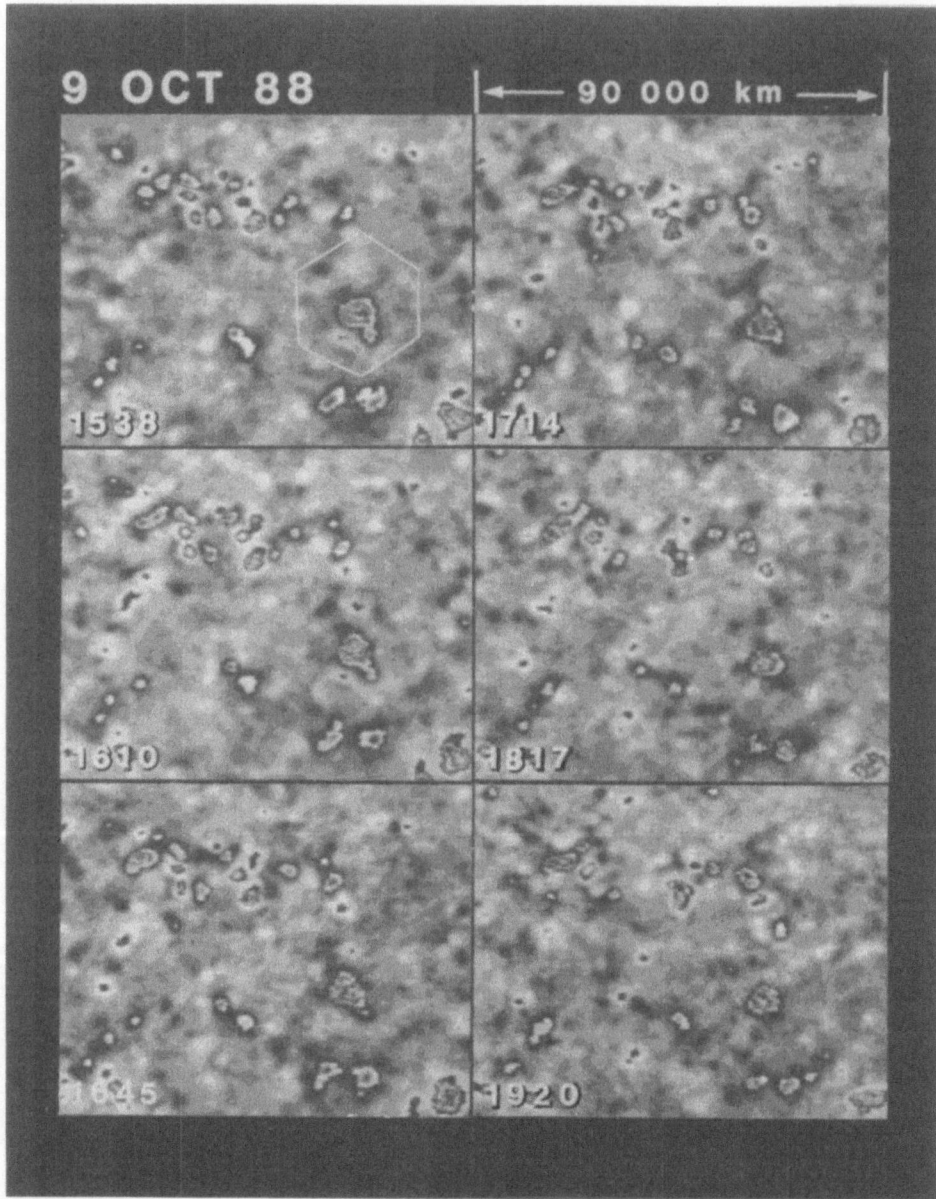

Figure 6. Within the white polygon, intranetwork magnetic field patches are seen to flow toward the area of network. The network fragment is at the vertices of several adjoining supergranule cells. In the upper left, a complex cluster of network magnetic fields of both polarities converge and cancel at another vertices between adjoining supergranule cells. The flux slowly reduces in this cluster throughout the observing day due to the cancellation of opposite polarity fields.

new ephemeral active regions and new active regions show the development of these secondary bipoles. Frazier (1972) has shown that the development and separation of these elementary bipoles, which he called 'magnetic knots', typically occurs under arch filament systems in Hα observations. Frazier (1972) and Vrabec (1974) found examples of the magnetic knots streaming into sunspots with speeds of approximately 0.3 km s when sufficient flux had accumulated for sunspots to form.

In the videomagnetogram observations collected at the Big Bear Solar Observatory, it can be seen that the elementary bipoles of both ephemeral regions and active regions often do not develop in the same orientation as the original pair or pairs of poles. If they develop with a reversed orientation from the original pair, they seem to be destined to cancel with previously emerged magnetic flux because the elementary poles separate from each other in the same way that first ephemeral region poles or active region poles move apart as their flux increases. Examples of secondary elementary bipoles are shown in the ephemeral active region and active regions illustrated respectively in Figures 7 and 8.

In Figure 7 the ephemeral region is first seen at 1732. The poles increase in magnetic flux and separate. Near the site of emergence of the original bipole, a secondary elementary bipole appears but its orientation is reversed from the original pair. The negative pole of the secondary bipole begins to cancel with the original positive pole and then becomes surrounded by the positive polarity fragments. After the complete cancellation of the negative elementary pole, the positive fragments coalesce into single unresolved pole. This positive pole then begins to cancel with the adjacent negative polarity network fragment. The predictable end of this sequence is that the negative pole of the ephemeral regions will remain and will become indistinguishable from any other network fragment.

The number of elementary bipoles increases for active regions of increasing ultimate area and magnetic flux. Examples of two small new active regions are shown in Figure 8. The region on the left has a whole cluster of elementary poles that are evolving and interacting. At the beginning of this sequence it is not clear which elementary poles originated at pairs. They seem to be randomly oriented. Some of the elementary poles of opposite polarity cancel and some of similar polarity merge. Overall the number of the mergers of similar polarity outnumber the elementary poles that cancel and the region grows.

The bipolar active region on the right has only one negative elementary pole at the beginning of this series; it is cancelling and disappears completely by 2003. Overall this bipolar active region has reached its maximum development. However that does not mean the flux is static. The negative pole is cancelling throughout this series with positive polarity magnetic flux of the other active region. At the same time, the positive polarity fragments are slowly increasing in magnetic flux at least until 2215. In addition, at the end of the observing day one more new elementary bipole appears above the primary negative pole in the last frame at 2314.

Small scale magnetic fragments are also seen during the decay of active regions. Numerous examples of small-scale cancelling fragments of magnetic flux are described and illustrated by Martin, Livi and Wang (1985) and Martin et al. (1989). They have shown that cancellation can account for the disappearance of either large quantities or all of the flux within some active regions.

142

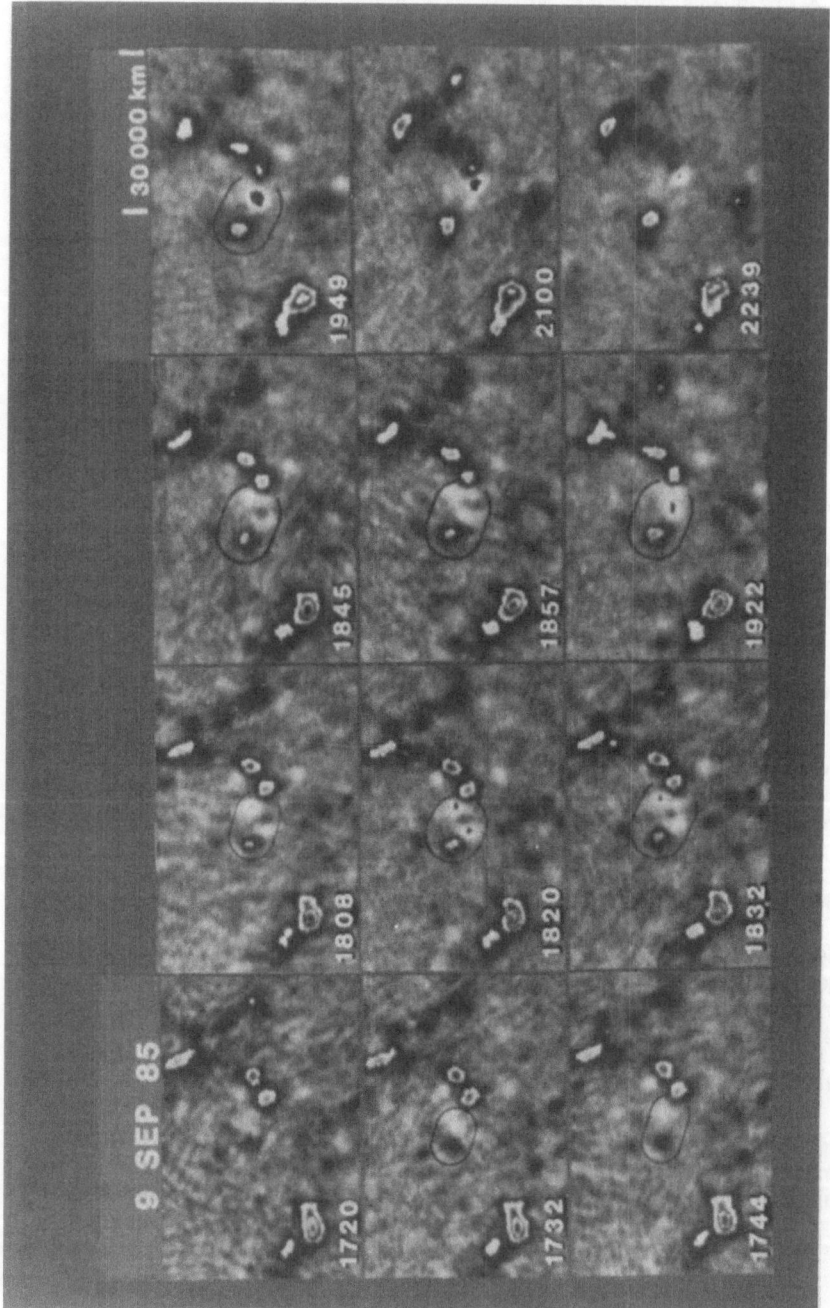

Figure 7. The initial poles of an ephemeral region are first seen at 1732. An elementary bipole appears between the initial poles at 1808. Note that the new elementary bipole is reversed in orientation from the poles of original bipole. The reversed negative pole cancels with the surrounding positive polarity patches until it completery disappears at 1922. Then the positive poles of the original and secondary bipoles coalesce into a single patch (1949) and thereafter cancel with the adjacent negative-polarity network. The surviving negative pole of the ephemeral region becomes part of the network.

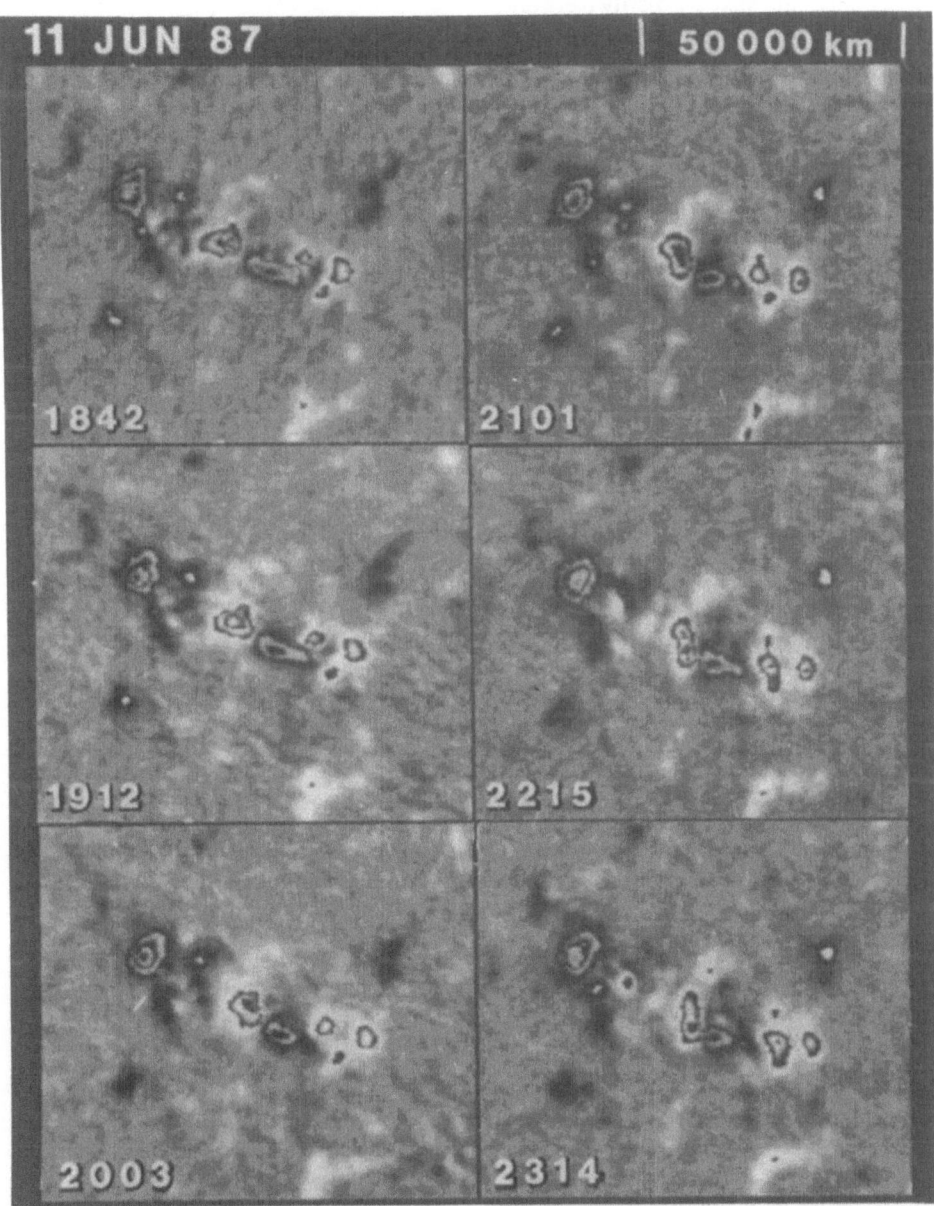

Figure 8. Two small active regions appear side-by-side. The one on the left is still growing as seen by the appearance of new elementary bipoles. The one on the right is cancelling at nearly the same rate that it is growing. Cancellation takes place where opposite polarity fields move together: withing the growing region, within the decaying region, and at the boundary between the two regions.

144

8. Origin of the Mixed-Polarity Network

Near the end of the day as shown in the lower half of Figure 3, 12% of the magnetic flux in the whole field of view was in new ephemeral regions, 42% in network fields and old active regions, and the balance of 46% of the flux was from the background intranetwork magnetic fields. The relative percentages of magnetic flux distributed among the different features depends on observational factors. In general the detectability of the intranetwork fields depends strongly on the number of integrations in the magnetograms. It also depends strongly on seeing and image stability which vary throughout the day and from day to day. Thus, the above percentages give us only a general idea of the proportion of magnetic flux among the differing classes of magnetic features. However, it is important to note that the intranetwork fields comprise about half of the magnetic flux in a quiet sun field of view where mostly mixed-polarity network is present. The amount of unipolar and mixed-polarity network varies greatly over the solar surface at any given time as well as over the solar cycle (Giovanelli, 1982). The number of ephemeral regions likewise varies with latitude and time during the solar cycle (Harvey, Harvey, and Martin, 1975; Martin and Harvey, 1979).

The mixed polarity network does not constitute new magnetic flux. It must therefore originate from either or both ephemeral regions or the intranetwork magnetic fields. The amount of ephemeral region flux and intranetwork flux that could contribute to the mixed polarity network depends not only on the ratio of the features in a given field-of-view but also on their mean survival time. Survival time in turn depends on the amount of time from initial appearance until encounter with adjacent features and on the average rate of cancellation with opposite polarity fields.

From this set of observations it can be seen that the intranetwork fields make a contribution to the mixed polarity network. The coalescence (and cancellation) of intranetwork fields which originate in adjacent cells are directly observable. Such coalesced intranetwork fields of the same polarity can become network elements by themselves or merge with other network fragments. An example is the cluster of positive polarity intranetwork fields at the extreme right side of the lower frame in Figure 3. Groups of intranetwork field of the same polarity are sometimes observed to add substantially to existing network elements. One of the network elements in the field was observed to double in size due only to coalescence with intranetwork fields. This example is the positive polarity intranetwork patches enclosed in a dashed polygon in the lower left corner of the upper frame in Figure 3.

The intranetwork clusters marked in Figure 3 are the largest such clusters that were observed to contribute to the network during this observing day. I estimate that the contribution to the mixed-polarity network from temporary clusterings of intranetwork patches is 10% or less. Since the ephemeral regions are the only other observed form of new flux in this field of view, I conclude that 90% or more of the mixed-polarity network must originate from the separated poles of ephemeral regions. This is consistent with the observed duration of the ephemeral regions. All 11 of the 11 ephemeral regions had traceable flux or were still growing at the end of the 7.5 hour observing interval. Hence, the survival time of the ephemeral region flux must be several times the observing interval. Since the observations cover slightly less than one-third of the observing day, about 3 times this amount of flux appears in the form of ephemeral regions each day in this field. If the ephemeral region flux has an average survival time of only 3.5 times the observing interval, then most or all of the mixed-polarity network could be accounted for by just the dispersed ephemeral region flux.

9. Interpreting the Observations of Cancelling Magnetic Flux

As seen in magnetograms of the line-of-sight component, the majority of magnetic fields on the both the quiet sun and in active regions disappear by cancellation of opposite polarity fields that encounter one another[1m. The characteristics of cancelling magnetic fields have been described in Martin (1984), Livi, Wang and Martin (1985), Martin, Livi, and Wang (1985), Wang et al. (1988) and Martin et al. (1989).

The disappearance of magnetic flux by cancellation can be interpreted as submergence or magnetic field reconnection (Zwaan 1985, 1987). Because ephemeral regions, as illustrated here, originate as bipoles but cancel with external features, I favor the interpretation as magnetic reconnection. Because the cancellations of ephemeral region poles with network and intranetwork fields have the same properties as cancellations between all other solar fields, (network fields, intranetwork fields and active region fields), one mechanism is sufficient to explain all cases of observed cancelling fields of opposite polarity.

A spectrum of regimes of magnetic reconnection have been discovered by Priest and Forbes (1986). I suggest that the type of reconnection associated with 'cancellation' is at the opposite end of the spectrum of reconnection regimes from the type of reconnection commonly associated with solar flares. Flare reconnection is rapid, leads abrupt energy release, and has no signature in present-day, photospheric magnetograms. In contrast, the type of reconnection resulting in 'cancellation' is slow, is associated with very weak, long-enduring enhancements in hydrogen-alpha filtergrams, and is a commonly observed phenomena in photopheric magnetograms of the line-of-sight component of magnetic fields. Because cancellation is associated with increasing magnetic field gradients (Wang et al., 1988), I suggest that cancellation is most easily associated with 'magnetic pile-up reconnection' among the magnetic reconnection regimes described by Priest and Forbes (1986). More work in observing or deducing the 3 dimensional magnetic structures at cancellation sites is needed to obtain a better understanding of this phenomenon.

10. Summary

The quiet sun is composed is composed of three classes of magnetic features: ephemeral active regions, network and intranetwork magnetic fields. It is deduced that the mixed-polarity network originates primarily from surviving separated poles of ephemeral active regions but some network elements are also observed to form from the coalescence of similar polarity patches of intranetwork magnetic flux that merge at the boundaries of supergranule cells.

During their growth phase, both ephemeral active regions and active regions are characterized by the appearance of secondary elementary poles which successively evolve between the initial poles of a new region. Although many elementary bipoles develop with approximately the same orientation as the original poles, some also develop with reversed or random orientations with respect to the original poles of a new region; elementary poles with reversed orientations from the original poles usually cancel with the original poles or other elementary bipoles.

Ephemeral active regions and active regions also cancel with adjacent intranetwork, network or active region magnetic fields. These external cancellations are the primary reason why the magnetic flux of the poles of ephemeral active regions are rarely balanced.

The observations of cancelling magnetic features are interpreted by the author as representing a type of magnetic reconnection described as 'magnetic pile-up reconnection' in the spectrum of reconnection regimes found and described by Priest and Forbes (1986).

Acknowledgements

This study was supported by the Air Force Office of Scientific Research under contract AFOSR-87-0023.

References

Chou, D. and Wang, H.: 1987, *Solar Phys.* **110**, 81.
Frazier, E.N.: 1972, *Solar Phys.* **26**, 130.
Giovanelli, R.G.: 1982, *Solar Phys.* **77**, 27.
Harvey, K.L.: 1989, in preparation.
Harvey, K.L. and Martin, S.F.: 1973, *Solar Phys.* **32**, 389.
Harvey, K.L., Harvey, J.W. and Martin, S.F.: 1975, *Solar Physics* **40**, 87.
Livi, S.H.B., Wang, J. and Martin, S.F.: 1985, *Australian J. Phys.* **38**, 855.
Martin, S.F. 1988, *Solar Phys.* **117**, 243.
Martin, S.F. and Harvey, K.L.: 1979, *Solar Phys.* **64**, 93.
Martin, S.F. Livi, S.H.B. and Wang, J.: 1985, *Australian J. Phys.* **38**, 929.
Priest, E.R. and Forbes, T.G.: 1986, *J. Geophys. Res.* **91**, 5579.
Stenflo, J.O.: 1989, Astron. Astrophys. Review 1, 3.
Tang, F., Howard, R.F. and Adkins, J. M.: 1984, *Solar Phys.* **91**, 76.
Vrabec, D.: 1974, in R.G. Athay (ed.), *'Chromospheric Fine Structure'*, IAU Symp. **56**, 201.
Wang, J., Shi, Z., Liu, J. Han, Feng and Liu, G.: 1989, presented at IAU Symp. 138, Kiev, 14-19 May 1989.
Wang, J., Shi, Z., Martin, S.F. and Livi, S.H.B.: 1988, *Vistas in Astronomy,* **31**, 79.
Zirin, H.: 1987, *Solar Phys.* **110**, 101.
Zwaan, C.: 1978, *Solar Phys.* **60**, 213.
Zwaan, C.: 1985, *Solar Phys.* **100**, 397.
Zwaan, C.: 1985, in R. Muller, *Lecture Notes in Physics,* Vol. **233**, Springer-Verlag, Berlin, p.263.
Zwaan, C.: 1987, *Ann. Rev. Astron. Astrophys.* **25**, 83.

High-Resolution Observations of Emerging Magnetic Fields and Flux Tubes in Active Region Photosphere

T. Tarbell, S. Ferguson, Z. Frank, R. Shine, A. Title, K. Topka
Lockheed Palo Alto Research Laboratory
3251 Hanover Street, Palo Alto, California 94304, USA

and G. Scharmer
Royal Swedish Academy of Sciences
Stockholm Observatory, S-133 00 Saltsjöbaden, Sweden

Introduction

On 29 September 1988, filtergrams of the solar photosphere with excellent resolution (0.3 to 0.5 arcsecond) were obtained at the Swedish Solar Observatory on La Palma, Canary Islands. An outstanding 2.5 hour run of digital filtergram observations was obtained, looking at a small area within an active region near disk center. On 6 August 1987, an 80 minute run of similar observations was obtained at the Vacuum Tower Telescope of the National Solar Observatory at Sacramento Peak. Digital and video movies have been made of Dopplergrams, magnetograms, line center, continuum, and white light images. Several examples of magnetic field emergence and formation of flux tubes can be studied in detail in the movies. The relationship between photospheric bright points, "filigree", the line center brightness, and the magnetic field has been established for individual images in analysis to date.

Observations

The data were collected using an evaluation model of the tunable filter systems developed for the NASA Solar Optical Universal Polarimeter (SOUP) and the Coordinated Instrument Package of the NASA Orbiting Solar Laboratory (OSL). The system (see Figure 1) consists of a high speed steering mirror for image stabilization, reimaging optics, a polarization analyser, a blocking filter wheel, a narrowband (70 mÅ) tunable filter, and a 1024 × 1024 CCD camera that uses a Texas Instruments uniphase detector. During periods of the best seeing, only data from a 512 × 512 subarea of the array (about 85 × 85 arcseconds) were recorded to increase the frame rate (4–5 seconds per frame). The target region on 29 September was dense plage adjacent to a large sunspot in NOAA active region 5168 at North 18, West 0. The

J. O. Stenflo (ed.), Solar Photosphere: Structure, Convection, and Magnetic Fields, 147–152.
© *1990 by the IAU.*

148

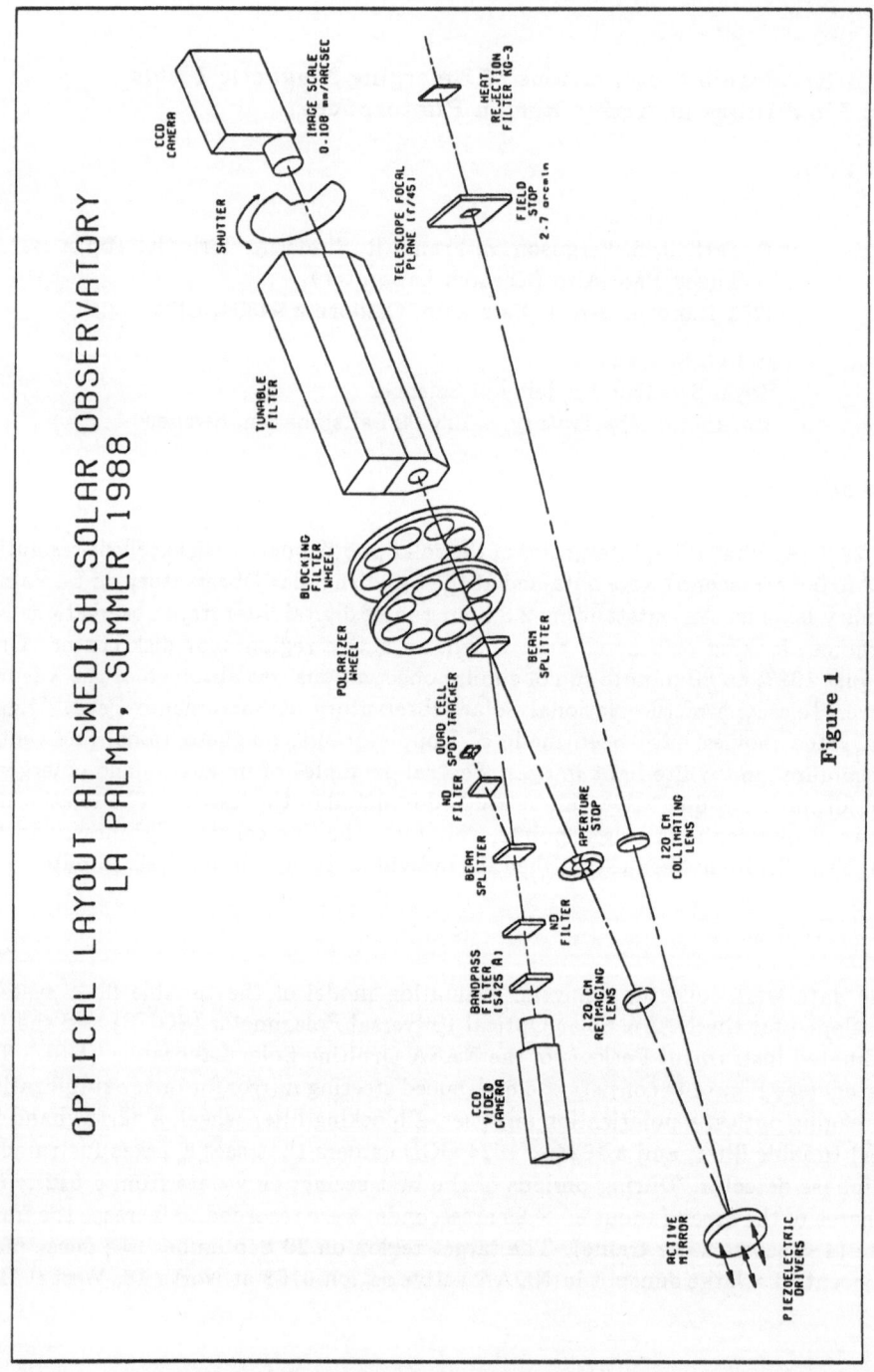

OPTICAL LAYOUT AT SWEDISH SOLAR OBSERVATORY
LA PALMA, SUMMER 1988

Figure 1

observing sequence had a 50 second cycle and consisted of images in RCP and LCP in the blue wing of Fe I 6302 for longitudinal magnetic measurements, four images spaced through Ni I 6768 for Doppler shift and line center intensity, and continuum images near one or both lines. At Sacramento Peak, we used the g=0 line Fe I 5576 instead of 6768 and included three Hα frames as well, with an overall cycle time of 80 seconds. The field-of-view included most of a small growing active region (NOAA 4835) at South 29, West 11.

Although the raw images make fascinating movies, intensive computer processing is needed to convert them into measurements of physical quantities with 0.3 – 0.5 arcsecond resolution (Title *et al.*, 1989a). Removal of spatial image distortions caused by atmospheric turbulence ("destretching") is the most difficult and time-consuming step (Topka *et al.*, 1986; November, 1986). With a suitable destretching procedure, it is possible to create time series of magnetograms, Doppler velocity, continuum, and line center intensity images for the entire 2.5 hour interval that are essentially simultaneous and aligned to a fraction of the spatial resolution.

Filigree and Magnetic Fields

Filigree are bright structures seen in the photosphere which are associated with fine scale magnetic fields (Dunn and Zirker, 1973). They range from points to linear structures, typically one or two arcseconds in length, one quarter arcsecond or less in width, and often "crinkled" on an arcsecond scale. Modern analysis of various spectroscopic observations (Stenflo, 1973; see reviews by Muller, 1985, and Stenflo, 1989; see Zirin, 1988, for a dissenting view) suggests that nearly all magnetic fields in the photosphere are in flux tubes too small to be observed directly, with kilogauss field strengths and diameters of order 100 km (0.14 arcseconds). It has been suggested (Dunn and Zirker, 1973; Mehltretter, 1974) that the filigree bright points are the elementary flux tubes, but some observations of very high quality appear to show the magnetic fields covering a much larger area (Simon and Zirker, 1974).

The 29 September data show many examples of filigree in the continuum and Ni I 6768 line center images. (See the review article by A. Title in this volume for examples of these images.) The magnetic fields outside sunspots have intricate fine structure down to the spatial resolution limit of the images. Some of our conclusions from viewing the movies and from detailed quantitative study of one time step of the 29 September data set are as follows.

(1) Magnetic regions have an "abnormal" granulation as seen in the continuum that is distinctly different from the granulation pattern seen in non-magnetic regions. Granules in the abnormal regions are smaller and the lanes between them are as well. In the photosphere above the abnormal granules, the vertical component of the velocity field is lower on average. Analysis of SOUP movies (Title *et al.*, 1989b) and the Doppler movie from Sacramento Peak (Tarbell *et al.*, 1988) show that this reduction occurs both in the 5 minute oscillations and in the granulation

velocities.

(2) The bright points observed in the Ni I 6768 line center are related to magnetic field structures. In weakly magnetic regions the bright points are cospatial with equally fine scale magnetic structures. In areas of higher magnetic flux, bright points lie within magnetic structures and are significantly smaller than the magnetic elements.

(3) Bright points in the Ni I 6768 line center are also usually seen as weak local maxima in the continuum intensity. In stronger magnetic areas, the continuum intensity is usually average or below average. The relation between line center and continuum varies considerably, depending on the magnetic flux, indicating varying intrinsic absorption depth of the Ni I line.

(4) In movies, both the bright points and the magnetic field structures evolve on comparable time scales. However, this does not mean that magnetic field disappears on a time scale of tens of minutes. Rather, magnetic field elements move and change their local configuration.

(5) The magnetic field appears to outline cellular structures of a wide range of sizes. A thresholded magnetogram resembles a fractal set, similar to those in Mandelbrot (1983). We can estimate the fractal dimension of this set from a plot of area versus length scale and find self-similar behaviour with a dimension of about 1.6 over the entire range considered, from 0.16 to 10 arcseconds.

We have made scatterplots of line center intensity versus continuum intensity, for pixels with magnetogram signals in selected bands. These show that the two intensities are correlated at all levels of the magnetogram, but the slopes and intercepts of the regression lines change in a complex way. Below 700 Gauss, the slope is much larger than in non-magnetic areas, indicating a smaller fractional line absorption depth in the magnetic structures. Above 700 Gauss, the slope decreases (line absorption increases), and most of the magnetic pixels are at average or below average intensity in both images. A more detailed study of the four-dimensional relationships between magnetogram signal, Doppler velocity, continuum, and line center intensity will be published elsewhere.

Emerging Magnetic Flux

Both Sacramento Peak and La Palma movies show many examples of tiny flux tubes emerging through the surface. We learned to recognize the signs of emerging flux in intensity and Doppler movies from 6 August 1987, and so it is now easy to find such events in the 29 September data, which have higher resolution and more uniform seeing. The quantitative discussion below is based on careful analysis of a few events in the 6 August movies (Tarbell *et al.*, 1989). The La Palma events are very similar in the movies but have not yet been studied quantitatively.

The first observable signal for the emergence of new magnetic flux is the formation of dark alignments seen in the continuum and line center. The dark alignments seen in our data are typically 1500 to 4000 km long and have lifetimes of 5 to 15 minutes. A strong upflow reaching 0.5 to 0.8 km/s may be seen for a few minutes in the dark alignment just after it appears. The upward velocity profile integrated over the duration of the event suggests a thickness of order only 100 km for the rising flux tube. Bright points form at one or both ends of the dark alignment a few minutes after it appears. Strong downflow reaching −1.0 km/s and new magnetic flux appear in the bright points. The downflow may persist for an hour or more; the mass flux integrated over this time period would empty a tube roughly 3000 km long at photospheric densities, consistent with the initial footpoint separation. In one event studied most carefully, the magnetic flux appears to increase gradually over a period of at least half an hour, instead of making an abrupt change when the bright point first appears. The observed emergence rate is about 3 x 10^{15} Mx/s. More events must be measured in the more uniform seeing of the 29 September movies to see if this result is confirmed. The footpoints separate at easily measurable horizontal speeds, sometimes as high as 3 km/s.

Zwaan and Brants (1985) have also seen many of these features in an emerging flux region. Theoretical studies of the emergence process have been published by Moreno-Insertis (1986) for the rise through the convection zone and by Shibata *et al.* (1989a, b) for eruption through the photosphere and expansion into the upper atmosphere. It is clear that a modest increase in resolution and uniformity of seeing would permit meaningful comparisons with MHD models. Continued observations of this type at excellent sites, perhaps with assistance from active optics, or balloon flights of the SOUP instrument could provide such observations in the near future.

ACKNOWLEDGMENTS. This work has been supported by NASA contracts NAS8-32805 (SOUP) and NAS5-26813 (OSL), by Lockheed independent research funds, and by the Royal Swedish Academy of Sciences. The observations were obtained at the Observatorio del Roque de los Muchachos of the Instituto de Astrofisica de Canarias and at the National Solar Observatory in Sunspot, New Mexico.

References

Dunn, R.B. and Zirker, J.B. (1973) *Solar Phys.*, **33**, 281.

Mandelbrot, B. (1983) "The Fractal Geometry of Nature" (W.H. Freeman, San Fransisco), pp. 308-9.

Mehltretter, P. (1974) *Solar Phys.*, **38**, 43.

Moreno-Insertis, F. (1986) *Astr. Ap.*, **166**, 291.

Muller, R. (1985) *Solar Phys.*, **100**, 237.

November, L.J. (1986) *Appl. Optics*, **25**, 392.

Shibata, K., Tajima, T., Matsumoto, R., Horiuchi, T., Hanawa, T., Rosner, R., and Uchida, Y. (1989a) *Ap. J.*, **338**, 471.

Shibata, K., Tajima, T., Steinolfson, R.S., and Matsumoto, R. (1989b) *Ap. J.*, in press.

Simon, G.W., and Zirker, J.B. (1974) *Solar Phys.*, **35**, 331.

Stenflo, J.O. (1973) *Solar Phys.*, **32**, 41.

Stenflo, J.O. (1989) *Astron. Astrophys. Rev.*, 1, 3.

Tarbell, T., Peri, M., Frank, Z., Shine, R., and Title, A. (1988) in "Seismology of the Sun and Sun-Like Stars," Domingo, V. and Rolfe, E.J. (eds.), European Space Agency SP-286, (ESTEC, Noordwijk, the Netherlands), 315.

Tarbell, T., Topka, K., Ferguson, S., Frank, Z., and Title, A. (1989) in "Proc. of Sac. Peak Summer Workshop on High Spatial Resolution Solar Observations", Luhe, O. von der (ed.), (NSO, Sunspot), in press.

Title, A., Tarbell, T., Topka, K., Cauffman, D., Balke, C., and Scharmer, G. (1989a) in "Physics of Magnetic Flux Ropes", Russell, C.T. (ed.), Am. Geophys. Union, in press.

Title, A.M., Tarbell, T.D., Topka, K.P., Ferguson, S.H., Shine, R.A., and the SOUP Team (1989b) *Ap. J.*, **336**, 475.

Topka, K.P., Tarbell, T.D., and Title, A.M. (1986) *Ap. J.*, **306**, 304.

Zirin, H. (1988) "Astrophysics of the Sun" (Cambridge Univ. Press, Cambridge), pp. 131-5.

Zwaan, C., and Brants, J.J. (1985) *Solar Phys.*, **95**, 3.

SMALL SCALE MOTIONS OVER CONCENTRATED MAGNETIC FIELD REGIONS OF THE QUIET SUN

DARA, H.C.[1], ALISSANDRAKIS, C.E.[2] and KOUTCHMY, S.[3]

[1]Research Center for Astronomy and Applied Mathematics, Academy of Athens, GR-10673, Athens, GREECE.
[2]Section of Astrophysics, Astronomy and Mechanics, Department of Physics, University of Athens, GR-15783, Athens, GREECE.
[3]Institut d' Astrophysique de Paris, Paris F-75014, FRANCE.

ABSTRACT. Using a time sequence of filtergrams in the magnetically sensitive $\lambda6103A$ CaI line (with circular polarization measurements), obtained with the SPO Vacuum Tower Telescope and the universal filter we mapped the line of sight velocity and the longitudinal magnetic field in three quiet solar regions. After elimination of the effect of the 5-minute photospheric oscillations we found in two regions of concentrated magnetic field no association with the line of sight velocity, while the third region was associated with small (<300 m/s) downflows.

1. INTRODUCTION

The magnetic flux tube is a key to a unified understanding of many solar phenomena (network, spicules, sunspots etc.). Since Howard and Stenflo (1972) inferred the existence of magnetic fluxtubes with strength of the order of KG and characteristic size of 100-300 km (Frazier and Stenflo, 1972, Stenflo, 1973) a lot of work has been carried out by many observers and theoreticians for their study. Flux tubes are believed to be associated to small (0".25) network elements, the filigree, inbedded in the intergranular lanes, although there are observations which show that the magnetic flux is confined in a larger area, of 1-3" (Koutchmy and Stellmacher,1978,Dara and Koutchmy, 1983). The question of compression and maintainance of the magnetic field to kG strengths is still unresolved since the magnetic pressure inside the tube is comparable to the gas pressure in the surroundings. The widely held view is that convective motions sweep the diffuse magnetic field to the supergranule boundaries, forming flux tubes of moderate strength. According to Parker (1978),the field is further compressed as a result of adiabatic cooling of the tube. Therefore, mass motion, in particular downdrafts, associated with magnetic structures, are believed to play an important role in the physics of flux tubes.

J. O. Stenflo (ed.), Solar Photosphere: Structure, Convection, and Magnetic Fields, 153–156.
© *1990 by the IAU.*

Velocities above 0,5 km s^{-1} have been reported initially by some observers (Stenflo,1973, Giovanelli and Slaughter, 1978; Wiehr, 1985) while more recent observations give small (<300 ms^{-1}) or zero velocities in regions of concentrated magnetic field (Stenflo and Harvey, 1985, Dara et al., 1987). In the latter paper we studied the motion in the vicinity of two regions of concentrated magnetic field in the quiet sun using a short (5.5 min) time series of spectra, and we found no convincing evidence of association of the magnetic field regions with downflows. In this paper we study the velocity and the magnetic field in three quiet regions using filtergrams, which permit the mapping over a two dimensional field of view.

2. OBSERVATIONS AND DATA REDUCTION

A long (1 hour 16 min) series of high resolution filtergrams in the wings and at the center of the magnetically 6103 CaI line (each wing simultaneously observed in two circular polarizations) were obtained with the Sacramento Peak Vacuum Tower telescope and the universal filter (UBF), on October 19, 1987. The filter bandwidth was 186 mA, while the scale of the image on the film was 10"/mm. For the separation of the two circular polarizations a Wollastone prism was used. The duration of each run was 32 sec and included two pictures in each of the wings, ±0.07 A from the center, and at the center of 6103A line, as well as two at the center of Hα. The field of view was 100 x 200" and the telescope was pointed near the center of the disk, at N18 W12.

Because of the important vignetting effect, flat field pictures were taken with the solar image out of focus. Photometric calibration was performed using a step wedge. The best of the two pictures at each wavelength was chosen for further processing with the fast microphotometer of the Sacremento peak Observatory. The micropho-tometer spot size was 0".7 and the step 0".3.

For the present analysis, we selected pictures from six runs, uniformly distributed in a time period of 5.3 min. Each frame was calibrated, corrected for the vignetting effect, and was rotated so that the orientation of all frames was the same.

The digital subtraction of the two pictures with opposite sense of circular polarization gives a signal proportional to the longitudinal magnetic field for each wing. The maps of the magnetic field were computed by averaging the magnetic signal from both wings. Moreover, the average of the two pictures in each wing gives the intensity, while subtraction of the intensity in the wings gives a signal proportional to the line of sight velocity. The magnetic and velocity signals were calibrated using a scan of the line profile obtained with the UBF.

A comparison of the six velocity maps shows that the effect of the photospheric oscillations is very strong and masks any flows associated with the granulation of the magnetic field. Therefore we computed average velocity maps over an interval of 5 minutes, where the effect of oscillations is strongly reduced. We also computed

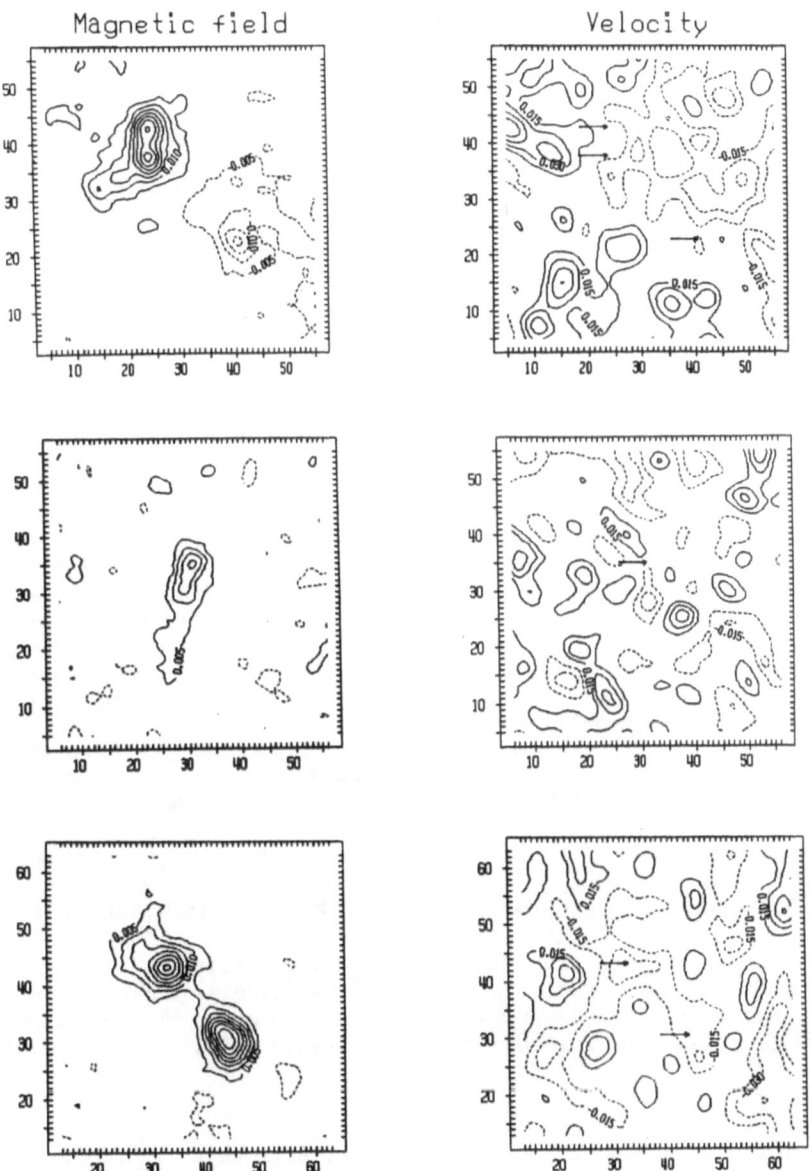

Figure 1. Magnetic field and velocity maps, averaged over 5 minutes. Dashed contours indicate negative values (downflows). Magnetic field contours are in steps of 0.5% contrast (∿40 Gauss) and velocity contours are in steps of 1.5% (∿100 m/sec). Tick marks are every 0.3". The arrows on the velocity maps show the position of the peaks of the magnetic field.

averages of the magnetic field maps in order to improve the signal to noise ratio.

3. RESULTS AND DISCUSSION

Three regions have been selected in the quiet sun for the study of small-scale motions and the calculation of the corresponding magnetic fields. Figure 1 shows the average maps of the longitudinal magnetic field and the corresponding line of sight velocity, after the elimination of the five minute oscillations.

The first region has a bipolar magnetic field and corresponds to a rosette in the center of Hα; the two opposite polarity regions are at a distance of 8", and the magnetic field is almost two times stronger in one polarity than on the other. The second region is unipolar and in the center of Hα appears as a bright point. One can clearly see that in these two regions the velocity corresponding to the magnetic features is practically zero. The third region is an extended, single polarity region with two well separated peaks, 5" apart. At the center of Hα it appears as a rosette with a bright point in its vicinity. In the velocity map of this region small downflows (<300 ms^{-1}) are observed. It is interesting to note that similar downflows on the velocity map are also observed in regions without any significant magnetic field; it is clear that, from the velocity maps alone, one cannot tell whether there are any concentrated magnetic field structures. This is in good agreement with our previous results (Dara et al., 1987).

It is evident that downflows cannot play a fundamental role in the confinement of small-scale magnetic fields, and any modelling of the tubes should take into consideration this observational result.

ACKNOWLEDGEMENTS

The analysis presented here was based on data taken with the Vacuum Solar Tower of the Sacramento Peak Observatory during the stay of two of the authors (H.D. and S.K). We are grateful to the director of the SPO for his kind hospitality and financial support, as well as to the staff for their assistance to our work. We are also much indebted to Laurence November for his valuable help.

REFERENCES

Dara, H.C. and Koutchmy, S.: 1983, Astron. Astrophys. 125,280.
Dara,H.C., Alissandrakis,C.E. and Koutchmy,S.:1987,Solar Phys. 109,19
Frazier, E.N. and Stenflo, J.O.: 1972, Solar Phys. 27, 330.
Giovanelli, R.G. and Slaughter, C.: 1977, Solar Phys. 57, 255.
Koutchmy, S. and Stellmacher, G.: 1978, Astron. Astrophys. 67, 93.
Parker, E.N.: 1978, Astrophys. J. 221, 328.
Stenflo, J.O.: 1973, Solar Phys. 32, 41.
Stenflo, J.O. and Harvey, J.W.: 1985, Solar Phys. 95, 99.
Wiehr, E.: 1985, Astron. Astrophys. 149, 217.

SMALL SCALE MAGNETIC STRUCTURES IN ACTIVE CENTERS

A. DOLLFUS
Observatoire de Paris
92195 MEUDON Principal Cedex
FRANCE

ABSTRACT. Active sunspot group n° 4567 was observed with high angular magnification on Sept. 2, 1984, using the instrument FPSS at Meudon Observatory. Within the sunspots umbrae, the longitudinal magnetic field shows a collection of small elements down to the limit of resolution of around 600 km. Between the trailing and leading spots, there were very sharp gradients, both in magnetization (450 gauss over 1000 km) and in Doppler velocity (350 m s^{-1} over 1 000 km). The line defined by these gradients was the location of a flare. Evershed driving forces may be responsible for building up these gradients and flare.

1. INSTRUMENT

The monochromatic solar telescope of Meudon Observatory, using the birefringent filter FPSS (Filtre Polarisant Solaire Sélectif) produces images of the solar surface with a spectral resolution of 2×10^{-5} (Dollfus et al. 1985). An optical design shifts the transmitted band from bottom to wings of a spectral line and to adjacent continuum, and selects the different states of polarization of the incident light. The images recorded in the different configurations are combined to produce new images of the Doppler motion, the magnetization, the line depth and the line width (Dollfus et al. 1986).

The images are obtained through a 27cm reflector directly pointed at the sun, in order to preserve a maximum telescopic image resolution. Our best magnetic images disclose an angular resolution on magnetic features of around 0,8 arc sec. (600 km at the solar surface).

2. MAGNETIC STRUCTURES IN SPOT UMBRAE

On September 2, 1984, the two small trailing spots of bipolar group n° 4567 (fig. 1), displayed the magnetic configuration enlarged in the figure 2. Into the umbrae, the magnetic field appears to be structured like a cluster of small elements, typically 600 km in diameter and smaller.

157

J. O. Stenflo (ed.), Solar Photosphere: Structure, Convection, and Magnetic Fields, 157–160.

158

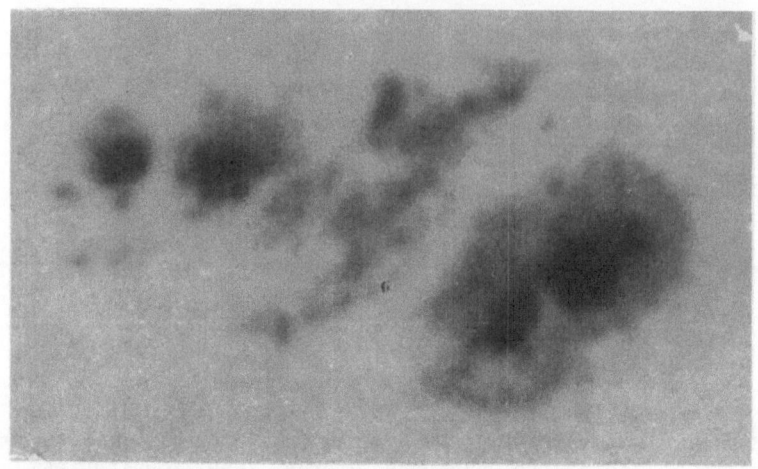

Figure 1 : Sunspot group n° 4567 on September 2, 1984 at 08:28 UT.
Field 110x70(arcsec)² centered at Long:140°, lat.=-10°. Meudon solar
telescope with FPSS. A. Dollfus.

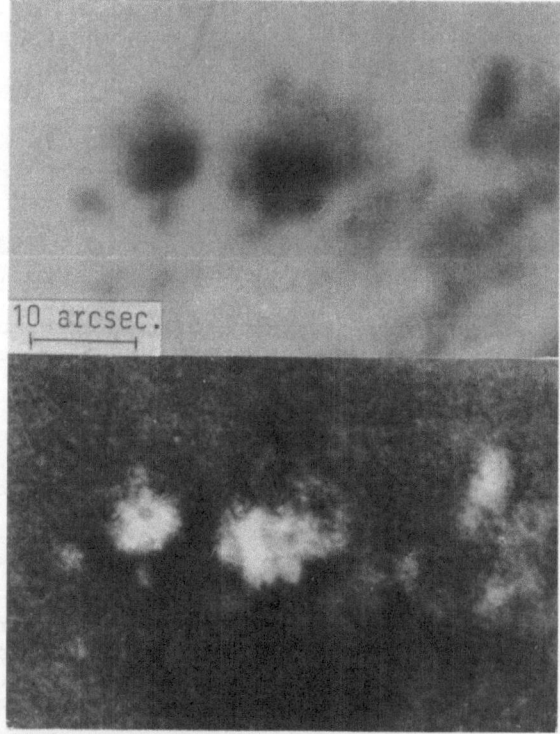

10 arcsec.

Figure 2 : Magnetic structures in the umbrae of the
two trailing spots of fig. 1. Positive polarity is
white. Line Fe I 6173.

A processing which avoids sharpness degradation due to pixelization and intensity discretization can enhance this structuration : the best negative photographic image taken in a wing of the line with a given circular polarization is copied by contact on another film which gives a positive image with a contrast of exactly 1.0. This image is then superimposed upon the best sharpest negative image taken with the other circular polarization component. The result is an image of the magnetization, with a minimum processing degradation. Now when the image centered upon the other is slightly shifted with excursions alternatively each side of the optimum position, the grey background, where there is no field, keeps

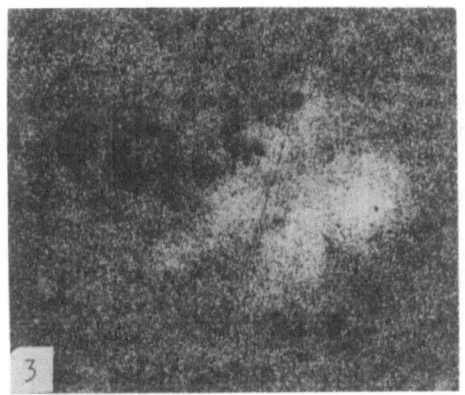 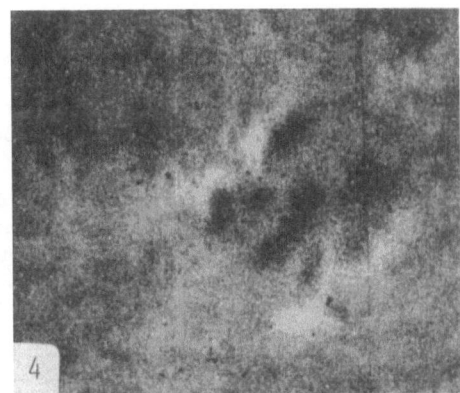

Figure 3 : Longitudinal magnetic field configuration around the spot group n°4567 of fig.1.
Line FeI 6173.

Figure 4 : Longitudinal velocity field configuration around the spot group n°4567 of fig.1.
Line FeI 6173.

a clotted aspect which fluctuates
randomly due to the graininess of
the emulsion. But on the spot umbrae,
these fluctuations are drastically
enhanced, because of a real magnetic
field structuration down to the limit
of the angular resolution.

3. SHARP GRADIENTS

Between the two spots and the
larger leading spot of opposite polarity (fig.1), there was a magnetic
neutral line. And across this line,
there were steep gradients (fig.3).
The longitudinal field gradient
reaches at least 450 gauss over 1000
km. At the same place, the longitudinal Doppler velocity image (fig.4)
exhibits at least 350 m s^{-1} over
1000km.

Evershed photospheric radial motions are observed around each of
the spot umbrae, particularly around
the larger spot. We speculate that
it is the forces driving the Evershed
motion which displace the photospheric

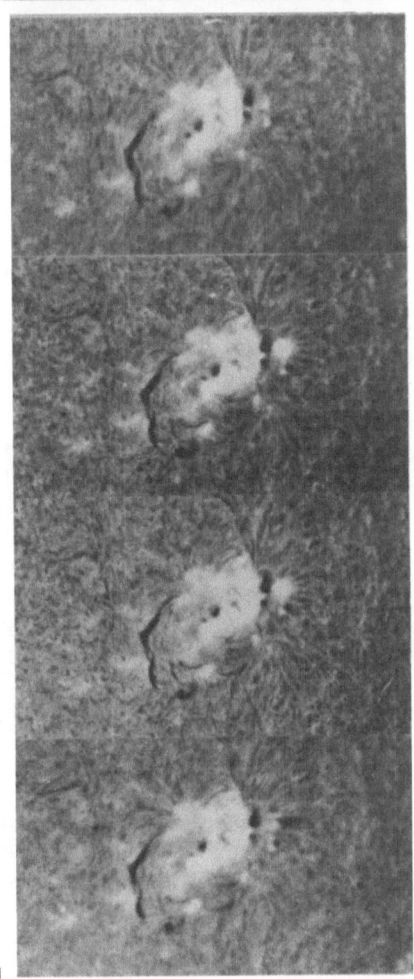

Figure 5 : Flare evolution of sunspot group
n°4567 on Sept.1,1984. Observed in Hα at Osservatorio Astrofisico di Catania by C.Bianco et al.
From top to bottom:10:50-10:15-10:36-10:40-10:45TU

material toward a same area between the two spots, antagonistically from both sides, and produce a concentration of material which have to escape laterally or vertically. This border appears as a dark cold lane in the white light image of fig. 1. Simultaneously some magnetization is moved accordingly and produces the magnetic concentration and gradient.

Around this area and one half hour later a two-ribbon flare occurred, at 10:15 UT. Recorded namely at Osservatorio Astronomico di Catania (fig. 5), this event imprinted on each side of the neutral magnetic and velocity line, as shown in fig. 6.

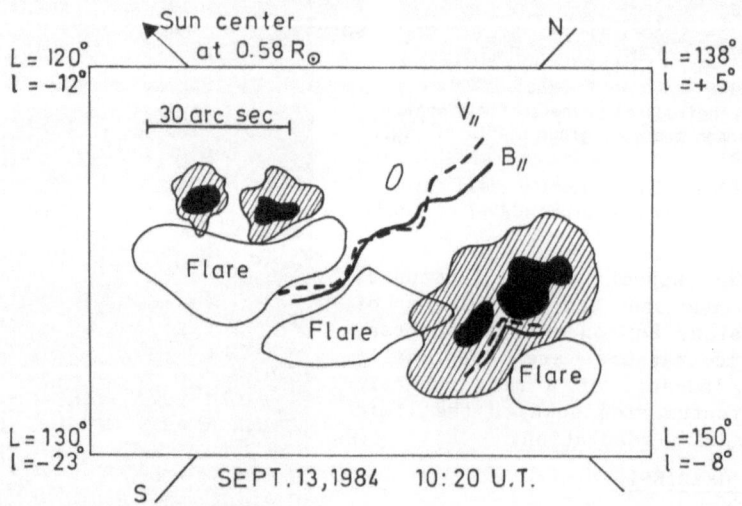

Figure 6 : Neutral magnetic line, neutral velocity line and flare configuration around sunspot group n°4567 on Sept.2, 1984.

Configurations with sharp gradients in motion and in magnetization are able to produce electric currents, responsible for flare occurrence. It is suggested that the flare eventually resulted at the expense of the Evershed driving forces.

Such a mechanism for flare occurrence may not be rare. The configuration required is seldom suitably position ed to be observable. Our bipolar group was observed at near 60° from disk center with the direction joi ning the spots of opposite polarities pointing precisely toward the disk center.

REFERENCES

Dollfus A., Colson F., Crussaire D. and Launay F. (1985).
    Ann. Astrophys. 151, 235-253.

Dollfus A., Crussaire D., Pernot E. and Lioure A. (1986).
    Compte-Rendus Acad. Sci. Paris 303, Série II, 153-158.

THEORETICAL ASPECTS OF SMALL-SCALE PHOTOSPHERIC MAGNETIC FIELDS

M. SCHÜSSLER [1]
Universitäts-Sternwarte
Geismarlandstr. 11
D-3400 Göttingen
Federal Republic of Germany

ABSTRACT. The state of theoretical description of small-scale concentrated magnetic fields in the solar photosphere (excluding oscillations and wave propagation) is reviewed with emphasis on work done since 1982. The processes which probably lead to the formation of strong fields (flux expulsion, convective collapse) are discussed in some detail and the present understanding of the subsequent (quasi-)equilibrium state is summarized. We consider in particular the magnetic and thermal structure of the basic magnetic flux concentrations (magnetic elements) and stress the importance of radiative transfer effects, e.g. the horizontal heat exchange with the surroundings and the effect of radiation from the hot bottom and walls on the upper layers. Velocity fields within and around magnetic flux concentrations are discussed with emphasis on shift and asymmetry of the observed Stokes V-profiles which have recently been understood in terms of a downflow in the immediate vicinity outside magnetic structures. Reconnection and instabilities are considered as possible destruction processes for magnetic elements.

1. Introduction

Most of the observable magnetic flux permeating the solar photosphere is organized in a hierarchy of structures which have a magnetic pressure comparable to the gas pressure in their apparently non-magnetic environment. Detailed analysis of spatially unresolved spectra (reviewed by Solanki in these proceedings, see also Stenflo, 1989) indicates the existence of a basic structure, the *magnetic element,* with a magnetic flux of a few times 10^{17} mx, a flux density of about 2000 Gauss and a diameter of less than 200 km (both at continuum optical depth unity within the magnetic structure). Magnetic elements comprise most of the flux in the magnetic network outside active regions and in plage areas. Larger structures like sunspots or pores are formed in the course of the eruption of new active regions. After the initial phases of magnetic flux emergence they sooner or later fragment into magnetic elements. The ubiquity of magnetic elements and the remarkable fact that they all basically share the same properties place these structures in the focus of observational and theoretical interest, even more so since they are suspected to play a crucial part in the solution of the long-standing problems of chromospheric/coronal heating and the acceleration of the fast solar wind. Apart from the implication for solar and stellar physics, small-scale

---

[1] Permanent address: Kiepenheuer-Institut für Sonnenphysik, Schöneckstr. 6, D-7800 Freiburg, Federal Republic of Germany

J. O. Stenflo (ed.), Solar Photosphere: Structure, Convection, and Magnetic Fields, 161–179.

magnetic flux concentrations in the solar (sub-)photosphere can be seen as an example for the formation of dissipative structures in systems far from thermal equilibrium.

Magnetic elements are too small to be individually studied in detail with presently existing spectroscopic instrumentation. Their tentative identification with small bright structures in the continuum and in spectral lines (Dunn and Zirker, 1973; Mehltretter, 1974; cf. review by Muller in these proceedings) has been supported by the analysis of spectral lines profiles (Schüssler and Solanki, 1988) and numerical model calculations (Spruit, 1976; Grossmann-Doerth et al., 1989a). Recently, it has been demonstrated by high-resolution magnetograms and filtergrams (Title et al., 1989) that magnetic structures coincide with bright features in the network while in plage regions the relation between brightness and magnetic structures seems to be more complicated. This is presumably due to the tendency of magnetic elements to collect into clusters in regions of large average magnetic filling factor (Knölker and Schüssler, 1988) where they strongly influence the granular motions and the convective energy transport.

The remarkable progress in our understanding of small-scale magnetic fields in the solar photosphere in spite of the resolution problem was made possible by the ingeneous use of indirect spectroscopic methods, the development of sophisticated instruments, most notably the Kitt Peak Fourier transform spectrograph/polarimeter (FTS), the theoretical study of basic physical processes in flux concentrations and magnetoconvection using simplified models, and the advent of comprehensive numerical simulations of MHD and radiative transfer.

Theoretical aspects of small-scale photospheric magnetic fields have been reviewed earlier by Meyer (1976), Weiss (1977), Parker (1979), Priest (1982), Spruit (1983), Spruit and Roberts (1983), Nordlund (1984,1985b,1986), Thomas (1985), Schüssler (1986, 1987a), and Spruit et al. (1989). This review concentrates on the developments which took place after the IAU-Symposium No. 102 in Zürich (Stenflo, 1983). A number of reviews in these proceedings is related to magnetic elements, namely those of Müller, Solanki and Title (observational aspects), Nordlund (interaction with convection), and Ryutova (waves and oscillations). In order to limit overlapping with the latter two contributions, magnetoconvection and the theory of waves in fluxtubes will not be discussed here. We shall focus our attention on the present state of theoretical understanding of the formation (Ch. 2) and destruction (Ch. 4) processes of magnetic elements and to the properties of their quasi-equilibrium state (Ch. 3).

2. Formation

Magnetic flux which is observed in the solar photosphere most probably has its origin in dynamo processes operating near the bottom of the convection zone from where it rises to the surface due to the combined effects of buoyancy and convective flows. Since the *average* field strength in an emerging flux region is rather high, a pre-eruption field strength of at least a few hundred Gauss in the uppermost layers of the convection zone must be assumed (Zwaan, 1978). The erupting flux transforms into strong flux concentrations in a matter of minutes since there is no evidence that the field strength in newly erupted active region is smaller. After the initial flux eruption we expect a dynamical state : Flux concentrations are formed, temporarily attain an equilibrium state and dissolve again while the magnetic flux is constantly moved around the changing pattern of granulation and supergranulation. On the other hand, at any instant from the beginning of the life of an active region to the dispersal of its flux in the network more than 90% of the magnetic flux which is observable through the analysis of circular polarization in spectral lines (see Stenflo, 1987, on the possible existence of a "turbulent" field with mixed polarity on very small scales) is in the form of magnetic elements with large field strength. Consequently, the formation and destruction processes must have a much shorter timescale than the lifetime of magnetic elements in equilibrium.

The remarkable result that all magnetic elements irrespective of being located in network or in active regions have similar thermodynamic and magnetohydrodynamic properties (e.g. Stenflo and Harvey, 1985; Zayer et al., 1989) which vary only weakly with increasing number density of flux concentrations (Solanki and Stenflo, 1984) indicates that they are formed by essentially the same process and reach a (quasi-)equilibrium which is determined by the local properties of the plasma within and around them. There are two mechanism which are held responsible for the concentration of magnetic flux into structures with large flux density: *Flux expulsion* and *convective collapse*. Although both processes are related we discuss them in somewhat artificial separation in order to emphasize the basic effects.

Convective motions in an electrically well-conducting plasma concentrate magnetic flux into structures with a local flux density much larger than its average value. This *flux expulsion* process (Parker, 1963; Weiss, 1966; Galloway and Weiss, 1981; Weiss, 1981a,b; Hurlburt et al., 1984; Hurlburt and Toomre, 1988) leads to a kind of "phase separation" between field-free convecting plasma and magnetic, almost stagnant regions. It has been suggested by Parker (1984) that in a stellar convection zone such a configuration is energetically favoured since it minimizes the interference of the magnetic field with the convective energy transport. In the case of the solar (sub-)photosphere, flux expulsion leads to a sweeping of magnetic flux into the intergranular downflow regions as demonstrated by high-resolution observations (Title et al., 1987) and numerical simulations (Nordlund 1983, 1986). The expulsion process works in essentially the same way for the vertical component of the *vorticity* as can be shown using the well-known formal identity of the equations describing the time evolution of vertical vorticity and magnetic field in the kinematical limit. This has the consequence that both magnetic flux and vertical vorticity are concentrated into the narrow downflow regions of granular convection (cf. Nordlund, 1985a,b) such that the magnetic flux concentrations become surrounded by rapidly rotating, descending whirl flows.

The back reaction of the magnetic field on the flow via the Lorentz force limits the flux density which can be achieved by flux expulsion to a value which is roughly given by the *equipartition* of magnetic and kinetic energy density. This limit may be modified by the effects of diffusion and compressibility (cf. Proctor and Weiss, 1982). For the case of the solar (sub-)photosphere, however, the equipartition limit (a few hundred Gauss) is rendered irrelevant by *thermal* effects. Since the horizontal flows of granular convection are responsible for both sweeping the magnetic field to the downflow regions and for carrying heat to those regions, the retardation of the flows by the growing magnetic field leads to a cooling of the magnetic region since the the radiative losses can no longer be balanced by the throttled horizontal flow. This cooling effect causes an increase of the magnetic field since the gas pressure in the magnetic region becomes smaller and it accelerates the downflow which gives rise to the *superadiabatic effect* (Parker, 1978): An adiabatic downflow in a magnetic flux tube which is thermally isolated from its surroundings leads to a cooling of the interior with respect to the superadiabatically stratified surroundings and a partial evacuation of the the upper layers ensues. Pressure equilibrium with the surrounding gas is maintained by a contraction of the flux tube which increases the magnetic pressure. In this way, the magnetic field can be locally intensified to values which are only limited by the confining pressure of the external gas.

It has been shown by a number of authors (Webb and Roberts, 1978; Spruit and Zweibel, 1979; Unno and Ando, 1979) that the superadiabatic effect in the case of a flux tube which is in magnetostatic and temperature equilibrium with its environment drives a *convective instability* in the form of a monotonically increasing up- or downflow. Consequently, the initial downflow due to the radiative cooling will be enhanced by this effect leading to an even stronger amplification of the magnetic field, a process which is often referred to as *convective collapse*.

While this convective instability of a flux tube with a weak magnetic field is theoretically well established, the results for strong fields and for the nonlinear evolution of the convective collapse give no unique picture. The claim of Spruit and Zweibel (1979) and Spruit (1979) that flux tubes with a strong enough magnetic field become stable with respect to convective collapse has been critized by Nordlund (1984) who argued that this result depends on the choice of boundary conditions: If the displacement of matter (or the fluid velocity along the tube) is not constrained to vanish at two fictitious endpoints of the tube, any adiabatic downward displacement leads to a state of lower energy and there is no stable equilibrium, irrespective of the field strength. This argument is supported by the linear results of Webb and Roberts (1978) who showed that the location of the lower (closed) boundary significantly influences the linear stability in a way that the stabilizing effect of the magnetic field decreases more and more as the location of the lower boundary is shifted deeper and deeper.

This dependence on the boundary conditions explains the discrepancy between the results of nonlinear simulations by Hasan (1983) and Venkatakrishnan (1983) and those performed later by Hasan (1984). In the first two papers constant internal gas pressure was assumed at the boundaries which thus were effectively "open". In these cases the convective instability evolved into a state of permanent downflow with high velocity, nearly independent of the strength of the initial field. In his subsequent paper, Hasan (1984) used closed boundaries, i.e. vanishing velocity at the end points. Now the instability was suppressed for strong enough magnetic field (in agreement with the linear results using the same boundary conditions) and the unstable configurations evolved into a state of stationary adiabatic oscillation.

With the exception of very special upper boundary conditions (e.g. Unno and Ando, 1979), the crucial point is the choice of the *lower* boundary condition (cf. Webb and Roberts, 1978; Hasan, 1986). A downflow leads to a gas pressure enhancement near a closed boundary and a local expansion of the flux tube in order to maintain equilibrium with the external gas pressure. For a weak field, this expansion is significant since only a moderate increase of the internal gas pressure can be balanced by a decrease of the magnetic field. Consequently, the vertical restoring force on the downflow due to the pressure enhancement is weak and the expansion of the tube provides space for the matter carried by the nearly unimpeded downflow. With a strong magnetic field, on the other hand, pressure balance with the exterior is readjusted by only a slight expansion and the internal gas pressure enhancement is fully available as restoring force in the vertical direction. However, if the flow is not constrained to vanish at the boundary or if the internal pressure is assumed to be constant there, the restoring force is much less effective and a strong magnetic field is not able to suppress the instability.

The question arises as to which boundary conditions should be used in the case of solar photospheric magnetic flux concentrations. Strictly speaking, a vanishing of the fluid displacement can reasonably be assumed only in a convectively stable stratification, i.e. if the bottom of the tube is placed *below* the convection zone. This has effectively been done by Webb and Roberts (1978, Ch. 6.2) by forcing the velocity perturbation to vanish at infinity and also by Spruit and Zweibel (1979) who put the boundary at the bottom of the convection zone. While Webb and Roberts (1978) found for a model with a constant temperature gradient that a strong field can only stabilize a small range of superadiabatic temperature gradients, Spruit and Zweibel (1979) showed that for their boundary conditions and a realistic convection zone a flux tube is stable if $\beta = 8\pi p/B^2 < 1.51$. Both calculations assume an equlibrium state with temperature in the flux tube being equal to the external temperature, i.e. a depth-independent value of β. The reason for the differing results lies in the small superadiabaticity of the deeper layers of a realistic convection zone while Webb and Roberts (1978) assume a constant superadiabaticity. However, the applicability of these linear results for strong fields to the real Sun is questionable for a number of reasons:

- The assumption of adiabatic changes is quite unrealistic because of the strong effects of radiation on the energy balance of the surface layers.
- Taking a depth-independent β implies unrealistically large values of the field strength in the deep layers: $3 \cdot 10^8$ Gauss at the bottom of the convection zone for $\beta = 1.5$.
- It is by no means obvious that the observed magnetic elements should maintain their identity as single flux tubes in deeper layers and the assumption of a vertical tube becomes inadequate at moderate depths of a few 10^3 km: Small flux concentrations become passive with respect to the convective flows due to the strong increase in density and will be severely distorted and fragmented (Schüssler, 1984a; 1987b).

In view of the observational results which imply that most of the magnetic flux at any given instant of time is in the form of magnetic elements approximately in hydrostatic equlibrium without a significant downflow (Stenflo and Harvey, 1985; Solanki, 1986) all of which have similar thermodynamic and magnetic properties, we conclude that this state cannot be determined by conditions deep within the convection zone but rather is controlled by the local conditions in or near the observable layers. In summary, although the linear results demonstrate the existence of the convective collapse mechanism they do not seem to be particularly relevant for the questions whether a stable state is reached by the instability and, if yes, which are its properties. The dynamics of the nonlinear evolution of the instability and its non-adiabatic character due to radiative effects have to be taken into account in order to quantitatively predict the result of a convective collapse.

The configuration resulting from the convective collapse must not necessarily be *globally* stable in order to represent a local quasi-equilibrium in the surface layers as implied by observations. The radiative cooling effect and the large superadiabaticity are restricted to the uppermost layers of the convection zone and the resulting unstable downflow will be localized in this region. The deeper layers are only slightly superadiabatic and thus almost neutrally stable but represent a large inertia because of the drastic increase of density with depth. We should therefore expect that the collapsing upper layers are stopped and reflected similar to a body colliding with another body of much larger mass at rest which is set into only very slow motion. It is presently unclear how strong the following upflow is. Nordlund's (1983, 1986) simulations do not show an upflow but their poor spatial resolution and strong numerical diffusion on the scale of the flux concentrations couple the magnetic regions artificially to the dynamics of the external downflow. The simulations of Hasan (1984, 1985) and Venkatakrishnan (1983, 1985) exhibit an upflow and a subsequent oscillation but they do not incorporate a precise treatment of the important effects of vertical radiative energy losses which are largely enhanced by the strong temperature dependence of the continuum opacity. If the upflow is strong enough it could possibly drive a spicule through the formation of shocks (cf. Suematsu et al., 1982; Hollweg, 1982; Sterling and Hollweg, 1988). However, it is well possible that the matter has lost so much energy through radiative losses that the upflow is weak or even non-existent. Eventually, this issue has to be settled by observation where the only indication of a convective collapse has been provided by Wiehr (1985; see, however, Solanki and Stenflo, 1986). It would be a major achievement if spectroscopic observations with high spatial resolution could follow the convective collapse of a single magnetic flux concentration.

Whichever is the detailed dynamic evolution of the convective collapse, in the uppermost layers of the convection zone it leads to a state near hydrostatic equlibrium. Due to the strong radiative losses during the collapse phase the gas in the flux concentration is now significantly cooler than the surrounding medium at equal depth. This temperature reduction is well capable of stabilizing the equilibrium of the upper layers with respect to further convective collapse (see Webb and Roberts, 1978) independent of the choice of particular boundary conditions in the linear stability analysis. Thus it is the non-adiabaticity of the collapse due to the radiative energy losses which may well be responsible for the establishment of a locally stable configuration while the increase in magnetic field strength only is of minor importance.

The thermal isolation with respect to convection is counteracted by radiation if the magnetic structure is so small that its horizontal optical depth becomes of order unity. Consequently, very small structures will always be kept at the temperature of the surrounding gas and therefore cannot undergo a convective collapse (Schüssler, 1986; Venkatakrishnan, 1986). Hasan (1986) has shown for a realistic solar atmosphere that the critical value of β for the onset of monotonic convective instability increases rapidly for decreasing diameter of the flux tube if lateral radiative energy exchange is taken into account.

It remains an open question whether magnetic elements are susceptible to overstability caused by horizontal radiative transfer (Roberts, 1976; Spruit, 1979; Hasan, 1985, 1986; Venkatakrishnan, 1985; Massaglia et al., 1989) if the energy losses by *vertical* radiative transfer are consistently taken into account. A proper treatment of radiation far beyond the limits of the diffusion equation or "Newton's law of cooling" and an adequate level of spatial resolution in numerical simulations is necessary in order to decide whether overstable oscillations are excited in a realistic model of a magnetic element. Even a crude inclusion of vertical radiative transfer (Venkatakrishnan, 1985) or the step from the radiative diffusion/relaxation time approach to the Eddington approximation (Massaglia et al., 1989; Hasan, 1989) led to drastic changes especially for the interesting case of a flux tube with a horizontal optical depth around unity.

3. Equilibrium

Observational results indicate that photospheric magnetic flux concentrations reach an equilibrium state whose properties are similar for most small-scale flux concentrations (Zayer et al., 1989) and depend only weakly on the filling factor, i.e. the fraction of the area covered with magnetic elements (Stenflo and Harvey, 1985; Solanki and Stenflo, 1984). The upper limit of 250 m·s^{-1} for an average flow *within* the magnetic structures determined from the absolute shift of the Stokes V-profile zero crossings of spectral lines in spatially and temporarily unresolved FTS spectra (Stenflo and Harvey, 1985; Solanki, 1986) indicates that the structures are approximately in (magneto-)hydrostatic equlibrium. This upper limit also excludes efficient vertical convective energy transport, for instance by large-amplitude overstable oscillations, since this would also lead to a significant average shift due to a correlation between intensity and velocity (e.g. Hasan, 1985).

These results justify the theoretical working hypothesis that, except for the formation and destruction phases, the basic properties of photospheric magnetic flux concentrations may be represented by prototype flux tubes or flux slabs in static or stationary equilibrium embedded in a non-magnetic environment. The theoretical objective is to describe this state self-consistently including force balance and dynamics, energy transport by radiation and flows, and interaction with the environment. Ultimately, this task calls for a comprehensive 3D time-dependent numerical simulation. However, this cannot by achieved with the computational facilities available at present or in the near future. For example, the simulations by Nordlund (1983, 1986) have a spatial resolution of about 250 km in the horizontal direction while a value of a few km is needed to resolve the boundary layer between a flux concentration and its surroundings which is crucial for a correct description of the energy balance. Consequently, the available 3D simulations may describe the average motion of an *ensemble* of magnetic elements in a time-dependent granular velocity field but they give no information on the dynamics and the properties of individual flux concentrations. So it is necessary to consider models restricted to one or two spatial dimensions which allow a much better spatial resolution.

Some guidance as to the assumptions going into these models can be taken from simple considerations. For example, the strong buoyancy force of a magnetic element keeps it essentially vertical in the (sub-)photospheric layers (Schüssler, 1986) such that the model of a vertical flux tube with a straight axis seems reasonable unless its diameter is much smaller than the scale height. Also the assumption of a thin boundary (current sheet) between the flux concentration and its environment is supported by estimates of the widths of resistive and viscous boundary layers (Schüssler, 1986). The cross-field flows due to finite resistivity are less than 10 m·s^{-1} for the photospheric regions while the drift velocities of neutral atoms is a few cm·s^{-1} at maximum (Hasan and Schüssler, 1985). Thus the effects of finite resistivity and incomplete ionization are irrelevant and the approximation of ideal magnetohydrodynamics is well justified for the photospheric layers.

3.1 MAGNETIC FIELD

In order to describe the structure of the magnetic field in a photospheric flux concentration a number of approaches has been used. *Static* models based on the approximation of slender fluxtubes have been presented by Ferrari et al. (1985) and Hasan (1988). Higher orders in the radial expansion procedure which to zeroth order gives the slender fluxtube approximation (cf. Roberts and Webb, 1978; Ferriz-Mas and Schüssler, 1989) and allow to include twisted

fields have been considered by Wilson (1977), Browning and Priest (1983) and Pneuman et al. (1986). The similarity approach was used by Osherovich et al. (1983) and Solov'ev (1984). 2D models of potential fields have been presented by Spruit (1976, 1977), Simon et al. (1983) and by Schmidt and Wegmann (1983, see also Jahn, these proceedings) who solved the free boundary problem consistently . Full 2D magnetostatic models with internal currents and current sheets have been published by Pizzo (1986), Steiner et al. (1986) and recently by Steiner and Pizzo (1989).

Time dependence is included in a number of approaches which aim either at describing dynamical processes (flows, waves, oscillations, shocks) or try to model the evolution of the flux concentration to a stationary state self-consistently. In this connection slender flux tubes have been assumed, among others, by Unno and Ribes (1979), Hollweg (1982), Hasan (1984, 1985), Venkatakrishnan (1983, 1984), Herbold et al. (1985), Ribes et al. (1985), Hasan and Schüssler (1985), Ferriz-Mas and Moreno-Insertis (1987), Ferriz-Mas (1988), Thomas (1988), Montesinos and Thomas (1989), and Degenhardt (1989). Higher orders of the radial expansion have been considered by Ferriz-Mas et al. (1989) and Anton (1989). Simulations in 2D slab geometry have been performed by Deinzer et al. (1984a,b), Knölker et al. (1988), Knölker and Schüssler (1988) and Grossmann-Doerth et al. (1989a). The path followed by slender flux tubes in prescribed cellular velocity fields meant to represent supergranular and granular flows has been investigated in a time-dependent calculation by Meyer et al. (1979) and Schmidt et al. (1985).

The height-dependence and the horizontal constancy of the magnetic field given by the slender flux tube approximation are in good agreement with results derived from FTS spectra in the visible and in the infrared (Zayer et al., 1989), a result which is further supported by comparison with 2D models (Knölker et al., 1988). Steiner and Pizzo (1989) have shown that an unrealistically large amount of heating or cooling in the photospheric layers of a flux tube would be necessary in order to significantly influence its shape. The effect of a twisted field on the equilibrium and shape of a flux tube has been investigated by Steiner et al. (1986) who found that magnetostatic equilibrium cannot be achieved if the azimuthal field component exceeds a critical value of about a third of the axial field strength at the base of the model. For statically allowed values of the twist the shape of the flux tube and the height where it merges with the neighbouring flux tubes for a given filling factor is not strongly affected by twisting the field (see also Pneuman et al., 1986).

3.2 THERMAL STRUCTURE

A variety of effects influences the energetics and the temperature structure of a photospheric flux concentration:

- Advection of heat by flow fields in the environment,
- lateral exchange of energy with the surroundings by radiative transport,
- reduction/suppression of convective energy transport by a strong magnetic field,
- vertical radiative loss, anisotropic radiation field.

A proper treatment of the radiative energy transport is crucial for quantitative modelling of the thermodynamics of magnetic elements since radiation largely determines the energy budget. Besides models which deal with specific aspects of the radiative energy transport (e.g. the lateral energy exchange by a relaxation time appoach) there is a number of investigations which attempt to include radiation more comprehensively. The comparatively simplest approach is to take a slender flux tube which is assumed to be optically thin, i.e. has a diameter smaller than the photon mean free path. In this case the mean intensity is mainly determined by the external medium and the temperature is nearly constant in horizontal planes (Ferrari et al., 1985; Kalkofen et al., 1986). It turns out that the temperature as function of *optical* depth for such models is 1000-2000 K higher than that given by semi-empirical models derived from Stokes V-profiles (Solanki, 1986; Keller, these proceedings). Consequently, optically thin flux tubes are not adequate to describe solar magnetic elements.

Other simplified approaches are the Eddington approximation for the case of a slender flux tube which has been used by Hasan (1988) and the 2D diffusion approximation (Spruit, 1976, 1977; Deinzer et al., 1984a,b). However, the anisotropy of the radiation field and the necessity to describe optically thick and optically thin regions equally well demands a full treatment of the radiative transfer by integration of the transport equation along many rays and angles. For the slab geometry and a grey atmosphere this has been incorporated in the time-dependent simulations of Grossmann-Doerth et al. (1989a) while Steiner (these proceedings) has included a non-grey radiative transport in magnetostatic models of cylindrical flux tubes. In a prescribed slab geometry resembling a flux concentration, Kalkofen et al. (1989) have determined a grey radiative equilibrium atmosphere which has qualitatively similar properties as the self-consistently determined models of Grossmann-Doerth et al. and Steiner.

The model calculations which incorporate a full radiative transfer revealed an important effect, i.e. the heating of the upper layers within the flux concentration by *radiative illumination:* The material in the region above optical depth unity of a partially evacuated magnetic element is bathed in the radiation from the hot bottom (with a temperature of more than 7000 K at $\tau_c = 1$). It therefore reaches an equilibrium temperature which is larger than that of the gas at the same height in the quiet atmosphere which "sees" radiation from a layer of optical depth unity at a temperature of about 6400 K. The result is the formation of a hot region with a temperature which is a few hundred degrees larger than that of the environment at equal *geometrical* depth. As an example, Fig. 1 shows the temperature distribution for the 2D slab model of Grossmann-Doerth et al. (1989a).

In the layers below $\tau_c = 1$ the suppression of convective energy transport leads to a temperature deficit in the flux concentration (with respect to the nonmagnetic gas at the same depth far away from the flux concentration) which reaches a maximum of nearly 3000 K. The resulting horizontal temperature gradient drains energy from the external medium via a lateral radiative heat flux directed into the magnetic element. This inflow of heat balances the vertical radiative losses and limits the temperature deficit to a level which still leads to a temperature excess compared to the quiet Sun at equal *optical* depth for vertical incidence of the line of sight. These results are in qualitative agreement with semiempirical models determined from Stokes V-profiles (Solanki, 1986; Keller, these proceedings, see his Fig. 2) without the necessity to incorporate any mechanical heating. Still, potentially important effects have to be included in the comprehensive model calculations, namely a more detailed treatment of spectral lines, particularly those of CO which may have an influence on the temperature profiles in the layers above $\tau_c \approx 10^{-4}$ (see Ayres, these proceedings; Hasan and Kneer, 1986; Massaglia et al., 1988).

The lateral influx of heat by radiation, together with the partial evacuation of the flux concentration and the strong temperature dependence of the opacity has the consequence that the magnetic elements become brighter than the quiet atmosphere when observed in the continuum with high spatial resolution. At a wavelength of 5000 Å, the slab model of

Grossmann-Doerth et al. (1989a) gives a value of about 1.6 of the intensity of the quiet Sun. Most of this excess intensity, however, is due to a *redistribution* of the heat flux: Energy has flown laterally into the flux concentration leading to a cooling of the exterior and the appearance of a darker region around the magnetic element. Only a small part of the heat flux disturbance propagates into the deeper layers where it is spread rapidly over the whole convection zone. In this way, the magnetic elements can act as a "heat leak" (Spruit, 1977, 1982). The amount of the net excess heat flux is difficult to predict precisely with presently available model calculations since it depends on the treatment of the convective energy transport and also on boundary conditions. However, both observation (Foukal and Fowler, 1984; Hirayama et al., 1985) and simulations (Deinzer et al., 1984b; Knölker et al., 1988) indicate that the net excess flux is likely to be small such that most of the large flux excess within the magnetic elements is compensated by a deficit in the surrounding non-magnetic atmosphere. Even though the excess flux is small for an individual magnetic structure, magnetic elements clustered in plage and network regions may well contribute significantly to the observed solar irradiance variations (cf. Foukal and Lean, 1987).

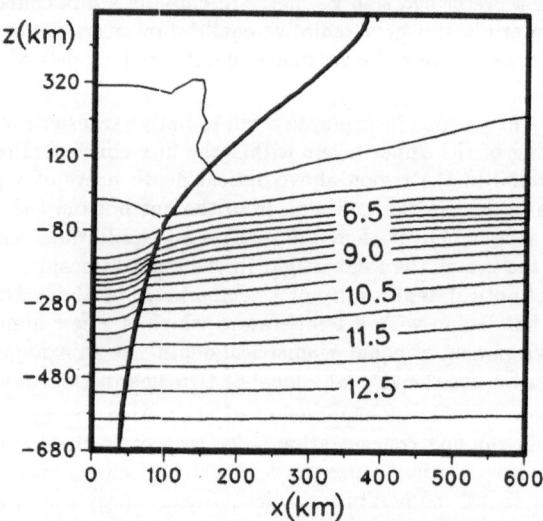

Fig. 1: Temperature distribution in the 2D slab model of Grossmann-Doerth et al. (1989a). The labels at the contour lines are given in units of 1000 K. Only half of the symmetric structure is displayed and the boundary of the flux slab is indicated by the thick line. In the deeper layers the magnetic structure is cooler than the environment at the same height due to the suppression of convective energy transport while in the upper layers it is hotter than the surroundings because of radiative illumination from the hot bottom.

The center-to-limb variation of the intensity contrast has been discussed in some detail in an earlier review (Schüssler, 1987a; for a different point of view see Schatten et al., 1986). Three effects contribute to the variation of the intensity contrast for finite inclination of the line of sight:

- The hot bottom of the magnetic element is obscured for rather small inclinations leading to the disappearance of bright points (see Müller and Roudier, 1984; Müller et al., 1989);

- the *bright wall* of the flux concentration becomes visible at large inclination angles and leads to a positive intensity contrast depending on the ratio between size and depth (Wilson depression) of the magnetic structure (Spruit, 1976);
- the hot upper regions of the magnetic elements overlap near the limb and lead to a sharp contrast increase (Steiner, these proceedings; see also Rogerson, 1961).

These effects are sufficient in order to understand the various observational results if spatial resolution, selection effects and the precise way of measurement are taken into account. High resolution observations near disc centre are mainly determined by the hot bottom of the magnetic elements, while the selection of individual bright "facular granules" (e.g. Muller, 1975) reveals the effect of the bright wall. The brightness evolution of individual faculae near the limb (Akimov et al., 1987), the observations of elevated faculae during eclipses (Akimov et al., 1982) and the measurements at the extreme limb (Chapman and Klabunde, 1982; Lawrence and Chapman, 1988) can be understood by the effect of overlapping hot regions.

3.3 DYNAMICS

The upper limit of 250 m·s^{-1} for the average velocities within magnetic elements determined from the observed V-profile zero-crossing shifts excludes strong systematic flows and large-amplitude overstable oscillations with intensity-velocity correlation. On the other hand, the large *width* of the V-profiles indicates the existence of "turbulence" with velocities of a few km·s^{-1} within magnetic elements. The nature of this velocity field is presently unknown but it is tempting to speculate that the various wave modes of a flux tube are excited by the interaction with convective flows and p-mode oscillations (Bogdan and Knölker, 1989; see also Roberts and Solanki, these proceedings) Flux tube oscillations and waves are discussed in more deatil by M. Ryutova elsewhere in these proceedings.

Another important indicator for dynamics associated with magnetic elements is the area and amplitude *asymmetry* of the observed Stokes V-profiles (Stenflo et al., 1984). Apart from atomic orientation (Kemp et al., 1984; Landi Degl'Innocenti, 1985) which seems inconsistent with the observed sign reversal of the asymmetry near the solar limb (Stenflo et al., 1987; Pantellini et al., 1988), a combination of magnetic field and velocity gradients along the line of sight appears to be the only reasonable explanation (Illing et al., 1975; Auer and Heasley, 1978; Sanchez-Almeida et al., 1988). However, flows *within* the magnetic structure in a physically realistic configuration (e.g. magnetic field decreasing with height) which reproduce the observed asymmetries lead to large shifts of the V-profile zero crossings which contradict the observations (Solanki and Pahlke, 1988).

Van Ballegooijen (1985) has suggested that an area and amplitude asymmetry of the V-profile may also be caused by a downflow *outside* but in the immediate vicinity of a static magnetic flux concentration: Since the magnetic field flares out with height, lines of sight at the periphery traverse static magnetic (upper part of the atmosphere) and downflowing non-magnetic (lower part of the atmosphere) regions. It has been shown by Grossmann-Doerth et al. (1988, 1989b) that quite generally such a configuration leads to asymmetric V-profiles with *unshifted* zero crossings. Solanki (1989) was able to demonstrate that the observed V-profile area asymmetries of many spectral lines can be quantitatively reproduced in this way. Furthermore, since the downflows are fed by horizontal flows directed towards the flux concentration this model at the same time accounts in a natural way for the sign reversal of the asymmetry shown by observations near the solar limb (Grossmann-Doerth et al., in preparation). Fig. 2 illustrates the geometry of the magnetic element and the surrounding flow field which leads to the formation of asymmetric V-profiles.

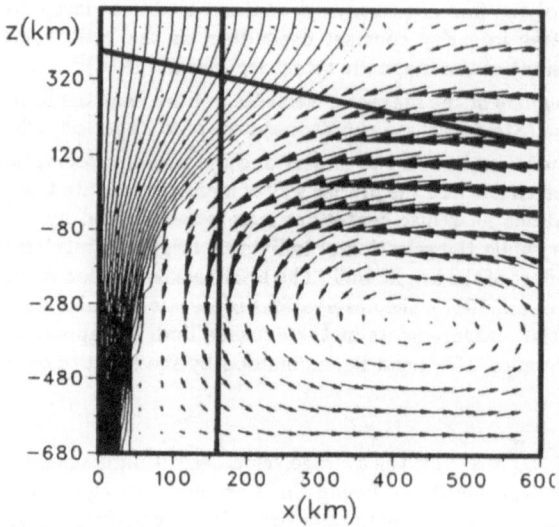

Fig. 2: Velocity and magnetic field structure of the slab model of Grossmann-Doerth et al. (1989a). Only half of the symmetric structure is displayed. While the interior of the magnetic element is almost static, a thermal circulation cell with a strong downflow (maximum velocity about 1.5 km·s^{-1}) and large horizontal velocities (up to 2 km·s^{-1}) has evolved in the non-magnetic environment. Two representative lines of sight are indicated which traverse static, magnetic regions and non-magnetic moving gas leading to asymmetric Stokes V-profiles. The vertical line "sees" a flow away from the observer while the inclined ray cuts a flow towards the observer such that the resulting asymmetries have different signs.

It seems as if the long-standing "V-profile dilemma" has found its resolution in the physically appealing concept of a magnetic element surrounded by a strong downflow in a cool environment. Photospheric magnetic flux is observed to be situated predominantly in the intergranular downflow regions (Title et al., 1987) which are also the site of the network bright points (Müller, 1983). The cooling effect of the magnetic elements on the surrounding gas supports and accelerates such a downflow (Deinzer et al., 1984b). The formerly enigmatic asymmetries thus constitute an important diagnostic tool for the structure of the flow field in the vicinity of magnetic elements. Additionally, the amplitude asymmetries may yield information about internal oscillations and waves (Solanki, 1989; Roberts and Solanki, these proceedings).

4. Destruction

The lifetime of individual magnetic elements is difficult to determine observationally. The simulations carried out by Nordlund (1983, 1986) show a continuous rearrangement of magnetic flux in the integranular lanes with a lifetime of the simulated magnetic structures (clusters of magnetic elements ?) determined by the timescale of the granular velocity field. However, due the low spatial resolution of his grid, the internal structure of the flux concentrations is unresolved and they are artificially coupled to the granular velocity field.

The resolution of the 3D simulations has to be increased by at least an order of magnitude before they can contribute to the solution of this problem.

Muller (1983) found a mean lifetime of network bright points near disk center of about 20 minutes, but it is not clear whether this represents also the life span of the underlying magnetic structure. However, if the observed bright point represents the hot bottom layers of magnetic elements as indicated by the results discussed in Ch. 3 its fading signals a major change in its structure, possibly its dissolution.

A crude estimate of a minimum lifetime of a magnetic element in strong-field form can be derived from the lower limit of 90% for the fraction of magnetic flux (excluding the "turbulent" flux) in strong-field form (Howard and Stenflo, 1972) and the timescale of the convective collapse of 2 to 5 minutes (Hasan, 1985; Nordlund, 1986). Allowing for a quick destruction of the magnetic element by an instability (see below) within one minute we find a minimum lifetime between 30 and 60 min for the quasi-equilibrium state.

Which processes can possibly destruct magnetic elements ? In regions of mixed polarity, *reconnection* is important (Spruit et al., 1987). The result of reconnection of two opposite polarity magnetic elements depends on the location of the reconnection point: If it is *below* the surface an ∪-shaped loop forms which floats upwards due to magnetic buoyancy. It arrives there with a low field strength since the strong decrease of density with height and mass conservation leads to a significant expansion of the rising flux tube. Such a process possibly is a source of intrinsically weak magnetic field and might be related to the "intranetwork fields" (Martin et al., 1985; Livi et al., 1985). If reconnection takes place *above* the surface it forms a loop which can be drawn below the surface due to the action of magnetic tension forces if the footpoint separation is less than a few scale heights (Parker, 1979, Ch. 8). Both possibilities lead to quite different observational signatures (see discussion in Spruit et al., 1989).

An individual flux concentration can be destroyed by dynamical processes, most efficiently by an instability. Besides the destabilizing influence of external flows related to the Kelvin-Helmholtz instability (e.g. Schüssler, 1979; Tsinganos, 1979), the interchange or fluting instability is most important (Parker, 1975). While pores and sunspots can be stabilized by gravity (Meyer et al., 1977), small flux concentrations are stable with respect to fluting if they are surrounded by a strong whirl flow (Schüssler, 1984b). Intermediate size structures cannot be stabilized by either effect which gives an upper limit for the size of magnetic elements depending on the maximum azimuthal velocity in an intergranular vortex. Such a structure is likely to form by advection of angular momentum towards the localized downflows by the familiar "bathtub effect". In fact, simulations of granular convection clearly show the formation of intense vortices (Nordlund, 1985a). On the observationally easier accesible mesogranular scale an example of such a vortical downflow has recently been observed (Brandt et al., 1988). Since the flux concentration cools its surroundings and thus enhances the converging downflows, magnetic structure and flow pattern can mutually stabilize each other: The strong thermal effects shape and stabilize the convective flow structure while this pattern, by means of advection of vorticity, stabilizes the magnetic element if its size is smaller than some critical value (Schüssler, 1984b). The observed deformation of granules around bright points (Muller et al., 1989) and the prolonged lifetime of the granular pattern in plage regions (Title et al., 1987) support this conjecture.

If for some reason, e.g. because of a major reorganisation of the pattern of convection, the supply of angular momentum becomes insufficient and the whirl decelerates, fluting instability sets in and the flux concentration is disrupted typically within the Alfvén transit time of less than a minute (Schüssler, 1986). The following evolution depends on the size of the fragments: If they are small enough (of the order of a few km), magnetic diffusion becomes relevant and the fragments tend to disperse into weak fields which may go through another flux expulsion/convective collapse cycle. Larger fragments may survive long enough to be reassembled by the granular flows and fuse into new flux concentrations without the necessity of another convective collapse.

In this way, a dynamical view of small-scale photospheric magnetic fields emerges. At any given time, most of the flux is in magnetic elements, but the individual elements sooner or later split into fragments. Small fragments diffuse and, together with rising U-loops, contribute to a weak-field component which partially becomes reconcentrated by flux expulsion and convective collapse. Larger fragments (with a diameter > 10 km, say) will rapidly heat up by radiation from the side leading to a decrease of the magnetic field strength. They are passive with respect to flows and may become severly distorted and inclined from the vertical direction before being assembled in integranular downdrafts to form new magnetic elements. All processes of splitting, diffusion, expulsion, collapse and accumulation of fragments operate in a timescale of minutes such that the majority of the flux at any given instant of time resides in the quasi-equilibrium strong-field form of magnetic elements.

5. Conclusions

The progress of our understanding of concentrated magnetic fields in the solar atmosphere achieved since the IAU-Symposium No. 102 in Zürich is considerable. The activity of research in this field has increased rapidly and a close interaction between theoretical and observational work has evolved. Comprehensive model calculations have been presented which reproduce the principal features of magnetic elements derived from observation without fine tuning of a large number of free parameters. These models begin to serve as tools for the diagnostics of solar magnetic structures by providing synthetic profiles of the full Stokes vector (I, V, Q, U) of spectral lines which can be directly compared with observations (e.g. Grossmann-Doerth et al., 1989a). The concept of small-scale fields consisting mainly of ensembles of similar structures (*magnetic elements*) which may be described by basic flux tube or flux slab models embedded in a non-magnetic environment has turned out to be remarkably successful. A consistent picture of magnetic elements begins to emerge from sophisticated analysis of the spatially unresolved FTS data which have unsurpassed spectral quality (in terms of resolution, noise and wavelength range) and the basic physical effects which have been revealed by comprehensive model calculations and analytical studies of idealized problems. Let us try to summarize this picture:

Expulsion of magnetic flux by strong horizontal granular flows leads to magnetic flux concentrations in the intergranular lanes. Radiative cooling and the large superadiabaticity of the uppermost layers of the convection zone cause a strong local intensification of the magnetic field by way of a partial evacuation (convective collapse). A quasi-equilbrium evolves which is characterized by the absence of systematic internal flows and the balance of the magnetic pressure by an internal gas pressure deficit. The magnetic flux density decreases with height in a way well described by the slender flux tube approximation.

This equilibrium is stabilized against further collapse by a temperature deficit of the layers below optical depth unity in the magnetic structure due to the suppression of convective energy transport. Heating by lateral influx of radiation, reduced density and the strong temperature dependence of the continuum opacity cause the magnetic element to be much hotter than the quiet atmosphere at equal *optical* depth and to reach a temperature above 7000 K at $\tau_c = 1$. Therefore, if observed with high spatial resolution near the center of the disc, the magnetic structure appears bright with a continuum intensity of about 1.5 times the value of the average Sun at 5000 Å. The "hot bottom and wall" of magnetic elements illuminate the upper layers of its atmosphere which becomes hotter than the environment even at equal geometrical depth. This contributes to the observed weakening of photospheric spectral lines.

The excess emission of magnetic elements is nearly compensated by an energy flux deficit in its environment such that only a small net excess flux is left. This cooling of the environment caused by lateral radiative energy flux into the magnetic element drives a thermal flow which supports, accelerates and stabilizes the granular downflows next to the magnetic structure. The external flows are responsible for the asymmetry of the observed Stokes V-profiles. Conservation of angular momentum leads to rapid rotation of these downflows which stabilizes the magnetic element with respect to the interchange/fluting instability.

The quasi-equilibrium state seems to be well represented by the FTS spectra. However, spectroscopic observations of individual magnetic structures with large spatial resolution are highly desirable in order to have an independent check of the methods which have been used to interpret the spatially unresolved FTS spectra. The challenging demand for high spatial and temporal resolution is compulsive for observational study of the formation and destruction processes, the dynamical interaction of the magnetic elements with convective flows, vortices and p-mode oscillations, the excitation and propagation of oscillations and waves within magnetic structures, and the interaction with other magnetic elements.

As far as theory is concerned, in contrast to some fashionable folklore existing and forthcoming numerical simulations do not make other approaches obsolete. The dynamics of motions and magnetic fields in the solar convection zone and atmosphere extends over huge ranges of temporal and spatial scales which in both case comprise more than ten decades. Since only a small part of these can be covered by a simulation, artificial boundaries have to be introduced, certain scales are ignored and others are includes only in a parametrized form. Such parametrizations can only be made in a sensible way if they are based on a sound understanding of processes which determine the properties of flows and fields on the scales which they attempt to describe.

The failure of 3D simulations to reproduce the differential rotation of the convection zone and the characteristics of the solar activity cycle has taught an important lesson: Unless we gain a better understanding of the effect of the small scales which then may lead to a reliable parametrization, comprehensive 3D simulations with low spatial resolution are potentially misleading. They are very helpful in drawing the attention to the relevant processes but these then have to be studied in detail in order to assess their general properties and consequences. This can be done by the analytical treatment of idealized problems and by simulations which gain spatial resolution at the expense of simplification, e.g. in dimensionality, of course keeping in mind that these approaches, due to their restrictions, might be misleading as well. So even in the case of solar granulation, a 3D time-dependent phenomenon par excellence, a 2D model (Steffen, these proceedings; Steffen et al., 1989) has apparently captured the basic physical situation (strong, localized, cool downflows and broad upflows, coupled by horizontal flows, advection and radiative transfer) and excellently reproduces the observed spectral and continuum features. Since such a model allows

176

a much higher spatial grid resolution, it is well suited to study the effect of small scales and sharp gradients which may severly compromise 3D simulations. To qualify this work as "wrong" only because it is 2D (as it has been explicitly done in a summary talk at this conference) reveals a striking ignorance. In reality, comprehensive 3D simulations and idealized/simplified approaches are complimentary: The simulations help to identify the relevant processes and allow us a glimpse at phenomena which observationally are hidden behind a curtain of unsufficient spatial resolution or optically thick material. They can guide us in picking the relevant pieces of physics to study in detail without falling into the trap of oversimplified or prejudiced concepts. An understanding of the physics governing these processes, of their general properties and the validity of their description in the simulation can only come from a detailed study in the spirit of theoretical physics.

References

Akimov, L.A., Belkina, I.L., Dyatel, N.P.: 1982, *Sov. Astron.* **26**, 334
Akimov, L.A., Belkina, I.L., Dyatel, N.P., Marchenko, G.P.: 1987, *Sov. Astron.* **31**, 64
Anton, V.: 1989, Dissertation, Universität Göttingen
Auer, L.H., Heasley, J.N.: 1978, *Astron. Astrophys.* **64**, 67
Bogdan, T.J., Knölker, M: 1989, *Astrophys. J.* **339**, 579
Brandt, P., Scharmer, G.B., Ferguson, S., Shine, R.A., Tarbell, T.D., Title, A.M.: 1988, *Nature* **335**, 238
Browning, P.K., Priest, E.R.: 1983, *Astrophys. J.* **266**, 848
Chapman, G.A., Klabunde, D.P.: 1982, *Astrophys. J.* **261**, 387
Degenhardt, D.: 1989, *Astron. Astrophys.*, in press
Deinzer, W., Hensler, G., Schüssler, M., Weisshaar, E.:
    1984a, *Astron. Astrophys.* **139**, 426
Deinzer, W., Hensler, G., Schüssler, M., Weisshaar, E.:
    1984b, *Astron. Astrophys.* **139**, 435
Dunn, R.B., Zirker, J.B.: 1973, *Solar Phys.* **33**, 281
Ferrari, A., Massaglia, S., Kalkofen, W., Rosner, R., Bodo, G.:
    1985, *Astrophys. J.* **298**, 181
Ferriz-Mas, A.: 1988, *Phys. Fluids* **31**, 2583
Ferriz-Mas, A., Moreno-Insertis, F.: 1987, *Astron. Astrophys.* **179**, 268
Ferriz Mas, A., Schüssler, M.: 1989, *Geophys. Astrophys. Fluid Dyn.*, in press
Ferriz Mas, A., Schüssler, M., Anton, V.: 1989, *Astron. Astrophys.* **210**, 425
Foukal, P., Fowler, L.: 1984, *Astrophys. J.* **281**, 442
Foukal, P., Lean, J.: 1988, *Astrophys. J.* **328**, 347
Galloway, D.J., Weiss, N.O.: 1981, *Astrophys. J.* **243**, 945
Grossmann-Doerth, U., Knölker, M., Schüssler, M., Weisshaar, E.: 1989a, in R.J. Rutten
    and G. Severino (eds.): *Solar and Stellar Granulation*, Kluwer, Dordrecht, p. 481
Grossmann-Doerth, U., Schüssler, M., Solanki, S.K.: 1988, *Astron. Astrophys.* **206**, L37
Grossmann-Doerth, U., Schüssler, M., Solanki, S.K.: 1989b, *Astron. Astrophys.*, in press
Hasan, S.S.: 1983, see Stenflo (1983), p. 73
Hasan, S.S.: 1984, *Astrophys. J.* **285**, 851
Hasan, S.S.: 1985, *Astron. Astrophys.* **143**, 39
Hasan, S.S.: 1986, *Mon. Not. Roy. Astron. Soc.* **219**, 357

Hasan, S.S.: 1988, *Astrophys. J.* **332**, 499

Hasan, S.S.: 1989, preprint

Hasan, S.S., Kneer, F.: 1986, *Astron. Astrophys.* **158**, 288

Hasan, S.S., Schüssler, M.: 1985, *Astron. Astrophys.* **151**, 69

Herbold, G., Ulmschneider, P., Spruit, H.C., Rosner, R.:
    1985, *Astron. Astrophys.* **145**, 157

Hirayama, T., Hamana, S., Mizugaki, K.: 1985, *Solar Phys.* **99**, 43

Hollweg, J.V.: 1982, *Astrophys. J.* **257**, 345

Howard, R.W., Stenflo, J.O.: 1972, *Solar Phys.* **22**, 402

Hurlburt, N.E., Toomre, J.: 1988, *Astrophys. J.* **327**, 920

Hurlburt, N.E., Toomre, J., Massaguer, J.M.: 1984, *Astrophys. J.* **282**, 557

Illing, R.M.E., Landman, D.A., Mickey, D.L.: 1975, *Astron. Astrophys.* **41**, 183

Kalkofen, W., Bodo, G., Massaglia, S., Rossi, P.: 1989, in R.J. Rutten
    and G. Severino (eds.): *Solar and Stellar Granulation*, Kluwer, Dordrecht, p. 571

Kalkofen, W., Rosner, R., Ferrari, A., Massaglia, S.: 1986, *Astrophys. J.* **304**, 519

Kemp, J.C., Macek, J.H., Nehring, F.W.: 1984, *Astrophys. J.* **278**, 863

Knölker, M., Schüssler, M.: 1988, *Astron. Astrophys.* **202**, 275

Knölker, M., Schüssler, M., Weisshaar, E.: 1988, *Astron. Astrophys.* **194**, 257

Landi Degl'Innocenti, E.: 1985, in H.U. Schmidt (ed.): *Theoretical Problems in High
    Resolution Solar Physics*, MPA 212, Max-Planck Institut für Astrophysik,
    München, p. 162

Lawrence, J.K., Chapman, G.A.: 1988, *Astrophys. J.* **335**, 996

Livi, S.H.B., Wang, J., Martin, S.F.: 1985, *Australian J. Phys.* **38**, 855

Martin, S.F., Livi, S.H.B., Wang, J.: 1985, *Australian J. Phys.* **38**, 929

Massaglia, S., Bodo, G., Kalkofen, W., Rosner, R.: 1988, *Astrophys. J.* **333**, 925

Massaglia, S., Bodo, G., Rossi, P.: 1989, *Astron. Astrophys.* **209**, 399

Mehltretter, J.P.: 1974, *Solar Phys.* **38**, 43

Meyer, F.: 1976, in R.-M. Bonnet, Ph. Delache (eds.): *The Energy Balance and Hydrody-
    namics of the Solar Chromosphere and Corona*, IAU-Colloq. No. 36, Nice, p. 111

Meyer, F., Schmidt, H.U., Simon, G.W., Weiss, N.O.: 1979, *Astron. Astrophys.* **76**, 35

Meyer, F., Schmidt, H.U., Weiss, N.O.: 1977, *Mon. Not. Roy. Astron. Soc.* **179**, 741

Montesinos, B., Thomas, J.H.: 1989, *Astrophys. J.* **337**, 977

Muller, R.: 1975, *Solar Phys.* **45**, 105

Muller, R.: 1983, *Solar Phys.* **85**, 113

Muller, R., Roudier, Th.: 1984, *Solar Phys.* **94**, 33

Muller, R., Roudier, Th., Hulot, J.C.: 1989, *Solar Phys.* **119**, 229

Nordlund, Å.: 1983, see Stenflo (1983), p. 79

Nordlund, Å.: 1984, in T.D. Guyenne and J.J. Hunt (eds.): *The Hydromagnetics of the
    Sun*, ESA SP-220, p. 37

Nordlund, Å.: 1985a, in H.U. Schmidt (ed.): *Theoretical Problems in High Resolution Solar
    Physics*, MPA 212, Max-Planck Institut für Astrophysik, München, p. 1

Nordlund, Å.: 1985b, in H.U. Schmidt (ed.): *Theoretical Problems in High Resolution Solar
    Physics*, MPA 212, Max-Planck Institut für Astrophysik, München, p. 101

Nordlund, Å.: 1986, in W. Deinzer et al. (eds.): *Small Scale Magnetic Flux Concentrations
    in the Solar Photosphere*, Abh. Akad. d. Wiss. Göttingen No. 38, Vandenhoeck &
    Ruprecht, Göttingen, p. 83

Osherovich, V., Flå, T., Chapman, G.A.: 1983, *Astrophys. J.* **268**, 412

Pantellini, F.G.E., Solanki, S.K., Stenflo, J.O.: 1988, *Astron. Astrophys.* **189**, 263

Parker, E.N.: 1963, *Astrophys. J.* **138**, 552

Parker, E.N.: 1975, *Solar Phys.* **40**, 291

Parker, E.N.: 1978, *Astrophys. J.* **221**, 368

Parker, E.N.: 1979, *Cosmical Magnetic Fields*, Clarendon, Oxford

Parker, E.N.: 1984, *Astrophys. J.* **283**, 343

Pizzo, V.J.: 1986, *Astrophys. J.* **302**, 785

Pneuman, G.W., Solanki, S.K., Stenflo, J.O.: 1986, *Astron. Astrophys.* **154**, 231

Priest, E.R.: 1982, *Solar Magneto-Hydrodynamics*, Reidel, Dordrecht

Proctor, M.R.E., Weiss, N.O.: 1982, *Rep. Progr. Phys.* **45**, 1317

Ribes, E., Rees, D.E., Fang, C.: 1985, *Astrophys. J.* **296**, 268

Roberts, B. : 1976, *Astrophys. J.* **204**, 268

Roberts, B., Webb, A.R.: 1978, *Solar Phys.* **56**, 5

Rogerson, J.B.: 1961, *Astrophys. J.* **134**, 331

Sanchez-Almeida, J., Collados, J., del Toro Iniesta, J.: 1988, *Astron. Astrophys.* **201**, L37

Schatten, K.H., Mayr, H.G., Omidvar, K., Maier, E.: 1986, *Astrophys. J.* **311**, 460

Schmidt, H.U., Simon, G.W., Weiss, N.O.: 1985, *Astron. Astrophys.* **148**, 191

Schmidt, H.U., Wegmann, R.: 1983, in B. Brosowski and E. Martensen (eds.): *Dynamical Problems in Mathematical Physics*, Lang, Frankfurt/Main, p. 137

Schüssler, M.: 1979, *Astron. Astrophys.* **71**, 79

Schüssler, M.: 1984a, in T.D. Guyenne and J.J. Hunt (eds.): *The Hydromagnetics of the Sun*, ESA SP-220, p. 67

Schüssler, M.: 1984b, *Astron. Astrophys.* **140**, 453

Schüssler, M.: 1986, in W. Deinzer et al. (eds.): *Small Scale Magnetic Flux Concentrations in the Solar Photosphere*, Abh. Akad. d. Wiss. Göttingen No. 38, Vandenhoeck & Ruprecht, Göttingen, p. 103

Schüssler, M.: 1987a, in E.-H. Schröter et al. (eds.): *The Role of Fine-Scale Magnetic Fields on the Structure of the Solar Atmosphere*, Cambridge University Press, p. 223

Schüssler, M.: 1987b, in B.R. Durney and S. Sofia (eds.): *The Internal Solar Angular Velocity*, Reidel, Dordrecht, p. 303

Schüssler, M., Solanki, S.K.: 1988, *Astron. Astrophys.* **192**, 338

Simon, G.W., Weiss, N.O., Nye, A.: 1983, *Solar Phys.* **87**, 65

Solanki, S.K.: 1986, *Astron. Astrophys.* **168**, 311

Solanki, S.K.: 1989, *Astron. Astrophys.*, in press

Solanki, S.K., Pahlke, K.-D.: 1988, *Astron. Astrophys.* **201**, 143

Solanki, S.K., Stenflo, J.O.: 1984, *Astron. Astrophys.* **140**, 185

Solanki, S.K., Stenflo, J.O.: 1986, *Astron. Astrophys.* **170**, 120

Solov'ev, A.A.: 1984, *Sov. Astron.*, **28**, 54

Spruit, H.C.: 1976, *Solar Phys.* **50**, 269

Spruit, H.C.: 1977, *Solar Phys.* **55**, 3

Spruit, H.C.: 1979, *Solar Phys.* **61**, 363

Spruit, H.C.: 1982, *Astron. Astrophys.* **108**, 348

Spruit, H.C.: 1983, see Stenflo (1983), p. 41

Spruit, H.C., Roberts, B.: 1983, *Nature* **304**, 401

Spruit, H.C., Schüssler, M., Solanki, S.K.: 1989, in A.N. Cox and W.C. Livingston (eds.): *The Solar Interior and Atmosphere*, University of Arizona Press, Tucson, in press

Spruit, H.C., van Ballegooijen, A.A., Title, A.M.: 1987, *Solar Phys.* **110**, 115

Spruit, H.C., Zweibel, E.G.: 1979, *Solar Phys.* **62**, 15

Steffen, M., Ludwig, H.-G., Krüß, A.: 1989, *Astron. Astrophys.* **213**, 371

Steiner, O., Pizzo, V.J.: 1989, *Astron. Astrophys.* **211**, 447

Steiner, O., Pneuman, G.W., Stenflo, J.O.: 1986, *Astron. Astrophys.* **170**, 126

Stenflo, J.O. (ed.): 1983, *Solar and Stellar Magnetic Fields: Origins and Coronal Effects*, IAU-Symp. No. 102, Reidel, Dordrecht

Stenflo, J.O.: 1987, *Solar Phys.* **114**, 1

Stenflo, J.O.: 1989, *Astron. Astrophys. Rev.* **1**, 3

Stenflo, J.O., Harvey, J.W.: 1985, *Solar Phys.* **95**, 99

Stenflo, J.O., Solanki, S.K., Harvey, J.W.: 1987, *Astron. Astrophys.* **171**, 305

Sterling, A.C., Hollweg, J.V.: 1988, *Astrophys. J.* **327**, 950

Suematsu, Y., Shibata, K., Nishikawa, T., Kitai, R.: 1982, *Solar Phys.* **75**, 99

Thomas, J.H.: 1985, in H.U. Schmidt (ed.): *Theoretical Problems in High Resolution Solar Physics*, MPA 212, Max-Planck Institut für Astrophysik, München, p. 126

Thomas, J.H.: 1988, *Astrophys. J.* **337**, 407

Title, A.M., Tarbell, T.D., Topka, K.P.: 1987, *Astrophys. J.* **317**, 892

Title, A.M., Tarbell, T.D., Topka, K.P., Cauffman, D., Balke, C., Scharmer, G.: 1989, preprint

Tsinganos, K.C.: 1980, *Astrophys. J.* **239**, 746

Unno, W., Ando, H.: 1979, *Geophys. Astrophys. Fluid Dyn.* **12**, 107

Unno, W., Ribes, E.: 1979, *Astron. Astrophys.* **73**, 314

Van Ballegooijen, A.A.: 1985, discussion remark in H.U. Schmidt (ed.): *Theoretical Problems in High Resolution Solar Physics*, MPA 212, Max-Planck Institut für Astrophysik, München, p. 177

Venkatakrishnan, P.: 1983, *J. Astrophys. Astron.* **4**, 135

Venkatakrishnan, P.: 1985, *J. Astrophys. Astron.* **6**, 21

Venkatakrishnan, P.: 1986, *Nature* **322**, 156

Webb, A.R., Roberts, B.: 1978, *Solar Phys.* **59**, 249

Weiss, N.O.: 1966, *Proc. Roy. Soc.* **A293**, 310

Weiss, N.O.: 1977, in E.A. Müller (ed.), *Highlights of Astronomy* **4**, 241

Weiss, N.O.: 1981a, *J. Fluid Mech.* **108**, 247

Weiss, N.O.: 1981b, *J. Fluid Mech.* **108**, 273

Wiehr, E.: 1985, *Astron. Astrophys.* **149**, 217

Wilson, P.R.: 1977, *Astrophys. J.* **214**, 611

Zayer, I., Solanki, S.K., Stenflo, J.O.: 1989, *Astron. Astrophys.* **211**, 463

Zwaan, C.: 1978, *Solar Phys.* **60**, 213

MODEL CALCULATIONS OF THE PHOTOSPHERIC LAYERS OF SOLAR MAGNETIC FLUXTUBES

OSKAR STEINER and J.O. STENFLO
Institut für Astronomie
ETH-Zentrum
8092 Zürich
Switzerland

ABSTRACT. Multi-dimensional radiative energy transport is coupled self-consistently to magnetohydrostatic solutions for fluxtubes with rotational symmetry. It is shown that the photospheric layers of plage and network fluxtubes are heated by radiation by as much as 300 K at equal geometrical height. The amount of heating depends on the density reduction within the tube. The results are compared with observational data and the most recent semi-empirical model.

1. Magnetohydrostatics and Radiative Transport

Magnetohydrostatic solutions of rotationally symmetric, vertically oriented magnetic fluxtubes in the solar photosphere are considered. The models should account for the basic physical processes of the tiny photospheric structures, which appear as bright points in the continuous spectrum of plages and network regions, and which are observed to have strong magnetic fields in the 1–2 kG range (Stenflo, 1973, 1989).

The magnetohydrostatic equation,

$$-\nabla p + \rho \mathbf{g} + \frac{1}{4\pi}(\nabla \times \mathbf{B}) \times \mathbf{B} = 0 \quad , \tag{1}$$

is solved using a finite difference scheme, allowing for fluxtubes having a sharp boundary (current sheet) where the magnetic field strength jumps from a finite value to zero. The method for solving (1) has been described in great detail in Steiner et al. (1986), who also provide an accuracy test.

As a next step, to obtain more realistic models for plage and network magnetic fluxtubes and to reduce the number of free parameters, we include an energy equation, which allows us to derive the temperature structures of such regions. We concentrate on the photospheric layers, since this is the height region most accessible to Stokes polarimetry. In this region radiative transfer is the dominant mode of energy transport. The radiative transfer equation is solved using cylinder coordinates under the constraint of radiative equilibrium:

$$(\mathbf{n} \cdot \nabla) \, \mathrm{I}(\mathbf{r}, \mathbf{n}, \nu) = \eta(\mathbf{r}, \nu) - \chi(\mathbf{r}, \nu) \, \mathrm{I}(\mathbf{r}, \mathbf{n}, \nu) \quad , \tag{2}$$

$$\int_0^\infty \chi(\mathbf{r}, \nu) \mathrm{S}(\mathbf{r}, \nu) \, \mathrm{d}\nu = \int_0^\infty \chi(\mathbf{r}, \nu) \, \mathrm{J}(\mathbf{r}, \nu) \, \mathrm{d}\nu \quad . \tag{3}$$

LTE is assumed with allowance for continuum scattering. The continuum opacities, scattering coefficients, and the mean molecular weight are computed using the statistical equilibrium and

J. O. Stenflo (ed.), Solar Photosphere: Structure, Convection, and Magnetic Fields, 181–184.

182

opacity code of Gustafsson (1973). Line blanketing is taken into account using opacity distribution functions (ODF) of Kurucz (1979). Couples of ODF's have been merged into one to save computer CPU-time. To avoid the inversion of the complete Λ-matrix the formal solution of the transfer equation

$$J_\nu(\mathbf{r}) = \Lambda_\nu(\mathbf{r}, \mathbf{r}')B_\nu(\mathbf{r}') \tag{4}$$

is only computed, using an accelerated λ-iteration to obtain iterative temperature corrections. This procedure is based on an operator perturbation technique as will be described in Steiner (1989). The method provides fast convergence also in high opacity atmospheres. This is greatly needed since radiative transfer within the tube may still be important even far below the $\tau = 1$ level, due to the inhibition of convective energy transport by the magnetic field.

The formal solution of the radiative transfer equation in cylinder coordinates has been computed using a modified version of the program CYL2D written by P. Kunasz, which is based on the short characteristic method of Kunasz and Auer (1988). This code reduces the 3D-transfer problem to an integration in several 2D-planes by exploiting the rotational symmetry.

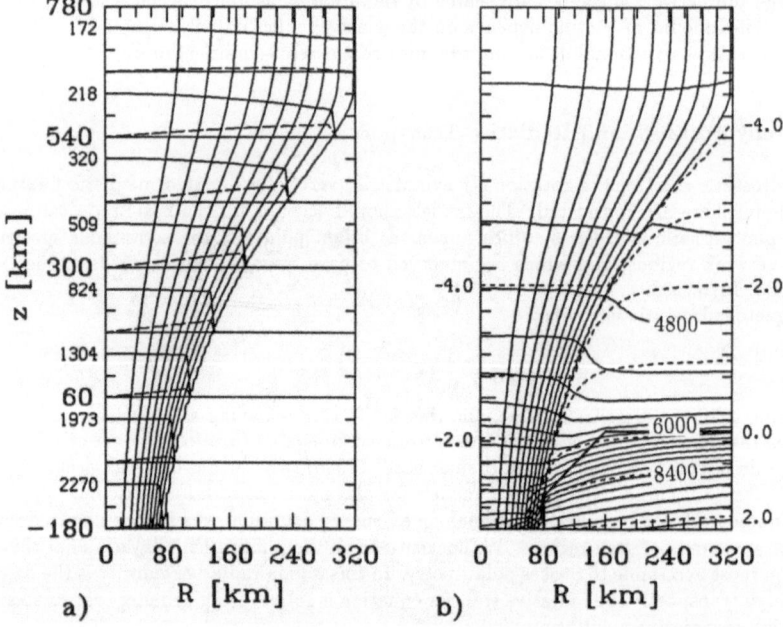

Figure 1. Model magnetic fluxtube representative of examples considered in the text. a) Field lines delineating the fluxtube which is separated from the surrounding field free plasma by a thin current sheet. Superimposed on the figure are horizontal curves showing the radial variation of B_z (solid lines) and B_r (dashed lines), each normalized to B_z at the axis, the value of which in Gauss is indicated to the left of each curve. b) Same fluxtube as in a) with contour lines of constant temperature (solid lines) and constant optical depth (dashed lines). log τ_{5000} is indicated to the left and right of the figure. The region below the heavy solid line is convectively unstable and its temperature has been prescribed (see text).

2. Magnetic Field and Temperature Structure

Fig. 1a shows representative magnetic field lines of a fluxtube with a radius of 100 km and a field strength of about 1600 G, at $z = 0$ (which refers to the height z at which $\tau_{5000} = 1$ in the undisturbed atmosphere [1]). Included in the figure are curves giving the field strength as a function of radius at several height levels. The field of the z component decreases from the fluxtube axis to the fluxtube boundary as a consequence of magnetic tension forces (Steiner and Pizzo, 1989). Because of the decreasing gas pressure the fluxtubes expand strongly with height and finally merge with neighbouring tubes into a uniform vertical weak field. The effect of the neighbouring fluxtubes is accounted for in the model by the choice of the boundary conditions for Equation (1).

Fig. 1b shows the temperature structure of the fluxtube of Fig. 1a together with iso-τ_{5000} lines. For the radiative transfer, Eq. (4), periodic boundary conditions have been used to take the influence of neighbouring fluxtubes into account. Since convective energy transport is not treated in the present model the temperature structure in the corresponding region (area below the heavy solid line in Fig. 1b) is prescribed, using the values of Grossmann-Doerth et al. (1988). The fluxtube interior is much less opaque than its surroundings because of its reduced density. It can be imagined as a hole in the solar surface through which radiation escapes more easily, causing an elevated temperature in the overlying photospheric layers.

Fig. 2a shows the variation of temperature with optical depth along the axis of fluxtubes having field strengths of 1600 (dot-dashed line) and 1500 Gauss (dashed line), as well as along the symmetry axis between the fluxtubes (solid line), which is very close to the temperature of the undisturbed atmosphere. The dotted line represents the semi-empirical model of Keller (1989). Note that the characteristic temperature depression around $\log \tau_{5000} = -2$ present in the semi-empirical model can not be explained by radiative effects. Fig. 2b shows for the same models the temperature as a function of geometrical height z.

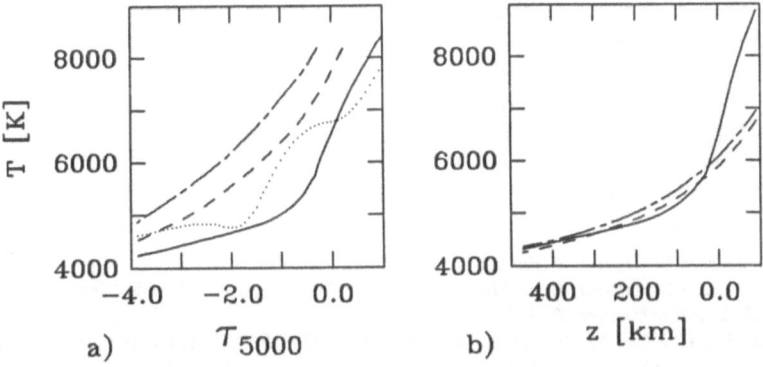

Figure 2. a) Temperature with optical depth along the symmetry axis between two neighbouring fluxtubes (solid line) and along the fluxtube axis for a tube with $B_z(z = 0) = 1600$ G (dot-dashed) and 1500 G (dashed line). The dotted line refers to the semi-empirical model of Keller (1989). b) Temperature curves as functions of geometrical height z.

---

[1] With 'undisturbed atmosphere' we mean the model atmosphere obtained with the same code as used for the fluxtube calculations but without magnetic fields. The resulting undisturbed or quiet atmosphere is very close to the solar model of Kurucz (1979).

184

3. Center to Limb Continuum Contrast

The solid line in Fig. 3 shows the calculated continuum contrast γ ($\lambda = 5000\,\text{Å}$) for values of $\mu = \cos\theta$ close to the solar limb and for a model with $B(z = 0) = 1500\text{G}$, $R(z = 0) = 100$ km, and a filling factor $f = 0.175$. The calculated contrast is an average over a region of about 1×1 arcseconds, simulating moderate resolution. The dot-dashed and the dashed lines are moderate resolution observations taken from Badaljan (1968) and Akimov et al. (1987), respectively. Three characteristics should be noted. Firstly, the maximum continuum contrast of the model occurs around $\mu = 0.2$ in agreement with observations. Secondly, the actual value of γ depends strongly on the filling factor. This can be readily understood if we imagine that with increasing f more and more fluxtubes appear in a given resolution element making it more or less uniformly bright. It is clear that the particular density and temperature structure of the fluxtube also plays a role. Thirdly, we draw attention to the sharp increase of γ very close to the limb, which is exclusively due to the radiative heating of the tube interior above $z \approx 0$. Increasing continuum contrasts at the extreme limb ($\mu \lesssim 0.1$) have been reported by several authors, for example by Akimov et al. (1987).

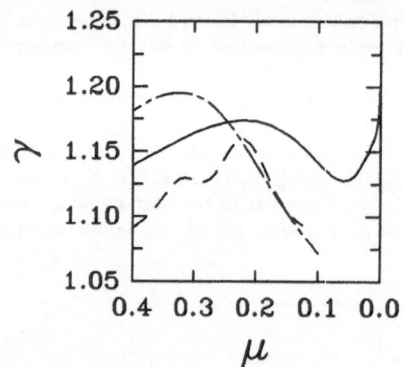

Figure 3. Center to limb variation of the continuum contrast ($\lambda = 5000\,\text{Å}$) derived from a model similar to that shown in Fig. 1 along with observed (dashed and dot-dashed lines) CLV contrasts.

ACKNOWLEDGEMENT. The authors are grateful to Dr. P. Kunasz for letting us use his yet unpublished radiative transfer code CYL2D.

References

Akimov, L.A., Belkina, I.L., Dyatel, N.P.: 1982, *Astron. Zh.* **59**, 552

Akimov, L.A., Belkina, I.L., Dyatel, N.P., Marchenko, G.P.: 1987, *Astron. Zh.* **64**, 126

Badaljan O.G.: 1968, *Soln. Dannye.* **5**, 105

Grossman-Doerth, U., Knölker, M., Schüssler, M., Weisshaar, E.: 1988, in *Solar and Stellar Granulation*, Proc. of the OAC–NATO Advanced Research Workshop, Capri, Italy, 1988

Gustafsson, B.: 1973, *Uppsala Astron. Obs. Ann.* **5**, No. 6

Keller, C.U.: 1989, *IAU Symp.* **138**

Kunasz, P.B., Auer, L.H.: 1988, *J. Quant. Spectrosc. Radiat. Transfer.* **39**, 67

Kurucz, R.L.: 1979, *Astrophys. J. Suppl. Ser.* **40**, 1

Steiner, O.: 1989, *A rapidly convergent temperature correction procedure using operator perturbation*, to be submitted

Steiner, O., Pizzo, V.J.: 1989, *Astron. Astrophys.* **211**, 447

Steiner, O., Pneuman, G.W., Stenflo, J.O.: 1986, *Astron. Astrophys.* **170**, 126

Stenflo, J.O.: 1973, *Solar Phys.* **32**, 41

Stenflo, J.O.: 1989, *The Astron. Astrophys. Rev.* **1**, 3

WAVE HEATING IN MAGNETIC FLUX TUBES

Wolfgang Kalkofen
Harvard-Smithsonian Center for Astrophysics
60 Garden Street, Cambridge, Massachussetts 02138

ABSTRACT: The solar chromosphere is identified with the atmosphere inside magnetic flux tubes. In the quiet sun, the layers of the low and middle chromosphere are heated by compressive waves with periods mainly between 2 min and 4 min. These long-period waves probably supply all the energy required for the heating of the quiet solar chromosphere.

1. INTRODUCTION

It is commonly assumed that if the chromosphere is heated by acoustic waves, as suggested by Biermann (1946) and Schwarzschild (1948), these waves must have periods shorter than the acoustic cutoff period. Consequently, practically all the work on chromospheric heating has been concerned with the so-called short-period waves (for a recent review, cf. Narain & Ulmschneider 1989). However, there is no observational evidence showing shock heating by such waves. This led Cram (1987) to claim that there is no evidence from ground-based or space-based facilities for *any* shock heating in the solar atmosphere. In contrast to this pessimistic view of acoustic heating of the solar chromosphere this paper asserts that the solar chromosphere below the 7000°K level is heated by compressive waves forming shocks and that their periods are near the acoustic cutoff period.

2. CHROMOSPHERIC MODEL AND OBSERVATIONS

There is a close correspondence between magnetic filaments and regions that are bright in the resonance lines of Ca^+ and the core of $H\alpha$ (Athay 1976; cf. also Foing & Bonnet 1984), implying dissipation of mechanical energy in regions of strong vertical magnetic field. This correlation between magnetic field and excess emission extends throughout the chromosphere and into the transition region between chromosphere and corona (Athay). And it is twofold: All regions of increased vertical magnetic field are bright in the cores of $H\alpha$ and the K line, and all regions that are bright correspond to peaks of vertical magnetic field strength (Zirin 1988). This one-to-one correspondence suggests that the chromosphere exists only in association with the magnetic field.

The small size of bright cell points (also called grains) and the pronounced tendency of brightenings to recur in the same location requires an organization of the waves that heat the medium by an underlying structure, most likely the small-scale, intense magnetic field.

The observations of chromospheric heating thus suggest a two-phase model of the outer solar atmosphere, consisting of gas inside magnetic flux tubes that is heated mechanically and of gas in the intertube medium that may receive little mechanical energy

J. O. Stenflo (ed.), Solar Photosphere: Structure, Convection, and Magnetic Fields, 185–188.

(Ayres 1981, Ayres & Testerman 1981, Ayres *et al.* 1986, Kalkofen 1989).

Observations of chromospheric oscillations show that waves with typically 3 min period dissipate energy in the chromosphere. The wave periods seen in the K line by Liu (1974) show periods mainly between 2 min and 4 min, with most periods longward of the acoustic cutoff period of about 3 min. These waves have been observed: (1) in Hα by Orrall (1966) and Bhatnagar & Tanaka (1972); (2) in the K line by Liu (1973, 1974), Cram (1974), and Damé (1984); (3) in the H line by Cram & Damé (1983) and Damé (1984); and (4) in the continuum by Yudin (1968) at 3.3 cm, by Simon & Shimabukuro (1971) at 3.3 mm and 3.5 mm, and by Lindsey & Roellig (1987) at 350 μm and 800 μm.

In the K line (Liu 1974) and in the H line (Cram & Damé 1983) the waves produce initially a symmetrical intensity enhancement in the far line wings, which arise deep in the photosphere. This feature propagates towards the line center, *i.e.*, outward in the atmosphere. When it reaches the line core, the profile becomes highly asymmetric: at the frequencies of the emission peaks K_2 and H_2, which are formed in the low chromosphere (*cf.* Fig. 1 of Vernazza *et al.* 1981, henceforth VAL81), the waves produce enhancements of only the blue peaks, K_{2v} and H_{2v}, typically by a factor of 3 at K_{2v}.

The size of the area that oscillates is larger than the size of the heated region. At the height of formation of the K_{2v} peak, the oscillations due to a wave extend over an area of the order of 2000 km to 4000 km, whereas the heated area is only 1500 km or less in diameter. This may imply that a typical flux tube diameter at the height of formation of the K_2 feature is of the order of 1500 km and that the atmosphere outside a tube oscillates in phase with the inside gas (*cf.* Defouw 1976).

3. WAVE ENERGY

Liu (1974) estimated the wave energy flux from the rms velocity amplitude and the phase speed of the waves, *i.e.*, the speed with which intensity enhancements in the K line propagate through the atmosphere, using an empirical model in order to relate the position of an intensity increase in the line profile to the position of the wave front in the atmosphere. Then, assuming that the waves were acoustic he calculated the group velocity and thus the energy flux. This flux, which may be uncertain by a large margin, falls short of the flux estimated by VAL81 (*cf.* their Table 29) by only a factor of about 2. Therefore, the estimate of the energy flux allows the conclusion that the waves observed in the K line may carry all the energy needed for the heating of the low and middle chromosphere.

In order to judge whether the waves observed in the H line carry enough energy to meet the needs of the chromosphere, Cram & Damé (1983) compared the average H line emission in their observations with the H line emission calculated from the empirical model of the average chromosphere (model C of Vernazza *et al.* 1981, henceforth VAL-C). They determined the observed emitted energy from the difference between the average profile of all their observations and the average profile of the lowest decile. Since the latter shows some evidence of heating, their estimate is less than the total dissipated mechanical energy. Their estimate gave 80% of the emission of the empirical model. Thus, one may conclude again that the heating of the layers in which the H line arises is consistent with heating by only these long-period waves.

Anderson & Athay (1989) determined the dissipation rate of the chromospheric heating mechanism from the empirical temperature structure of the model VAL-C using a model atmosphere code. In addition to the familiar radiation terms present in radiative equilibrium the energy equation included a source term representing mechanical energy input. They matched the temperature curve of the empirical model in the layers of the temperature plateau below about 7000°K with a constant flux divergence per unit mass of $dF/\rho dx = 4.5 \times 10^9$ erg g^{-1} s^{-1}. In order to deduce the nature of the heating mechanism they assumed dissipation by acoustic waves, which implied a velocity amplitude of about half the sound speed, a result that is consistent with heating by shock waves. As further support for their hypothesis of acoustic wave heating they could point to the value of the microturbulent velocity in the same layers, which is approximately equal to the sound speed in these layers. Thus, the energy input into the chromosphere in the region of the temperature plateau below about 7000°K is consistent with the assumption of heating by compressional waves alone.

Weak-shock theory (*cf.* Ulmschneider 1970, Bray & Loughhead 1974) permits an estimate of the wave period and the Mach number of the waves for shocks that have reached their limiting strength, where the growth of the wave amplitude due to the outward propagation in a gravitationally stratified gas is balanced by the decay of the amplitude due to dissipation. The theory is not directly applicable to the tube waves that heat the chromosphere since they are not plane and the shocks are not weak. But using the theory for a rough estimate of wave properties, the dissipation rate found by Anderson & Athay implies that the shocks are fairly strong, with a Mach number of nearly 2, and that the wave period is about 1 min, not much different from the observed periods. These results thus are broadly consistent with heating of the chromosphere in the layers of the temperature plateau by compressional waves with periods near the acoustic cutoff period.

4. WAVE PERIODS

Acoustic waves in a gravitationally stratified medium propagate in the vertical direction only if their period is shorter than the acoustic cutoff period (*cf.* Schatzman & Souffrin 1967), approximately 3 min in the upper solar atmosphere. In deriving this result from the linearized equations of hydrodynamics in a gravitational field (but without a magnetic field) it is assumed that a wave phenomenon may be described in terms of a largely inert background medium and a disturbance. Thus, the density and pressure perturbations are first-order quantities, and the energy flux of the wave is a second-order quantity.

With the result by Anderson & Athay for the flux divergence per gram in the region of the chromospheric temperature plateau we can check whether the basic assumptions on which the cutoff period rests are valid for the waves that heat the solar chromosphere. We find that each pulse of a 200 s wave dissipates an energy of about 1 eV per particle. This exceeds the thermal energy per gram of the chromospheric gas, which is approximately 1/2 eV per particle. Thus, for these waves, the wave energy is a zeroth-order quantity in the hydrodynamic equations, and hence the linearization is not allowed and the cutoff condition does not apply. There is therefore no *a priori* reason why the chromosphere should not be heated by waves with periods near the acoustic cutoff period.

5. SUMMARY

The chromosphere is identified with the atmosphere inside magnetic flux tubes. In the quiet sun below the 7000°K level, *i.e.,* the layers traditionally referred to as the low and middle chromosphere, the gas is heated by compressional waves with periods mainly between 2 min and 4 min. The heating mechanism in the high chromosphere, where hydrogen is completely ionized, may be different from that in the lower layers.

The energy dissipated by each wave pulse is comparable to the thermal energy of the chromospheric gas. Hence the cutoff condition that limits propagation of low-amplitude disturbances to waves with periods shorter than the cutoff period does not apply to these large-amplitude waves.

Estimates of the energy supplied to the low and middle chromosphere suggest that the 3 min waves can supply all the energy that is needed to heat the chromosphere to the observed temperatures. This conclusion is based on (1) the estimated energy flux in the compressional waves, which matches the observed chromospheric emission of the model VAL-C within a factor of about two, (2) the close agreement between the excess energy emitted by the H line in response to the dissipation by the observed compressional waves and the H line cooling rate in the empirical model VAL-C, (3) on the estimate of the mechanical dissipation rate per unit mass implied by the empirical temperature curve, which is consistent with heating by acoustic waves, and (4) on the fact that weak-shock theory can approximately account for the properties of the observed compressional waves. The weight of this evidence implies that the chromosphere is heated only by these long-period waves and that no other heating mechanism makes a significant contribution to the energy input of the chromosphere.

I thank Thomas R. Ayres, Franz-Ludwig Deubner, Jan Stenflo and Harold Zirin for comments on the chromospheric model. This work was supported by NASA grant NAGW-1568.

REFERENCES

Anderson, L. S. & Athay, R. G. 1989, *Astrophys. J.*, **336**, 1089.
Athay, R. G. 1976, *The Solar Chromosphere and Corona*, Reidel Pub. Co., Dordrecht, Holland.
Ayres, T. R. 1981, *Astrophys. J.*, **244**, 1064.
Ayres, T. R. & Testerman, L. 1981, *Astrophys. J.*, **245**, 1124.
Ayres, T. R., Testerman, L. & Brault, J. W. 1986, *Astrophys. J.*, **304**, 542.
Bhatnagar, A. & Tanaka, K. 1972, *Solar Phys.* **24**, 87.
Biermann, L. 1946, *Naturwiss.*, **25**, 161.
Bray, R. J. & Loughhead, R. E. 1974, *The Solar Chromosphere*, Chapman and Hall, London.
Cram, L. 1974, in: I.A.U. Symp. No. 56, *Chromospheric Fine Structure*, R. G. Athay ed., Reidel, Dordrecht Holland, 51.
―――――― 1987 in: *Cool Stars, Stellar Systems, and the Sun*, J. L. Linsky and R. E. Stencel eds., Springer Verlag, Berlin, 123.
Cram, L. E. & Damé, L. 1983, *Astroph. J.* **272**, 355.
Damé, L. 1984, in: *Small-Scale Dynamical Processes in Quiet Stellar Atmospheres*, S. L. Keil ed., Sacramento Peak, p54.
Defouw, R. J. 1976, *Astroph. J.* **209**, 266.
Foing, B. & Bonnet, R. M. 1984, *Astrophys. J.*, **279**, 848.
Kalkofen, W. 1989, *Astroph. J., subm.*
Lindsey, C. & Roellig, T. 1987, Astroph. J. **313**, 877.
Liu, S.-Y. 1973, *Solar Phys.* **31**, 127.
―――――― 1974, *Ap. J.* **189**, 359.
Narain, U. & Ulmschneider, P. 1989, *Space Sci. Rev.*, subm.
Orrall, F. Q., 1966, *Astroph. J.* **143**, 917.
Schatzman, E. & Souffrin, P. 1967, *Ann. Rev. Astron. Astroph.*, **5**, 67.
Schwarzschild, M. 1948, *Astrophys. J.*, **107**, 1.
Simon, M. & Shimabukuro, F. I. 1971, *Astroph. J.* **168**, 525.
Ulmschneider, P. 1970, *Solar Phys.* **12**, 403.
Vernazza, J. E., Avrett, E. H. & Loeser, R. 1981, *Astroph. J. Suppl.*, **45**, 635, (**VAL**).
Yudin, O. I. 1968, *Sov. Phys. – Doklady*, **13**, 503.
Zirin, H. 1988, *Astrophysics of the Sun*, Cambridge University Press, Cambridge, UK.

IV. MAGNETOHYDRODYNAMICS OF THE PHOTOSPHERE

SOLAR MAGNETOCONVECTION

A. Nordlund
Copenhagen University Observatory
Øster Voldgade 3
1350 Copenhagen K
Denmark

R. F. Stein
Dept. of Physics and Astronomy
Michigan State University,
East Lansing, MI 48823
U.S.A.

ABSTRACT. As a prelude to discussing the interaction of magnetic fields with convection, we first review some general properties of convection in a stratified medium. Granulation, which is the surface manifestation of the major energy carrying convection scales, is a shallow phenomenon. Below the surface, the topology changes to one of filamentary cool downdrafts, immersed in a gently ascending isentropic background. The granular downflows merge into more widely separated downdrafts, on scales of meso-granulation and super-granulation.

The local topology and time evolution of the small scale, kilo Gauss, network and facular magnetic field elements are controlled by convection on the scale of granulation. The topology and time evolution of larger scale magnetic field concentrations are controlled by the hierarchical structure of the horizontal components of the large scale velocity field. In sunspots, the small scale magnetic field structure determines the energy balance, the systematic flows and the waves. Below the surface, the small scale structure of the magnetic field may change drastically, with little observable effect at the surface. We discuss results of some recent numerical simulations of sunspot magnetic fields, and some mechanisms that may be relevant in determining the topology of the sub-surface magnetic field. Finally, we discuss the role of active region magnetic fields in the global solar dynamo.

1. Introduction

With few exceptions (some information from helioseismology measurements), we can only observe the surface manifestations of phenomena in the solar convection zone. We must deduce by a combination of physical intuition, simple models, and numerical simulations what goes on beneath the solar surface. The intrinsically three-dimensional nature of flows and magnetic fields in the turbulent solar plasma renders the use of analytical techniques

J. O. Stenflo (ed.), Solar Photosphere: Structure, Convection, and Magnetic Fields, 191–211.
© *1990 by the IAU.*

and overly simplified geometries questionable. Rather, we must confront the complex geometries, and use numerical techniques and available computer resources as our tools.

In this review, we summarize recent progress in efforts to understand solar convection and magnetoconvection. In Section 2, we summarize recent results on the topology of solar convection. In Section 3, we report on numerical simulations of sunspot umbrae, and discuss the structure of sunspots below the visible surface. In Section 4, we discuss network and facular magnetic fields, and in Section 5 we briefly discuss the role of active region magnetic fields in the solar dynamo process.

2. Convection Topology

2.1. SURFACE MANIFESTATIONS

Granulation is observationally defined as a well correlated brightness and velocity pattern (e.g. Bray et al. 1984). Some characteristic surface properties are: size scale fractions of *Mm* to a few *Mm*, time scales minutes to half an hour, asymmetry between bright and dark connectivity, asymmetry in the time evolution ("arrow of time"), and a large brightness contrast.

Meso-granulation was first observed as a weak signal in spatially filtered Doppler shift and brightness (November *et al.*, 1981). Its characteristic properties are: size scales of a few *Mm* to 10 *Mm*, time scale hours, horizontal advection of granulation and magnetic fields, weak temperature contrast, weak vertical velocity field. Meso-scale flows are most clearly revealed by auto-correlation tracking of small scale features (granulation) (Title *et al.* 1989, November & Simon 1988), which measures the advection produced by larger scale flows. Meso-scale flows also have noticeable effects on granule growth and on the distribution of granule sizes (cf. the review by Müller 1989, and this volume).

Super-granulation was first observed as a cellular pattern in the horizontal velocity field (Leighton *et al.*, 1962). Its characteristic properties are: size scales of 20 *Mm* to 50 *Mm*, time scale days, organization of the magnetic field into cells, very weak (undetectable) temperature contrast, and very weak vertical velocity field. Super-granular flows are also measurable with the auto-correlation tracking technique. A practical limitation is the size of CCD chips which, together with the resolution required to track individual granules, determines the area coverage.

Observational techniques and limitations (spatial and temporal filters, resolution) often influences the classification into separate phenomena. The original definition of the meso-granulation scale (November *et al.* 1981) was made using spatial filters which excluded smaller and larger scales. The observations demonstrated clearly that "there was something there", but could not accurately address the question of how "well defined" or "separated" the meso-granulation and supergranulation scales are. Simon & Leighton (1964) used a definition of supergranulation size based on the distance to secondary maxima in auto-correlation spectra. It is not obvious how the distribution of such distances maps onto a distribution of horizontal velocity power as a function of wave number. Simon & Leighton found a broad distribution of distances to secondary maxima, ranging from 20 to 50 Mm (cf. their Fig. 4). They adopted as the diameter of supergranulation cells the mean of the secondary maxima measurements, and gave the error as the standard deviation of the mean. This perhaps somewhat arbitrary definition of the size has been quoted ever since as the size of supergranulation cells.

With the autocorrelation tracking technique, it is, in principle, possible to accurately determine power spectra of horizontal velocities, and thus address the question of separation of scales. Since the motions are stochastic in nature, time series which cover a large number of cells, in space and/or time, are necessary to avoid obtaining power spectra which reflect individual cells. For reasons discussed below, it is doubtful on theoretical grounds that there is actually a separation between meso- and super-granulation scales. An observational clarification of this point would be most valuable, and we hope that new observations from La Palma (Scharmer *et al.*, present and future) and SOHO (Scherrer *et al.* 1989) will provide decisive data.

2.2. SURFACE TOPOLOGY: GRANULATION

A qualitative understanding of the granulation phenomenon has emerged from numerical simulations of convection in the surface layers of the Sun (Nordlund 1982, 1983, 1984abc, 1985abcd, Lites *et al.* 1989, Stein & Nordlund 1989) and other stars (Nordlund & Dravins, 1989). One of the main conclusions is that granulation is a shallow *surface phenomenon*, and that the topology below the surface is qualitatively different. To a large extent, the granulation pattern is a manifestation of the interaction of convection with a radiating surface, and the structure of the photosphere is established as a balance between competing convective and radiative processes.

Thermal energy is carried to the surface by advection and released into radiative flux in a thin (50 - 100 km) cooling layer. An ascent velocity of some 2 kms^{-1} is necessary to sustain the radiative losses at the surface. The photosphere has a stratification which is *strongly sub-adiabatic*; $\Delta \ln T$ (≈ 0.4) $\ll \Delta \ln P \times \nabla_{ad}$ ($\approx 5 \times 0.4 = 2.0$). Thus, unless a significant radiative heating occurred throughout the photosphere, the temperature of the upper photosphere would be much lower. The radiative heating is due to re-absorption of a small, but energetically significant fraction of the radiation, and the detailed temperature structure of the photosphere is the result of a fierce competition between convective (expansion) cooling and radiative heating. As a result, there are large temperature fluctuations on a small scale in the solar photosphere.

Dynamically, the main factors that determine the shape and evolution of the granulation pattern are *advection*, and *buoyancy braking*. The horizontal cellular outflows advect properties (including the flow pattern itself). The result is a tendency for cells to grow horizontally. The competition between neighboring cells in different phases of growth leads to the non-stationary, chaotic evolution of the granular pattern with time. For given vertical velocities, the horizontal velocities grow linearly with the horizontal size, and thus the pressure fluctuations that drive the horizontal velocities grow quadratically with the horizontal size (cf. Nordlund 1982, Hurlburt *et al.* 1984). This leads to excess densities and hence buoyancy breaking, especially in the centers of large granules which, when otherwise allowed to grow undisturbed, often develop dark centers surrounded by a ring of bright, expanding material ("exploding granules", cf. Bray *et al.* 1984, Section 2.3.7).

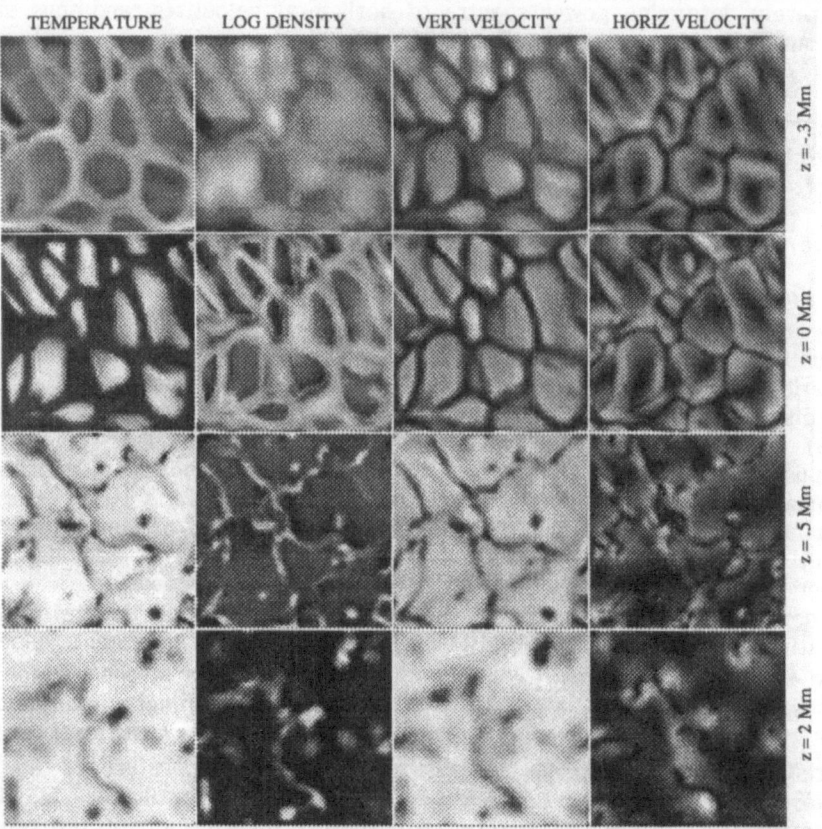

Figure 1. Composite plot, showing temperature, density, and the vertical and horizontal velocity amplitudes in horizontal planes at four different depths in a numerical model of granulation and meso-granulation (Stein & Nordlund, 1989). The horizontal size of the model is 6×6 *Mm*.

2.3. SUBSURFACE TOPOLOGY

Fig 1. illustrates the topology in horizontal planes at, above, and below the surface. Note that below the surface, the horizontal topology changes qualitatively in just a few hundred kilometers. This is a depth interval which is only a fraction of the horizontal cell size. The horizontal topology changes from one with descending gas in connected intergranular lanes to one with isolated spots of descending material. In three dimensions, the topology is *intermittent*, with vertically oriented *filaments* of rapidly descending, entropy deficient material, immersed in a background of slowly ascending, nearly isentropic material. It should be noted that there is no clear cell structure in the vertical direction.

Motions below the surface are nearly adiabatic and anelastic; evolution is mainly by advection. Properties of a fluid element at a given time and place are given by the "sum of the histories" of its constituent parcels. Therefore, test particles are useful in understanding the evolution. Given a record of velocities u(t), particles may be traced forwards and backwards in time. Fig. 2 shows an example of such trace plots.

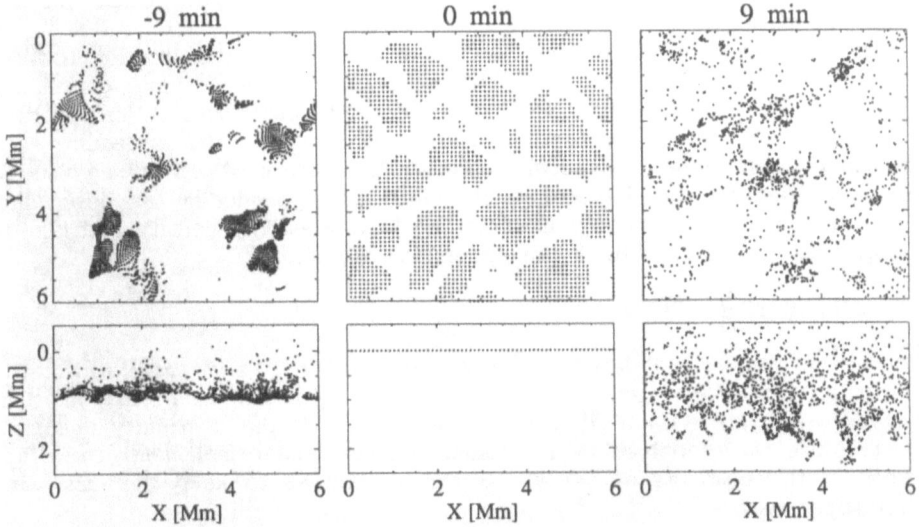

Figure 2. Trace particle plots, showing the location of selected test particles at three different times. The two mid panels show the location at a reference time, with trace particles at all grid points with ascending velocities in the plane $z = 0$. The two panels to the left show the location of these test particles nine minutes earlier, and the two panels to the right show the location nine minutes after the reference time.

As discussed by Stein & Nordlund (1989), the sub-surface topology is a consequence of the density stratification. Because of the strong density stratification, fluid ascending / descending only a few Mm (the size of a cell) must expand / contract orders of magnitude. Because of mass conservation, most *ascending* gas must overturn within a density scale height, while most *descending* gas becomes engulfed in overturning gas, and keeps descending many scale heights. In other words, only a small fraction of the ascending material at any one depth ascends many scale heights, and only a small fraction of the descending material at any one depth overturns within a scale height.

Below the radiating surface, non-adiabatic effects are negligible; the motion is almost *adiabatic*. Furthermore, the rapid expansion of ascending material wipes out entropy inhomogeneities, and thus ascending material is very nearly *isentropic*. Consequently, overturning gas below the surface is also nearly isentropic. This is the reason for the change of topology below the surface; entropy deficient gas in the interconnecting lanes is rapidly replaced with entropy neutral gas below the surface. All entropy deficient material from the surface eventually ends up in the vertices between cells.

It is important to realize that the surface is the only source of entropy fluctuations (except for similar effects at the lower boundary of the convection zone), and thus the surface is also the ultimate cause for the driving of motions.

The fact that the convective flux is directed from the interior towards the surface is perhaps somewhat misleading in this connection; descending cool material and ascending hot material both contribute to a positive convective flux, and because of mass conservation, the ascending and descending mass fluxes are equally large. However, most of the convective flux is carried by the descending filaments, and dynamically the descending filaments are also dominating; most of the kinetic energy density and kinetic energy flux is associated with the descending filaments.

2.4. ANELASTIC MOTION

Below the surface, Eulerian density changes are very small; i.e., the continuity equation acts basically as a *constraint* on the motion; $\partial \ln \rho / \partial t = -\nabla \cdot (\rho \mathbf{u}) \approx 0$. With this *anelastic* form of the continuity equation, the pressure is determined by a Poisson equation and complements the other forces in the equation of motion in such a way as to keep the mass flux divergence free.

Using the Poisson equation for the pressure, one may show that localized (small scale) buoyancy fluctuations lead to localized pressure fluctuations, hence localized motion. On the other hand, large scale buoyancy fluctuations lead to pressure fluctuations who's relative amplitudes vary little with height (cf. Nordlund 1985, Sect. 2.6). In other words, for large scale perturbations, the atmosphere moves locally up/down as a whole, with local stratifications in near hydrostatic equilibrium. As a consequence, large scale components of the horizontal velocity field do not have much vertical shear.

For a horizontally Fourier decomposed velocity field, the anelastic continuity equation implies

$$\frac{u_{\text{vert}}}{u_{\text{hor}}} \approx kH_{\rho u_z} \qquad (1)$$

where k is the horizontal wave number, and $H_{\rho u_z}$ is the scale height of the vertical mass flux.

2.5. LARGER SCALE MOTIONS

In the interior of the convection zone, a more gradual change of the topology occurs. The downflows that originate from intergranular lanes at the surface merge into fewer, more widely separated filamentary downdrafts. The horizontal velocities of the large scale flows advect the small scale structure sideways to produce this merging. As a result, the horizontal scale of the velocity field increases with depth. Conversely, it is the merging smaller scale filaments which provide the entropy fluctuations that drive the larger scale flows.

The small density scale height in the surface layers dictates that the energy carrying convection cells (which must have vertical velocities in excess of some 2 kms^{-1}) must not be larger than a few Mm near the surface (cf. Eq. (1) above). At larger depths, larger cell sizes are allowed, and according to the discussion above, larger scales are indeed driven by the merging of smaller scale filaments from the surface layers.

The simulations have demonstrated this only for scales marginally larger than the surface granulation, but presumably the same mechanism works for still larger scales, including the supergranular scale. The general scenario then is one where the merging of downdrafts on granular scales drives flows on meso-granular scales at a depth of a few Mm below the surface, and the merging of meso-granular downdrafts drives flows on supergranular scales at depths of some 10 - 20 Mm. Presumably, flows on even larger scales (traditionally called giant-cells) are driven at even larger depths, by merging supergranular downdrafts.

As mentioned earlier, the pressure fluctuations which drive the horizontal components of large scale velocity fields extend over a height range comparable to or larger than the horizontal size of the fluctuations. Since the aspect ratio (ratio of horizontal size to distance from the surface) of these flows is larger than unity, this implies that the horizontal velocity fields of larger scale flows extend up to the surface. The numerical simulations indicate similar vertical and horizontal rms amplitudes below the surface. Accordingly, the distribution of surface horizontal velocity amplitude with horizontal size reflects the dependence of vertical velocity amplitudes on depth below the surface.

There is no obvious reason why, in this process, certain distinct scales should be favored. The often mentioned helium ionization zones, which extend over quite a range in depth, centered at some 5 and 15 Mm, has no particular relevance in this scenario, except for reducing the adiabatic temperature gradient, and hence somewhat reducing the density scale height.

The effect is a change in the density scale height, but only of a few tens of percent, and is hardly likely to have much effect on the distribution of horizontal velocity amplitude with horizontal size.

2.6. GLOBAL CONVECTION ZONE FLOWS

We expect the strong asymmetry between ascending and descending motions to prevail on all scales, including scales comparable to the depth of the convection zone. The merging downdrafts with entropy deficient gas are likely to extend throughout the convection zone. Turbulence in the downdrafts induces mixing with surrounding, entropy neutral fluid, but because of the general convergence of descending gas, only a small fraction actually turns over into ascending gas. At the very bottom of the convection zone, the downdrafts hit the interface to the stable region below the convection zone. In this layer, descending gas is diverging, ascending gas is converging, and overturning gas is reheated by radiative diffusion or by mixing. The situation is to some extent the reverse of that at the surface. However, at some distance above the lower boundary, ascending gas again must be expanding, and the analogy to the surface layers is lost.

It would seem that the global dynamics of the convection zone must be strongly influenced by the asymmetry between descending and ascending gas, and that models that do not take this asymmetry properly into account may easily fail to produce realistic result. This is a likely cause for the failure of current numerical global convection zone models (Gilman & Miller 1986, Glatzmaier 1987) to predict differential rotation properties and dynamo action consistent with the observations.

The observational evidence (cf. Libbrecht 1988, and other references in the same proceedings) indicates that the differential rotation of the solar convection zone to a first approximation is constant along radii, rather than constant on cylinders (as predicted by the Taylor-Proudman theorem). This shows that, in some sense, the solar convection zone is "vertically stiff"; i.e., ascending and descending material tend to conserve angular speed, rather than angular momentum.

If the topology of the global convection zone is intermittent, with localized filamentary downdrafts immersed in a gently ascending background, then it might be possible to work out simplified models of the differential rotation, where the exchange of angular momentum between ascending and descending flow components is estimated from drag and mass exchange between the filamentary downdrafts and the gently ascending background. As discussed above, the mass exchange between the ascending and descending components is largely determined by the vertical mass flux scale height, through the continuity equation.

3. Magneto-Convection

We now turn our attention to convection in the presence of a magnetic field, starting with the extreme case of sunspot umbrae.

3.1. UMBRAL SIMULATIONS

We have recently performed a preliminary simulation of the central parts of a sunspot umbra (Nordlund & Stein 1989). In this simulation, a vertically homogeneous magnetic field with a strength of 0.2 T (2 kG) is superimposed on a snapshot from a granulation simulation. This adds a constant pressure everywhere in the atmosphere, but produces no additional forces. Hence the initial condition is self-consistent, although somewhat artificial. The strong magnetic field rapidly quenches the convection.

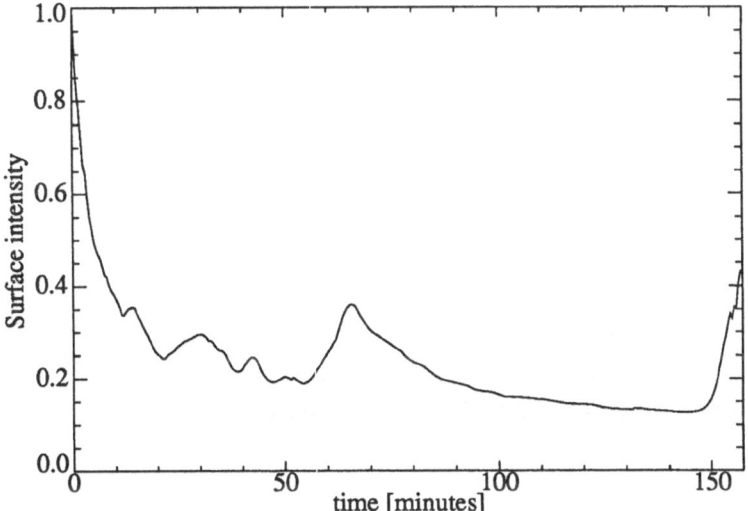

Figure 3. The average surface intensity in an umbra simulation, relative to the initial photospheric surface intensity, as a function of time.

Fig. 3 shows the average surface intensity as a function of time for this simulation. Because convection is suppressed, there is nothing to compensate for the strong radiative losses at the surface, and the surface begins to cool rapidly. Within 5 minutes, the surface radiation intensity is less than half the nominal one. Due to the rapidly decreasing surface flux, and the exponentially increasing heat capacity per unit volume (because the visible surface descends), the rate of cooling slows down, and it takes approximately one hour to reach a surface intensity of 20 %. We regard our initial condition as rather artificial, but the slow cooling, and the formation of a dark umbra "in place", may correspond to a particular umbra formation process, with no accompanying convergence of pores and small scale magnetic features, observed by Zirin (1987).

Fig. 4 shows the corresponding evolution of the horizontally averaged temperature, as a function of time and depth. The very steep temperature drop near $\tau = 1$ is obvious, and one may follow the descent of this surface as a function of time by noting that the vertical distance between grid lines is approximately 50 km. Note that the $\tau = 1$ surface descends approximately

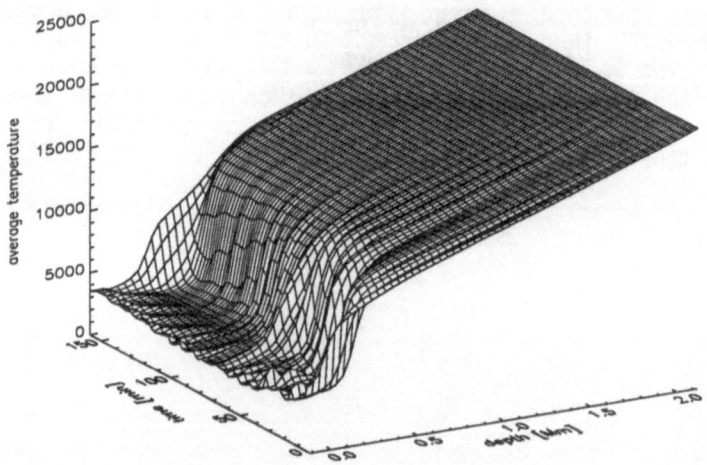

Figure 4. The horizontally averaged temperature in the umbra simulation, as a function of time and depth. The total time covered is about two and a half solar hours. The fine structure in the atmosphere in the first part of the simulation is an artefact of the initial conditions.

400 km in about one hour.

Some of the variation of the surface intensity with time visible in Fig. 3 is due to a vertical buoyancy oscillation that was present in the initial model, and which continues, driven by inertia. However, the two dominant peaks are due to two episodes of convection, which carry heat up to the surface and thus increase the surface radiative flux. Fig. 5 shows the horizontally averaged convective flux, as a function of time and depth. Note that, for brief periods, the convective flux just below the surface exceeds the nominal *photospheric* surface flux. During the first episode, the convective flux peaks at just over 100 % of the nominal solar flux, but during the second episode it exceeds 300 % of the nominal solar flux. The convective flux is localized to a shallow layer centered on the steepest part of the temperature distribution, and serves to slightly flatten the temperature profiles displayed in Fig. 4. This increases the surface temperature and hence the surface radiative flux.

The effect on the surface radiative flux is only about 15 % during the

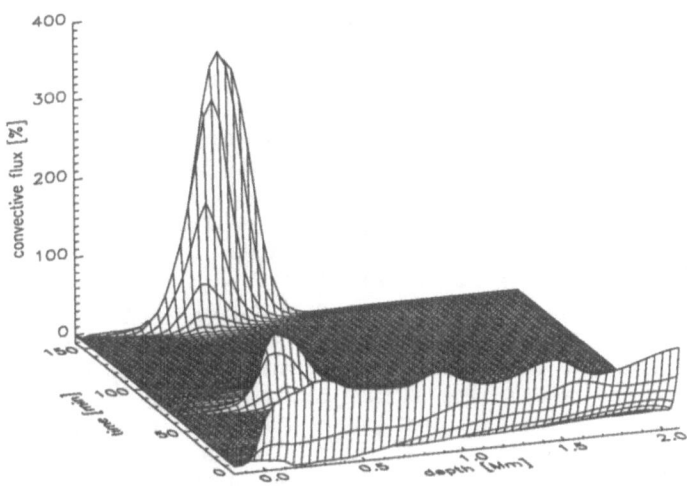

Figure 5. The average convective flux in the umbra simulation, as a function of time and depth. The convective flux that is present in the initial snapshot rapidly dies away. Later, after about one solar hour, and after about two and a half solar hours, short episodes of convection develop in a narrow surface layer.

first episode, which may be surprising considering the convective flux of over 100 % just below the surface. However, the heat capacity of the surface layers is large, and the huge divergence of the convective flux only leads to a rather insignificant flattening of the average temperature profile. During the second episode, the convective flux is large enough - and its duration long enough - to significantly heat the surface layers. This results in an upward displacement of the surface temperature drop, with a corresponding increase in the surface radiation intensity, which reaches about 40 %.

The topology of the heat flow is similar to that of ordinary granulation, with cells of ascending and expanding gas which push the magnetic field aside. The cells are roundish, not space filling, and their size taper off with height, because of the increasing dominance of the magnetic pressure. The surface brightness pattern has bright edges on the round cells, because of the transparency of the plasma in the surrounding, strong magnetic field. As illustrated by Fig. 5, the flow pattern is shallow.

3.2. UMBRAL STRUCTURE

The qualitative features of an umbra's vertical structure may be deduced from simple considerations of pressure and energy equilibrium. To a first approximation, sunspot umbrae are similar to cool stellar atmospheres with inhibited convection. At lower temperatures the opacity is smaller, so the pressure at $\tau = 1$ is larger. Inhibition of convection reduces the convective heat flux, so the $\tau = 1$ layer is cooled by radiation. Since the convective flux is small, the atmosphere is close to radiative equilibrium, and the temperature rises rapidly below the surface until it reaches the interior adiabat. The bottom of the steep temperature gradient is an important "pivot" (transition) point in the structure of the umbra.

In fact, the "pivot point" may be defined as the place where the radiative flux becomes a small fraction of the surface flux. Above this point, there is approximate radiative equilibrium, because the time scale to approach radiative equilibrium decreases rapidly with height. Near the pivot point, there is a significant divergence of radiative flux, which must be balanced either by the divergence of another (e.g. convective) energy flux, or else by a local loss of thermal energy. In the latter case, the loss of thermal energy implies that the pivot point descends with time. Its rate of descent is determined by the ratio of the surface radiative energy loss to the energy density per unit volume at the pivot point. The pivot point must lie rather close to the visible surface ($\tau = 1$), since the gas rapidly becomes very opaque below the surface. The depth of the pivot point is basically the same as the "Wilson depression"; i.e., the height difference between $\tau = 1$ in the umbra and in the surrounding photosphere. Thus, if there is negligible convective flux in the deep umbra, the umbra surface descends; i.e., the Wilson depression increases with time. If and when convection becomes sufficiently effective to compensate for the surface energy losses, the umra ceases to descend. The vertical position of the umbra at any one time is determined by how long it has been cooling, and how much heat has been replenished by convection. The shape of the temperature profile is qualitatively the same inside and outside the umbra, with a deep, nearly isentropic part, separated from the cool optically thin atmosphere by a steep temperature drop just below the visible surface.

In the external photosphere, the isentropic region extends all the way up to just below the surface of the photosphere. Inside the umbra the visible surface and steep temperature gradient are depressed, as discussed above. Thus, there is a depth interval between the surface of the photosphere and the surface of the umbra where the temperature inside the spot is much smaller than the temperature in the surrounding photosphere. As a consequence, the internal gas pressure drops much more rapidly with height in this interval than the external gas pressure. The gas pressure difference, which controls the strength and topology of the magnetic field at the edge of the spot, thus obtains a characteristic shape. From its small subsurface value, the difference increases rapidly with height near the pivot point, because of the drop in the interior temperature. When the internal gas pressure is negligible compared to the external pressure, the pressure difference is essentially equal to the external pressure, and hence decreases exponentially with height.

Below the pivot point, the pressure difference is equal to the difference between two exponentially increasing pressures. If the temperature is equal inside and outside, then the *relative* pressure difference is nearly constant with depth. This implies a nearly constant plasma β ($\beta = P_{gas}/P_{mag}$). The

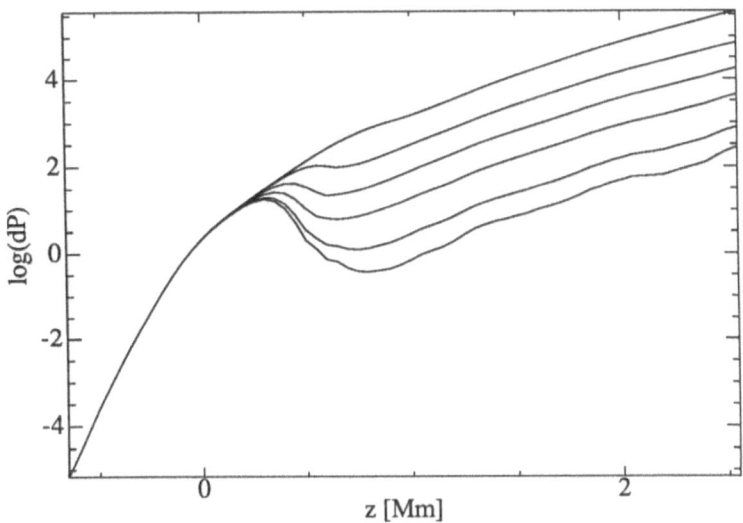

Figure 6. The pressure difference between the interior of the umbra and the external convection zone, after about two hours of simulated solar time. The six curves correspond to (from bottom to top) vertical displacements of the subsurface umbra, relative to the surrounding convection zone, of 10, 20, 50, 100, 200, and 500 *km*, respectively.

pressure difference may also be parametrized in terms of a relative displacement of the inside with respect to the outside.

Fig. 6 shows the gas pressure difference, as a function of height, for 6 different assumed height differences (10, 20, 50, 100, 200, and 500 *km*). Note the characteristic shape, with a pressure difference maximum near the pivot point. Only in the (unrealistic) case with a 500 *km* displacement is there no maximum near the pivot point. This is because, for this particular case, the internal gas pressure is a small fraction of the external pressure at all depths. We may conclude that, for reasonable values of the vertical displacement, the pressure difference has a characteristic shape, with a maximum near the pivot point.

When trying to fit the detailed distribution of field strength in sunspots, Jahn (1989) empirically deduced pressure difference profiles of just this shape. He also found it necessary to introduce volume currents in the penumbral region of his spot model, to satisfy constraints from measurements of penumbral magnetic fields.

3.3. SUNSPOT PARAMETERS

The pressure difference profile, together with the total magnetic flux of the spot, determines the umbra flux density, and hence the size and topology of the spot. Since the shape of the pressure profile is nearly universal, the main characteristics of an idealized, symmetric, spot is determined uniquely by three parameters; the total flux, the depth of the pivot point (Wilson depression), and the (nearly constant) relative pressure difference in the subsurface layers. In principle, the subsurface entropy difference between the inside and the outside enters as a fourth independent parameter (van Ballegooijen, 1982), but small entropy differences do not significantly influence the structure near the surface.

The meaning of the first two parameters is clear, but what is the physical significance of the third one? Obviously, the relative pressure difference in the subsurface layers may be changed by pushing matter up or down the flux tube. In the Sun, this must be controlled by conditions in the deepest part of the flux rope, since that is where most of the mass in the flux rope is located. There may also be couplings to what happens in other part of a topologically connected structure. Globally, the total mass within a flux structure must be approximately conserved, if perhaps in a "leaky" way, depending on how coherent the flux structure is. In the most naive picture, pushing matter down at one place will push it up somewhere else.

When considering sunspots as part of the global solar magnetic field, such geometrical constraints have interesting implications that may be relevant for the behavior of active regions and for the solar dynamo process (cf. the next section). The relative pressure difference parameter might control the "looseness" of a spot or spot group, and might be what determines the systematic umbra intensity dependency on cycle phase discovered by Maltby and co-workers (Albregtsen & Maltby 1978, 1981; cf. also Maltby *et al.*, 1986). This is also the parameter who's evolution may be controlling the break-up of a spot. When the subsurface pressure difference decreases, the sub-surface field expands, and the spot may become unstable and break up.

The relative pressure difference at depth may also control the formation of spots and pores in an emerging flux region. Until the surface layers have cooled sufficiently, the surface pressure difference may not be large enough to hold a spot together at the surface, even if it is substantial some distance below the surface. An aggregate of flux is formed, with pores and faculae, loosely held together by the pressure difference at depth, but with insufficient pressure difference at the surface to form large spots. As the individual pores cool off, the surface pressure difference increases, and large spots form by merging of smaller pores and spots.

Morphological changes of the global magnetic field can change the relative pressure difference. If a flux structure is bent over backwards, because of differential rotation, it tends to become shorter along the bottom. Conservation of mass requires that matter be pushed up in the flux structure, so the sub-surface gas pressure difference drops. The surface flux concentration will no longer be held together below the surface, and spots will start to dissolve. This may be what happens in the following polarity of an active region, if the surface rotates more slowly than the bottom of the convection zone. Conversely, the leading part of a flux structure would be stretched out along the bottom, which would increase the sub-surface pressure difference, and hence tend to stabilize spots in the leading part of active regions, as is observed.

The scenario works if the bottom of the convection zone rotates faster than the surface. Recent helioseismology measurements (e.g., Libbrecht, 1988) indicate that this is indeed the case, with the radially "stiff" mapping of the surface differential rotation through the top part of the convection zone turning over into more rigid rotation, near the lower boundary of the convection zone. Stenflo (1989ab) independenty deduced such a rotation profile by noting that rotation rates measured by correlating surface patterns over one or several rotation periods are systematically larger than the rotation rates for individual magnetic features, measured over shorter time periods, near the central meridian (Snodgrass, 1983).

4. Network and Facular Magnetic Fields

Network and facular magnetic fields consist of large numbers of small magnetic field structures, with magnetic field strengths of the order of 1 - 2 kG. Hence, their magnetic pressures are comparable to the photospheric gas pressure. They are similar to the larger pores and spots, but their sizes are typically smaller than can be resolved with present instruments. The local topology and time evolution of such structures are controlled by convection on the scale of granulation. The magnetic flux is concentrated in intergranular lanes. Their interiors become evacuated, because the surface radiative losses cannot be balanced by advection of entropy across the field lines. The flux concentrations are in quasi-static pressure equilibrium with surrounding evolving granules, and the flux concentrations "creep" into newly formed intergranular lanes. (Nordlund 1985d, 1986; Nordlund & Stein 1989)

Channeling of the radiative flux into thin flux structures may explain the brightness of small scale flux concentrations, as compared to darker larger scale flux concentrations such as pores and sunspots (Spruit 1976, Spruit & Zwaan 1981). The flux structures, which are perhaps better represented by slabs than by flux tubes, are separated from the surrounding granulation by a very thin boundary layer (Deinzer et al. 1984ab, Knölker et al. 1988, Grossmann-Doerth et al. 1988). Enhanced radiative heating and suppressed convective cooling of the upper photosphere may explain the relatively hot upper photospheric layers deduced from the temperature weakening of Stokes V profiles (Stenflo, 1975; Solanki & Stenflo 1984, 1985; Stenflo et al. 1987).

The topology and time evolution of larger clusters of small magnetic field concentrations is influenced by convection on larger scales: meso-granulation and super-granulation. Most of the field is swept to the boundaries of supergranulation cells, and local auto-correlation tracking of granules shows that the horizontal motion of small magnetic elements agrees with the horizontal motion of granules (Simon et al. 1988). The supergranular velocities are similar in the quiet sun and in the enhanced network (Wang 1989).

In plages, supergranulation cells are no longer visible, and horizontal velocities are significantly suppressed (Title et al. 1989). This may be related to a qualitative change of topology, where the magnetic field fills

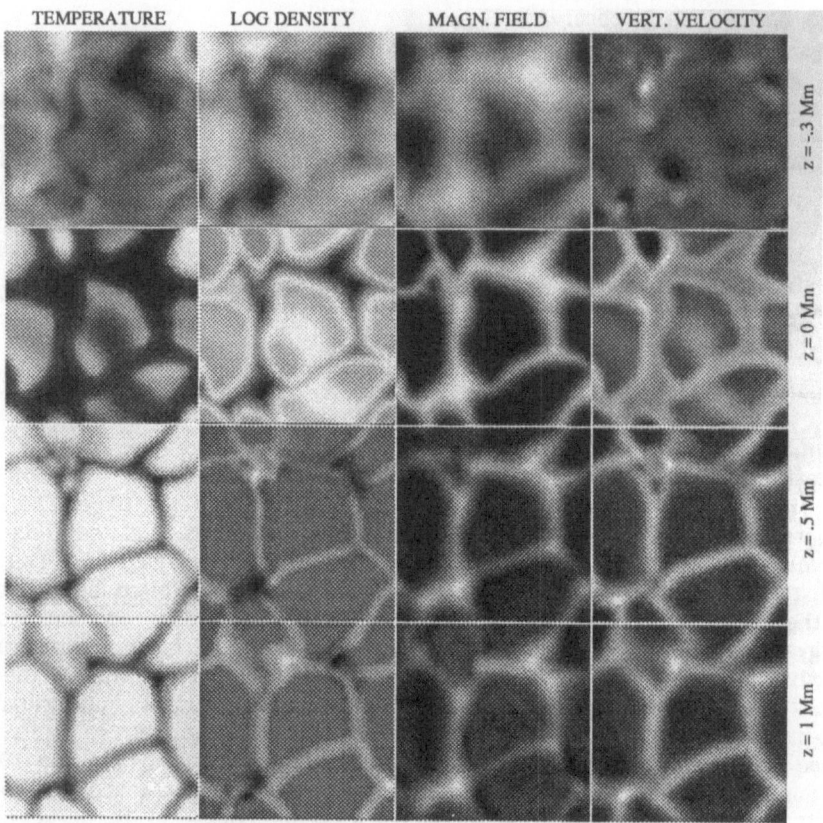

Figure 7. Composite plot, showing temperature, density, and the vertical components of the magnetic field and the velocity field, in horizontal planes at four different depths in a numerical model of the interaction of granulation and a (rather strong) facular magnetic field (Nordlund & Stein, 1989). The horizontal size of the model is 3×3 *Mm*.

essentially all available intergranular lanes, and hence becomes topologically connected in the horizontal plane (Nordlund & Stein, 1989). Such a magnetic flux topology inhibits large scale horizontal velocities, at least near the solar surface. The horizontal topology of a case with a 500 *G* average vertical field is illustrated in Fig. 7 (analogous to Fig. 1). Note that the surface (granulation) topology is visible over a larger range in depth, compared to the case in Fig. 1. Also, as illustrated by Fig. 1 of Nordlund & Stein (1989), the granules become more roundish, and their life times increase, relative to ordinary granules. This is because the nearly vertical sheets of magnetic field stabilize the convection patterns, by preventing the interaction of flows from neighboring granules.

Schrijver (1989) has suggested a simple mechanism, based on the topology of the paths available for horizontal flux transport, to explain the rather sharp boundary between a plage and the surrounding network. It may well be, however, that plages are intrinsically coherent. If they are the surface manifestations of subsurface flux ropes with a relative gas pressure difference sufficient to hold the flux together at depth, but insufficient to bring the flux together into sunspots, then the relatively well defined boundary may simply reflect the boundary of the subsurface flux rope. Near the surface, the flux rope is "frayed", with the individual strands concentrated to kilo Gauss strength by the surface effects discussed above. However, below the surface, the flux rope may retain some of the coherent structure it undoubtedly had when emerging through the surface as the following polarity part of an emerging flux region.

5. Active Regions and the Solar Dynamo

Active regions are often considered passive consequences of the solar dynamo; the queer and intricate surface manifestation of subsurface dynamo action. However, a scenario for a "topological solar dynamo" may be constructed where active region magnetic fields play an important part in the dynamo process, and the detailed connectivity of the global solar magnetic field is essential to the dynamo process (Nordlund, 1989). The "topological dynamo" is a classical "$\alpha - \omega$ dynamo", but complemented with detailed suggestions for the topology of the α and ω parts of the process. This scenario is able to explain a number of observed features of the solar cycle and solar active regions. A short summary of the main features is given here:

A flux rope breaking off from an azimuthal magnetic flux system at the bottom of the convection zone, and rising towards the surface, experiences a systematic coriolis force due to the persistent expansion of ascending plasma. The sense of the coriolis force is contrary to the sense of rotation; i.e., the rising part of the flux rope tends to rotate its leading polarity towards the equator and its following polarity away from the equator. This tendency is counteracted by the tension force along the flux rope, which tends to keep the flux rope in alignment with the main azimuthal flux system. The well known tendency for active region magnetic fields to have a slight inclination to the equator is most likely a consequence of this effect which, of course, is nothing else than what is usually called the "α-effect" in dynamo theory. In general terms, the α-effect is responsible for generating a poloidal component of the magnetic field, from an originally azimuthal field component.

The other ingredience in a classical $\alpha - \omega$ dynamo is the "ω-effect", which regenerates an azimuthal component, of opposite sign to the original one, from the poloidal component. In general terms, the ω effect is of course a consequence of the differential rotation of the solar convection zone, acting on the poloidal component of the global solar magnetic field. Note, however, that the "poloidal" part generated by the weak tilt of active region magnetic fields is localized to the active region. Part of the "topological dynamo" scenario is a suggestion for the actual topology of the "ω-effect", which involves the distortion of active region magnetic fields by differential rotation.

As an active region flux rope breaks through the surface, the crest of the loop extends into the corona. We know observationally that reconnection is efficient in the corona. On the time scale of days, the original connectivity between leading and following polarity is lost. We may therefore regard the original flux rope as effectively severed above the surface, with the following and leading polarity crossections acting as two "loose" ends of the original flux rope.

As discussed in the previous section, the leading polarity end of the flux rope, with flux concentrated into a few spots, is "dragged" along by the faster rotation at depth. The following polarity end is bent over backwards by the differential rotation, and forms a diffuse (plage) crossection with the surface. As a consequence, the latitudinal position of the leading polarity remains well defined and stable, while the following polarity, because it is bent over backwards, more easily drifts in latitude. The observed poleward drift of the following polarity implies that the dispersed plage area hauls the bottom field along as a heavy rope trailing polewards around along the bottom. This leads to "unwinding" of the bottom part of the following polarity from the original azimuthal flux system, and the winding up of new azimuthal flux in the opposite direction at high latitudes. At the same time, the leading flux is being wound up on the equatorward side of the azimuthal flux system, which leads to an equatorward migration of the original azimuthal flux system. This process keeps going as long as the surface rotates slower than the bottom of the convection zone. When the original azimuthal flux system reaches latitudes where the radial differential rotation vanishes (cf. Libbrecht 1988, Fig. 2), the "unwinding" process wins over the "winding" process, and the original azimuthal flux system eventually vanishes. In the mean time, a new flux system with opposite polarity has formed at high latitudes, and a new cycle begins.

6. Concluding Remarks

We have stressed, throughout this paper, the importance of considering the three-dimensional topology of solar magnetic fields and flows. Because of the enormous pressures and densities in the deep convection zone, relative to the observable surface, much of the large scale and long time behavior of surface phenomena is likely to be controlled from below. The near adiabatic and anelastic nature of subsurface flows, and the connectivity of magnetic fields, provide important constraints on the behavior of the subsurface flows and magnetic fields. We can only observe the surface manifestations of these subsurface phenomena, and we must deduce the behavior below the surface by indirect means.

As we have illustrated here, numerical simulations may play an important role in this process, by providing examples of the complicated behavior, and of the three-dimensional topologies involved. Numerical simulations are most useful when set up to directly model the behavior of solar phenomena, by using three-dimensional and unbounded geometries, by accurately modeling surface radiative transfer effects, and by using realistic equations of state and absorption coefficients. One argument for this detailed approach is to be able to directly compare with observed surface phenomena. Another important reason for using a realistic simulation is to avoid introducing spurious effects, e.g., at non-penetrative or reflecting boundaries.

However, given results of such detailed and realistic simulations, qualitative interpretation of the often very complicated spatial and temporal behavior is vitally important, in order to identify the essential physical processes, and to be able to draw more general conclusions from the numerical models, which are necessarily limited in temporal and spatial extent and resolution.

Solar physics is presently enjoying a renaissance, with new instruments and clever observational techniques, together with supercomputer numerical simulations, providing the impetus for rapid progress in our understanding of physical processes on the Sun. The development of new earth- and space-based instrumentation (LEST, SOHO, OSL), and the continual improvement of computer hardware and software, will supply even more powerful tools to assist our venture.

Acknowledgements

A.N. gratefully acknowledges support from the Danish Natural Science Research Council and the Danish Space Board. R.F.S. thanks NASA for support under grant NAGW 1695. They both would like to acknowledge stimulating discussions with Axel Brandenburg, Chris Durrant, Göran Scharmer, Henk Spruit, Alan Title, and Juri Toomre.

REFERENCES

Albregtsen, F., Maltby., P., 1978, *Nature*, **274**, 41.

Albregtsen, F., Maltby., P., 1981, *Solar Phys.* **71**, 269.

Bray, R.J., Loughhead, R.E., Durrant, C.J., 1984, *The Solar Granulation*, Cambridge University Press.

Deinzer, W., Hensler, G., Schüssler, M., Weisshaar, E., 1984a, *Astron. Astrophys.* **139**, 426.

Deinzer, W., Hensler, G., Schüssler, M., Weisshaar, E., 1984b, *Astron. Astrophys.* **139**, 435.

Gilman, P.A., Miller, J., 1986, *Astrophys. J. Suppl.* **61**, 585

Glatzmaier, G.A., 1987, in *The Solar Internal Angular Velocity: Theory, Observations and Relationships to the Solar Magnetic Fields*, eds. B.R. Durney and S. Sofia, Reidel, Dordrecht, p. 263.

Grossmann-Doerth, U., Schüssler, M., Solanki, S.K., 1988, *Astron. Astrophys.* **206**, L37.

Hurlburt, N. E., Toomre, J. and Massaguer, J. M. 1984, *Astrophys. J.* **282**, 557.

Jahn, K., 1989, *Astron. Astrophys.* (in press).

Knölker, M., Schüssler, M., Weisshaar, E., 1988, *Astron. Astrophys.* **194**, 257.

Leighton, R.B., Noyes, R.W., and Simon, G.W., 1962, *Astrophys. J.* **135**, 474.

Libbrecht, K.G., 1988, *Proc. Symp. Seismology of the Sun and Sun-like Stars, Tenerif, Spain, 26-30 September 1988*, ESA SP-286, p. 131.

Lites, B.W., Nordlund, Å., and Scharmer, G.B. 1989, in *Proceedings NATO Advanced workshop on Solar and Stellar Granulation*, eds. R.J. Rutten and G. Severino, Kluwer Academic Publishers, Dordrecht, p. 349.

Maltby, P., Avrett, E.H., Carlsson, M., Kjeldseth-Moe, O., Kurucz, R.L., Loeser, R., 1986, *Astrophys. J.* **306**, 284.

Müller, R., 1989, in *Proceedings NATO Advanced workshop on Solar and Stellar Granulation*, eds. R.J. Rutten and G. Severino, Kluwer Academic Publishers, Dordrecht, p. 9.

Nordlund, Å., 1982, *Astron. Astrophys.* **107**, 1.

Nordlund, Å, 1983, in "Solar and Stellar Magnetic Fields: Origin and Coronal Effects", ed. J.-O. Stenflo, *IAU Symp.* **102**, 79.

Nordlund, Å, 1984a, in *Small Scale Processes in Quiet Stellar Atmospheres*, ed. S.L. Keil, Sacramento Peak Observatory, Sunspot, N.M. 88349, p. 174.

Nordlund, Å, 1984b, in *Small Scale Processes in Quiet Stellar Atmospheres*, ed. S.L. Keil, Sacramento Peak Observatory, Sunspot, N.M. 88349, p. 181.

Nordlund, Å, 1984c, in *The Hydromagnetics of the Sun, Proc. 4th European Meeting in Solar Physics*, ESA SP-220, p. 37.

Nordlund, Å, 1985a, *Solar Phys.* **100**, 209.

Nordlund, Å, 1985a, in *Problems in Stellar Spectral Line Formation Theory*, eds. J.O. Beckman and L. Crivellari, Reidel, Dordrecht, p. 215.

Nordlund, Å, 1985c, in *Proc. MPA/LPARL Workshop on Theoretical Problems in Solar Physics*, MPA 212, p. 1.

Nordlund, Å, 1985d, in *Proc. MPA/LPARL Workshop on Theoretical Problems in Solar Physics*, MPA 212, p. 101.

Nordlund, Å, 1986, *Abh. der Akad. der Wissensch. in Göttingen*, **38**, 83.

Nordlund, Å, 1989, (in preparation).

Nordlund, Å. and Dravins, D., 1989, *Astron. Astrophys.* (in press).

Nordlund, Å., Stein, R.F., 1989a, *Proceedings NATO Advanced workshop on Solar and Stellar Granulation*, eds. R.J. Rutten and G. Severino, Kluwer Academic Publishers, Dordrecht, p. 453.

November, L.J., Toomre, J, and Gebbie, K.B., 1981, *Astrophys. J.* **245**, L123.

November, L.J., Simon, G.W., 1988, *Astrophys. J.* **333**, 427.

Scherrer, P.H., Hoeksema, J.T., Bogart, R.S., Walker, Jr., A.B.C., Title, A.M., Tarbell, T.D., Wolfson, C.J., Brown, T.M., Christensen-Dalsgaard, J., Gough, D.O., Kuhn, J.R., Leibacher, J.W., Libbrecht, K.G., Noyes, R.W., Rhodes, Jr., E.J., Toomre, J., Zweibel, E.G., Ulrich Jr., R.K., 1989, ESA SP-1104, p. 25.

Schrijver, C.J., 1989, *Solar Phys.* (in press).

Simon, G.W., and Leighton, R.B., 1964, *Astrophys. J.* **140**, 1120.

Simon, G.W., Title, A.M., Topka, K.P., Tarbell, T.D., Shine, R.A., Ferguson, S.H., Zirin, H., and the Soup team, 1988, *Astrophys. J.* **327**, 964.

Snodgrass, H.B., 1983, *Astrophys. J.* **270**, 288.

Solanki, S.K., Stenflo, J.O., 1984, *Astron. Astrophys.* **140**, 185.

Solanki, S.K., Stenflo, J.O., 1985, *Astron. Astrophys.* **148**, 123.

Spruit, H.C., 1976, *Solar Phys.* **50**, 269.

Spruit, H.C., Zwaan, C., 1981, *Solar Phys.* **70**, 207.

Stenflo, J.O., 1975, *Solar Phys.* **42**, 79.

Stenflo, J.O., 1989a, *Astron. Astrophys.* **210**, 403.

Stenflo, J.O., 1989b, *Astron. Astrophys. Review* **1**, 3.

Stenflo, J.O., Solanki, S.K., Harvey, J.W., 1987, *Astron. Astrophys.* **171**, 305.
Stein, R.F. and Nordlund, Å, 1989, *Astrophys. J. (Letters)* **342**, L95.

Title, A.M., Tarbell, T.D., Topka, K.P., Ferguson, S.H., Shine, R.A., and the SOUP team, 1989, *Astrophys. J.* **336**, 475.

van Ballegooijen, A.A., 1982, *Astron. Astrophys.* **106**, 43.

Wang, H., 1989, *Solar Phys.* (in press).

Zirin, H., 1987, *Solar Phys.* **114**, 239.

RESULTS FROM 2-D NUMERICAL SIMULATIONS OF SOLAR GRANULES

M. STEFFEN, D. GIGAS, H. HOLWEGER, A. KRÜSS, H.-G. LUDWIG
Institut für Theoretische Physik und Sternwarte der Universität Kiel
Olshausenstrasse 40
D-2300 Kiel 1
Federal Republic of Germany

1 Calculations

We have carried out detailed numerical simulations of solar granular convection cells of different horizontal dimension. The calculations account for the basic physics of compressible convection, including the ionization of H I, He I and He II, and H_2 molecule formation as well as non-local, multi-dimensional radiative transfer (grey approximation in LTE). A more detailed description of the simulations has been given by Steffen and Muchmore (1988) and by Steffen et al. (1989).

Recently, the scheme for the computation of the radiation field has been improved to give a better angular resolution and to make sure that numerical integration of the rate of radiative energy exchange, $\mathrm{div} F_{\mathrm{rad}}$, over the model volume gives the same radiative surface flux as obtained directly from integration of the radiative transfer equation along the ray system. This greatly improves the accuracy of the numerical energy conservation, flux errors being of the order of a few per cent, as can be checked easily for steady state models. This is a satisfactory result in view of the steep gradients occurring in the models and the presence of a narrow layer where the energy transport changes from primarily convective to essentially radiative.

2 Steady State Models

Steady state solutions are found for model diameters below a critical upper limit. This limit was roughly 2000 km with the earlier version of the code, while it has not yet been determined with the new radiative transfer scheme, which gives steeper temperature gradients and larger horizontal temperature differences between hot and cool parts of the flow.

We always find a strong downdraft at the axis of symmetry with maximum velocities of the order of 6 km/s, which is surrounded by a broader ring-like upflow with lower velocity ($\approx$ 3 km/s). At the side walls we find another downflow. The convective velocity field extends considerably into the stable atmospheric layers. The rms value of the vertical velocity decreases approximately exponentially with height. The corresponding scale height is of the order of 200 km, the exact value being a function of cell size.

Granulation generates large horizontal temperature differences, reaching typically 4500 K about 100 km below the visible surface (τ=1). Equally remarkable, the calculations produce very steep vertical temperature gradients at the top of the ascending part of the flow (up to 80 K/km) where the hot gas cools rapidly due to efficient radiative energy losses (typically 10^{10}–10^{11} erg/g/s). The steep temperature gradient in concert with the recombination of

213

J. O. Stenflo (ed.), Solar Photosphere: Structure, Convection, and Magnetic Fields, 213–216.
© *1990 by the IAU.*

hydrogen produces a density inversion in this region, i.e. a layer of higher density lies on top of gas with lower density. In the cool intergranular downflows, however, density increases monotonically with depth.

In the overshooting layers the temperature fluctuations change sign. Here the rising gas is cooler than the surrounding sinking parts of the flow. This behavior is a consequence of the penetration of convective motions into stably stratified atmospheric layers. The uppermost layers of the steady state models is essentially in radiative equilibrium; the temperature is almost constant with height as expected for a grey radiative atmosphere. Horizontal temperature fluctuations are insignificant at heights $> 300\,\mathrm{km}$ above $\tau = 1$. For a detailed study of calculated steady state velocity and temperature fields as a function of horizontal cell size see Steffen et al. (1989).

From detailed radiative transfer calculations we find the rms intensity contrast, δI_{rms}, of the 2-dimensional intensity pattern to range typically between 14 and 16 % in the continuum at 5000 Å. This value seems in reasonable agreement with observational evidence (e.g. Bray et al., 1984). Towards smaller granular scales the amplitude of the horizontal intensity fluctuations decreases considerably (Steffen et al., 1989). The same is true for the variation of contrast toward the limb. We find a monotonic, roughly linear decrease of δI_{rms} with μ.

As has been demonstrated elsewhere (Steffen, 1987), line bisectors computed from the larger steady state models are in excellent agreement with observations. It must be noted, however, that the convective velocity field of the stationary models is not sufficient to fully explain the observed line broadening.

3 Non-stationary Models

Simulations with model diameters exceeding a critical limit never reach a steady state, not even asymptotically. They are truly instationary. Increasing the model diameter beyond this limit, the time evolution of the flow becomes more and more chaotic. It may be considered as a continuous splitting and merging of granules under the constraints imposed by the cylindrical symmetry. Significant changes occur on time scales of the order of 10 minutes (approximately one turnover time), comparable to typical granular lifetimes.

The magnitude of the velocity and temperature fluctuations in the deep photospheric layers is similar to that found in the steady state models. We obtain a time average of the rms intensity contrast that is not significantly different from that found in the steady state models, maximum values at individual instants of time not exceeding 20 %. The decrease of contrast towards the limb is, however, significantly less steep than in the steady state models. This is partly due to the somewhat different thermal structure of the photospheric layers. Furthermore, the considerable temporal variations of the emergent mean intensity translate into an additional contribution to the intensity contrast which varies only slightly across the solar disk.

Due to the time dependence of the flow topology the upper photosphere is much more affected by the convective motions than in case of steady state flows. Stochastic up- and downward motions of considerable amplitude are found in the line formation layers. Spatial and temporal averaging of this type of velocity field will probably result in a height dependence of the rms vertical velocity that shows a local minimum somewhere above the $\tau = 1$ level with an increase toward higher layers. Furthermore, the non-stationary velocity field should provide sufficient line broadening to explain observations. These points will need further study.

Test calculations have shown that the main effect on the line bisector is to shift it back and forth without seriously distorting its shape; the bisectors more or less keep their C-shapes as the flow evolves in time. We never found an inverted C-shape. Therefore, the line bisectors obtained from a time average of the simulations is similar to that derived from comparable steady state models.

It seems evident that the non-stationary models give a more realistic description of the real solar photosphere than steady state flows. In view of the results presented above we have to conclude that the dynamical and thermal structure of the solar photosphere is much different from the situation suggested by flux-constant mixing-length models. In particular, the upper photospheric layers are probably not horizontally homogeneous and their thermal structure is not exclusively determined by radiative equilibrium but also by dynamical phenomena.

4 Oscillations in Stationary Convection Cells

As mentioned above, all but the largest model granules studied with the earlier version of the code finally reach a stationary state. Superposed on the convective flow, almost sinusoidal vertical oscillations with a period of about 4 min are ubiquitous. A preliminary report has been given by Steffen (1988). The oscillation period depends only slightly on model diameter, increasing from about 200 s to 265 s over the diameter range 200–2000 km.

In all models the velocity amplitude increases with height, typically from about $\pm 100\,\text{m/s}$ at the lower boundary to several times that value in photospheric layers. The increase of amplitude with height is less steep than $1/\sqrt{\rho}$. All layers oscillate in phase. Together, this behavior is compatible with that of evanescent waves. The velocity oscillations are associated with oscillations of temperature that have a non-monotonic height dependence, attaining maximum values of typically $\pm 50\,\text{K}$ in the upper part of the granules where the transition from convective to radiative energy transport occurs.

These oscillations are remarkably persistent. In most models the amplitude remains constant over many periods; some models show a slow decrease, others an increase towards an asymptotic value which never exceeds 500 m/s. Extensive test calculations have shown that the basic properties of the oscillations are insensitive to details of the numerical scheme such as time step, grid spacing, or SGS viscosity. In addition, the vertical extent of the simulation volume turned out to be not critical.

Which physical properties of the simulated granules determine the frequency, ω_{sim}, of these oscillations? The striking independence of cell geometry seems to rule out an explanation in terms of a resonant cavity. More likely, ω_{sim} is determined by intrinsic properties of the convective flow. Our hypothesis is that ω_{sim} is related to the acoustic cut-off frequency, ω_{ac}, in the upper part of the granule close to $\tau{=}1$, where both the convective temperature gradient and the temperature oscillations attain their maximum, and the mass density has a local minimum. We suspect that it is here that the driving oscillator may be located. Indeed the value of ω_{sim} ($\approx 26\,\text{mHz}$) is found to closely coincide with ω_{ac} in all model granules. As another test of this hypothesis the dependence of ω_{sim} on gravity was investigated. We have run various simulations for solar type stars, keeping T_{eff} unchanged but varying $\log g$ by up to ± 0.1. Indeed ω_{sim} turned out to be strictly proportional to gravity. This supports our view that ω_{sim} is closely related to ω_{ac}, which, for an isothermal atmosphere, is given by $\omega_{\text{ac}} = \gamma g/2a$.

Our simulations imply that a stationary convective flow is able to generate oscillatory power in the five minute band, most probably through some kind of overstability mechanism. It must be expected that this contributes to the excitation of solar p-modes, in addition to stochastic interaction with instationary convection. Furthermore, an efficient damping mechanism must be active which keeps the amplitude in the linear regime. We suspect that radiative transfer at photospheric levels is the cause, but this has still to be studied in detail. It should be noted that the non-stationary simulations show basically the same kind of oscillations, superimposed on the stochastic convective velocity field.

5 Simulations for other Stars

Currently attempts are under way to extend our simulations to stars other than the Sun. As a first application the atmosphere of an early A-type main-sequence star of spectral type A0 V has been choosen. The atmospheric structure of such stars is generally considered to be well understood: convective energy transport is believed to be completely unimportant in the atmospheres of these stars, which are therefore described in terms of static, plane-parallel models.

A different picture came to light in the numeric simulations undertaken so far (see also Gigas, 1988, 1989). Although convective energy transport is still unimportant ($\approx 0.5\%$ of the total flux) in comparison to the energy transported by radiation, mainly vertical gas flows are encountered with velocities considerably larger than the predictions of mixing-length theory. Maximum flow velocities amount to a few hundred meters per second in deep atmospheric layers below the hydrogen ionization zone; similar to the solar case they increase towards smaller, spectroscopically accessible depths up to values of ≈ 2 km/s. The temporal power spectrum shows a dominant period of ≈ 1070 s, a secondary peak of ≈ 220 s, as well as higher frequency "noise". Like in the solar case, the main period seems to be related to the acoustic cut-off period in the upper part of the computational domain.

Corresponding temperature fluctuations show a maximum of $\approx \pm 500$ K in the region close to the hydrogen ionization zone. Contrary to the velocity field, pressure and temperature fluctuations appear to be in antiphase in the upper and in the lower part of the computational domain with a common phase shift of $\approx \pi/2$ relative to the velocity field.

The presence of such oscillatory velocity fields in the atmospheres of early-type stars may provide a satisfactory explanation for the microturbulence parameter ξ, which has to be employed in abundance analyses of such stars even if deviations from LTE are taken into account. Typically, microturbulence values of ≈ 1.0–2.0 km/s are required, which is in satisfactory agreement with the results encountered in our simulations so far. We are planning to extend these computations to stars of other effective temperature and gravity.

6 References

Bray, R.J., Loughhead, R.E., Durrant, C.J.: 1984, in: *The Solar Granulation*, Cambridge University Press, 2nd ed.

Gigas, D.: 1988, in: *The Impact of Very High S/N Spectroscopy on Stellar Physics*, Proc. IAU Symposium No. 132, eds. G. Cayrel de Strobel, M. Spite, p. 395

Gigas, D.: 1989, in: *Solar and Stellar Granulation*, eds. R.J. Rutten, G. Severino, Kluwer, p. 533

Steffen, M.: 1987, in: *The Role of Fine-Scale Magnetic Fields on the Structure of the Solar Atmosphere*, eds. E.-H. Schröter, M. Vazquez, A.A. Wyller, Cambridge University Press, p. 47

Steffen, M.: 1988, in: *Advances in Helio- and Asteroseismology*, Proc. IAU Symposium No. 123, eds. J. Christensen-Dalsgaard, S. Frandsen, p. 379

Steffen, M., Muchmore, D.: 1988, *Astron. Astrophys.* **193**, 281

Steffen, M., Ludwig, H.-G., Krüß, A.: 1989, *Astron. Astrophys.* **213**, 371

WAVES AND OSCILLATIONS IN THE NON-MAGNETIC PHOTOSPHERE

Franz - Ludwig Deubner

Institut für Astronomie und Astrophysik
der Universität Würzburg
Am Hubland, D-8700 Würzburg

1. Introduction

The solar photosphere is the outer boundary of the cavity which defines the spectrum of the global p-modes. It is also the lowest layer that permits direct observation of the boiling mixture of motions which eventually create the complex patterns apparent in the chromosphere and corona. Studies of the dynamical behaviour of this layer are therefore of paramount importance for an overall understanding of the dynamic sun.

Although the basic hydrodynamic processes occurring in and close to the "quiet" photosphere were described more than two decades ago, the verification of its detailed dynamic behaviour by observation created a complex puzzle some important pieces of which have only lately fallen into their proper places, while others, surprisingly, are still being added.

It is impossible, in the frame of this review, to give a fair and full account of the immense volume of observational accomplishments in the field of solar atmospheric dynamics achieved since, say, the classical report by Noyes (1967), when the 5-min oscillations were not yet recognized as a phenomenon of the solar interior. The interested reader is referred to more recent reviews by Frandsen (1988), and by Deubner et al. (1984).

After the discovery of the global coherence of the solar p-modes motions and waves in the visible atmosphere were studied for three major purposes, apart from Helioseismology:

(1) The heating of the outer layers of the atmosphere; in the quiet sun acoustic and gravity waves are still considered prime candidates for carriers of non-thermal energy, at least in the low chromosphere. Yet, a conclusive quantitative account of the contributions from various competitive MHD and non-MHD processes is not available. However, this challenging topic is outside the scope of this article.

(2) Investigations of stellar convection; measurements of motions and brightness distributions of granulation, meso- and supergranulation reveal basic physical properties of convection, against which theory and models need to be checked, before they can be confidently applied to the invisible interior of any star.

(3) The main emphasis of this review will be on spectral diagnostics of oscillations and waves. Power, phase and coherence spectra serve to define the character of a wave - propagating or evanescent, standing or running - as they can be easily related to the relevant dispersion relations, and to explore the physical properties of the medium that supports oscillatory motions. The latter aspect is particularly important for our understanding of the structure of the dynamical atmosphere.

217

J. O. Stenflo (ed.), Solar Photosphere: Structure, Convection, and Magnetic Fields, 217–228.

In the following chapter we shall very briefly review recent observational work describing the interaction of granulation and waves. Chapters 3 to 5 will center around some unresolved issues arising from new results of observational photospheric oscillation studies. It appears that the k-ω spectra of phase and coherence, while improving the wave diagnostics dramatically, are also good for intriguing surprises.

2. Low frequency waves and convection

The art of modelling solar (and stellar) convection by simulation calculations (Nordlund, 1984; Stein et al., 1989; Steffen et al., 1989) has now reached a degree of perfection, that permits direct comparison of observable parameters (brightness contrast, r.m.s. velocities, line width variations, bisectors etc.) with corresponding values derived from the models (Dravins et al., 1981; Wöhl and Nordlund, 1985; Steffen, 1989). However, there are still severe geometrical restrictions to those models; owing to the limited memory space in computers the volume studied in 3-D simulations typically includes no more than a few granules, and consequently the discussion of larger scale structures such as mesogranulation is mostly based on similarity arguments. Due to these circumstances there is (fortunately) still much room for fundamental observational contributions.

To begin with the larger scale phenomena, there is first the simple question, whether mesogranulation indeed constitutes a distinct regime of convective motions separated from granulation and from supergranulation by well defined gaps in the spatial velocity or intensity power spectra. Some evidence in favour of such a picture was recently presented and discussed by Deubner (1989), who finds enhanced power in the velocity and brightness distribution in a range of scales between 5" and 10"; but a firm conclusion on this issue may require an investigation of two-dimensional data. In fact, this seemingly simple question entails a rather difficult observational task: High spatial resolution (and very good seeing!) is needed as well as good wave number separation, i.e. a large field of view; since the velocity signal changes its structure and amplitude with distance from the disc center due to perspective, one must probably be satisfied with the rather low contrast brightness signal for the low wave number section of the spatial power spectra; the duration of the observation should comply with the typical time scales of the larger spatial scale phenomena (at least several hours), and with the necessity to separate them in frequency space from waves and oscillations.

A second question concerns the dynamics of solar convection near the photosphere, and of mesogranulation in particular. If mesogranulation is indeed convective in origin, as is generally assumed, the vertical flow pattern observed by November et al. (1981) should be correlated with the brightness distribution, as in the case of granules. Mesoscale brightness fluctuations have been observed as early as 1932 by Strebel and Thüring, and again by Koutchmy and Lebecq (1986), and by Oda (1984); a high degree of correlation with the flow pattern is confirmed by Deubner's (1989) analysis. The temporal phase differences observed between velocity fluctuations at different levels of the atmosphere, and between velocity and brightness fluctuations in various spectral lines as function of the spatial scale testify strongly for the convective origin of the observed quasi cellular motions at all spatial scales in the photosphere, and of mesogranules in particular.

The mesoscale horizontal velocities inferred from our power spectra are of the order of 750 m/s. They are in good agreement with the r.m.s. values derived from high resolution observations of granule proper motions (November et al., 1987; Brandt et al., 1988). This comparison justifies the previous assumption, that the observed proper motions represent in fact the horizontal bulk velocities of small scale convective elements. The observed vertical motions are fairly strong (~300 m/s) and obviously sufficient to suppress, impede or accelerate the evolution of individual granules, as seen in white light movies (Brandt et al., 1989). From Deubner's study follows also, that the observed ratio of vertical to horizontal flow velocity of convective motions depends nearly linearly on the spatial scale, being on the order of 1 for granules, 2.6 for mesogranules and about 20 for supergranules. In cylindrical geometry the continuity equation does indeed suggest such a linear scaling law.

A third area of interest is the overshoot region and the interaction of convection with the stable layers on top. The early observational finding by Evans and Catalano (1972), namely the inversion of the granular brightness contrast at a hight of approximately 100 km, has been confirmed by many observers, and is clearly present also in the simulations of Nordlund (....) and Steffen et al. (1989) as a consequence of the adiabatic expansion of overshooting elements into the stably stratified atmosphere. This is not the whole story, however, and the actual distribution of brightness and velocity relates in a very complex way to the convective pattern underneath. In a statistical study of the intensity fluctuations and Doppler shifts measured in individual high spatial resolution spectrograms, Nesis et al. (1988) find a rapid decay of the spatial correlation of the observed structures with height, suggestively described as a "loss of memory". An unknown independent velocity field is invoked to explain the observational results. Let us have a look at the kind of motions which might perturb a quasi stationary velocity and brightness distribution.

If these disturbances somehow formed a coherent pattern they should become apparent in a Fourier decomposition of the velocity or brightness data with regard to frequency and wavenumber, i.e. in k-ω diagrams of power or temporal phase lag. The SOUP experiment on the Spacelab 2 shuttle flight (Title et al., 1986) detected f-modes up to $l = 3500$ (1."7). In this wavenumber regime confusion with the velocity and brightness distribution of granulation is very likely, even if at this l value the fundamental mode period (~200 s) is considerably shorter than the granular lifetime. At lower frequencies (v≤4 mHz), and for $l \geq 1400$ Deubner and Fleck (1989) have demonstrated the presence of propagating internal gravity waves in the photosphere by comparing the phase lag between the Doppler shifts measured at different heights in the atmosphere. Locally coherent evanescent oscillations are present in several other areas of the k-ω diagram as we shall discuss in Chapter 3 and 4.

The relation between any of these wave fields and granular or mesogranular motions has not been studied in detail yet. Therefore, the question which of these motions and corresponding brightness fluctuations are indeed generated independently from the convective processes, and which ones occur always in the wake of granules can not be answered from the observational point of view at the present time.

3. Evanescent waves of various kinds

In an adiabatic stratified atmosphere, theory predicts a phase difference between velocity and temperature fluctuations of 90° in the evanescent regime, and an asymptotic approach to 0° at frequencies beyond the acoustic cut off. With thermal relaxation taken into account the phase in the evanescent regime tends to higher values, and should become 180° in the fully isothermal case. This behaviour was first studied extensively both observationally and in fundamental theoretical investigations by Schmieder (1976, 1977, 1978), and later by Lites and Chipman (1979), who confirmed Schmieder's observations.

Within certain limits, to be discussed in Chapter 5, those observations are in accordance with theoretical calculations for frequencies higher than ~ 2.5 mHz. But, in Lites and Chipman's (1979) Figure 1, a surprising phase discontinuity is observed at approximately 2.2

mHz which is clearly visible also in every single photospheric V-I phase spectrum displayed in Staiger's (1985) thesis. We present a collage of Staiger's observations in Figure 1.

Gravity waves can not be responsible for the discontinuity and the negative phases found between 1 and 2 mHz, because the V-I phases of gravity waves are rather similar to those of evanescent oscillations at the observed frequencies (Mihalas and Toomre, 1981). Superposition of evanescent waves with granular impulsive motions was therefore suggested as an alternative explanation (Lites and Chipman, 1979), although the discontinuity appears most conspicuously in the phase spectra of lines formed near the temperature minimum, and the absolute value of the maximum phase difference (50° - 60°) depends very little on the height of the line forming layer.

Decomposition of the phase difference spectra with respect to frequency *and* wavenumber (Deubner and Fleck, 1989, Figure 6) revealed that the highest positive phase values are connected to fluctuations observed at wavenumbers with a spatial scale larger than 5". The presence of convective motions at this scale (corresponding to the mesogranular pattern discovered by November et al., 1981) has indeed been demonstrated by Deubner (1989); however, the case for this particular regime of convective motions causing the strange phase discontinuity did not appear very convincing.

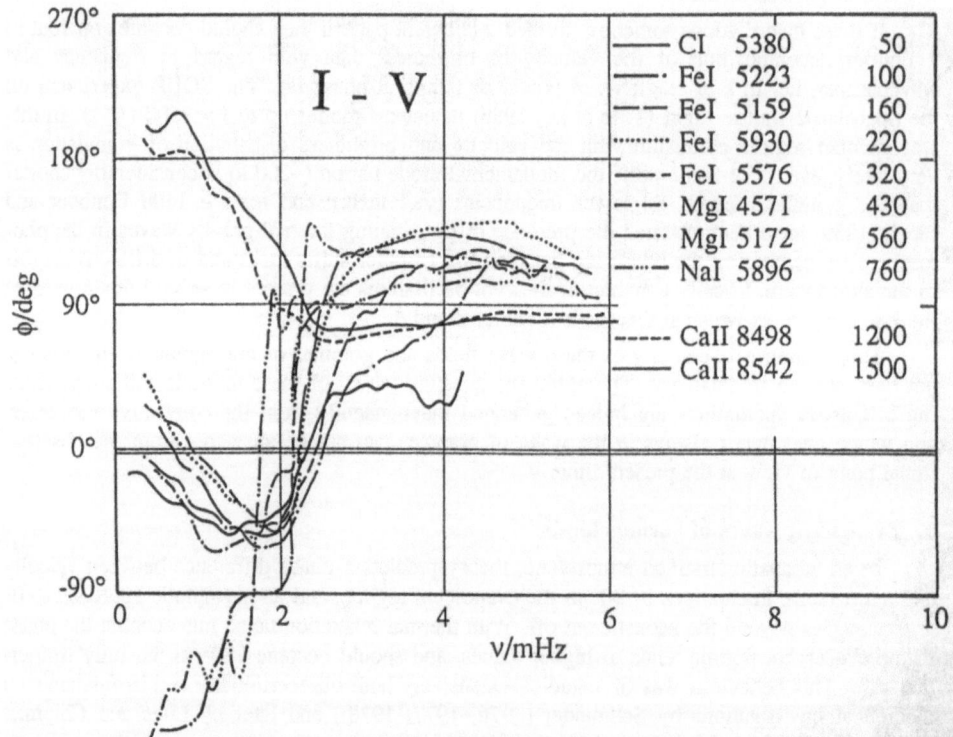

Fig. 1. Brightness - velocity phase spectra compiled from Staiger's (1985) thesis work. Positive phase indicates that brightness leads upward velocity. Approximate line formation heights are given in km.

At this stage, Marmolino and Severino (1990) presented theoretical phase difference diagrams in the k-ω plane, based on an isothermal model atmosphere with adjustable thermal relaxation. These diagrams indicated clearly, that the observed discontinuity is not related to the presence of large scale convective motions, but follows directly from the dispersion relation of waves in a stratified atmosphere. In fact, there are *two* such discontinuities present in the k-ω diagram, enclosing a trough with a triangular base (in the double logarithmic presentation of Marmolino and Severino, this conference) dubbed the "missing piece of cake" for its shape.

The position of the two discontinuities (i.e. the walls of the trough) correspond (a) to the linear dispersion relation of Lamb waves (horizontally propagating sound waves with $\omega=kV_s$) for the low frequency flank, and (b) to the dispersion relation of divergence free waves ($\omega^2=gk$) for the high frequency flank. In the latter case the locus of the discontinuity is characterized by a minimum of the pressure disturbation; the other discontinuity occurs where the vertical velocity component of the wave field has a minimum. According to this theoretical diagram we should compare the observed discontinuity in the frequency spectra (Figure 1) with case (b), since only here the phase jump goes in the right direction. The inverse slope of the phase at lower frequencies (<1.5 mHz) would then have to be associated with case (a).

Can the counterparts of the "missing piece" be seen directly in observed k-ω diagrams? The aforementioned Figure 6 in Deubner and Fleck (1989) does not fully match the theoretical diagram, because the data are only one-dimensional in the spatial coordinate, k_x. In Figure 2 we present the results of a recent study in the form of k-ω diagrams, where the horizontal wavenumber k was deduced from two-dimensional data. Now the p-mode ridges can be well recognized in the crosspower diagram in the lower left. The lowest one is the f- or divergence free mode. The Lamb waves are indicated by the slanting straight line. V-I phases are coded by two sets of grey levels which differ by 60° as indicated in the right column next to the phase diagrams.

In the high wavenumber regime, the phase diagrams are obviously dominated by noise and seeing effects, pulling the phase towards zero. Below the Lamb line the phase is negative (approximately -30°), and positive (approximately +90°) directly above it, in qualitative agreement with the theoretical model of a nearly adiabatic gas. Averaging of all spatial scales yields a gradually increasing contribution of positive phases with increasing frequency. This explains the smooth transition from positive to negative values in the 0 to 1.5 mHz range of the one-dimensional frequency spectra derived for the photospheric lines in Figure 1. (Figures 1 and 2 have opposite sign conventions!) The f-mode on the other hand appears in the diagram as a *ridge* with a distinct negative (~-90°) phase rather than as a *cliff*. The counterpart of the high frequency edge of the piece of cake apparently is not formed by the dividing line between "gravity like" and "pressure like" waves, but by the low frequency envelope of the power distribution of the low order p-modes. Again, averaging of the spatial coordinate readily explains the existence of a sharp ridge at ~2.2 mHz in the frequency spectra.

Alas! This finding does not concur in all aspects with theoretical expectation. It appears that more realistic simulations of atmospheric waves, including the global modes, are necessary to reproduce the observational results.

Our data indicate the presence of solar noise power in the range of frequencies occupied by low order global p- and g-modes. It is difficult to estimate its true amplitude because of the superposed effects of instrumental trends and seeing revealed by the coherence spectrum in Figure 2. Nevertheless, one needs to be aware of the possibility, that these helioseismologically most important modes are contaminated by a kind of noise, which can not be avoided by space experiments.

222

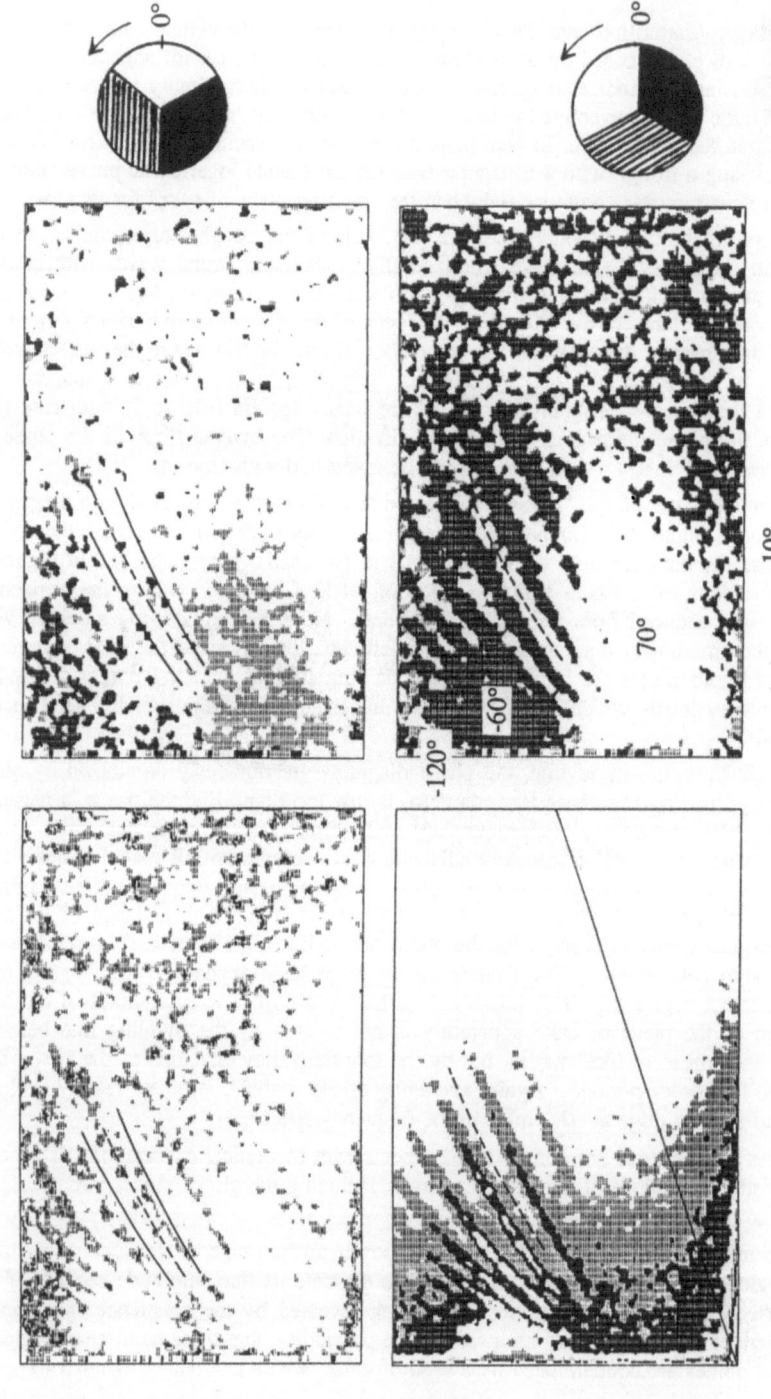

Fig. 2. Diagnostic diagrams of brightness and velocity fluctuations in the FeI 5576 line. *Lower left:* V*I crosspower. *Lower right:* V-I phase (grey level phase coding as indicated on the right hand). Positive phase indicates upward velocity leading brightness. *Upper right:* V-I phase (with different coding as indicated). *Upper left:* V*I coherence (darker shading indicates lower coherence). The acoustic cut-off frequency is near the upper edge of each frame. The full curves follow p-mode ridges, the dashed curve follows the valley between p_1 and p_2

4. More evanescent waves

Another reason for continued efforts to better simulate atmospheric waves can be seen next to the region of the phase diagram we have just discussed. As we shall discuss in the next chapter, the low order p-mode ridges exhibit phases which are in satisfactory agreement with theoretical expectation for evanescent waves in an adiabatic gas, namely close to -90°, in Figure 2. However, side by side with the p-mode ridges, the phase observed in the gaps is drastically different, and probably equal to the value of +90° assumed within the trough as discussed in Chapter 3. This is very clearly visible between the f -, p_1- and p_2-modes. With higher order the modes begin to merge, and due to the limited resolving power in our observations the gap between p_2 and p_3 is the last one that shows the effect.

We might wonder - since there is much less power in between the ridges - whether the effect is just an illusion due to increased noise in the data. However, incoherent random noise in the phase diagram would have a quite different appearence in our display, with speckles of all three grey scales evenly distributed, whereas the interridge regions as well as the adjacent bottom of the "trough" are fairly uniform in phase.

We have tried to substantiate this statement by computing the coherence of the velocity and brightness fluctuations. The result is displayed in the upper left part of the Figure, decreasing coherence being indicated by darker grey scales. In agreement with our judgement based on the phase displays the coherence is large in the trough and in the adjacent interridge regions, as it is on the ridges themselves. Only at the boundary between the ridge and the gap region the coherence drops sharply for obvious reasons. This can be seen e.g. between the p_1 and p_2 ridges.

If the phase anomaly in between the ridges is not an artefact of either the data analysis or seeing (image motion) which we don't believe after several numerical tests we made, we feel encouraged to speculate about a possible solar origin of the weak coherent signal.

It is very unlikely, that the weak continuum under the ridges is caused by short range impulsive events, because in such a case (if the time lag between velocity and brightness is different from zero) one would expect a linear phase dispersion as function of frequency. This is not observed. The only "conservative" assumption left at the moment appears to be the existence of a different type of evanescent waves whose progressive part carries energy downward. A similar type of solution had once been invoked by Hill et al. (1978) to account for the discrepancy between p-mode amplitudes observed as Doppler effect on the disc and the limb oscillations studied with the SCLERA instrument. Here we talk about oscillations being scattered in the inhomogeneous upper atmosphere, with their downward wave fronts directed arbitrarily, which are then no more globally coherent, and therefore uniformly distributed in wavenumber space. In a recent study of the 5-min oscillations Stebbins and Goode (1987) find occasionally a downward phase speed, at certain positions on the sun, especially where the photospheric amplitude is low. It is most interesting to note, that preliminary numerical simulations carried out by Severino and Marmolino (priv. communication) do indeed indicate a V-I phase behaviour that is different from the one observed in the upward solution.

With the present data further hints to the solution of this problem might be expected from studies of the vertical amplitude profile of motions observed between the ridges. A rather more direct approach is obviously to check the cross spectra of velocities measured at different hight levels, and their phases. Unfortunately no such data are available at this moment which have sufficient resolution in both wavenumber and frequency.

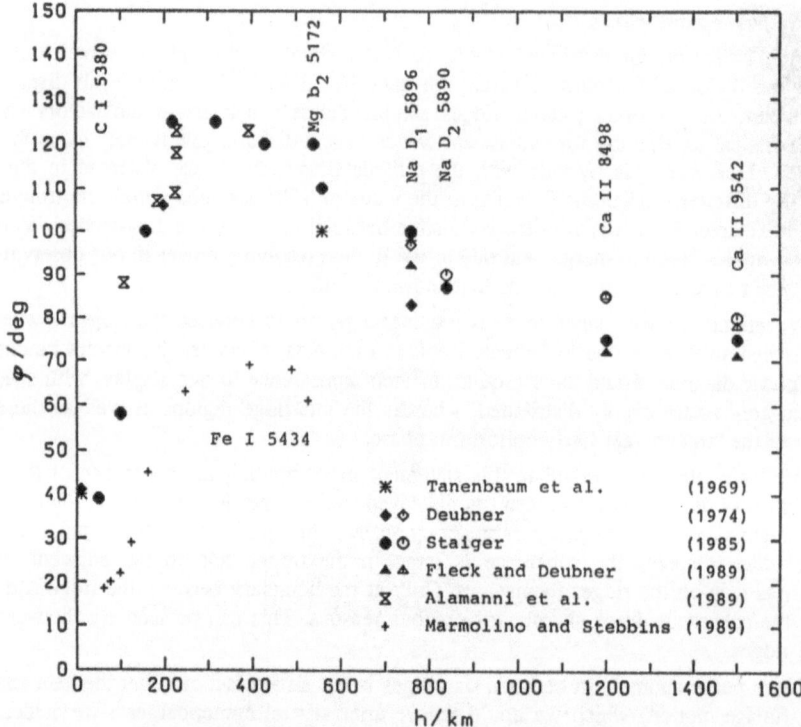

Fig. 3. Phase lag between brightness and velocity in the evanescent frequency range (5.5 mHz: open diamonds, dots and triangles; 3.3 mHz: all other symbols). Positive phase indicates brightness leading upward velocity. See text for the FeI 5434 results.

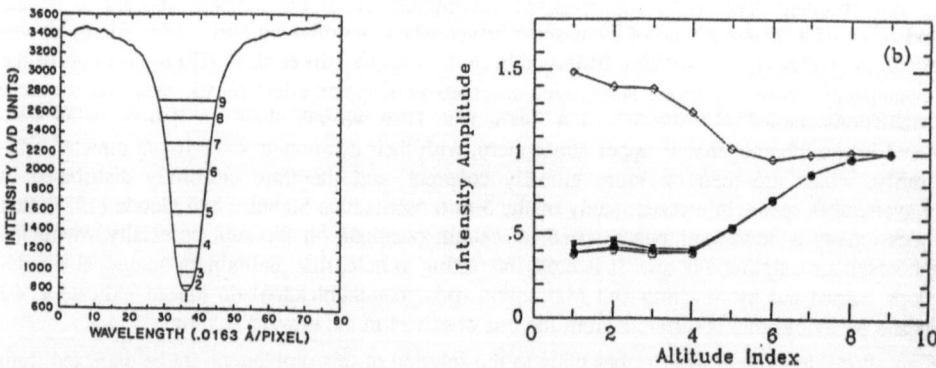

Fig. 4. a: Photometric profile of the FeI 5434 line, with 9 intensity levels at which the V-I phases printed in Fig. 3 with a "+" were determined. Courtesy R.Stebbins and Ph.R.Goode. **b:** Relative brightness amplitudes of p-mode oscillations in the FeI 5434 line (diamonds: observations; other symbols: various theoretical simulations) as a function of height. The altitude index refers to the levels indicated in Fig. 4a. Courtesy C.Marmolino and R.T.Stebbins.

5. The height dependence of the Velocity-Intensity phases of solar p-modes

From the theory of evanescent waves we recall that phase values between 90° and 180° are predicted in an atmosphere with thermal relaxation. The exact value depends on the relaxation time τ and on the frequency ω. Above the temperature minimum values close to 90° (for adiabatic conditions) are expected which should increase continuously to higher values as the photosphere (which is more nearly isothermal) is approached. As we can see in Figure 1, a maximum is reached at a height of about 220 km above $\tau_{5000}=1$ in the FeI 5930 line; but closer to the photosphere the V-I phase value decreases again. This was already noted long before and has since been studied by several observers whose results are compiled in Figure 3 (with references therein). The Figure evidences a rapid decrease of the phase to values close to 40° in the lowest 200 km.

It should be kept in mind, that *in the evanescent regime* phase differences observed between intensity fluctuations at different heights in the atmosphere are almost entirely due to the height dependence of the radiation effects acting on the I signal; since the differential phase lag of the corresponding velocity signals across the full range of heights covered in our diagram is less than 10°, it is also possible to deduce the differential phase lag of the intensity signals from the difference of the corresponding V-I phases.

Recent measurements obtained by Lindsey and Roellig (1987) in the infrared continuum ($\lambda = 350\mu$ and 800μ) are in excellent agreement with the V-I phases in Figure 3, bearing in mind that the infrared signals are emitted at levels corresponding to the line forming layers of MgI 5172 and NaI 5896 respectively. No special damping mechanism has to be assumed at the higher level to explain the observed vertical phase lag of 25° to 35°. The value matches closely the corresponding phase lags of the brightness fluctuations in the visual lines at the frequencies between 3 and 6 mHz, as may be seen in Figures 1 and 3. On the contrary, the phase lag observed at the upper level is a consequence of radiative damping at the lower level.

Good agreement exists also with the observations of Jimenez et al. (1989), who report average I-V phases between -140° and -145° at 3.3 mHz. It has to be considered, that their brightness signal is measured in the continuum, i.e. in the lowest photosphere, and is compared with the velocity signal of the potassium resonance line at 770 nm. With the sign convention of Figure 3 (upward velocity positive) the corresponding phase difference is 35° to 40°, i.e. close to the phase lag measured in the CI 5380 line.

Recent theoretical work on the dynamics of solar p-modes in the atmosphere has been published by Marmolino and Stebbins (1989) and - along the lines of the classical work of Schmieder - by Alamanni et al. (1989) and Cavallini et al. (1987). Neither one of these contributions has been successful in explaining the decrease of V-I phases below 90° in the photosphere as a property of the p-modes. As in the case of the "missing piece of cake" it was suggested that superposition of the evanescent wave field with convective motions would bias the phase (Schmieder, 1976). Since the phenomenon occurs indeed only close to the photosphere, the argument appears far more convincing in the present context. By the way, the phase values for the FeI 5434 line in Figure 3, taken from the work of Marmolino and Stebbins (1989), have been deduced by averaging the phase signal in the frequency range from 1.3 to 6.3 mHz, which includes a large portion of the 2 mHz trough in the phase spectrum, causing a systematic underestimate of these values.

To substantiate or disprove the superposition hypothesis we have constructed k-ω phase diagrams such as Figure 2 for other spectral lines, CI 5380 and FeI 5576. The search for other sources of a coherent signal with a small V-I phase was unsuccessful in either wavelength; there is just not enough power in the evanescent range at any other spatial scale than that of the 5-min oscillations. More importantly, the phases measured in the very ridges follow the same trend with height in the atmosphere, and correspond very closely to those found previously in the frequency spectra, as plotted in Figure 3.

Now, that the observational homework seems to be finished, the solution of the riddle of the p-mode phases can be safely delegated to theory where another problem is already waiting: In their study of the dynamics of p-mode oscillations in the atmosphere Marmolino and Stebbins (1989) have pointed out that the height dependence of the intensity amplitude is seriously at variance with theoretical predictions of all models studied so far (see Figure 4). Rather than the expected decrease of the amplitude of brightness fluctuations, a significant rise of the amplitude with height is observed on the sun; if the simulated signal is calibrated with the observed amplitude at the lowest photospheric level, the difference is as large as a factor of 3 - 4 at a height of 500 km.

In Figure 5 we have qualitatively depicted the effects on the V-I phase and on the brightness amplitude we have just discussed. V_- symbolizes upward velocity (blue shift), which lags temperature T by 90° in the adiabatic case. If brightness was a good proxi of temperature its phase vector (dashed arrow) would point towards the long dashes with increasing numbers for increasing non-adiabaticity. The limiting case evidently would be an isothermal medium with a phase angle of 180° between V_- and T. In order to obtain a phase angle smaller than 90°, as observed, opacity effects must be important in the photosphere and have to be taken into account in the model calculations.

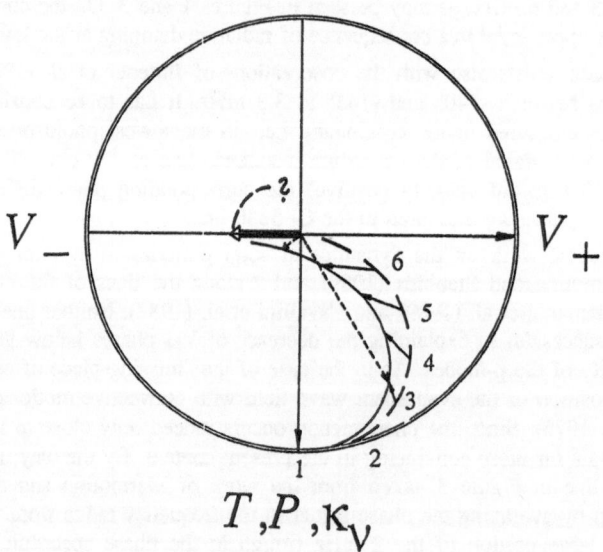

Fig. 5. Schematic phase diagram illustrating the effects of non-adiabaticity and opacity on velocity - brightness phase lags; see text.

6. Conclusion

Recent theoretical and observational studies have provided an improved understanding of the interaction of convection (in particular granulation) and photospheric oscillations and waves. The hazy picture of the vertical structure of the overshoot region previously derived from coherence studies based on individual high spatial resolution spectra becomes clearer through the analysis of time series, which reveal the existence of coexisting gravity and compression free wave fields, presumably excited by the granules.

On the other hand, the penetration of the global p-mode oscillations into the visible atmosphere, and the whole range of evanescent waves is less well understood than we thought it was. Existing models do not yet reproduce the phase relations between velocity and brightness observed in the low photosphere, nor are the phases in the vicinity of the f- and the low order p-modes in accordance with the customary picture. Apparently we have to regard these regions in the k-ω diagram as "white spots" which still need a detailed survey by observation, but at the same time we need urgently a more realistic theoretical modelling.

Acknowledgements

I would like to express my sincere gratitude to the many colleagues who responded to my request of material for this review, and to J. Laufer and B. Fleck who provided valuable practical support at various stages of the research that was incorporated in this report.

References

Alamanni,N., Bertello,L., Cavallini,F., Ceppatelli,G., Righini,A.: 1989, *Astron.Astrophys.* (submitted)

Brandt,P.N., Scharmer,G.B., Ferguson,S., Shine,R.A., Tarbell,T.D., Title,A.M.: 1988, *Nature* **335**, 238

Brandt,P.N., Scharmer,G.B., Ferguson,S., Shine,R.A., Tarbell,T.D., Title,A.M.: 1989, in *High Spatial Resolution Solar Observations* (O. von der Lühe, ed.) Sacramento Peak, Sunspot N.M. (in print)

Cavallini,F., Ceppatelli,G., Righini,A., Alamanni,N.: 1987, *Astron. Astrophys.* **173**, 161

Deubner.F.-L.: 1974, *Solar Phys.* **39**, 31

Deubner,F.-L.: 1989, *Astron.Astrophys.* **216** , 259

Deubner,F.-L., Endler,F., Staiger,J.: 1984, in *Oscillations as a Probe of the Sun's Interior* (G.Belvedere, L.Paterno, eds.) *Mem.Soc.Astron.Ital.* **55**, 147

Deubner,F.-L., Fleck,B.: 1989, *Astron.Astrophys.* **213**, 423

Dravins,D., Lindegren,L., Nordlund,Å.: 1981, *Astron.Astrophys.* **96**, 345

Evans,J.W., Catalano,C.P.: 1972, *Solar Phys.* **27**, 299

Fleck,B., Deubner,F.-L.: 1989, *Astron.Astrophys.* (in print)

Frandsen,S.: 1988, in *Advances in Helio- and Asteroseismology* (J. Christensen-Dalsgaard, S.Frandsen, eds.) D.Reidel, Dordrecht. p. 405

Hill,H.A., Rosenwald,R.D., Caudell,T.P.: 1978, *Astrophys.J.* **225**, 304

Jimenez,A., Pallé,P.L., Roca Cortés,T., Andersen,N.B., Domingo,V., Jones,A., Alvarez,M., Ledezma,E.: 1989, in *Seismology of the Sun and Sun-like Stars*, Puerto dela Cruz, Tenerife (in print)

Koutchmy,S., Lebecq,D., 1986, *Astron.Astrophys.* **169**, 323

Lindsey,C., Roellig,T.: 1987, *Astrophys.J.* **313**, 877

Lites,B.W., Chipman,E.G.: 1979, *Astrophys.J.* **231**, 570

Marmolino,C., Severino,G.: 1990, in *Solar Photosphere: Structure, Convection, and Magnetic Fields* (J.O.Stenflo, ed.)

Marmolino,C., Stebbins,R.T.: 1989, *Solar Phys.* (submitted)

Mihalas,B.W., Toomre,J.: 1981, *Astrophys.J.* **249**, 349

Nesis,A., Durrant,C.J., Mattig,W.: 1988, *Astron.Astrophys.* **201**, 153

Nordlund,Å.: 1984, in *Small Scale Dynamical Processes in Quiet Stellar Atmospheres* (S.Keil, ed.) Sunspot N.M. p.181

November,L.J., Toomre,J., Gebbie,K.B., Simon,G.W.: 1981, *Astrophys.J.* **245**, L 123

November,L.J., Simon,G.W., Tarbell,T.D., Title,A.M., Ferguson,S.: 1987, in *Theoretical Problems in High Resolution Solar Physics II* (G.Athay,D.S. Spicer, eds.) NASA Conf.Publ. **2483R**, 121

Noyes,R.W.: 1967, in *Aerodynamic Phenomena in Stellar Atmospheres* (R.N.Thomas, ed.) New York, Academic Press. p.293

Oda,N.: 1984, *Solar Phys.* **93**, 243

Schmieder,B.: 1976, *Solar Phys.* **47**, 435

Schmieder,B.: 1977, *Solar Phys.* **54**, 269

Schmieder,B.: 1978, *Solar Phys.* **57**, 245

Staiger,J.: 1985, *Thesis* (Freiburg i.Br.)

Stebbins,R., Goode,Ph.R.: 1987, *Solar Phys.* **110**, 237

Steffen,M.: 1989, in *Solar and Stellar Granulation* (R.J.Rutten, G.Severino, eds.) Kluwer Acad. Publ., Dordrecht. p.425

Steffen,N., Ludwig,H.G., Krüß,A.: 1989, *Astron.Astrophys.* **213**, 371

Stein,R.F., Nordlund,Å., Kuhn,J.R.: 1989, in *Solar and Stellar Granulation* (R.J.Rutten, G.Severino, eds.) Kluwer Acad. Publ., Dordrecht. p.381

Strebel,H., Thüring,B.: 1932, *Zs.Ap.* **5**, 348

Tanenbaum,A.S., Wilcox,J.M., Frazier,E.N., Howard,R.: 1969, *Solar Phys.* **9**, 328

Title,A., Tarbell,T., Simon,G., and the SOUP team: 1986, *Adv.Space Res.* **6**, 253

Wöhl,H., Nordlund,Å.: 1985, *Solar Phys.* **97**, 213

WAVES AND OSCILLATIONS IN MAGNETIC FLUXTUBES

M.P. RYUTOVA
Institute of Nuclear Physics
630090 Novosibirsk, 90
USSR

ABSTRACT. According to observational data solar magnetic fields have a pronounced filamentary structure. Theoretical investigations of plasmas containing structured magnetic fields, including the study of the properties of these structures and their interactions with associated gas flows, are of great importance for our understanding of the basic processes in the solar atmosphere, whose structure and dynamics are dominated by magnetic fields. In the present review theoretical models of thin magnetic fluxtubes and their behaviour in the ambient plasma are discussed.

1. INTRODUCTION

A situation when the magnetic field is concentrated into randomly distributed bundles of field lines is often met in laboratory and space plasmas. It is well known that all the solar magnetic fields, from the convection zone to the heliosphere, have a pronounced filamentary structure (cf. Stenflo, 1989 [1] and References therein). In the photosphere the magnetic field is concentrated in almost vertical, thin (about 300 km) fluxtubes, usually widely separated from each other, with field strengths of the order of 1-2 kG. In sunspots intense (3-4 kG) fluxtubes are assumed to be tightly packed. The isolated fluxtubes are as a rule localized at the supergranular cell boundaries, extending from the subsurface regions through the photosphere and chromosphere, where they form a great variety of magnetic structures. These structures generally have longitudinal dimensions much larger than the transverse ones, and are exposed to the action of a constantly "booming" atmosphere, which results in the generation of different kinds of waves and oscillations. For a better insight into the active processes in the solar atmosphere, such as field concentration, the processes of energy transfer from the lower to the upper layers of the atmosphere, the processes of energy storage and release, preflare and flare processes, etc., the properties of both isolated tubes and ensembles of them should be analysed. Besides its significance for astrophysical objects, such a study is also of interest from the point of view of general physics, due to the wealth of wave processes in such structures.

J. O. Stenflo (ed.), Solar Photosphere: Structure, Convection, and Magnetic Fields, 229–249.

In general, the studies of the problem show a rapid progress. There should be mentioned the essential contributions by P.R.Wilson, L.E.Cram, E.Parker, W.Unno, E.R.Priest, N.Weiss, J.Hollweg, H.Spruit, B.Roberts, M.Schüssler and others (see [2] and References therein). But main problems raised by observational evidence are not yet understood, though numerous theories that address these problems have been proposed. The physics of magnetic fluxtubes remains today of great interest. In the present paper I will give a survey of some works which are devoted to the studies of oscillations of magnetic fluxtubes. I will not try to apply these results to explain any particular observational data but there is a hope that the theoretical results which will be presented can give at least some reasonable frame for the analysis of particular situations at the Sun.

As it was mentioned above, the permanently booming atmosphere, in particular, motions in convective zone excite several types of oscillations of fluxtubes. We shall consider long-wave oscillations whose wavelength $\lambda = 1/k$ is much larger than the radius of magnetic fluxtube R: kR << 1. Just these oscillations are most readily excited by large-scale plasma motions and have a relatively low damping rate. Among these oscillations the most important modes are the two shown in Fig.1: a) the bending (kink) oscillations which are actually the dipole mode corresponding to the azimuthal wavenumber m = ±1, and b) axisymmetric sausage mode (m = 0).

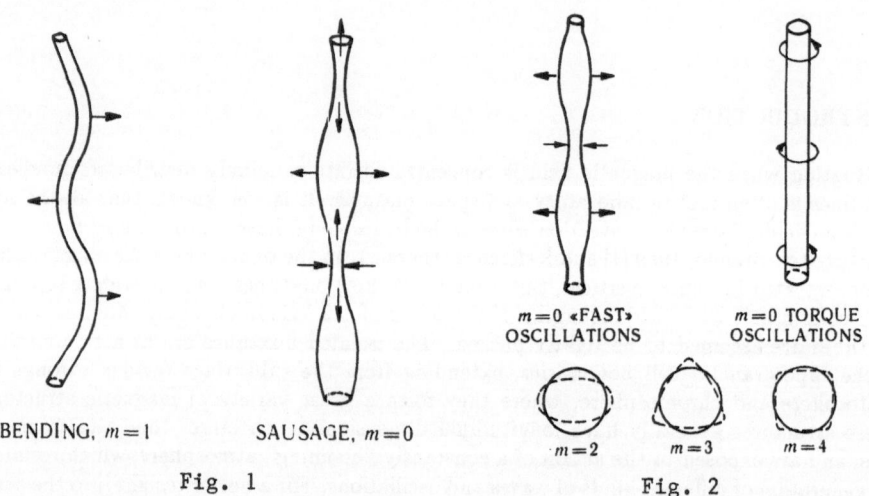

$m=0$ «FAST» OSCILLATIONS

$m=0$ TORQUE OSCILLATIONS

BENDING, $m=1$ SAUSAGE, $m=0$

$m=2$ $m=3$ $m=4$

Fig. 1   Fig. 2

For both types of oscillations the frequency scales linearly with wavenumber.

The bending oscillations are some analogue of Alfvén waves. As the external plasma participates in the motion in the vicinity of a tube via the "added mass" effect its density enters their dispersion relation (Ryutov & Ryutova, 1976):

$$c_b = \frac{\omega}{k_z} = \frac{a}{\sqrt{1 + \rho_e/\rho_i}} \quad . \tag{1}$$

Here subscript "i" refers to the tube interior, while "e" - to the external plasma; ρ is the plasma density, $a = B/\sqrt{4\pi\rho_i}$ is the Alfvén velocity.

The sausage mode is specific quasi-longitudinal oscillation of fluxtube in which a compression (expansion) of a plasma inside the tube is compensated by the decrease (increase) in the longitudinal magnetic field due to a corresponding change in the cross section of fluxtube, so that the sum of gas-kinetic and magnetic pressures is almost not perturbed. Thus the plasma parameters outside the tube have little influence on their dispersion relation (giving the corrections of the order of $(kR)^2$) (Defouw, 1976):

$$c_T = \frac{\omega}{k_z} = \frac{a \; s_i}{\sqrt{a^2 + s_i^2}} \quad . \tag{2}$$

Here $s = \gamma P/\rho$ is a sound speed (γ is a specific heat ratio).

Some less important modes are shown in Fig. 2. There are high frequency fast oscillations which are analogue of fast magnetosonic waves. As their frequency is extremely high (of the order of a/R) they can hardly be excited and experience fast radiative damping. The $m = 0$ torsional oscillations are just Alfvén wave, but their amplitude is very small (of the order of $V_{conv.} \cdot R/L$, where L is convective cell size and $V_{conv.}$ is characteristic velocity in convective zone) and also are of less interest. The oscillations with the higher azimuthal mode numbers $m = 2; 3,...$ are very weakly coupled with the large scale motions of medium since the matrix elements which determine this coupling contain a small parameter $(kR)^{|m|}$.

It should be noted that for theoretical investigation of the properties of fluxtubes and their ensembles (including the model of a spot as a cluster of intense fluxtubes) the concept of a thin fluxtube has proved to be very fruitful and justified. All mentioned above relates to the thin fluxtube model. Even in this concept the physics of plasma containing fluxtubes is very rich.

There exist specific damping mechanisms consisting in the radiation of secondary acoustic waves into external plasma by oscillating fluxtube (Ryutov and Ryutova, 1976), and in resonance excitation of Alfven waves with continuous spectrum in the region where the phase velocity of oscillations is close to local Alfvén velocity (Ryutova, 1977, Ionson, 1978). These effects contribute to the heating of upper layers of atmosphere. The gravity leads to the changing of the character of waves, which become dispersive. Another effect is that in the simple case of isothermal atmosphere amplitude of oscillations increases exponentially with height which leads to the development of   strong nonlinear effects in tube environment (Spruit, 1981). In stratified atmosphere there appears the frequency cutoff which leads to the existence of propagating (above cutoff frequency) and evanescent (below cutoff frequency) waves (Defouw,1976; Roberts and Webb, 1978,1979). An evanescent wave can be regarded as the quasistatic response of a tube to the slow changes in the region of wave excitation. If the wave is generated impulsively at the base of a tube it propagates with height in the form of a wave front with velocity of corresponding wave mode. Behind the wave front is trailed a wake oscillating at the cutoff frequency (Rae and Roberts, 1982). For the compressive case the formation of shocks both from the wave front and from the oscillating wake takes place (Hollweg, 1982).

When fluxtube parameters have smooth radial dependence there appear very peculiar evolution of the radial mode structure of tube waves: it becomes more and more spiky at higher altitudes and respectively the dissipative processes become important, which results in a faster damping of the wave. The longitudinal dependence of the energy flux becomes nonexponential. Statistical analysis of the case when tube experiences random oscillations caused by photospheric convective motions shows that even "white noise" of convective zone can result in a temporal brightening of the tube region at definite height (Ryutova, 1989).

Very interesting physics is brought about by taking into account the presence of shear flow along the fluxtubes. The plasma flows with different velocities inside and outside fluxtube gives rise to the qualitatively new effects: the appearance of negative energy waves, reversal of the sign of radiative damping, the development of explosive instability at nonlinear stage, and the development of linear hydrodynamic instability similar to instability of tangential discontinuity (Ryutova, 1988).

The picture of nonlinear effects in structured magnetic fields even at the stage which we have today (with a great work yet to be done) is very rich. Roberts and Mangeney (1982) have shown that when gravity is ignored in a slab geometry the Benjamin-Ono type solitons can appear. After this paper, solitary waves in a slab and cylindrical tube are extensively studied. Roberts (1985) showed that slow surface waves are governed by nonlinear integrodifferential equation which possesses the soliton solution. Edwin and Roberts studied Benjamin-Ono-Burgers type equation and concluded that the estimation of the damping rate of corresponding solitary wave enables these waves to propagate from lower layers to upper chromosphere unharmed. Molotovshchikov and Ruderman (1987) obtained the equations describing the long nonlinear sausage waves (both slow and body) in fluxtube in magnetic environment. Note that under the conditions when negative energy waves are excited in the system the solitary waves with explosively growing amplitude can appear.

Another class of nonlinear effects is connected with the backward effect of longwave oscillations on plasma which consists in the generation of secondary plasma flows and electric currents. These phenomena lead to macroscopic effects which can play an essential role in the dynamics of structured magnetic fields and which, in principle, can be observed. The effect of secondary plasma flows (Ryutova, 1986) is similar to the effect of "acoustic" or "quartz wind" in usual hydrodynamics but its picture in MHD is more complicated. At the propagation of longwave oscillations along the fluxtube there arise stationary vortex flows in the plane perpendicular to the magnetic field and upward mass flows along the field. The main effect of the "magnetosonic wind" is that the fluxtube is vanishing "diffusively" or is splitting into thinner independent tubes. If the absorption of oscillation energy is determined mostly by one of the plasma components then besides the generation of "magnetosonic wind" it is accompanied by the excitation of currents. The effects mentioned above are connected with the oscillations of fluxtubes. It is worth mentioning that there exists another mechanism of current drive and, respectively, of a generation of magnetic fields which is not necessarily connected with fluxtube oscillations. Generally speaking the absorption of the momentum of usual acoustic waves generated in solar atmosphere results in a transfer of the momentum to plasma electrons and ions. It is shown that in a collisional case of solar chromosphere where damping of acoustic waves is caused by the nonlinear effects (formation of weak shocks) quite strong currents and magnetic fields are generated (Ryutov & Ryutova, 1989).

The problem connected with the properties of ensemble of magnetic fluxtubes is explored much less than the properties of separate fluxtube. Ryutov and Ryutova (1976) studied the propagation of sound waves in a plasma containing an ensemble of randomly distributed magnetic fluxtubes and found that even in the absence of any dissipative effects (viscosity, thermal conductivity, Ohmic losses) sound waves are absorbed due to the effect similar to Landau damping and consisting in the resonance excitation of fluxtube oscillations. In a much longer time than the time during which the energy of outer motions is transferred into the energy of fluxtube oscillations magnetic tubes release their energy in the form of secondary sound waves into the upper layers of atmosphere. The contribution of noncollinearity of fluxtubes into the resonant absorption and resonant scattering of sound waves is found. Quite recently Bogdan (1989) studied the resonance scattering of sound waves by fluxtubes paying the most attention to the interaction of solar p-mode with fluxtubes and indicating the importance of this effect as a diagnostic probe of the structure of solar magnetic flux concentrations (see also Bogdan & Zweibel, 1985, 1987; Bogdan & Cataneo, 1989). Ryutova & Persson (1984) studied dispersion properties of a plasma containing small scale random inhomogeneities when density and magnetic field change by the order of unity at a length small compared to the wavelength (the model of a spot as a cluster of intense fluxtubes). Unlike the case of homogeneous plasma the wave propagation in such system is accompanied by vortex motion of plasma having the same scale as that of the nonhomogeneities. The main effect which takes place in such a clusters of fluxtubes is the enhanced dissipation of MHD waves caused by large local gradients of temperature, velocity, etc.

Of course, the present review is far from complete and in addition to References presented here I end the Introduction with some more of them. Namely, the papers of N.O.Weiss (1981), Hasan & Schüssler (1985), Ferriz-Mas (1988), Hollweg (1987), Hollweg et al. (1989), Seehafer

(1988), Henoux & Somov (1987), and earlier reviews by Spruit (1981), Thomas (1985) and Priest (1988).

2. LINEAR EFFECTS

2.1. Some basic properties of fluxtube oscillations

The study of fundamental modes of fluxtube oscillations is based on the linearized MHD-equations complemented by the equilibrium condition of fluxtube in unperturbed state

$$P_i(r) + \frac{B^2(r)}{8\pi} = P_e . \tag{3}$$

Here P_i and P_e are gas-kinetic pressures inside and outside the tube and B is the magnetic field inside the tube. The condition (3) is written for cylindrical fluxtube embedded in a field free medium. For the perturbations proportional to $\exp(-i\omega t + ikz + im\varphi)$ (we use the cylindrical coordinates with z-axis coinciding with tube axis) in the absence of gravity the phase velocities of two fundamental modes, kink (m = 1) and sausage (m = 0), have a form (1) and (2), respectively.

When gravity is included there appears a buoyancy force which changes the character of tube oscillation.

For isothermal case neglecting the compressibility of medium the equation for kink oscillations can be written as (Spruit, 1981a,b):

$$\frac{\partial^2 \xi}{\partial t^2} = c_b^2 \frac{\partial^2 \xi}{\partial z^2} + g\left(\frac{\rho_i - \rho_e}{\rho_i + \rho_e}\right) \frac{\partial \xi}{\partial z} ; \tag{4}$$

ξ is the transverse displacement of the tube (which is much larger for these oscillations than the vertical displacement of the plasma inside the tube). The solution of Eq. (4) gives for the amplitude

$$\xi \sim e^{-i\omega t + ikz + \frac{z}{4H}} \tag{5}$$

and for the frequency

$$\omega^2 = c_b k^2 = \frac{c_b^2}{4H} \tag{6}$$

where H is a pressure scale height and $\omega_c^b = \frac{c_b}{4H}$ is the cutoff frequency.

The same holds for the sausage oscillations. The corresponding equations written for the vertical component of velocity which for these oscillations in contrast with the kink mode is much larger than the transverse one, have a form (Defouw):

$$\rho \frac{\partial v_z}{\partial t} + \rho v_z \frac{\partial v_z}{\partial z} \simeq - \frac{\partial P}{\partial z} + \rho g$$

$$\frac{\partial}{\partial t} \left(\frac{\rho}{B}\right) + \frac{\partial}{\partial z} \left(\frac{\rho u}{B}\right) = 0 . \tag{7}$$

These equations together with the equilibrium condition of a thin fluxtube (3) give for the isothermal atmosphere the following disper-sion relation:

$$\omega^2 = c_T^2 k^2 + (\omega_c^T)^2 \tag{8}$$

with the cutoff frequency

$$\omega_c^T = \frac{c_T}{4H} \left(9 - \frac{8}{\gamma} + \frac{16(\gamma - 1)}{\gamma^2} \frac{s^2}{a^2}\right)^{\frac{1}{2}} .$$

As to solution for the velocity amplitude it has the same form as (5). (Appropriate studies for the arbitrary temperature profile $T(z)$ see Roberts and Webb (1978); for the case of variable specific heat ratio and the mean molecular weight see Spruit and Zweibel (1979).

Thus, the dispersive nature of both modes in isothermal atmos-phere is analogous to that of an acoustic wave propagating vertically upward. The waves with the frequencies $\omega < \omega_c$ are evanescent. For pro-pagating waves the frequency cutoff works as a filter which lets through the thickness of atmosphere only those waves among the great variety of waves excited in convective zone whose frequency is larger than ω_c. Since for bending oscillations the cutoff is quite low these waves are expected to reach chromosphere with growing amplitudes (in accordance with Eq. (5))(Spruit, 1981a). Rae and Roberts have shown that if disturbances of fluxtube are impulsively generated at its base, then the wave propagation occurs and the existence of cutoff manifests itself in the formation of an oscillating wake. The wake is oscillating with the tube frequency.

The above conclusions are valid for isothermal atmosphere. In a non-isothermal atmosphere fluxtube oscillations may become unstable. For a thin tube embedded in a convective zone the instability is driven by the buoyancy force in much the same way as in ordinary convection (Webb & Roberts, 1978; Spruit, 1979; Spruit & Zweibel, 1979). The sta-bility criterion (actually Schwarzschild's criterion) is modified by the presence of magnetic field. Namely, the magnetic field exerts the stabilizing influence and constrains the unstable motion to a flow along the tube. If the instability sets in as a downdraft it leads to a new equilibrium state of higher field strength, but a lower total energy. The magnetic energy is increased at the expense of the gravita-tional energy of plasma within the tube.

One more example of changing the character of tube oscillations is that when there is a sharp discontinuity in the vertical temperature distribution - the situation typical for chromosphere and the transition

zone. The pressure balance condition means that in this region a steep
drop of plasma density inside the tube occurs providing the condition
for the wave reflection and for the formation of resonance structure in
the open fluxtube. This means that in the region of the temperature
jump one can expect quite a large oscillation level. Since the lower
point of fluxtube experiences random oscillations caused by the convective motions, besides the study of boundary value problem one needs to
analyse the behaviour of the spectral density of tube oscillations. The
appropriate analysis shows that with a considerable probability the
amplitude of tube oscillations can be 3 or 4 times larger than its
average value which by itself, in the region of temperature jump becomes
much larger than the amplitude of motions in convective zone (Ryutova,
1989). In the observational data such events can manifest themselves
as a temporal brightening of tube region. The same holds for another
mechanism of the formation of 1-D resonator in open fluxtube which is
connected with the rapid radial broadening of the tube at some altitude
caused by the loss of radial equilibrium. As is well known, such phenomenon occurs when the plasma temperature outside the tube is less than
that inside. The rapid broadening gives rise to a growing tube inertia
which results in the "pinning" of the tube at the altitude of the loss
of equilibrium.

Up to now, we considered the fluxtubes homogeneous in their cross
section. In the next section we will see that the radial dependence of
tube parameters leads to a number of new effects. In the same section
we consider the influence of compressibility of medium.

2.1. Anomalous and radiative damping of fluxtube oscillations

The allowance for a smooth profile of plasma parameters over the tube
radius brings about the specific (and strong) damping of oscillations
which is caused by the resonance between the phase velocity of oscillations and the local Alfvén velocity. Mathematical form of corresponding
equation is of the type of Rayleigh equation in usual hydrodynamics
with the coefficient of the higher derivative approaching zero in
singular point. Beginning with the paper of Timofeev (1970) such type
of equation and corresponding effects, in particular Alfvén resonance,
in laboratory plasma is studied in detail. The effect of Alfvén resonance appeared to be very important in physics of coronal loops and
heating processes (Ionson,1978; Hollweg,1979; Heyvaerts and Priest,
1983; see also Sudan and Similon, 1988). Just for magnetic fluxtube oscillations this effect was studied by Ryutova (1977) who considered
longwave bending (kink) oscillations in an incompressible limit. In
this case the MHD-equations come to a single equation for current
function ψ :

$$\frac{\partial}{\partial r} \left(\rho - \frac{k^2 B^2}{4\pi\omega^2} \right) r \frac{\partial\psi}{\partial r} - \left(\rho - \frac{k^2 B^2}{4\pi\omega^2} \right) \frac{\psi}{r} = 0 \qquad (9)$$

(current function relates to the velocity components by $v_r = \frac{1}{r}\frac{\partial\psi}{\partial\varphi}$ and

$v_\varphi = \frac{\partial \psi}{\partial r}$). For the arbitrary radial profiles of density and magnetic field the quantitative evaluation of the damping rate is not quite simple. The eigenvalue problem can evidently be solved for the model of a tube which is homogeneous everywhere except the narrow (but finite) boundary layer of the width εR, where the Alfvén and sound speeds are linear functions of radius (ε is a small parameter. For this model the damping rate is proportional to ε (qualitatively, if ε is of the order of R, the damping rate becomes comparable with the frequency):

$$\nu_{res} = \frac{\pi \varepsilon}{4} \frac{\rho_i}{\rho_i + \rho_e} \tag{10}$$

Note that the anomalous damping, by itself, does not transform the energy of oscillations to the plasma heating. It just results in the concentration of the initially smooth eigenfunction near the resonant point, or, in other words, to a conversion of oscillation energy to the Alfvén continuum.   But then, of course, the usual dissipation mechanisms, like viscosity and others, turn on, and this strongly oscillating distribution damps out.

   Returning to the case of a really smooth radial dependence of fluxtube's parameters note that the weakly damped wave which could be described in terms of a smoothly varying radial eigenfunction does not exist any more. As in a real situation the radial distributions in most cases should be smooth, the question arises of whether under such conditions the bending oscillations of fluxtube can still be an agent responsible for the energy transfer from the underlying surface to upper chromosphere. To answer this question one should study a boundary value problem for the tube excited in its footpoint. The appropriate analysis in a most general case of a profile without any small parameter shows (Ryutova, 1989) that at higher altitudes the radial structure of the perturbation which was smooth near the excitation point, becomes more and more spiky at higher altitudes. The characteristic radial scale length Δr diminishes inversely proportional to z:

   $\Delta r \sim Ra/\omega z$.

Respectively, due to the presence of these small scale structures the dissipative processes (viscosity, thermal conductivity, etc.) become more and more important at larger z. The longitudinal dependence of the energy flux of the oscillations becomes nonexponential. The corresponding estimates give

$$Q \sim \rho \nu \frac{\omega^2 \xi^2}{(\Delta r)^2} ,$$

where Q is a volume density of a power released by viscous dissipation (ν is a kinematic viscosity coefficient). As Δr diminishes with height,

Q is growing with z. At some altitude z* reaches maximum, and then rapidly decreases - just because the amplitude of oscillation decreases. Thus, the heating power has a very characteristic shape with a pronounced maximum at some altitude z*. For comparison the dotted line in Fig. 4 shows Q for usual exponentially damping wave. The estimate of the altitude z* is

$$z^* \sim \frac{a}{\omega} \left(\frac{R^2 \omega}{\nu} \right)^{1/3}$$

Thus, one can conclude that even in the case of a smooth radial profile of plasma density and magnetic field the bending waves can transfer energy from underlying surface to upper chromosphere.

Let us proceed now to another important feature of fluxtube oscillations connected with the compressibility of medium. Ryutov and Ryutova (1976) have shown that in compressible medium oscillating fluxtube can be a source of secondary acoustic waves. The frequency of the radiated acoustic wave is obviously equal to that of the fluxtube oscillations. The same holds for the longitudinal wavenumber k. The radial wavenumber of acoustic wave is determined from the dispersion relation and is equal to:

$$k_r = \frac{1}{s_e} \left(\omega^2 - k^2 s_e^2 \right)^{\frac{1}{2}}$$

Here s_e is the sound speed in the external plasma.

The radiation occurs if k_r is real, that is if the phase velocity ω/k of flux tube oscillations is larger that s_e. In the opposite case the acoustic waves are evanescent, and their presence does cause only a small change of phase velocity of tube oscillations. So, for the case when k_r is real radiative damping rate of an arbitrary mode (except the axisymmetrical one) is as follows (Ryutov and Ryutova,1976):

$$\frac{\alpha_{rad}^{(m)}}{\omega} = \frac{\pi}{|m|!(|m|-1)!(1+\eta)} \left(\frac{Rk}{2} \right)^{2|m|} \left[\frac{2}{\gamma(1+\eta)} - 1 \right]^{|m|} .$$

Respectively for bending oscillations we have

$$\frac{\alpha_{rad}^{(1)}}{\omega} = \frac{\pi}{1 + \rho_i/\rho_e} \left(\frac{Rk}{2} \right)^2 \left[\frac{2}{\gamma(1 + \rho_i/\rho_e)} - 1 \right] .$$

For sausage oscillations (Ryutova, 1981):

$$\frac{\alpha_{rad}^{(0)}}{\omega} = \frac{\pi}{2} \left(\frac{Rk}{2} \right)^2 \frac{s_i^6}{s_e^2(a^2 + s_i^2)^2} .$$

The damping rate contains in all the cases a small parameter $(kR)^2$ and thus is relatively small. This smallness is a reflection of the fact that the frequency of tube oscillations is small as compared to the eigenfrequency of its radial acoustic oscillations. For short wavelengths the damping rate becomes large, of the order of frequency. This is another reason why just the longwave oscillations are of the most interest.

So, approaching the real conditions of solar atmosphere we have taken into account the gravity effect, the radial inhomogeneity of fluxtube and the compressibility of medium. The next step is taking into account the presense of longitudinal mass flows along the magnetic structures.

1.3. Effect of sheared flows

As it was mentioned in Introduction, the presence of longitudinal flow of a plasma in the vicinity of fluxtube gives rise to qualitatively new effects (Ryutova, 1988). First of all, when the velocity of shear flow exceeds a certain threshold value there arise negative energy waves in the system which can become unstable due to various energy absorption processes including nondissipative mechanisms of damping. In the case of fluxtubes just because of their specific feature described above the instability can occur due to the radiation of secondary sound waves and due to anomalous damping in Alfvén resonance region. When the velocity exceeds the second threshold there arises the linear hydrodynamic instability similar to the instability of the tangential discontinuity (TD) in MHD or Kelvin-Helmholz instability in hydrodynamics. Note, that according to observational data the plasma flows usually with different velocities inside and outside the magnetic structures are observed in all the magnetized regions in solar atmosphere.

In the coordinate system where the substance inside the tube is at rest, while the flow velocity outside it equals u and is directed along the z-axis, the linearized MHD-equations together with the equilibrium condition (3) lead to the following set of equations for small oscillations of magnetic fluxtube in the presence of shear flow:

$$i\delta\mathcal{P} = \rho(r)\frac{\Omega^2(r)(s^2+a^2)-k^2s^2a^2}{\eta^2 - k^2s^2}\left[\frac{1}{r}\frac{\partial}{\partial r}r\frac{v_r}{\Omega(r)} + \frac{im}{r}\frac{v_\varphi}{\Omega(r)}\right]$$

$$\frac{\partial\delta\mathcal{P}}{\partial r} = i\rho(r)[\Omega^2(r) - k^2a^2]\frac{v_r}{\Omega(r)} \tag{11}$$

$$\frac{im}{r}\delta\mathcal{P} = i\rho(r)[\Omega^2(r) - k^2a^2]\frac{v_\varphi}{\Omega(r)} .$$

Here $\delta\mathcal{P} = \delta p + b_zB/4\pi$ is the total pressure perturbation (all perturbed quantities are assumed to be proportional $\exp(-i\omega t + ikz + im\varphi)$), and

$$\Omega(r) = \omega - ku(r).$$

The set (11) describes all the types of linear oscillations of magnetic fluxtube. (For u = 0 they were described above). To illustrate the properties of negative energy waves we consider the bending oscillations. In this case the set (11) is reduced to the following equation for the displacement vector (cf. Eq. (4)):

$$\rho_i \frac{\partial^2 \vec{\xi}}{\partial t^2} = -\rho_e \left(\frac{\partial}{\partial t} + \frac{\partial}{\partial z} \right)^2 \vec{\xi} + \frac{B^2}{4\pi} \frac{\partial^2 \vec{\xi}}{\partial z^2} \ . \tag{12}$$

The dispersion relation following from (12) has a form:

$$\omega^2 + \frac{1}{\eta} (\omega - ku)^2 - k^2 a^2 = 0 \tag{13}$$

where $\eta = \rho_i / \rho_e$. From (13) we have

$$\frac{\omega}{k} = \frac{1}{1 + \eta} \left\{ u \pm \sqrt{\eta [a^2 (1 + \eta) - u^2]} \right\} \ . \tag{14}$$

Before describing the instability of negative energy waves let us examine the region, where

$$u > a \sqrt{1 + \eta} \ . \tag{15}$$

This is a region of TD-instability which can be referred as a "coarse" one since its growth rate is comparable with the frequency when **u** exceeds the threshold by a factor of two. This instability must play an essential role in different astrophysical objects where there are high speed streams along the magnetic field. In particular, the excitation of solar fluxtube oscillations is thought to be caused by shaking of tube's footpoint in convective zone. The frequency of these oscillations is very low, which presents the problem in an attempt to explain the energy transfer from photosphere to upper layers. The TD-instability leads to existence of another mechanism of excitation of oscillations magnetically structured media which is independent of the motions in the base of fluxtube and which can work far from convective zone and moreover, in any region with the strong mass flow. So, if in some region a fluxtube is "blown out" by the upward flow the oscillations excited here propagate further upwards. The frequency of these oscillations can be much higher than the inverse time of re-arrangement of granulation picture.

In the region $u > a \sqrt{1 + \eta}$ when system is still stable with respect to TD-instability the instability of negative energy waves can occur. For the lower branch (sign "minus") in dispersion relation (14) the energy density has a form

$$W = \pi R^2 k^2 \rho_e \xi^2 (\eta a^2 - u^2) \frac{x}{x + u} \tag{16}$$

where $x = \sqrt{\eta [a^2 (1 + \eta) - u^2]}$.

It is seen that the wave energy becomes negative at

$$u > a \sqrt{\eta} \qquad (17)$$

Therefore, within the interval

$$a\sqrt{\eta} < u < a\sqrt{1 + \eta} \qquad (18)$$

there can exist the instability of NEW caused by any of dissipative processes. In other words, dissipative effects lead to loosing the energy of NEW and, hence, to the growth of their amplitude. Note, that the possibility of the existence of negative energy waves in nonequilibrium plasma for the first time was pointed out by Kadomtsev et al. (1964); on the negative energy waves in hydrodynamics see, for example, Ostrovsky et al. (1986).

It is remarkable, that for magnetic fluxtubes this instability can occur even in the absence of any usual dissipative processes due to their specific features mentioned above. First, the instability takes place due to the damping in the Alfvén resonance layer. Recently, similar results have been obtained by Hollweg et al. (1989), who studied the effects of velocity shear on the resonance absorption of incompressible MHD surface waves and indicated the importance of this effect for the development of turbulence in regions of strong velocity shear. I would like to emphasize that the strong instability takes place not only for axisymmetrical waves but also for bending waves (see below). Second, in the case of compressible medium the instability occurs due to the radiation of secondary sound waves. Note again, that the development of NEW-instability leads to the reversal of the sign of radiative damping and to the growth of the fluxtube oscillations amplitude. This is possible in two cases: when fluxtube oscillation has a negative energy, while a radiated sound wave has a positive one, or when former has a positive energy, while a radiated wave has a positive one. For example, the conditions under which bending oscillations of positive energy radiate negative energy sound waves are as follows:

$$a > s_e/\sqrt{\eta} \,, \qquad u > s_e + \sqrt{a^2 - \frac{1}{\eta} s_e^2} \,. \qquad (19)$$

As to sausage oscillations the outer flow has a weak influence on them; in particular, their energy remains positive in the presence of flow. So that the instability in this case can be caused by the radiation of negative energy sound waves. The necessary condition has the form:

$$c_T = u - s_e \sqrt{1 + k^2/k} > 0$$

which is possible when fulfilling the requirement

$$u > c_T + s_e \,.$$

The most important feature of negative energy waves is that their presence leads to enlargement of classes of instabilities and one can expect a very vigorous nonlinear activity in a system. A few of them we describe in the next section.

2. NONLINEAR EFFECTS

2.1. Explosive instability

In the system where the waves with opposite signs of energy are simultaneously present the nonlinear explosive instability can develop. This instability was first considered by Dikasov et al. (1965) and illustrated by the waves with random phases. Later on, it was analysed by Coppi et al. (1969) for a triplet of coherent waves.

As is well known, the main feature of this instability is that the amplitudes of interacting waves achieve infinitely large values for a finite period of time. As an example let us consider an explosive instability for a triplet of waves propagating along the fluxtube.

Let wave 1 be of a sausage type, thus having a positive energy, (as was mentioned above, external flows have a weak influence on this mode), wave 2 - of bending type with negative energy, that is propagating against the flow and having k_2 negative, and wave 3 - of bending type with positive energy ($k_3 > 0$). The bending wave can be either linearly polarized, that is be a mixture of waves with $m = 1$ and $m = -1$, or circularly polarized in opposite directions.

The matching conditions can be satisfied if shear flow velocity exceeds some critical value equal to

$$u_c^{exp} = \frac{c_T}{1 + \eta} + [a^2\eta - c_T^2 \frac{\eta(\eta^2 + 3\eta + 3)}{1 + \eta^2}]^{\frac{1}{2}} . \tag{20}$$

It is easy to verify that the value (20) is below the limit of hydrodynamic instability and just gets into the interval corresponding to the existence of negative energy waves determined by (18). So, under the condition

$$a\sqrt{\eta} < u < u_c^{exp}$$

one can expect a very vigorous nonlinear activity in system. So, if only one wave (say wave 1) is excited in the system at the initial moment of time and the amplitudes of two other waves are determined by thermal noise , then at the initial stage of development of the explosive instability these amplitudes increase exponentially. The characteristic growth rate by the order of magnitude is equal to $k_1 v_{1\sim}$, where $v_{1\sim}$ is the velocity amplitude of tube's boundary in slow oscillations. In a time of the order of several inverse growth rates when the amplitudes of all three waves become the quantities of the same order, there begins the power growth of the amplitudes of all interacting waves according to the law

$$v_\sim \sim \frac{1}{t - t_0}$$

and the amplitudes of all three waves achieve infinitely large values in a time t_0, which in our case is also of the order of $(k_1 v_1)^{-4}$. Of course the assertion made above is formal to some extent: higher-order nonlinear processes can limit the growth of amplitudes at a finite level. So, the further development of explosive instability depends on the character of these nonlinear processes as well as the dispersive properties of system. There can be, for example, "real explosion", that is fast ($\sim t_0$) release of energy stored in a system; the stabilization of instability with appearance of solitary wave; gas of solitons; solitons with explosively growing amplitudes; shocks; turbulence, etc. The detailed analysis of these processes is the future problem. What is well known today is that nonlinear equations describing fluxtube oscillations possess soliton-type solutions which will be shortly presented in the next section.

2.2. Solitons in magnetic fluxtubes

Eq . (7) for sausage together with the equilibrium condition (3) form hyperbolic   equations which possess the shock solution. The allowance of the compressibility of medium leads to dispersion of the wave. If the nonlinearities responsible for the creation of shocks are counter-balanced by the dispersive effects the solitons can appear. Roberts (1985) has shown that sausage waves in a thin cylindrical fluxtube satisfy the nonlinear integrodifferential equation which is analogous to the equation which was got by Leibovich (1970) for the description of water waves on a vortex and which has a solitary solution. Earlier such an equation for a slab geometry was obtained by Roberts and Mangeney (1982):

$$\frac{\partial v_z}{\partial t} + c_T \frac{\partial v_z}{\partial z} + \alpha_1 v_z \frac{\partial v_z}{\partial z} + \alpha_2 \frac{1}{\pi} \frac{\partial^2}{\partial z^2} \int_{-\infty}^{\infty} \frac{v_z(s,t)}{s-z} ds = 0 \qquad (21)$$

where

$$\alpha_1 = \frac{[(\gamma + 1) a^2 + 3s_e^2] a^2}{2(s_e^2 + a^2)^2} \; ; \; \alpha_2 = \frac{1}{2} \frac{\rho_e}{\rho_i} \left(\frac{c_T}{a}\right)^3 x_0 c_T \; .$$

Here $2 x$ is the width of the magnetic slab.

Eq. (21) is the Benjamin-Ono type and the corresponding solution is

$$v_z(z,t) = \frac{v_0}{1 + (z - st)^2/1^2}$$

where the amplitude v_0, speed s and scale   of soliton are related by

$$s = c_T + \frac{1}{u} \; , \qquad 1 = 4\alpha_2/v_0 \alpha_1 \; .$$

This solution corresponds to a swelling in the slab which propagates with a speed s. The cross section of a slab depends on v_z as

$$A = A_0 (1 + \frac{c_T}{a} v_z) .$$

Nonlinear analysis including dissipative effects lead to equation of the Benjamin-Ono-Burgers type (Edwin and Roberts, 1986). The estimates of the decay rate of corresponding solution show that solitons can propagate through the nonadiabatic atmosphere and to emerge at higher altitudes.

2.3. "Magnetosonic flows"

As was described above, in plasma containing structured magnetic fields a great variety of wave processes take place. This leads to very interesting nonlinear effects analogous to the effect of "acoustic flows" or "quartz wind" in usual gasdynamics (see, for example, Nyborg, 1965 and References therein). But in magnetohydrodynamics the picture of these effects is much more rich (Ryutova, 1986). Namely, at the propagation of oscillations along the fluxtubes secondary plasma flows and currents are excited inside the tube as well as outside it. There are two main reasons leading to these effects: the action of ponderomotive force on plasma (this mechanism is not connected with the absorption of oscillations and with any dissipative processes) and absorption of momentum and angular momentum of oscillations propagating along the tubes (nonzero angular momentum can be transferred by the circularly polarized bending oscillations). The absorption of angular momentum causes the rotational mass flow around the fluxtube and the absorption of longitudinal momentum can lead to the upward mass flow. If the absorption is provided mostly with one of the plasma components (the specific of situation depends on the damping mechanism), then it is accompanied by the excitation of currents which can distort the initial magnetic field.

The velocity field of generated secondary flows is described by the following equation

$$\rho \frac{\partial \vec{u}}{\partial t} + \rho \nu \nabla^2 \vec{u} = -\nabla (P + \frac{B^2}{8\pi}) + \vec{f} \tag{22}$$

where ν is kinematic viscosity and $\vec{f}$ is ponderomotive force acting on the unit volume of plasma. $\vec{f}$ has a following form:

$$\vec{f} = - < \tilde{\rho} \frac{d\tilde{\vec{v}}}{dt} > + \frac{1}{c} < [\tilde{\vec{j}}\tilde{\vec{B}}] > . \tag{23}$$

Here tilde marks perturbed quantities defined by linearized MHD-equations corresponding to one of the modes (bending or sausage). One can see from equation (12) that convective motions can arise only if

$$\text{rot } \vec{f} \neq 0 .$$

Otherwise, the ponderomotive force leads only to an insignificant redistribution of plasma density and magnetic field. The magnitude of $\vec{f}$ becomes especially large in Alfvén resonance layer where quite   strong convective motions can arise.

These secondary ("magnetoacoustic") flows can bring about different evolution regimes of magnetic fluxtubes which depends on the correlation properties of the wave trains propagating in the tube and on the parameters of fluxtube itself. If we consider a case of long coherent train with the duration T much in excess of the time of establishing of viscous flow ($\tau_\nu \sim R^2/\nu$), the density and magnetic field become constant along the lines of generated motion, and the fluxtube is "splitting" into four independent tubes. In the opposite case of a sequence of mutually incoherent (including the plane of polarization) short ($T \ll \tau_\nu$) wave trains, a kind of stochastic motion is induced, which results in a diffusive broadening of a tube and ends up in a complete "dissolving" of fluxtube in the ambient plasma.

As it was mentioned the effects described above exist even in the absence of any mechanism of absorption. In the presence of absorption z-component of ponderomotive force directed along the magnetic field appear.

Other mechanism of generation of magnetosonic flows and in this case also the current drives is connected just with the absorption of the oscillation energy. Formally, the expression for f has the same form as Eq. (23), but now the terms directly connected with the absorption are taken into account. Besides the absorption due to the usual dissipative processes for fluxtubes nondissipative mechanisms of absorption described above are quite efficient. For example, in the case of resonance damping with the damping rate (10) the energy of bending oscillations is pumping into the resonant layer where the dissipation occurs. Respectively, the whole momentum of oscillations is transferred to plasma in the narrow layer, causing the strong upward mass flow along the magnetic field. The corresponding force can be estimated as follows

$$f_z \sim \alpha \, \frac{B^2}{4\pi} \, \frac{|\tilde{v}|^2}{a^2 R} \quad .$$

Now, if the absorption is provided mostly in electron (or ion) component of plasma the generation of currents inside the tube as well as outside occurs. As in the case of secondary flows, the absorption of momentum leads to the current drives along the fluxtubes and the absorption of angular momentum leads to the generation of azimuthal currents.

3. THE ENSEMBLES OF FLUXTUBES

This part of paper deals with the effects accompanying the propagation of long-wave MHD-oscillations in a plasma containing small-scale inhomogeneities, namely, the ensembles of magnetic fluxtubes. In the first two sections we consider the ensemble of fluxtubes far removed from each other, while in the third section we discuss briefly the tightly settled fluxtubes.

3.1. Resonance damping of sound waves in system of fluxtubes

The system of fluxtubes has an interesting feature: the longwave acoustic oscillations in the system are damped due to the effect similar to Landau damping which consists in the resonance excitation of bending waves propagating along the fluxtubes (Ryutov and Ryutova, 1976). The bending oscillations propagate along the fluxtube separate with the phase velocity defined by Eq. (1) and if the angle θ between the direction of sound propagation and fluxtube direction satisfies the condition

$$s = c_b \cos \nu \tag{27}$$

then the resonant transfer of sound wave energy to the energy of fluxtube oscillations takes place.

Since in general the density and the magnetic field inside the different fluxtubes are different, so the velocity changes from one fluxtube to another, and hence, at each propagation angle there are tubes for which the condition (27) is satisfied and which absorb the energy of sound waves.

We define the distribution function of fluxtubes $f(r,\eta)$ over their radius R and parameter $\eta = \dfrac{\rho_i}{\rho_e}$ (we consider that R is much less than the average distance between fluxtubes L which is much less than wavelength: $R \ll L \ll \lambda$):

$$\sigma = \int_0^\infty \!\!\! \int dRd\eta \; f(R,\eta)$$

where σ is the total fraction of volume occupied with fluxtubes ($\sigma \sim \dfrac{R^2}{L^2} \ll 1$). The damping rate of sound oscillations has a form:

$$\nu_s = \pi ks \sin^2\theta \left\{ \begin{array}{ll} g(\eta_0), & \eta_0 > 0 \\ 0 & , \eta_0 < 0 \end{array} \right. ; \; \eta_0 = \frac{2\cos^2\theta}{\gamma} - 1 \tag{28}$$

where we introduce the function

$$g(\eta) = \int_0^\infty f(r,\eta) \; dR \; .$$

If we make the natural assumption that ρ_i changes with respect to ρ_e not more than by several times from one tube to another, then $g(\eta_0) \sim \sigma$, and for damping rate we can write the following estimate:

$$\nu_s \sim \sigma ks$$

which holds only for those values of θ, where $\eta_0 > 0$, i.e. where $\cos\theta > (\gamma/2)^{1/2}$. For monatomic gas the corresponding region is quite narrow, $\theta \le 5°$, so that it is essential to take into account the noncollinearity of separate fluxtubes. Let $h(\vec{n})$ be the distribution function of fluxtubes over the directions. $\bar{n}$ is the unit vector directed along the tube's axis. We can normalize $h(\bar{n})$ as follows:

$$\sigma = \int h(\bar{n}) \, d\sigma \, ,$$

$d\sigma$ is the element of solid angle. Then for isotropic distribution of fluxtubes the damping rate of sound oscillations has a form

$$\frac{\nu_s}{ks} = \frac{1}{4\sqrt{2}} \frac{\sigma\gamma \, [2 - \gamma(1 + \eta)]}{\sqrt{\gamma(1 + \eta)}} \, .$$

So magnetic fluxtubes absorb the energy of outer motions at the photosphere level during the time of the order of ν_s^{-1}. Then the process of energy accumulation goes on, and in a time α_{rad}^{-1} which is much greater $(\alpha_{rad}^{-1} \gg \nu_s^{-1})$, fluxtubes release their energy due to radiation of secondary acoustic waves at the higher layers of atmosphere.

3.2. Scattering of sound waves by fluxtubes

The detailed theory of scattering of sound waves by magnetic fluxtubes (Ryutov and Ryutova, 1976) shows that there are resonances in scattering due to the existence of weakly damped natural oscillations of fluxtube. The main contribution to damping is of dipole mode, i.e. the bending oscillations. The damping rate of acoustic wave with frequency ω has a form

$$\nu_{scatt} = \int \frac{f(R,\eta)}{\pi R^2} \, \beta(\eta,R,\omega) \, d\eta d\omega dR \qquad (29)$$

where β is some function containing the resonance denominators

$$\beta \sim \frac{\Omega^2}{(\omega - \Omega)^2 + \alpha_{rad}^2} \, ; \quad \Omega = k \, s\sqrt{\frac{2}{\gamma(1 + \eta)}}$$

For the narrow region of $\quad = \quad$ one can perform integration over η and R resulting

$$\nu_{scatt} = \pi s k \, \sin^2\theta \cdot g(\eta_0)$$

which formally coincides with (28). We imply that $\eta_0 > 0$; in the opposite case the scattering becomes nonresonant and damping rate becomes much smaller: $\nu_{scatt} \sim \sigma ks(kR)^2$.

Note that if the condition $(kR)^2 < \sigma$ is satisfied the scattering is negligible.

In the case of comparatively short waves ($R \ll \lambda \ll L$) the damping is completely determined by scattering. In this case the damping rate ν_s appears to be smaller than α_{rad} and this means that direct scattering of primary acoustic wave into the secondary sound waves without the preliminary accumulation of energy in natural oscillations of fluxtubes takes place. In other words for short waves one has $\nu_s = \nu_{scatt}$, where ν_{scatt} is defined by Eq. (29).

Recently resonant scattering of sound waves by slender fluxtubes was studied by Bogdan (1989), who emphasized the importance of this effect for the diagnostics of the structure of magnetic flux concentrations across the solar surface. This may be possible because near the resonance the elastic scattering cross section (per unit length of the fluxtube) is comparable to the wavelength of the incident solar p mode, which for slender fluxtubes may exceed both the geometrical and non-resonant scattering cross sections by several orders of magnitude.

3.3 Enhanced dissipation by tightly packed fluxtubes

Until now we have discussed the problems connected with isolated fluxtubes and ensembles of fluxtubes widely separated from each other. In the solar atmosphere the situation when fluxtubes are tightly packed seems quite realistic, so the question arises what the effects are that accompany the propagation of long-wave MHD oscillations in a plasma with random inhomogeneities, when the density, temperature, and magnetic field change by a large amount over a distance small as compared with the wavelength. Intuitively it seems that the large-scale perturbations will propagate in quite the same way as in a homogeneous plasma, with the difference that the density, velocity, etc., will be replaced by some averaged quantities, and that intuitive conclusions should turn out to be more or less correct. It is however far from easy to find an explicit procedure for even the derivation of the averaged equations, especially for a real 3-D problem.

There are a few cases in which one can write the averaged equations in explicit form (Ryutova and Persson, 1984). It is interesting that such systems reveal some less obvious properties. One is that unlike the case of a homogeneous plasma, the wave propagation in a plasma with random (and strong) inhomogeneities is accompanied by a vortex motion of the plasma. These motions have the same scale as the fluxtube cross sections. Besides, because of the large gradients of velocity and temperature that are associated with tightly packed fluxtubes, enhanced dissipation of MHD waves can take place. The dissipation rates associated with both the viscosity and thermal conductivity are $(\lambda/R)^2$ (with $\lambda \gg R$) times larger as compared with the homogeneous case. Ohmic dissipation gives only a minor contribution to this effect. Thus although dissipative effects are believed to be small in the solar atmosphere, these effects may strongly influence the dynamics of inhomogeneous regions.

REFERENCES

1. Stenflo, J.O. (1989) The Astronomy and Astrophysics Review 1, 3.
2. Priest, E.R. (1982) Solar Magnetohydrodynamics, Dodrecht, D.Reidel Publ. Co.
3. Ryutov, D.D. & Ryutova, M.P. (1976) Sov. Phys. JETP, 43, 491.
4. Defouw, R.J. (1976), Ap.J. 209, 266.
5. Ryutova, M.P. (1977) Proc. of XIII Int. Conf. on Ionized Gases,p. 859.
6. Ionson, J.A. (1978) Ap.J. 226, 650.
7. Spruit, H.C. (1981a) Astr. Ap., 98, 155.

8. Roberts, B. & Webb, A.R. (1978) Solar Phys. 59, 249.
9. Webb, A.R. & Roberts, B. (1979) Solar Phys. 64, 77.
10. Rae, I.C. & Roberts, B. (1982) Ap. J. 256, 761.
11. Hollweg, J.V. (1982) Ap. J. 257, 345.
12. Ryutova, M.P. (1989) Proc. of AGU Chapman Conf. on Physics of Magnetic Flux Ropes (in press).
13. Ryutova, M.P.(1988), JETP, 94, 138.
14. Roberts, B. & Mangeney, A. (1982) Mon. Not. Roy. Astr. Soc. 198,7p.
15. Roberts, B. (1985) Phys. Fluids, 28, 3280.
16. Molotovshchikov, A.L. & Ruderman, M.S. (1987) Solar Phys.
17. Ryutova, M.P. (1986) Proc. of Joint Varenna-Abastumani Int. School on Plasma Astrophys., ESA, p.71.
18. Ryutov, D.D. & Ryutova, M.P. (1989) Proc. of AGU Chapman Conf. on Physics of Magnetic Flux Ropes (in press).
19. Bogdan, T. (1989) Ap. J. (in press).
20. Bogdan, T & Zweibel, E.G. (1985) Ap. J. 298, 867.
21. Bogdan, T. & Zweibel, E.G. (1987) Ap. J.318, 888.
22. Bogdan, T. & Cattaneo, F. (1989) Ap. J. (in press).
23. Ryutova, M. & Persson, M. (1984) Physica Scripta, 29, 353.
24. Weiss, N.O. (1981) J.Fluid Mech., 108, 247; 273.
25. Hasan, S.S. & Schüssler, M. (1985) Astr. Ap. 151, 69.
26. Ferriz-Mas, A. (1988) Phys. Fluids, 31, 2583.
27. Hollweg, J. (1987) Ap. J. 317, 918.
28. Hollweg, J., Yang, G., Cadez, V.M. & Gakovic (1989) Ap.J.(in press).
29. Seehafer, N. (1988) Proc. Int. Workshop on Reconnection in Space Plasma, Potsdam, p. 5.
30. Henoux, J.C. & Somov, B.V. (1987) Astr. Ap., 185, 306.
31. Spruit, H.C. (1981) in The Sun as a Star, ed. S.Jordan (Washington, NASA SP-450) p. 385.
32. Thomas, J.H. (1985) Proc. Solar Optical Telescope Workshop, "Theoretical Problems in Solar Physics", Munich.
33. Priest, E.R. (1988) Proc. of Joint Varenna-Abastumani Int. School on Plasma Astrophysics, ESA, p. 189.
34. Spruit,H.C. (1981b) Astr. Ap. 102, 129.
35. Roberts, B. & Webb, A.R. (1978) Solar Phys. 56, 5.
36. Spruit, H.C. & Zweibel, E.G. (1979) Solar Phys. 62, 15.
37. Spruit, H.C. (1979) Solar Phys. 61, 363.
38. Timofeev, A.V. (1970) Usp. Fiz. Nauk, 102, 185.
39. Hollweg, J. (1979) Solar Phys. 62, 227.
40. Heyaeverts, J. & Priest, E. (1983) Astr. Ap. 117, 220.
41. Sudan, R. & Similon, P. (1988) Proc. of Joint Varenna-Abastumani Int. School on Plasma Astrophysics, ESA, p. 63.
42. Sakurai, T. & Granik, A. (1984) Ap. J. 177, 404.
43. Ryutova, M.P. (1981) Sov. Phys. JETP, 53(3), 529.
44. Kadomtsev, B.B., Mikhailovsky, A.B. & Timofeev, A.V. (1964)JETP, 47, 2267.
45. Ostrovsky, L.A., Rybak, S.A. & Tsimring, L. (1986) Usp. Fiz. Nauk, 150, 417.
46. Dikasov, V.M., Rudakov, L.I. & Ryutov, D.D. (1965) JETP, 48, 913.
47. Coppi, B., Rosenbluth, M.N. & Sudan, R. (1969) Ann. Phys.55, 207.
48. Leibovich, S. (1970) J. Fluid Mech. 51, 625.

ON THE 5-MINUTE PHOTOSPHERIC OSCILLATION AND ITS MODELING

C. MARMOLINO[1] and G. SEVERINO[2]
[1]*Dipartimento di Scienze Fisiche dell'Università*
Mostra d'Oltremare pad. 19
I - 80125 Napoli, Italia
[2]*Osservatorio Astronomico di Capodimonte*
Via Moiariello 16
I - 80131 Napoli, Italia

1 Introduction

The 5-minute oscillations are standing waves trapped in subphotospheric cavities, whose evanescent part affect the profiles of photospheric lines.

The models developped to reproduce the signatures of these oscillations on spectral lines are not completely satisfactory. In fact
i) computed $I - V$ phases in very weak lines forming below h $\sim$ 100 km are greater than those observed, $\geq 90°$ against $\leq 60°$ (e.g. Deubner, 1989);
ii) computed I oscillations in the line wings are stronger than those in the line core, but the opposite is observed (Marmolino and Stebbins, 1989, their Fig. 3);
iii) computed and observed frequencies disagree significantly for higher degree modes which are confined in the very surface layers (Christensen-Dalsgaard et al., 1985).

To solve point iii) the use of an improved equation of state appears promising (Christensen-Dalsgaard et al., 1988). How this could help in the cases i) and ii) has not yet been investigated.

The theoretical results i) and ii) are based on an oscillation model which considers linear perturbations propagating in a compressible medium, assumed to be a perfect gas stably stratified, and damped by radiation according to the Spiegel's formula (Marmolino and Stebbins, 1989). The resulting vertical velocity V, temperature perturbation T and pressure perturbation P are, in an atmosphere characterized by a constant scale height H, sound velocity c, and radiative decay time τ_r:

$$V \equiv v_z = v_o exp(\frac{z}{h_\pm} + i(\omega t - k_x x \mp k_z z)) \tag{1}$$

251

J. O. Stenflo (ed.), Solar Photosphere: Structure, Convection, and Magnetic Fields, 251–254.
© *1990 by the IAU.*

$$T \equiv \frac{\delta T}{T} = \frac{(\gamma - 1)\omega(1 + i\alpha r)}{(\omega^2 - c^2 k_z^2)(1 + \alpha^2 r^2)}(\pm k_z + i(\frac{1}{h_\pm} - \frac{c^2 k_z^2}{\gamma \omega^2 H})) \cdot V \qquad (2)$$

$$P \equiv \frac{\delta P}{P} = \frac{\gamma \omega(1 + i\alpha r)}{(\omega^2 - c^2 k_z^2)(1 + \alpha^2 r^2)}(\pm k_z + \alpha(\frac{1}{h_\pm} - \frac{1}{H}) + i(\frac{1}{h_\pm} - \frac{1}{\gamma H} \mp \alpha k_z)) \cdot V \qquad (3)$$

with $h_\pm = 2H/(1 \pm 2Hk_i)$, $\alpha = 1/(\gamma \omega \tau_r)$, $r = (\gamma \omega^2 - c^2 k_z^2)/(\omega^2 - c^2 k_z^2)$, and $k_z + i k_i$ the complex vertical wavenumber; in the double signs the upper and lower one corresponds to upward and downward phase propagation respectively.

In an effort to specify the origin of the discrepancies between theory and observations, we started a review of the model analysing its forecasts on a wider area of the $k - \omega$ diagram than that one where the 5-minute oscillations are confined. This work has revealed itself fruitful leading to the identification of *well defined areas with different phase relations in the evanescent part of the $k - \omega$ diagram*.

The preliminary results we present here refer to waves whose progressive part carries energy upwards all over the $k - \omega$ diagram.

2 Results

A contour map of the $T - V$ phase differences shows clearly the existence of two regimes in the evanescent part of the diagnostic diagram (Fig. 1b):
zone a, where $90° \leq \phi_{TV} \leq 180°$ and most of the 5 minute oscillations are confined; $\phi_{TV} = 90°$ without radiative damping;
zone b, where $\phi_{TV} \leq 0°$; $\phi_{TV} = -90°$ without radiative damping.

The walls of zone b are sharp and mark a jump of almost $\pm 180°$ in the phase (exactly $180°$ in the adiabatic case). Radiation damping fixes also the phase offset, $0°$ with damping, $-90°$ without. The walls coincide with the lines labelled L_m and f on the $k - \omega$ diagram in Fig. 1a, which have the equations $k_z^2 = \omega^2/c^2 \cdot (\gamma^2 \omega^2 \tau_r^2 + \gamma)/(\gamma^2 \omega^2 \tau_r^2 + 1)$ and $\omega^2 = g k_z$, respectively, where g is the solar surface gravity.

The equation for L_m is the real part of the dispersion relation for pure acoustic waves in the presence of radiative damping (e.g. Mihalas and Mihalas, 1984 their Eq. 101.11). On this line the denominator of the Eqs. (2) and (3) gets its minimum values and the topology of the surface representing the T and P amplitudes relative to V over the $k - \omega$ diagram is dominated by the peaks occuring along it (Fig. 1c for the T perturbation). In absence of damping the denominator of Eqs. (2) and (3) is zero, reduces to the dispersion relation of *Lamb* waves (curve L in Fig. 1a), and corresponds to vanishing vertical velocity.

The equation for f defines the *fundamental mode*. On this line and for the waves carrying energy upward the T amplitude vanishes for $k_z \leq 1/2H$ (Eq. 2). This behaviour is clearly seen both in the $T - V$ phase jump and in the contour maps involving the T amplitude (Figs. 1c,d).

The differentiation between the two evanescent zones is lost in the $P - V$ phase differences, since the phase jump on the *fundamental mode* disappears and $\phi_{PV} \sim -90°$ with and without damping in both the zones.

The evanescent area at $k_z \leq 1/2H$ is also characterized by a general increase of the amplitude of P relative to T. This relative increase is strengthened by radiation damping which reduces significantly the T amplitudes and smooths only P amplitudes.

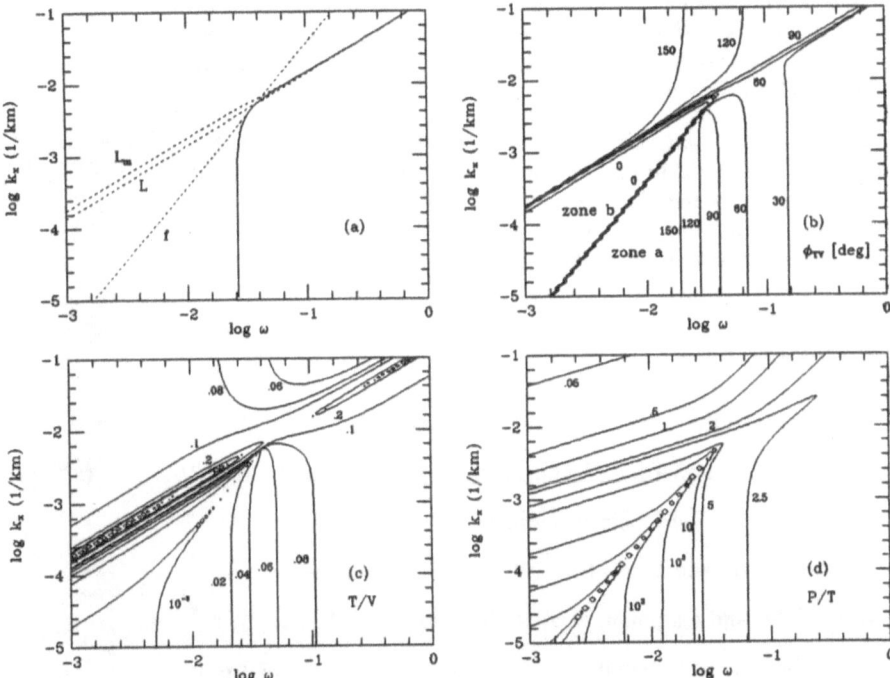

Figure 1. *window (a)*: k-ω diagram for $T = 4900°K$, in presence of radiative damping, $\tau_r = 15s$. Solid line corresponds to $k_z^2 - k_i^2 = 0$; dashed lines are the *fundamental mode*, labelled f, the *modified Lamb* waves, L_m, and the *adiabatic Lamb* waves, L. *window (b)*: contour map of the $T - V$ phase differences; *window (c)*: contour map of the T amplitude relative to V; *window (d)*: contour map of the P amplitude relative to T.

3 Conclusions

Analysing the forecasts of the photospheric oscillation model on the $k - \omega$ diagram we found that the *Lamb* waves appear to be modified by the radiative damping, and that these modified *Lamb* waves and the *fundamental* mode are *dividing lines in the phase relations* between the different perturbations. Along these lines the vertical velocity and the temperature perturbation get their minima respectively.

The transition between zone a, of the ordinary 5 minute modes, and zone b, confined between *fundamental* mode and *Lamb* waves, has probably already been observed in the phase spectra of Schmieder (1976), Lites and Chipman (1979), Staiger (1984), and Deubner and Fleck (1989).

Pressures higher than temperatures are a possible source of discrepancy between theory and observation for the 5-minute modes. In fact since each perturbation has an amplitude increasing with height, oscillations stronger in the line wings than in the line core, as provided by theory, should imply core oscillations leaded by temperature and wing oscillations leaded by pressure.

We are completing this work considering waves whose progressive part carries energy downward. The model has to be improved allowing for complex horizontal wavenumber to define properly the modified *Lamb* waves. Finally, we plan to compute theoretical phase spectra for specific lines, which are directly comparable with observations, including the atmospheric temperature stratification and transfer effects.

References

Christensen-Dalsgaard, J., Gough, D. O. and Toomre, J.: 1985, *Science*, **229**, 923

Christensen-Dalsgaard, J., Dappen, W. and Le Breton, J.: 1988, *Nature* , **336**, 634

Deubner, F.-L.: 1989, this conference

Deubner, F.-L. and Fleck, B.: 1989, *Astron. Astrophys.*, in press

Lites, B.W. and Chipman, E.G.: 1979, *Astrophys. J.*, **231**, 570

Marmolino, C. and Stebbins R. T.: 1989, *Solar Phys.*, submitted

Mihalas, D. and Mihalas, D.W.: 1984, *Foundations of Radiation Hydrodynamics*, Oxford University Press, New York

Schmieder, B.: 1976, *Solar Phys.* **47**, 435

Staiger, J., Schmieder, B., Deubner, F.-L. and Mattig, W.: 1984, *Mem. Soc. Astron. Ital.*, **55**, 147

CLASSIFICATION OF MAGNETOATMOSPHERIC MODES IN SUNSPOT UMBRAE

S.S.HASAN
Indian Institute of Astrophysics
Bangalore 560034,India

Y.SOBOUTI
Physics Department
Shiraz,Iran

ABSTRACT. We examine the wave modes in a sunspot umbra. Assuming a stratification, based on a model atmosphere in a sunspot, the normal mode spectrum is determined. The modes are classified using a scheme based on a Helmholtz decomposition of the displacements into l(longitudinal) and t(transverse) components. In certain cases these can be related to the usual fast and slow MHD waves. We compute the theoretical eigenfrequencies and note the existence of umbral oscillations with periods in the range 2-3 min, which are interpreted as slow and mixed modes. The frequencies of the Alfvén waves are also calculated. It is suggested that these modes might also have been observed.

1. Introduction

Oscillations in the umbrae of sunspots have been widely reported (e.g. Beckers and Schulz,1972; Balthasar and Wiehr,1984; Lites and Thomas,1985; Lites,1986; Abdelatif et al.,1986;Thomas et al. ,1987 and Gurman,1987). The periods of these oscillations typically lie in the range 2-3 min, although larger periods of some 300-400 s (Bhatnagar et al.,1972; Soltau et al.,1976 and Balthasar et al.,1987) or smaller ones around 100 s (Schröter and Soltau,1976) cannot be ruled out.

A number of theoretical investigations have been carried out to examine the nature of wave propagation in sunspot umbrae (e.g. Uchida and Sakurai,1975; Scheuer and Thomas,1981; Thomas and Scheuer,1982 and Leroy and Schwartz,1982). The interpretation of the observed modes is still not established. Scheuer and Thomas (1981) suggested that umbral oscillations are essentially fast waves trapped in a photopheric cavity. On the other hand, Zhugzhda et al. (1983) and Gurman and Leibacher (1984) have argued in favour of a slow mode, trapped in a chromospheric cavity above the temperature minimum. The aim of the present analysis is to look carefully into this question. Our plan is first to calculate the normal modes of a model sunspot atmosphere with a uniform vertical magnetic field. A classification of these modes is then attempted, based upon a technique developed earlier by us (Hasan and Sobouti,1987; henceforth HS). We also discuss the connection between the modes, classified according to our scheme, and the conventional MHD modes of an unstratified medium.

2. Equations and Method

2.1 EQUILIBRIUM UMBRAL MODEL

In order to compute the wave modes in an umbra, we require an equilibrium model which mimics a real atmosphere in a sunspot. We selected the core umbral model M of Maltby et al. (1985) (kindly provided to us by T. Abdelatif). Beneath the photosphere, the atmosphere is matched to a convection zone model. Figure 1 shows the temperature variation with z (depth) in a sunspot. The convention used is that z increases into the Sun and z=0 corresponds to optical depth unity.

J. O. Stenflo (ed.), Solar Photosphere: Structure, Convection, and Magnetic Fields, 255–258.

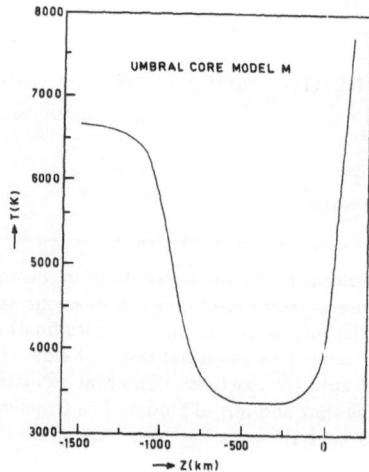

Figure 1. Temperature as a function of z in a sunspot umbra (based on the core umbral model M of Maltby *et al.*).

2.2 LINEARIZED EQUATIONS

We assume that a uniform vertical magnetic field is embedded in the model atmosphere. The normal modes can be determined by solving the wave equation

$$\rho\frac{\partial^2 \xi}{\partial t^2} = -\nabla\delta p + \mathbf{g}\delta\rho + \frac{1}{4\pi}(\nabla \times \delta\mathbf{B}) \times \mathbf{B} \tag{1}$$

where $\delta\rho, \delta p, \delta\mathbf{B}$ denote perturbations in density, pressure and magnetic field respectively and ξ is a small displacement of a fluid element. Equation (1) can be recast into an energy integral, which lends itself to a variational treatment (for details see HS). Assuming a time dependence of the form $e^{i\omega t}$, the normal frequencies can be calculated using a Rayleigh Ritz variational method.

2.3 MODE CLASSIFICATION

In order to classify the modes, we decompose ζ, where ζ is a linear displacement, using a modified form of Helmholtz' theorem (Sobouti 1981), as follows

$$\zeta = \zeta_l + \zeta_t + \zeta_a \tag{2}$$

where the various components can be expressed in terms of scalar functions χ as

$$\zeta_l = -\nabla\chi_l \quad, \quad \zeta_t = \frac{1}{\rho}\nabla \times \nabla \times (\hat{z}\chi_t) \quad \text{and} \quad \zeta_a = \nabla \times \nabla \times \nabla \times (\hat{z}\chi_a)$$

The irrotational vector ζ_l is associated with longitudinal motions, whereas the remaining two are essentially solenoidal transverse displacements. It turns out that ζ_a can always be identified with Alfvén waves. We expand ξ as a linear combination of ζ , so that

$$\xi = \sum_i Z_i\zeta_i$$

where Z_i are proportionality constants and $i = l, t$ or a. For a uniform field, it is found that the Alfvénic motions get decoupled from the rest and can be treated separately.

3. Results

Equation (1) was solved using a Rayleigh Ritz method (see HS for details). The boundary conditions used were $\xi_z = d\xi_x/dz = 0$ at $z = 1000$ km and $\xi_x = d\xi_z/dz = 0$ at $z = -90$ km. A constant vertical field $B = 2000$ G, corresponding to $\beta = 0.84$ at $z = 0$ was used. We first consider the solutions for the l and t modes.

3.1 l AND t MODES

Figure 2 shows the $\omega - k$ (diagnostic) diagram for umbral oscillations, where k is the horizontal wave number. For a fixed value of k, a number of solutions exist satisfying the boundary conditions. These correspond to the normal mode frequencies or harmonics, which form a discrete spectrum. The curves depict the variation of ω with k for fixed order. We also classify the modes by comparing the magnitudes of the l and t components of ξ. Thus, in a l mode $|\xi_l| \gg |\xi_t|$ Open and closed circles correspond to t and l modes respectively, whereas the squares denote mixed modes. In the latter, the first letter denotes the components which as larger magnitude. Physically, the l and t modes can be crudely related to the conventional MHD modes. Above the photosphere, the Alfvén speed is much greater than the sound speed. In this case, the fast mode is essentially a transverse mode (t type) and the slow mode is mainly longitudinal (l type) for parallel propagation to the field. For perpendicular propagation, however, the fast and slow modes are of the l and t types respectively.

3.2 ALFVÉN MODES

The frequencies of the Alfvén modes were also calculated. Since these are independent of the horizontal wave number, we present the results in tabular form, for different mode orders only. Table 1 gives the frequencies and corresponding periods for the lowest five modes.

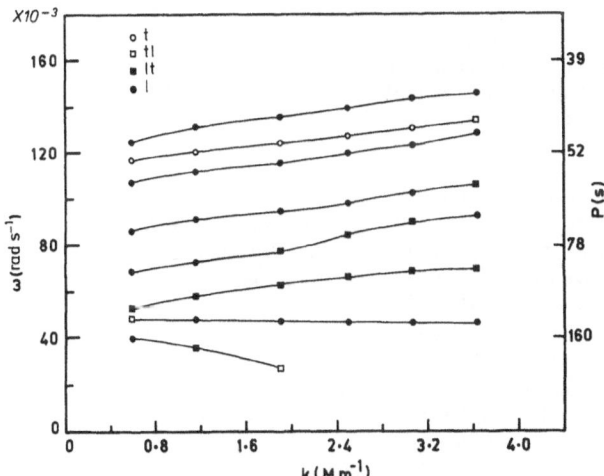

Figure 2. Variation of frequency ω with k in a sunspot umbra, assuming a uniform vertical magnetic field of 2000G. Open and filled circles denote t and l modes respectively, whereas open and filled squares correspond to mixed modes tl and lt respectively.

TABLE 1. Alfvén modes in a sunspot umbra

n	$\omega(s^{-1})$	$P(s)$
1	0.025	250
2	0.098	64
3	0.160	38
4	0.230	27
5	0.310	2

4. Discussion and Conclusions

We now consider the interpetation of oscillations with periods around 180 s. Assuming an umbral radius of some 3000 km, we find $k = 1.3$ Mm^{-1} (i.e., for the lowest order mode which has a vanishing radial displacement at the umbral boundary). The corresponding frequencies are 180 s and 130 s for the lowest two modes, and are of type lt and l respectively. For parallel propagation above the photosphere, the 130 s mode resembles a slow wave, whereas the 180 s modes is a mixed mode, which has a dominant slow component. It should, however, be noted that the nature of the modes changes with k (although the period is comparatively insensitive to k). We also find that oscillations in the 2-3 min range are unlikely to be of the Alfvénic kind. It is conceivable that the large period (around 300 s) oscillations observed for example by Balthasar et al. (1987), might possibly be Alfvén waves in the photosphere. Owing to the rapid increase of the Alfvén speed with height, these waves are unlikely to to reach chromospheric levels, since they suffer strong reflection as they propagate upwards.

In conclusion, we have theoretically calculated the spectrum of umbral oscillations. We suggest that the observed modes in the 2-3 min range are of the slow or mixed type. The higher period oscillations may possibly be of the Alfvén type.

References

Abdelatif,T.E.,Lites,B.W.,and Thomas,J.H.(1986),Astrophys. J. **311**,1015.

Antia,H. and Chitre,S.M. (1979),Solar Phys. **63**,67.

Balthasar,H. and Wiehr,E. (1984),Solar Phys. **94**,99.

Balthasar,H.,Küveler,G.,and Wiehr,E. (1987),Solar Phys. **112**, 39-48.

Beckers,J.M. and Schulz,R.B. (1972),Solar Phys. **27**,61.

Bhatnagar,A.,Livingston,W.C.,and Harvey,J.W.(1972),Solar Phys. **27**,80.

Gurman,J.B. and Leibacher,J.W. (1984), Astrophys. J. **283**, 859.

Gurman,J.B. (1987),Solar Phys. **108**,61.

Hasan,S.S. and Sobouti,Y. (1987),Mon. Not. R. astr. Soc. **228**,427.

Leroy,B. and Schwartz,S.J. (1982),Astron. Astrophys. **112**,84.

Lites,B.W. and Thomas,J.H. (1985),Astrophys. J. **294**,682.

Lites,B.W. (1986),Astrophys. J. **301**,922.

Maltby,P.,Avrett,E.H.,Carlsson,M.,Kjeldseth-Moe,O.,Kurucz,R.L. and Loeser,R. (1986), Astrophys. J. **306**,284.

Scheuer,M.A. and Thomas,J.H. (1981),Solar Phys. **71**,21.

Schröter,E.H. and Soltau,D. (1976),Astron. Astrophys. **449**,463.

Sobouti,Y. (1981), Astron. Astrophys. **100**,319.

Soltau,D.,Schröter,E.H.,and Wöhl,H. (1976),Astron. Astrophys. **50**,367.

Thomas,J.H.,Lites,B.W.,and Gurman,J.B. (1987), Astrophys. J. **312**,457.

Uchida,Y. and Sakurai,T. (1975),Publ. Astron. Soc. Japan **27**,259.

Zhugzhda,Y.D. (1979),Sov. Astron. **23**,42.

Zhugzhda,Y.D., Locan, V., and Staude, J. (1983) Solar Phys. **82**, 369.

THE OBSERVATIONAL SIGNATURE OF FLUX TUBE WAVES AND AN UPPER LIMIT ON THE ENERGY FLUX TRANSPORTED BY THEM

S.K. SOLANKI and B. ROBERTS
Department of Mathematical Sciences
University of St Andrews
St Andrews, KY16 9SS
Scotland

Abstract. The influence of undamped linear longitudinal tube waves on Stokes V profiles is considered. A rough upper limit is set on the energy flux transported by such waves through the photosphere. It is found that this upper limit is larger than the flux in the quiet sun. However, due to the small filling factor of the magnetic elements, the total luminosity of flux tube waves is unlikely to be larger than that of acoustic waves when averaged over the whole sun. Therefore, probably both kinds of waves contribute to chromospheric heating. However, the derived upper limit does not rule out that flux tube waves can significantly enhance the chromospheric brightness in active regions and the supergranular network where the magnetic filling factor is large.

1. Introduction

Flux tube waves are among the main contenders for transporting the energy required to heat the outer solar atmosphere. However, no observational limits have so far been set on the amount of energy which such waves transport. Due to the small size of the flux tubes (below the best current spatial resolution), no direct observations of locally excited waves exist (but see Giovanelli et al., 1978, for observations of globally excited modes). The indirect evidence is restricted to observations of line broadening (Solanki, 1986) and the modelling of the Stokes V asymmetry (Solanki, 1989). It is, therefore, not possible to directly determine the amount of energy transported by such waves, but we can use the line broadening to set an upper limit. First we briefly describe the general influence of tube waves on the widths of a selected set of spectral lines, since no previous study of this effect exists. We then go on to set a rough upper limit on the energy flux carried by tube waves by comparing with observed line profiles.

2. Summary of Calculations

In calculating the waves we follow the theory developed by Roberts and Webb (1978, 1979) combined with the following assumptions: 1) The calculations are linear, 2) the thin tube approximation is used, 3) radiative damping is neglected, and 4) the gas within the flux tube is uncoupled from that in its surroundings. A differential equation for the velocity must then be solved, after which the rest of the linear perturbations in atmospheric quantities produced by the wave (e.g. pressure, temperature, density, magnetic field strength) can be derived via the thin tube equations.

The radiative transfer is carried out numerically. For the investigation of basic influences of tube waves on line profiles five hypothetical spectral lines have been chosen. No. 1: Fe I with equivalent width $W_\lambda = 55$ mÅ in the quiet sun and excitation potential $\chi_e = 0$ eV; No. 2: Fe I, $W_\lambda = 15$ mÅ, $\chi_e = 0$ eV; No. 3: Fe

J. O. Stenflo (ed.), Solar Photosphere: Structure, Convection, and Magnetic Fields, 259–262.

I, $W_\lambda = 110$ mÅ, $\chi_e = 0$ eV; No. 4: Fe I, $W_\lambda = 55$ mÅ, $\chi_e = 4$ eV; No. 5: Fe II, $W_\lambda = 55$ mÅ, $\chi_e = 3$ eV. All lines have a Landé factor of unity. They have been chosen such that they cover a wide range of W_λ and χ_e. All calculations are for solar disk centre, so that only Stokes I and Stokes V need be considered. We concentrate here on Stokes V. Both single- and multi-ray (1.5-D) radiative transfer calculations have been carried out, but the influence of the dimension of the model on the widths of Stokes V profiles is minimal, so that we shall not differentiate between the two in the following.

3. Influence of Longitudinal Tube Waves on Line Widths

If we suppress the temperature fluctuations caused by the wave, i.e. consider an isothermal wave, then the widths of all lines are affected similarly. In this case the line width v_D can be written as: $v_D^2 \approx (\delta \cdot v_a(z))^2 + v_{D_0}^2$, where v_{D_0} is the line width in the absence of the wave, $v_a(z)$ is the wave amplitude at the "height of formation", z, of the line and δ is a multiplicative factor which takes into account that the wave profile is closer to a sine wave than to a Gaussian. δ differs appreciably from unity for practically all waves, suggesting that a Gaussian turbulence broadening model (as used by, for example, Solanki, 1986) will give a different (larger) limit on the flux tube wave energy flux. For waves with sufficiently long wavelengths δ becomes independent of the line (macroturbulence limit), whereas for short wavelength waves it varies considerably from line to line, depending on the sensitivity of the particular line to microturbulence. Also, for isothermal waves, standing and propagating waves show the same effect on the line width.

If we now let the temperature vary with phase, then standing waves still broaden the lines in the same manner as before (due to the phase difference of 90° between temperature and velocity fluctuations), while the influence of propagating waves is greatly changed (phase difference of approximately 180°). Fig. 1 shows v_D vs. $v_a(z = 0)$ for isothermal (Fig. 1a) and moderately non-isothermal (Fig. 1b) propagating waves with a wavelength of 300 km and a period of approximately 80 s. Note that the two weaker low excitation Fe I lines (lines No. 1 and 2) are very strongly affected by the temperature fluctuations, while the others show little or no influence on the line widths. The behaviour of the line widths of the three low excitation lines can be easily understood if we keep their different strengths in mind and follow the weakening and strengthening of the lines in the down- and upflowing phases, respectively. Fig. 1 carries a clear message: as long as we do not know how strong the thermal fluctuations associated with the tube waves are on the sun, it is wise to use a temperature insensitive spectral line for setting a limit on the energy flux transported by such waves.

4. Upper Limit on the Wave Energy Flux

The energy flux carried by a tube wave can be written as

$$F(z) = \langle c_T(z)\rho(z)\,v^2(z)\rangle \approx \frac{1}{2}c_T(z)\rho_0(z)v_a^2(z), \tag{1}$$

where c_T is the tube speed (i.e., the propagation speed of a tube wave) and ρ is the gas density (ρ_0 = time averaged gas density). We derive the upper limit to v_a from a fit to the line widths of Stokes V profiles observed with a Fourier transform spectrometer (FTS) at Kitt Peak. The data have been described by Stenflo et al. (1984). The flux tube model describing the undisturbed atmosphere has been taken from Solanki (1986). In accordance with the conclusion of Sect. 3, we use the Fe II line at 5197.6 Å whose width is unaffected by the thermal fluctuations due to the wave.

From a variety of calculations, using different waves, we arrive at the following preliminary upper limit for the energy flux transported by longitudinal tube waves F_{Tube}:

$$F_{Tube}(5197.6 \text{ Å}) \lesssim 3 \times 10^9 \text{erg s}^{-1} \text{cm}^{-2}. \tag{2}$$

Fe II 5197.6 Å is formed in the lower photosphere, so that F_{Tube} refers to that height. This limit applies to both an active plage region and an enhanced network region.

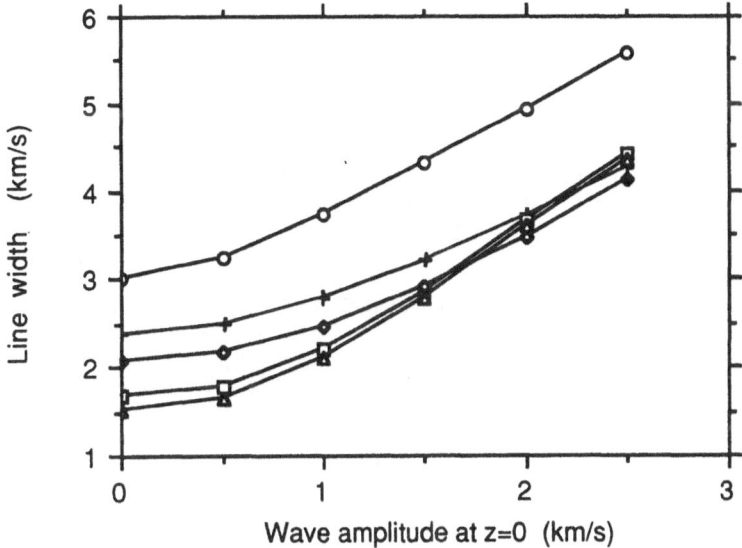

Fig. 1a   Line width of Stokes V in km s^{-1} vs. the wave amplitude at $z = 0$ (corresponding to $\tau_{5000} = 1$ in the quiet sun) for isothermal waves with wavelengths of 300 km and periods of around 80 s. Squares: line 1, triangles: line 2, circles: line 3, diamonds: line 4, plusses: line 5.

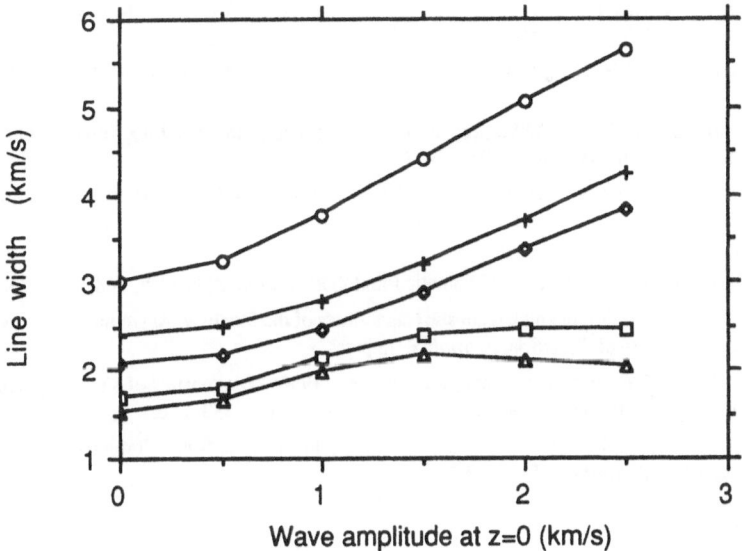

Fig. 1b   The same for moderately non-isothermal waves.

At first sight, the upper limit (2) appears quite large. It is much larger than the recent estimate of Keil and Mosman (1989) for acoustic waves in the quiet sun, who find $F_{\text{Acoustic}} \approx 1 \times 10^8$ erg s^{-1} cm^{-2} after correcting the power upwards by about 2 orders of magnitude to take into account radiative transfer effects. Also, at first sight the upper limit to the tube wave energy flux appears ample to heat the chromosphere, which requires 4×10^6 erg s^{-1} cm^{-2} in the quiet sun (Withbroe and Noyes, 1977). However, it should be borne in mind that in contrast to the quiet sun values the flux tube values are *upper limits*. Furthermore, flux tubes cover only a small fraction of the solar surface and, therefore, in order to obtain an "area averaged" value of the limit on the tube wave energy flux it is necessary to multiply the value given in Eq. (2) by the global solar magnetic filling factor. As an optimistic estimate of the global filling factor we take 1%. We are then left with an upper limit of the wave luminosity in flux tubes which is of about the same order as that in the quiet sun, and may be less.

We can reach two main conclusions from the above: 1) Since approximately equal amounts of energy are transported by both kinds of waves, it appears likely that the chromosphere is heated by a *combination* of tube waves and acoustic waves. Our investigation, therefore, supports the notion of a "basal flux" of acoustic waves which heat the "quiet" chromosphere, first derived from a study of late-type stars by Schrijver (1987). 2) Since magnetic features can cover 10–20% of the surface area of active regions, flux tube waves potentially carry sufficient energy into the chromosphere above active regions and the network to be primarily responsible for the excess heating seen there (radiative losses are 2×10^7 erg s^{-1} cm^{-2} in active regions; cf. Withbroe and Noyes, 1977), and so give rise to the "patchiness" of chromospheric Ca II H and K emission seen in spectroheliograms in these lines.

References

Giovanelli, R.G., Livingston, W.C. and Harvey, J.W. (1978) "Motions in solar magnetic tubes. II. The oscillations", *Solar Phys.* 59, 49–64.

Keil, S.L. and Mosman, A. (1989) "Observations of High Frequency Waves in the Solar Atmosphere", in *Solar and Stellar Granulation*, Proc. NATO Advanced Research Workshop, R. Rutten and G. Severino (Eds.), Reidel, Dordrecht, in press

Roberts, B., and Webb, A.R. (1978) "Vertical Motions in an Intense Magnetic Flux Tube.", *Solar Phys.* 56, 5–35.

Roberts, B., and Webb, A.R. (1979) "Vertical Motions in an Intense Magnetic Flux Tube. III. On the Slender Flux Tube-Approximation", *Solar Phys.* 64, 77–92.

Schrijver, C.J. (1987) "Magnetic Structure in Cool Stars. XI. Relations Between Radiative Fluxes Measuring Stellar Activity, and Evidence for Two Components in Stellar Chromospheres", *Astron. Astrophys.* 172, 111–123.

Solanki, S.K. (1986) "Velocities in Solar Magnetic Fluxtubes", *Astron. Astrophys.* 168, 311–329.

Solanki, S.K. (1989) "The Origin and Diagnostic Capabilities of the Stokes V Asymmetrie Observed in Solar Faculae and the Network", *Astron. Astrophys.* in press

Stenflo, J.O., Harvey, J.W., Brault, J.W. and Solanki, S.K. (1984) "Diagnostics of solar magnetic fluxtubes using a Fourier transform spectrometer", *Astron. Astrophys.* 131, 333–346.

Withbroe, G.L and Noyes, R.W. (1977) "Mass and Energy Flow in the Solar Chromosphere and Corona", *Ann. Rev. Astron. Astrophys.* 15, 363–387.

MAGNETIC FLUX CONCENTRATION BY SIPHON FLOWS IN ISOLATED MAGNETIC FLUX TUBES

John H. Thomas
Department of Mechanical Engineering, Department of Physics and
Astronomy, and C. E. K. Mees Observatory
University of Rochester
Rochester, New York 14627, U.S.A.

Benjamin Montesinos
Department of Theoretical Physics
University of Oxford
1 Keble Road
Oxford OX1 3NP, U.K.

ABSTRACT. Siphon flows along arched, isolated magnetic flux tubes, connecting photospheric footpoints of opposite magnetic polarity, cause a significant increase in the magnetic field strength of the flux tube due to the decreased internal gas pressure associated with the flow (the Bernoulli effect). These siphon flows offer a possible mechanism for producing intense, inclined, small-scale magnetic structures in the solar photosphere.

In order to produce magnetic flux tubes in the solar photosphere with the observed field strengths of 1000 - 1500 G, the interior of the flux tubes must be substantially evacuated. The observed magnetic field strengths are very near the limiting maximum value $B_p = (8\pi p_e)^{1/2}$ of a totally evacuated flux tube confined by the external gas pressure p_e; thus, a mechanism that produces a substantial evacuation of the flux tube is required. One such mechanism, first proposed by Parker (1978), is the convective collapse of a vertical flux tube in the superadiabatic layer just below the photosphere. This mechanism has been studied extensively (see, for example, Webb and Roberts 1978; Spruit and Zweibel 1979; Spruit 1979; Hasan 1985; Venkatakrishnan 1985; Nordlund 1986; and the recent reviews by Solanki 1987, Stenflo 1989, and Thomas 1989). Here we discuss another possible mechanism, namely, siphon flows in an arched, isolated magnetic flux tube (Thomas 1984, 1988).

Earlier work on siphon flows, beginning with Meyer and Schmidt (1968), dealt exclusively with the case of an effectively rigid, embedded flux tube in the limit of low plasma beta, appropriate for conditions in the solar corona or chromosphere (see the review by Priest 1981). For conditions in the solar photosphere or convection zone, with plasma beta of order unity, the flux tube cannot be considered rigid; its cross-sectional area and magnetic field strength must change in response to changes in internal gas pressure induced by the siphon flow in order to maintain lateral pressure balance with the surrounding atmosphere. The reduced internal gas pressure associated with the siphon flow (the Bernoulli effect) results in an increased magnetic field strength and a reduced cross-sectional area of the flux tube.

The critical speed for siphon flows in a thin isolated flux tube is the so-called *tube speed* $c_t = [c^2 a^2/(c^2+a^2)]^{1/2}$, where c is the internal sound speed and a is the Alfvén speed.

J. O. Stenflo (ed.), Solar Photosphere: Structure, Convection, and Magnetic Fields, 263–266.

(The tube speed is the speed of propagation of a longitudinal, or sausage-mode, wave along a thin isolated magnetic flux tube; it is smaller than either the sound speed or the Alfvén speed.) Flows with speeds less than c_t (subcritical flows) or greater than c_t (supercritical flows) along an isolated flux tube are qualitatively equivalent to subsonic or supersonic flows along a rigid, embedded flux tube. In an arched flux tube, the flow can undergo a smooth transition from subcritical to supercritical speed only at the top of the arch where the tube is locally horizontal.

The maximum speed of the siphon flow depends on the height of the arch and the initial velocity at the upstream footpoint. If the arch height is small enough or the initial velocity is low enough, the flow remains subcritical throughout. If the maximum velocity is less than the characteristic speed $c_1 = [c^2 a^2/(2c^2+a^2)]^{1/2} < c_t$, then the flow accelerates and the tube expands with increasing height all along the upstream leg of the arch, and the flow decelerates and the tube contracts with decreasing height all along the downstream leg of the arch. The expansion of the tube with height is less than it would be without the flow, however, due to the decreased internal pressure caused by the flow (the Bernoulli effect). If the maximum velocity exceeds the characteristic speed c_1 but is still less than c_t, the flow accelerates and the tube expands up to the point where $v = c_1$ and beyond that point the flow accelerates but the tube contracts with increasing height in the upstream leg of the arch. This produces a point of local maximum cross-sectional area, or "bulge point," in the tube at the point where $v = c_1$. In the downstream leg of the arch the flow decelerates symmetrically and there is another bulge point. For such a flow the Bernoulli effect is so strong that it more than compensates for the decreasing pressure with height outside the tube and causes the tube to contract with height above the bulge point.

If the arch height is increased or the initial velocity at the upstream footpoint is increased, we reach a critical flow in which $v = c_t$ at the top of the arch. In this case the flow can accelerate smoothly to supercritical velocity in the downstream leg of the arch. The flow velocity continues to increases while the cross-sectional area decreases (and the magnetic field strength increases) with decreasing height along the downstream leg of the arch. However, this supercritical flow requires a very small internal gas pressure at the downstream footpoint which will not be the case in the solar photosphere. For a higher value of the "backpressure" at the downstream footpoint, the flow will decelerate abruptly and the cross-sectional area will increase abruptly across a standing "tube shock" at some point along the downstream leg of the arch. The critical flow is "choked," in the sense that a decrease in the pressure at the downstream footpoint will not increase the mass flow rate along the tube; disturbances cannot propagate upstream along the supercritical section of the flow.

We have carried out extensive numerical computations of siphon flows in arched, isolated flux tubes in the limits of isothermal flow (Thomas 1988) and adiabatic flow (Monetsinos and Thomas 1989). Siphon flows are capable of increasing the magnetic field strength of an isolated flux tube up to as much as 80 or 90% of the limiting vacuum value and thus can, in principle, produce flux tubes with strengths 1500 G or more in the solar photosphere. In general, adiabatic flows produce greater magnetic flux concentration than isothermal flows (Montesinos and Thomas 1989). Radiative transfer in a real photospheric flux tube will produce conditions somewhere in between isothermal and adiabatic flow.

The large-scale equilibrium shape of a flux tube arch containing a steady siphon flow is determined by a balance among the buoyancy force, the net magnetic tension force, and the inertial force associated with the siphon flow along curved streamlines (the centrifugal force). We have calculated the equilibrium shape of the flux tube arch for a wide variety of siphon flows (Thomas and Montesinos 1989) and compared it with the shape of a static arched flux tube first determined by Parker (1975). In general, the presence of a siphon flow requires that the arch be more highly curved in order that the net magnetic tension force can balance the additional effect of the centrifugal force. With increasing flow speed, the footpoints of the arch move closer together relative to the height of the arch. Provided there is some mechanism for anchoring the ends of the flux-tube arch somewhere in the convection zone, the intensified part of the flux tube at the top of the arch might be held in mechanical equilibrium in the photosphere for a time much longer than the rise time associated with the magnetic buoyancy alone.

A totally isolated flux tube arch is highly curved and the distance between the footpoints in the photosphere is at most equal to about six times the density scale height of the surrounding atmosphere. This configuration would produce closely spaced intense magnetic elements of opposite polarity in the photosphere. However, more widely spaced footpoints are possible if the arch reaches up to the overlying magnetic canopy. In this case, the upper part of the arch can extend horizontally over a considerable distance, being held down against its buoyancy force by the magnetic pressure of the canopy.

We suggest that siphon flows in arched magnetic flux tubes may be the cause of some of the intense magnetic fields observed in the solar photosphere outside sunspots. In these cases the observed magnetic field will be inclined to the vertical, with the inclination depending on the height of formation of the spectral line. Observations do indicate that a significant fraction of the photospheric flux tubes are moderately inclined (10° or more) and that a smaller fraction may be highly inclined (Solanki 1987).

For an intense magnetic element associated with a flux tube containing a siphon flow, there will be either an upflow or a downflow inside the magnetic element, depending on whether it is associated with the upstream or the downstream footpoint of the flux tube. This flow will be persistent and thus should be observable, in contrast to the downflows associated with the convective collapse mechanism, which are only transient events and thus would be difficult to observe. Early observations of photospheric flux tubes were interpreted as indicating downflows within many flux tubes, but more recent observations of higher resolution are interpreted as indicating that downflows (or upflows) inside photospheric flux tubes are very weak or absent (see the reviews by Solanki 1987 and Stenflo 1989). Even the best observations to date fail to resolve individual flux tubes, however, so the evidence that there is no systematic flow inside a photospheric flux tube is not conclusive, and in any case may not apply to all photospheric flux tubes. Model calculations of the effect of siphon flows on observed line profiles are needed to help settle this issue.

The siphon flow mechanism at least has the advantage of solving one well-known problem associated with the early reports of downflows. Because all of the mass flux associated with the downflow is provided by an equal mass flux associated with the upflow at the other photospheric footpoint of the arch, there is no need to expect that the mass would have to be taken from the corona, in which case the corona would be totally drained in a matter of minutes by the many photospheric flux tubes.

This work was supported in part by the National Aeronautics and Space Administration under grant NSG-7562.

References

Hasan, S. S. 1985, *Astron. Astrophys.*, **143**, 39.
Meyer, F., and Schmidt, H. U. 1968, *Zeits. Angew. Math. Mech.*, **48**, 218.
Montesinos, B., and Thomas, J. H. 1989, *Astrophys J.*, **337**, 977.
Nordlund, A. 1986, in *Small Scale Magnetic Flux Concentrations in the Solar Photosphere*, ed. W. Deinzer, M. Knölker, and H. H. Voigt (Göttingen: Vanderhoeck & Ruprecht), pp. 83-102.

266

Parker, E. N. 1975, *Astrophys. J.*, **201**, 494.
Parker, E. N. 1978, *Astrophys. J.*, **221**, 368.
Priest, E. R. 1981, in *Solar Active Regions*, ed. F. Q. Orrall (Boulder, CO: Colorado Associated University Press), p. 213.
Solanki, S. K. 1987, in *Proc. Tenth European Regional Astronomy Meeting of the IAU. Vol. 1: The Sun*, ed. L. Hejna and M. Sobotka, Publ. Astron. Inst. Czechoslovakia Acad. Sci., p. 95.
Spruit, H. C. 1979, *Solar Phys.*, **61**, 363.
Spruit, H. C., and Zweibel, E. G. 1979, *Solar Phys.*, **62**, 15.
Stenflo, J. O. 1989, *Astron. Astrophys. Rev.* **1**, **3**.
Thomas, J. H. 1984, in *Small-Scale Dynamical Processes in Quiet Stellar Atmospheres*, ed. S. L. Keil (Sunspot, NM: National Solar Observatory), p. 276.
Thomas, J. H. 1988, *Astrophys. J.*, **333**, 407.
Thomas, J. H. 1989, in *The Physics of Magnetic Flux Ropes*, ed. C. T. Russell, E. R. Priest, and C. Lee (American Geophysical Union), in press.
Thomas, J. H., and Montesinos, B. 1989, preprint.
Venkatakrishnan, P. 1985, *J. Astrophys. Astron.*, **6**, 21.
Webb, A. R., and Roberts, B. 1978, *Solar Phys.*, **59**, 249.

ELECTRIC CURRENTS IN THE ATMOSPHERE OF THE SUN

V.I. ABRAMENKO, S.I. GOPASYUK, M.B. OGIR
Crimean Astrophysical Observatory
334413, p/o Nauchny, Crimea, USSR

ABSTRACT. The structure of the magnetic field, proper motions of sunspots and electric currents have been studied and related to the evolution of sunspot groups. Further the height variations of the magnetic fluxes and electric currents in active regions have been explored.

1. Introduction

The appearance of an active region on the sun and the evolution of the magnetic field is determined mainly by plasma motions in the subphotospheric layers. These motions induce electric currents and hence change the exterior magnetic field.

The vector magnetograph was a new step in the investigations of the structure of solar magnetic fields and electric currents (Stepanov and Severny, 1962; Nikulin, 1967). Using observational data on the total vector of the magnetic field, Severny (1965) discovered electric currents in the atmosphere of the Sun.

The evolution of the magnetic field of an active region is closely associated with the variations of the electric currents, which may lead to different types of plasma instabilities, in particular flares.

2. The observational data

We have studied the observational data of the total vector of the magnetic field for two active regions: 21.–26.10.1968 (longitude $\ell = -29°$ to $+37°$; latitude $\varphi = 16°$N) and 8.–14.06.1969 ($\ell = -23°$ to $+52°$; $\varphi = 14°$S). The first active region was rather stable, the second was decaying.

The magnetic field was recorded in the Fe I $\lambda5250.2$ Å line with the vector magnetograph of the Crimean Observatory (Nikulin, 1967). We have calculated the vertical component of the electric current density j_z using the transverse components of the observed field.

3. Global electric currents and vortex motions of the plasma

To obtain a better picture of the electric current structure and the current-generating part

J. O. Stenflo (ed.), Solar Photosphere: Structure, Convection, and Magnetic Fields, 267–271.
© 1990 by the IAU.

of the magnetic field, we have displayed in Abramenko and Gopasyuk (1987) the vector of the observed transverse field $H_\perp$ as two components, one along the vector of the potential transverse field, and the other perpendicular to it. The structure of the potential field was computed from the H_z component by solving a Neumann boundary value problem. The component perpendicular to the potential field depends only on the electric current, which determines the orientation of the observed transverse field vector relative to the potential field (see Figure 1.). The arrows reveal the presence of two vortex structures. One is twisted mainly in the clockwise direction and covers the whole leading part of the active region, while the second is twisted in the anti-clockwise direction. The vortex structure of the field confirms the presence of large-scale electric currents extended over the total areas of the leading and following parts of the active region. The current is flowing upwards in the leading part, downwards in the following part, independent of the sign of the leading polarity, and wherever the position of the active region is (northern or southern hemisphere). The values of the upflowing and downflowing currents are the same, approximately 2×10^{12} A . The equality of the currents in the leading and following parts suggests that the current is closed and quasi-stationary.

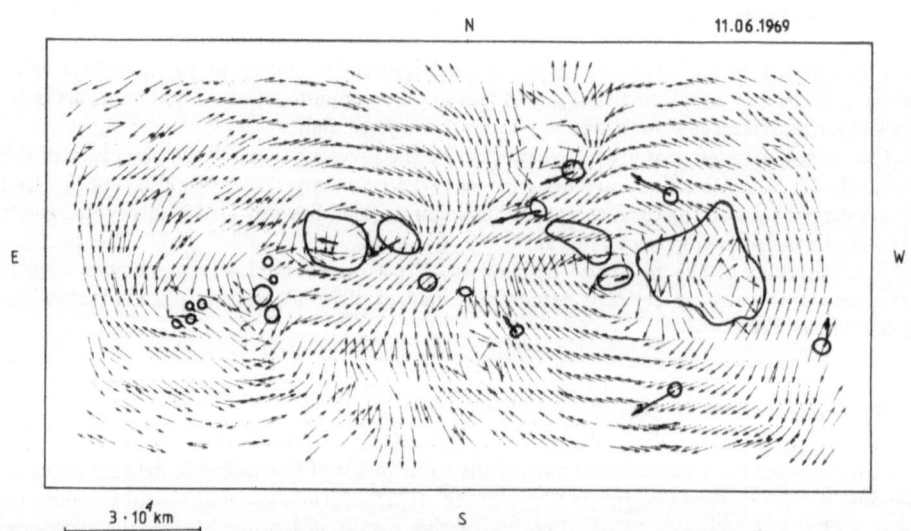

Figure 1. Direction of the azimuth angles (thin arrows) of the component of the observed transverse magnetic field vector that is perpendicular to the transverse field vector computed with the potential approximation. The thick lines show the position of the spots, with their proper motions indicated by thick arrows.

On the maps of the vortex structure of the magnetic field we have plotted the sunspots and indicated their trajectories by arrows. The main spots are located near the centres of the vortex structures. Their proper motions are small. The spots situated at significant distances from the centre of the vortex have large proper motions (up to 200 m s^{-1}). The trajectories of the spots in general coincide with the direction of the observed component of the transverse field vector that is plotted in Figure 1. This suggests that the proper motion of sunspots and the orientation of the non-potential component of the transverse field may have a common origin, both being determined by plasma motions not only at the photospheric level, but also in the subphotospheric layers.

Since in the leading part of the active region the vector of the observed transverse field $H_\perp$ was turning mainly in the clockwise direction and vice-versa in the following part, and the spots were moving along these vortex structures, we have come to the conclusion that there were two plasma vortices in these active regions: one covering the whole leading part, the other covering the following part of the active region. The rotation of the plasma in each of these vortices corresponded to the direction of the magnetic field of the global electric current (Abramenko et al., 1988 c). The vortex motions of the plasma occupied a rather extensive region from the level of spot formation in the deep layers upward to the photosphere. However, as follows from Figure 1, there exist several regions where the direction of the non-potential field component does not coincide with the main direction of the vortex. Probably this is a consequence of the circumstance that local disturbances are superposed on the global vortex structure. Our results indicate the presence of processes that twist the whole system of magnetic structures which form an active region.

4. Local electric currents and the evolution of spot groups

A system of local small-scale current structures are superposed on the global electric current. Using each individual record, we have calculated the average over the area of the active region of the local electric current density $\bar{j}_z$ in the upward and downward directions.

The calculations were made for both sunspot groups. Though these groups were at different stages of evolution, their data appeared to be aligned along the same straight line. The current density $\bar{j}_z$ (A km^{-2}) and the area S of the group (in square degrees) turned out to be connected by the relation (Abramenko et al., 1987)

$$\bar{j}_z = 74\,S + 950.$$

The correlation coefficient between $\bar{j}_z$ and S is 0.88. This relation shows that the spots in an active region appear only when the average current density exceeds some threshold value, approximately 10^3 A km^{-2}. Furthermore, with increasing area of the spots, the density of the current in the active region changes in synchrony.

From the maps of the observed field we can calculate the energy of the transverse magnetic field of the active region corresponding to a volume with the height of unity, $W = 1/8\pi \int_S H_\perp^2 \, dS$, and the value of the total current I_z (the sum of the local currents in the active region in the given direction). The day-to-day variations of the value of the observed field energy W show a correlation with the variations of the total current I_z in the active region (Abramenko et al., 1989). The observational data of both active regions gave a common dependence, which is approximately linear. It is not excluded, however, that each active region has its own dependence. This relation indicates that at least a considerable part of the photospheric transverse field is determined by currents that flow in the photosphere. In fact, if only a small part of the transverse field were produced by photospheric electric currents, its variations against the background magnetic field of the active region would be negligible, and would not have shown a relation with the variations of the photospheric electric currents.

The storage of energy due to the local electric currents in the active region significantly exceeds 10^{32} erg (Abramenko et al., 1989). The decay of the active region leads to dissipation of the local currents, and the field generally tends to become more potential (Abramenko et al., 1988 a).

5. Variations of the electric current and magnetic field with height

We have calculated the electric currents using the structure of the transverse potential field and the orientation of the Hα fibrils (Abramenko et al., 1988 b). To avoid projection effects on the Hα fibrils, we have used observations near the central meridian. The calculated currents nicely display not only the structure, but also the evolution of the currents in the photosphere of the active region. This enables us to obtain information concerning the currents not only in the photosphere, but also at the level of the Hα fibrils.

The structure of the potential field was calculated at different levels in the atmosphere up to 50'000 km. Assuming that the deviation of the real transverse field vector from the potential one does not change with height, and is given by the orientation of the Hα fibrils, we have calculated the currents at different levels. The data analysis showed that when going from the photosphere upwards, the main current structures on the whole remain unchanged, and can be observed during at least three days.

The magnetic fluxes were calculated at the same elevations by using the vertical component of the potential field. To compare the height variations of the magnetic flux Φ_z and total electric current I_z observed on different days with each other, we normalized them to their values in the photosphere. The photospheric value of the total current was computed using the transverse component of the potential field and the angular deviation of the potential field from the observed transverse field in the photosphere. The results of these calculations for three days of observations are illustrated in Figure 2. We see that the total electric current decreases faster than the magnetic flux of the potential field (Abramenko et al., 1988 b).

A potential model gives the most rapid decrease of the field with height as compared with all other models. The circumstance that the total current decreases much faster than the magnetic flux indicates that the magnetic structures tend to become more potential with height. This agrees with the results reported by Poletto et al. (1975) that the morphology of the coronal structures observed in soft X-ray, and the magnetic field structures computed from the observed H_z component in the photosphere assuming a potential field, look very similar.

6. Conclusion

The study of electric currents provides us with exceptional information about the processes taking place in active regions. The currents largely determine the structure and value of the transverse magnetic field at the photospheric level, as well as the evolution of the active region. However, when going to the upper levels of the solar atmosphere, the value of the electric current decreases significantly faster than the magnetic flux. Therefore, if at some level in the corona electric currents are present, they are most likely determined by transient processes associated with different deformations of the coronal magnetic field.

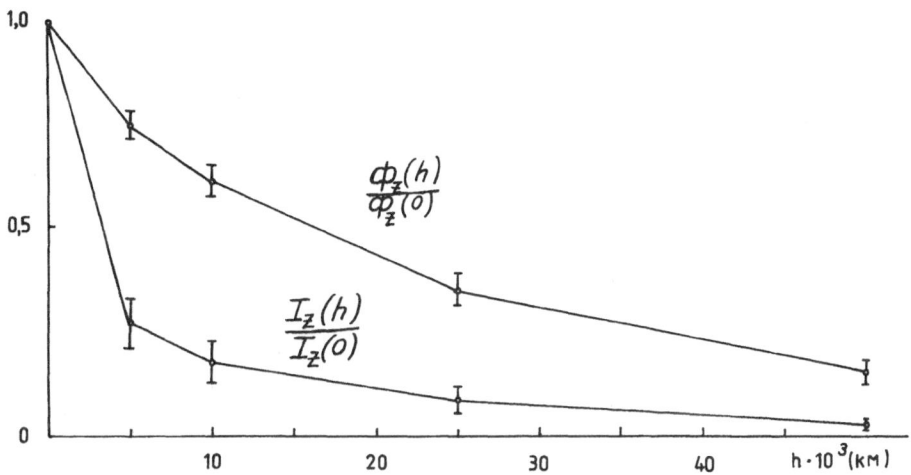

Figure 2. Relative height variations of the magnetic flux and electric current.

References

Abramenko, V.I. and Gopasyuk, S.I. (1987) 'A system of electric currents and magnetic field structure in the active region', *Izv. Krymsk. Astrofiz. Obs.* **76**, 147-168.

Abramenko, V.I., Gopasyuk, S.I., and Ogir, M.B. (1987) 'Electric currents and active region evolution', *Solnech. Dannye* **6**, 73-79.

Abramenko, V.I., Gopasyuk, S.I., and Ogir, M.B. (1988 a) 'Evolution of the active region, its current systems and flare activity', *Izv. Krymsk. Astrofiz. Obs.* **78**, 151-171.

Abramenko, V.I., Gopasyuk, S.I., and Ogir, M.B. (1988 b) 'The determination of the electric currents regarding the vertical component of the magnetic field and Hα fibrils', *Izv. Krymsk. Astrofiz. Obs.* **80**, 97-105.

Abramenko, V.I., Gopasyuk, S.I., and Ogir, M.B. (1988 c) 'Plasma motions and electric currents in the active region', *Izv. Krymsk. Astrofiz. Obs.* **81**.

Abramenko, V.I., Gopasyuk, S.I., and Ogir, M.B. (1989) 'Electric currents and mangetic field loop structures of the active regions on the Sun', *Izv. Krymsk. Astrofiz. Obs.* **82**.

Nikulin, N.S. (1967) 'Simultaneous registration of the main parameters of a magnetic field by means of a magnetograph', *Izv. Krymsk. Astrofiz. Obs.* **36**, 76-86.

Poletto, G., Vaiana, G.S., Zombeck, M.V., Krieger, A.S., and Timothy, A.F. (1975) 'A comparison of coronal X-ray structures of active regions with magnetic fields computed from photospheric observations', *Solar Phys.* **44**, 83-99.

Severny, A.B. (1965) 'A study of the magnetic field and electric currents of unipolar sunspots', *Izv. Krymsk. Astrofiz. Obs.* **33**, 34-79.

Stepanov, V.E. and Severny, A.B. (1962) 'A photoelectric method for measurements of the magnitude and direction of the solar magnetic field', *Izv. Krymsk. Astrofiz. Obs.* **28**, 166-193.

GENERATION OF MAGNETIC FIELDS AND ELECTRIC CURRENTS IN THE SOLAR PHOTOSPHERE

J.C.HENOUX* and B.V.SOMOV°
* Observatoire de Paris, DASOP, URA326,
92195 Meudon Principal, FRANCE,
◇ Lebedev Physical Institut, Leninskii Prospect 53,
SU-117924 Moscou USSR

ABSTRACT. Velocities of electrons, ions and neutrals are computed in the three-fluid approximation for an axisymmetrical magnetic field. By prescribing a radial dependence of the velocity of neutrals in agreement with a downflow, the radial dependence of the magnetic field energy density is derived for a given set of values of the magnetic field at the central and external boundaries. Flux-tube cooling by advection of ionization energy is found to be significant. Vortices in the low photosphere could produce significant electric power and DC current intensity along the coronal magnetic lines of forces. The velocities of neutrals, the size and the number of flux-tubes required to power flares in plage regions, are estimated.

1. Introduction

Some authors suggested that sunspots and active regions could be superficial phenomena originating in the photosphere and at the top of the convective zone (Bumba (1987a,b), Akasofu (1984a,b,c). Inward and downward motions below the photosphere have been considered as a mean of cooling and powering the sunspot dynamics by Schatten and Mayr (1985). Ambroz (1987) found that Active Regions are formed in places where the global circulation display vorticity. And Martres et al. (1973) related sunspot formation and disappearance to the direction of rotation of the vortices. As shown below, ambipolar diffusion in the weakly ionized photosphere creates electric DC currents and the kinetic energy of the neutral component could be converted into electrical and magnetic energy as long as convection maintains the flow (Spicer, 1982).

2. Particles velocities in an axisymetric magnetic field

Assuming steady state, the balance of the forces that act per unit volume on particules k of charge q_k is expressed as

$$n_k m_k \mathbf{V}_k . \nabla \mathbf{V}_k = n_k \sum_l \mathbf{F}_{l,k} + q_k n_k (\mathbf{E} + \mathbf{V}_k \wedge \mathbf{B}) - \nabla P_k + n_k m_k \mathbf{g}. \qquad (1)$$

$\mathbf{F}_{l,k}$ is the friction force acting on particule k due to particules l, and $\mathbf{F}_{l,k} = m_k \nu_{kl}(\mathbf{V}_l - \mathbf{V}_k)$. $\mathbf{B}$ and $\mathbf{E}$ are the magnetic and electric field vectors. $\mathbf{V}_k$, n_k and m_k are the velocity, density and mass of particle k. ν_{kl} is the coefficient of

J. O. Stenflo (ed.), Solar Photosphere: Structure, Convection, and Magnetic Fields, 273–277.

friction between particles k and l. In the case where one of the species is a neutral atom the subscript n is omitted in the friction coefficient. A velocity field $\mathbf{V^c}$ in the convective zone can create by continuity an electric field $\mathbf{E}$ in the photosphere (Hénoux and Somov, 1987). Assuming the magnetic field to be vertical in the convective zone, the electric field and plasma velocity in the convective zone are related by $\mathbf{E} = -\mathbf{V^c} \wedge \mathbf{B}$.

Restraining equation (1) to the horizontal components of the forces and using the relation between the velocity in the convection zone and the electric field we write

$$n_k \sum_l \mathbf{F}_{l,k} - q_k n_k (\mathbf{V}_k - \mathbf{V^c}) \wedge \mathbf{B_z} = -\nabla P_k - n_k m_k \mathbf{V}_k . \nabla \mathbf{V}_k + q_k n_k \mathbf{V}_{\mathbf{z},k} \wedge \mathbf{B}, \quad (2)$$

Equation (2) leads to a system of 6 linear equations that can be solved for the radial and azimuthal components of the velocities of ions, electrons, and neutrals refered by the subscripts r and θ. These components can be expressed as a function of V_θ^c, V_r^c, $V_{z,e}$, $V_{z,i}$, the radial pressure gradients and the radial and azimuthal components of the inertial term $n_k m_k \mathbf{V}_k . \nabla \mathbf{V}_k$. Formulæ are given in Hénoux and Somov (1988,1989). In what follows, we kept in the inertial terms only the dominant contribution of neutrals and assumed $B_r = 0$. Derived from the expression of the ion and electron velocities, the azimuthal and radial current densities are

$$j_\theta = (j_z B_\theta + \frac{\partial P_n^*}{\partial r})/B_z \tag{3}$$

$$j_r = -\frac{1}{B_z} n_n m_n \frac{V_{r,n}}{r} \frac{\partial r V_{\theta,n}}{\partial r}, \tag{4}$$

where j_z is the vertical current density and P_n^* is defined by

$$P_n^* = P_n + \frac{1}{2} n_n m_n V_{r,n}^2 - \frac{1}{r} V_{\theta,n}^2. \tag{5}$$

Assuming that $V_r^c = 0$, the radial velocities of neutrals and ions can be written as

$$V_{r,i} = -\frac{j_\theta B_z}{\sigma B_z^2} + I_{r,i} = \frac{1}{2\mu\sigma B_z^2} \frac{\partial B_z^2}{\partial r} + I_{r,i}, \tag{6}$$

$$V_{r,n} = V_{r,i} - \frac{1}{\alpha_s} \frac{\partial P_n^*}{\partial r}. \tag{7}$$

Here σ is the electric conductivity parallel to B, $\alpha_s = n_e(m_i \nu_i + m_e \nu_e)$ and

$$I_{r,i} \simeq -\frac{n_n m_n}{n_e e B_z} \frac{V_{r,n}}{r} \frac{\partial r V_{\theta,n}}{\partial r}. \tag{8}$$

In equation (3) $j_z B_\theta \geq 0$ and a positive pressure gradient always generate an azimuthal current and a radial magnetic field gradient. The azimuthal current is associated with a radial motion of ions, and with the neutrals moving inwards relatively to ions. Equation (4) shows that the loss of angular momentum of neutrals leads to the circulation of radial currents. These radial currents would generate vertical currents. It is outside the scope of this paper to compute self consistently the radial dependence of the angular momentum. From the expression of $V_{\theta,n}$ it can be shown that, except for constant B, $\partial r V_{\theta,n}/\partial r = 0$ is not a solution . We shall just assume that $r V_{\theta,n}$ decreases to zero inside the flux-tube radius. In the next section we relate the velocity of neutrals to the magnetic field gradient and to the creation of strong magnetic fields from a preexisting weak field. Then we study how vortices can lead to the circulation of radial and vertical electric currents.

3. Magnetic field concentration

From Ampère law and equation (3) we derive:

$$\frac{\partial P_n^*}{\partial r} = -(\frac{1}{2\mu}\frac{\partial B^2}{\partial r} + j_z B_\theta) \tag{9}$$

Defining c_α as $j_z B_\theta/(\partial P/\partial r)$ and using (6), (7), (8) and (9) we obtain

$$V_{r,n}[1 + \frac{\rho_{n,i}}{x\omega_i}\frac{1}{r}\frac{\partial r V_{\theta,n}}{\partial r}] = \frac{1}{2\mu}(\frac{1}{\sigma}\frac{\partial Log B_z^2}{\partial r} + \frac{1}{\alpha_s(1 + c_\alpha)}\frac{\partial B_z^2}{\partial r}), \tag{10}$$

where ω_i is the ion gyrofrequency, $\rho_{n,i} = m_n/m_i$ and $x = n_e/n_n$. By prescribing the radial dependence of the velocity field of neutrals, the radial dependence of the magnetic field can be derived. We assumed that, at $r \leq r_0$, $V_{r,n} = V_0 r/r_0$. This, together with the requirement of mass conservation ($div \rho V = 0$), imply a downflow inside the cylinder of radius r_0 with a velocity $V_{z,n} = 2V_0/r_0 H(1 - e^{z/H})$ ($z \leq 0$). The radial dependence of the magnetic field inside the flux-tube is given by:

$$\mu V_{r,n} r[1 + \varepsilon] = \frac{1}{\sigma} Log[\frac{B^2(r)}{B^2(0)}] + \frac{1}{\alpha_s(1 + c_\alpha)}[B^2(r) - B^2(0)], \tag{11}$$

where $\varepsilon = 2\Omega/x\omega_i$ and $\Omega = V_{\theta,n}/r$. In practice $\varepsilon \ll 1$. Assuming a magnetic field of 1000 Gauss at the center of the flux-tube and using the values of the conductivity and collisional frequencies computed by Kubat and Karlicky (1986), the radial dependence of B_z in the photosphere, at the optical depth $\tau_{5000} = 1$ can be computed. Defining the radius r_0 of the flux-tube as the radial distance where $B = 5$ Gauss leads to $V_0 r_0 = 1.2 \ 10^5$ m^2 s^{-1}. We must point out that r_0 is more an estimate of the scale-length, for the radial dependence of the magnetic field, than the exact value of the flux-tube radius r_t.

The inward radial flow of neutrals requires the existence of a pressure gradient that can be stabilized or amplified by advection of ionization energy. Assuming $r_t = r_0$, the extra power required to ionize the flux of neutrals falling inside the flux-tube per unit area is

$$\frac{dE}{dt} = \frac{2}{r_0^2}(r_0 V_0) H n \chi, \tag{12}$$

where χ is the ionization potential of Hydrogen and H is the length of the tube in the photosphere. Using the value of $V_0 r_0$ found precedently, assuming $r_0 = 40$ km, $H = 500$ km, $n = 2 \ 10^{17}$ cm^{-3} we obtain $dE/dt = 2.5 \ 10^{10}$ ergs cm^{-2} s^{-1} that is about one third of the radiation emittance of the solar surface. Consequently the resulting cooling is important and could maintain a significant pressure gradient.

4. Radial DC electric currents

Assuming no charge accumulation the vertical electric current density j_z is related to the radial current density j_r by $\partial j_z/\partial z = -2j_r/r$. Using equation (4), the integration of this equation over the flux-tube cross-section and height H, leads to the following expression of the total vertical current intensity J_z

$$J_z = J_0 + 4\pi n_n m_n \int_0^{r_t} \int_0^H \frac{V_{r,n}}{B_z r} \frac{\partial r V_{\theta,n}}{\partial r} dr\, dz. \tag{13}$$

Assuming $r_t = r_0$ and $V_{r,n} = V_0 r/r_0$ we derive

$$J_z \simeq J_0 + 4\pi \frac{n_n m_n}{B_z} H V_{\theta,n}(r = r_t) V_{r,n}(r = r_t). \tag{14}$$

Similar result is obtained by assuming that $r_0 < r_t$ and that $V_{\theta,n}$ goes to zero over a distance $\leq r_0$. Total current intensities from 10^{10} to 10^{11} Amperes are generated for photospheric velocities of 2.10^2 m s^{-1} and 6.10^2 m s^{-1} associated with horizontal magnetic field scale lengths of respectively a km to a few hundred of m ($B_z = 100$ Gauss). However due to the high conductivity of the subphotospheric layers, most of the current shall flow in these layers. Assuming that the flux-tube extends vertically from the corona to the subphotospheric layers the current J_c flowing into the corona is

$$J_c = J_M \frac{R_b}{R_a}, \tag{15}$$

where J_M is the maximum intensity that would flow into the corona in the absence of any subphotospheric shunt and R_a and R_b are respectively the resistances of the flux-tube above and below the photosphere. From Kubat and Karlicky we estimate R_b/R_a to be close to 0.1. This reduces J_c to 10^9 - 10^{10} Amperes.

The azimuthal force acting on neutrals is $F_\theta = -j_r B_z$. The work made by this force per unit of time is $dW/dt = -j_r B_z V_{\theta,n}$. From this expression and from equation (4) the total power that is taken out of the kinetic energy of neutrals in the dynamo region to generate electric currents can be derived and is

$$P_v = -\pi n_n m_n \int_0^{r_t} \int_0^H \frac{V_{r,n}}{r} \frac{\partial (r V_{\theta,n})^2}{\partial r} dr\, dz. \tag{16}$$

The same assumptions as for deriving equation (14) lead to

$$P_v = -\pi n_n m_n H r_t V_{r,n}(r = r_t) V_{\theta,n}^2(r = r_t). \tag{17}$$

Due to the shunt of the subphotospheric layers the power available in the corona P_c is only one tenth of P_v.

The total intensities observed in active regions and the total power required in order to power one powerful flare (10^{24} Joules) per week, in an area of 10^8 km^2 (radius R

$= 5600$ km), are of about 10^{11} Amperes and 10^{18} watts. These values can be reached in plages if many concentrated flux-tubes are present in the flaring area. About 10^6 flux-tubes of cross section ≈ 1 km^2, surrounded by vortices with velocities close to 2. to 5. 10^2 m s^{-1}, are required to give a power of 10^{18} watts. This imply a filling factor of only 10^{-2}. The net total D.C. Current intensity J_c depends of the sign of V_θ and, if there is a net loss of angular momentum over the active region, it could easily be as high as requested. The requirements on the filling factor would be relaxed in quiet sun area.

5. Conclusion

We have shown that, in the hypothesis of axial symmetry, any mechanism that could create a radial pressure gradient, in a region with a low preexisting magnetic field, would increase the magnetic energy density by creating azimuthal currents. By a cooling of the flux-tube by advection of ionisation energy, the initial radial pressure gradient could be increased. This effect may lead to the formation of concentrated magnetic flux-tubes. The annihilation of the kinetic energy of vortices in the low photosphere could produce significant electric power and current intensity in the corona.

6. References

Akasofu, S.I. (1984a) 'Vortical distribution of Sunspots', Planet.Space Sci., Vol.33, No.3, 275

Akasofu, S.I. (1984b) 'Sunspot pair formation by the photospheric dynamo process' Planet.Space Sci., Vol.32, No.10, 1257

Akasofu, S.I. (1984c) 'An essay on Sunspots and Solar Flares', Planet.Space Sci., Vol.32, No.11, 1469

Ambroz, P. (1987) 'The global horizontal circulation on the Sun', Bull. Astron.Inst. Czechosl., 38, 110

Bumba, V. (1987a) 'Magnetic Fields of the Sun and Stars', 10th European Regional Astronomy Meeting of the IAU, The Sun, Vol.1,3(1987)

Bumba, V. (1987b) 'Does the solar dynamo act in Sunspot Groups', 10th European Regional Astronomy Meeting of the IAU, The Sun, Vol.1, 59

Hénoux, J.C. and Somov, B.V. (1987) 'Generation and structure of the electric currents in a flaring activity complex', Astron. & Astrophys., 185, 306

Hénoux, J.C. and Somov, B.V. (1988) 'Electrodynamic conversion of energy: Magnetic field amplification in the solar photosphere' Adv. in Space Res., in press

Hénoux, J.C. and Somov, B.V. (1989), submitted to Astron. & Astrophys.

Kubát, J. and Karlický, M. (1986) 'Electrical conductivity in the solar photosphere and chromosphere, Bull. Astron. Inst. Czechosl., 37, 155

Martres, M.J., Soru-Escaut, I., Rayrole, J. (1973) 'Relationship between some Photospheric Motions and the Evolution of Active Centers', Solar Phys., 32, 365

Schatten, K.H. and Mayr, H.G. (1985) 'On the maintenance of Sunspots: An ion hurricane mechanism', Ap.J., 299, 1051

Spicer, D.S. (1982) 'Magnetic energy storage and conversion in the solar atmosphere', Space Science Reviews, 31, 435

V. LARGE-SCALE STRUCTURE AND DYNAMICS

GLOBAL EVOLUTION OF PHOTOSPHERIC MAGNETIC FIELDS

V. I. Makarov
Kislovodsk Station of the Pulkovo Observatory, Kislovodsk, 357741,
USSR

K. R. Sivaraman
Indian Institute of Astrophysics, Bangalore 560034,
INDIA

ABSTRACT

The main features concerning the evolution of the large scale photospheric magnetic fields derived from synoptic maps as well as from
H-alpha synoptic charts are reviewed. The significance of a variety
of observations that indicate the presence of a high latitude component as a counterpart to the sunspot phenomenon at lower latitudes
is reviewed. It is argued that these two components describe the
global magnetic field on the sun. It is demonstrated that this
scenario is able to link many phenomena observed on the sun (coronal
emission, ephemeral active regions, geomagnetic activity, torsional
oscillations, polar faculae and global modes in the magnetic field
pattern) with the global magnetic activity.

1.   INTRODUCTION

Since the discovery of magnetic fields in sunspots by Hale (1908)
solar magnetic fields have been an area of investigation both with
countinously improving observing techniques as well as methods of
theoretical modelling. The important role of the magnetic fields
in every aspect of solar activity and variability has now been fully
recognised. Although most of the flux observed on the sun appears
in the form of tiny fragments associated with strong fields, these
structures organise themselves into large scale global patterns
that evolve over the time scale of the solar cycle. Magnetic fields
on the sun manifest themselves on several length scales and change
on several time scales.

i.   Large scale fields with mean field values in the neighbourhood
of 1 gauss and time scale of the order of a few rotation periods
of the sun.

ii.   Intermediate scale fields which occur in the form of sunspots
with high values of fields (~ 2000 gauss).

iii. Small scale structures which have sizes of a few arc seconds
(even sub-arc size elements are present) with strong fields residing

J. O. Stenflo (ed.), Solar Photosphere: Structure, Convection, and Magnetic Fields, 281–295.

in them.

In this review, we shall be concerned with the first type, namely, the large scale fields.

The existence of solar magnetic fields outside of sunspots was first detected by H.W.Babcock and H.D.Babcock (1955). Their magnetograms of the sun showed the following features:

I. Solar magnetic field consists of regions of strong (10^3 Gauss) as well as regions of weak diffuse fields (~ 1 Gauss).

II. The regions of strong fields are related with active regions and the sunspots, while the regions of weak fields consist of large unipolar structures.

III. At the boundaries of unipolar regions where the radial component of the magnetic field is zero, prominences are observed. Over the solar disc these appear as H-alpha dark filaments.

IV. The poles have a weak but significant measurable fields all the time and the polarity of these fields at the poles reversed near the epoch of the maximum activity (1957-1959).

Following this, Babcock (1961) enunciated his classical model of the solar cycle. But a more fundamental problem which remains unsolved pertains to the origin and support of the solar magnetic field or the solar dynamo. Different theoretical groups at work on this have constructed several models. Although none of them is satisfactory, enough observations do not also exist today which can pick out or eliminate any of the models in preference to others with any confidence.

Further progress in the studies of solar magnetism is marked by the commencement of regular observations of the photospheric magnetic fields at the Mt.Wilson Observatory in 1959 (Howard 1967; Howard et al. 1967), at the Crimean Astrophysical Observatory (Severny 1966), at the Kitt Peak National Observatory and Stanford Observatory in 1976 and at Sayan Observatory at Sibizmir (Grigoriev et al. 1983). The main features that emerged from the synoptic maps constructed from the Mt.Wilson magnetograms for the period 1959-1980 (Bumba and Howard 1965; Howard and LaBonte, 1981) are the following:

I. Active regions break up into fragments of weak fields that coalesce to form global patterns of unipolar magnetic field regions. These slowly expand, are stretched by differential rotation and drift polewards forming the polar fields.

II. The solar equator is not the polarity division line for the background fields as in the case with sunspots.

III. The sunspot latitudes are characterised by fields of the preceding polarity while, the polar fields are composed by field flows of the following polarity which migrate towards the poles with a velocity of ~ 10 m sec^{-1}.

IV. The total magnetic flux on the sun changes only by a factor of 3 from the minimum to the maximum epochs of activity.

2. DYNAMICAL FEATURES OF EVOLUTION

During the last few years, considerable information concerning the dynamics of large scale fields have been derived both from the full disc magnetogram as well as the H-alpha synoptic charts. The latter form a good proxy for the magnetograms. The H-alpha filaments which are neutral dividing lines between two regions of opposite polarity can be used as a good tool to study the dynamical features associated with the evolution of global magnetic field pattern (McIntosh 1979; Makarov et al. 1983). The migration of the filament bands is represented most conveniently by plotting their mean latitude positions rotation wise, on a latitude – time diagram Fig.1 (Makarov et al. 1983). On this diagram the filament bands are seen to migrate pole-wards continously at speeds of ~5 m sec^{-1}. Topka et al. (1982) demostrated that the poleward drift of the filament bands (and hence of the unipolar regions) is not by diffusion, but the surface magnetic fields are transported to the poles by poleward meridional flows. Also, it is seen that these unipolar regions always migrate poleward and at no time show an equatorward drift (Makarov, 1984). The polemost filament shows a dramatic increase in its poleward motion attaining speeds of 15-40 m sec^{-1} simultaneous with the steep rise in sunspot activity. The speed of the poleward migration seems to be related

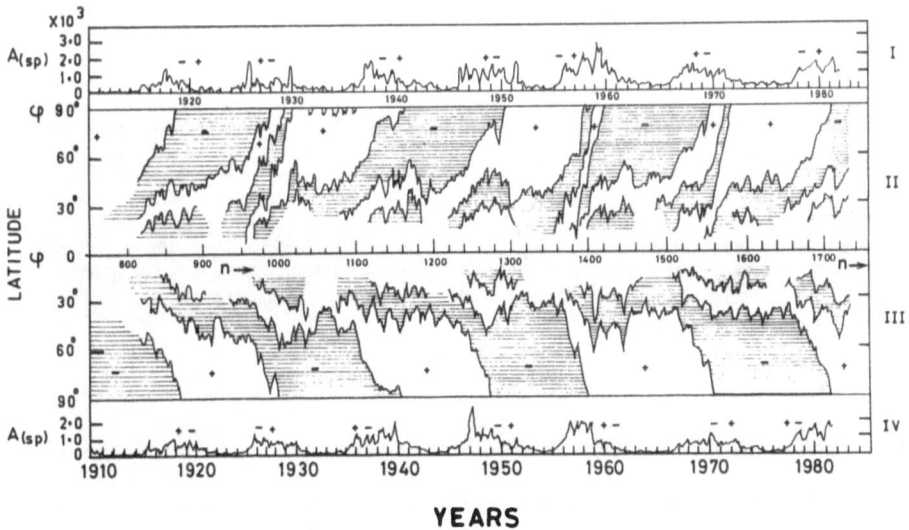

Figure 1. Boxes II and III show the migration trajectories of the neutral lines of the large scale magnetic field form $H\alpha$ synoptic charts for the period 1910–1982. + and − stand for the polarity signs of the magnetic field in the conventional way. n is the number of the Carrington rotation. Boxes I and IV show the plots of daily sunspot areas $A_{(sp)}$ in millionths of the visible hemisphere.

with the strength of the cycle concerned. The polemost filament reaches the pole first and causes the reversal of the polar field. The polemost filament in both the hemispheres do not reach the respective poles simultaneously and in such situations the polar reversal at one of the poles takes place earlier than at the other. The sun exhibits the same polarity on both the poles (monopole) till such time, the polemost filament in the second hemisphere has reached the pole and causes the reversal there. This is the picture when a single fold reversal takes place in both the hemispheres (eg. years 1920, 1940, 1950 and 1980 in Fig.1). There are instances when a three fold reversal occurs in either of the hemispheres. In such cases all the three filament bands (Fig.1) travel to the respective poles one after the other and cause a three fold reversal. Such three fold reversals took place in the northern hemisphere alone in 1930, 1960 and 1970 and in the southern hemisphere alone in 1885 and 1910 (Fig.2). The phenomenon of a three fold reversal in both the hemispheres has not been observed any time during the last 115 years (Fig.2).

Years / Cycles	1856-1878 / 10 - 11	1879-1901 / 12 - 13	1902-1923 / 14 - 15	1924-1944 / 16 - 17	1945-1964 / 18 - 19	1965-1985 / 20 - 21	1986- / 22 -
N	↓	↑ ↓	↑ ↓	↑↓↑ ↓	↑ ↑↓↑	↑↓↑ ↓	
S	↑	↑↓↑ ↑	↑↓↑ ↑	↓ ↑	↓ ↑	↓ ↑	

Figure 2. Hale's 22-year cycles and the reversal of the sun's magnetic field in the odd and even 11-year cycles in the northern and southern hemispheres. Three vertical arrows in one group (↑↓↑) represent a three-fold reversal, while a single vertical arrow (↑ or ↓) represents a one-fold reversal.

3. FORMATION OF UNIPOLAR MAGNETIC REGIONS

Both the magnetogram data as well as the H-alpha synoptic charts show that the polar magnetic fields during any cycle are built up and maintained by the continous arrival of discrete f-polarity regions. These regions originate in active region latitudes and migrate towards the poles. This picture can be illustrated better with the help of Fig.1 for any cycle particularly in the southern hemisphere where the single fold reversal make the illustration easier. During cycle 20 (1964-1974) the following part of active centres in the southern hemisphere was of south polarity (-), whereas the polarity at the pole was positive (+) from 1960 to 1971, while the reversal took place. Fig.1 shows something more than this. It can be seen that these negative unipolar regions were formed at latitudes > 30° from as early as 1957, although their poleward migration started only as late as 1966-1967. Thus, the zone of negative polarity that formed in cycle 19 determined the polarity of the high latitude field in cycle 20 during the period 1970-1981. This picture would lead to the interpretation that this zone with the new fields (-) was formed out of the p-polarity of the active regions starting with 1957 as, these regions could not have been formed out of the regions of cycle 20, which are yet to appear on the sun at least at these latitudes (> 30°). It may be that the f-polarity regions of cycle 20 aded fresh unipolar regions to those already formed from the p-polarity of the earlier cycle. This process is most obvious for cycle 18 (1944-1954) in Fig.1.

4. EVOLUTION OF FIELDS AT HIGH LATITUDES

The solar cycle has been defined traditionally from the spot counts or areas of spots as the interval between two successive minima giving an average value of ~11 years. But a number of observations show that the activity begins at high latitudes soon after the reversal of the polar fields and a few years before the first appearance of the spots of the new cycle. These observations are the following:
i.   Coronal emission observation in the 5303A line.
ii.   Ephemeral active regions (ERs).
iii. Geomagnetic activity.
iv.   Torsional oscillations.
v.   Polar faculae
vi.   Global modes in the magnetic field pattern.

4.1. Coronal Emission in 5303A line

It is known that the coronal emission intensity in 5303A line is a good index of the magnetic field. In the 1950s the coronal observers (Waldmeier 1957; Trellis 1963) noticed in their data that a zone of enhanced emission in 5303A line makes its appearance at high latitudes in each hemisphere several years before the commencement of the "classical" sunspot cycle. This is in addition to the strong emission component present during the years of the sunspot cycle that always mathes with the butterfly diagram of sunspots. The high latitude bands are brought out conspicuously when the standard deviation of the coronal emission σ_{5303} is used as the emission index rather than the mere average emission values. This index is very useful for detecting particularly the newly emerging magnetic flux regions. The plot of the isovalues of the quantity $\sigma_{5303} - \overline{\sigma}_{5303}$ ($\overline{\sigma}_{5303}$ is the latitude average of σ_{5303} over 20° intervals) for years 1944-1974 shows the two components clearly, when freed of the background emission (Leroy and Noens, 1983). The high latitude components appear immediately after the polar reversals and drift towards the

286

poles (Fig.3) (Makarov et al. 1987a). The two latitude components
partly overlap in time and thereby extend the duration of the coronal
activity to 16-18 years. Results of Altrock (1988) for cycle 21
and of Bumba et al. (1989) for the period 1965-1986 show the two
latitude components in each hemisphere. The latter data even suggest
a possible connection between the two components unlike the results
of Leroy and Noens (1983).

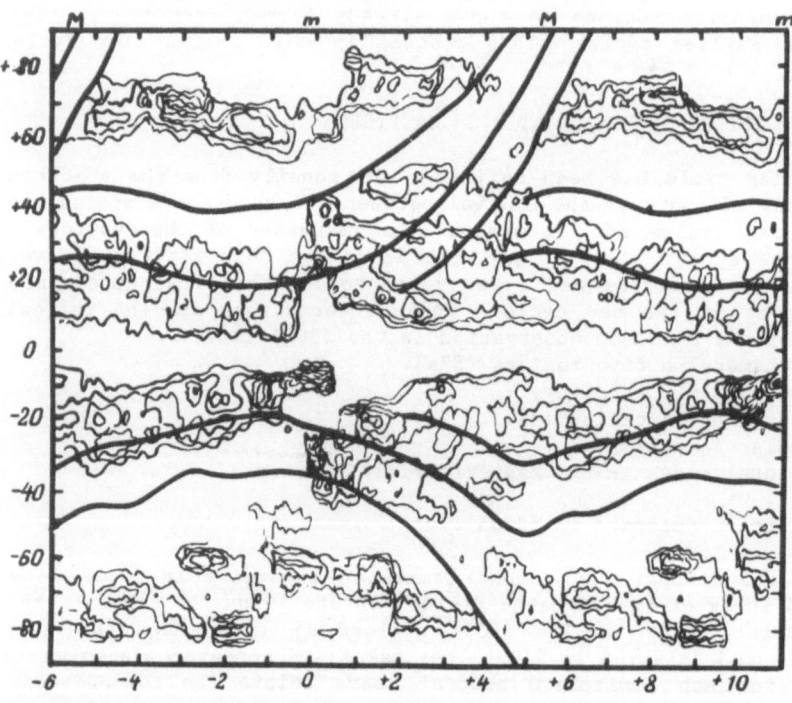

Figure 3. Map of isovalues of the quantity $\sigma_{5303} - \bar{\sigma}_{5303}$ (Leroy and Noens, 1983). σ_{5303} is
the standard deviation of the coronal intensity at each latitude over periods of about 1 year,
and $\bar{\sigma}_{5303}$ is the latitude average of σ over 20° intervals. The thick lines are the migration
trajectories of neutral filament bands for cycle 20 (Makarov, Leroy, and Noens, 1987a). m is
the minimum epoch of the solar cycle, M the maximum epoch. The vertical axis gives the
solar latitude in degrees.

4.2. Ephemeral Active Regions (ERs)

The properties of ERs and their evolution in relation to the solar cycle have been studied by Martin and Harvey (1979) for the 20th cycle. The ERs occur in the form of tiny bipoles. They identified a high latitude band of ERs in 1973 and 1975 in each hemisphere. Their important finding is the detection of a significant number of ERs contained in the high latitude bands that exhibited a polarity orientation dictated by Hale's law appropriate to the next cycle (cycle 21) rather than to cycle 20, while the low latitude ERs showed the polarity of cycle 20. Thus the ERs belonging to two successive cycles coexist for more than 3 years, with the high latitude component leading the low latitude component. These findings have been confirmed from further analysis of subsequent data for cycle 21 by K.Harvey (Wilson 1988).

4.3. Geomagnetic Activity

Another evidence for the high latitude component of activity is provided by Legrand and Simon (1981) who formulated a pattern for the global solar activity of 16-18 years duration from their study of geomagnetic activity in relation to the solar cycle. They noticed geomagnetic activity related to the solar cycle arises from two components: one, the high speed wind streams originating from the coronal holes at high latitudes and the other, related to the sunspot activity. The high latitude component makes its appearance shortly after the polar field reversal and coexists with the low latitude component of activity with an overlap of 6-7 years, thereby extending the duration of solar cycle to 16-18 years.

4.4. Torsional Oscillations (TO)

The torsional oscillation refer to the alternating latitude bands of faster than average and slower than average rotation present on the sun discovered by Howard and LaBonte (1980) from the Mt.Wilson velocity data. According to them, the torsional waves (two per hemisphere) start from either pole once in 11 years before the sunspot maximum and travel to the equator in the course of 18-22 years. The epoch of appearance of sunspots at $\pm 40°$ latitudes coincides with the arrival of the faster band of the torsional wave at these latitudes and the TO merge with the butterfly diagram then on. The TO is considered as a signal representing the propagation of magnetic activity on the sun. If the time of travel of TO from the pole to the equator on the solar hemisphere represents the duration of the solar cycle, then the latter turns out to be ~ 18 years. Such an extended duration although, hypothesised independently by Legrand and Simon (1981) from the solar wind stream studies and by Martin and Harvey (1979) from the ERs, this concept gained strength only after the TO came to be known.

4.5. Polar Faculae

After every polar reversal, regions above $\pm 40°$ latitudes show polar faculae. These can be seen in white light images of the sun as well as in the Ca II K line spectroheliograms. The number of polar faculae present on the sun follow a cyclic variation with the period of 11 years which differ in phase with the sunspot numbers by $\sim 90°$ (Sheeley 1976). The most significant results that have emerged from a study of the evolution of polar faculae are the following:

i. The faculae appear a few months after the polar revrsal. During the deep solar minimum periods, about 900 faculae can be identified on the solar disc and over a third of these occur as bipoles, some of them aligned in the E-W direction. The unipolar faculae have a polarity identical to that of the background field, while in the case of the bipolar faculae the preceding polarity is identical to that of the background field. The polarity orientation of these bipoles is opposite to that of the spots of the same cycle (Hale's law), but identical to the polarity orientation for bipolar spots of the next following cycle for the hemisphere concerned (Makarov and Makarova 1984, 1987). In other words, while faculae of the (N+1) cycle make their appearance at high latitudes, the activity of the preceding cycle (N) is still present at lower latitudes in the sunspot phase. This is similar to the behaviour of ephemeral active regions.

ii. The faculae appear first at latitude zones 40°-60° and the zones of appearance migrate slowly and reach high latitudes 70°-80° as the cycle progresses (Fig.4) (Makarov and Sivaraman 1986; Makarov et al. 1987b).

iii. The new cycle shows up first as faculae at high latitudes and leads the sunspot phenomenon by 5-6 years. Each of these has a duration of 11 years, but occur at separate latitudes and displaced from each other in time by 5-6 years within a 22 year magnetic cycle (Fig.4). Thus the conventional solar cycle, which is defined as the duration of the butterfly patterns based on the spot number counts, describes only that part of the activity relating to the cycle that occurs within the $\pm 40°$ latitude zones; whereas, if the activity at latitudes $> 40°$ is also taken into account then the solar activity cycle starts from the appeerence of the faculae and lasts till the end of the butterfly diagram. The duration of this global cycle turns out to be 16-18 years (Makarov and Sivaraman 1989).

If we now compare the coronal emission pattern as that of Leroy and Noens (1983) with the faculae distribution, the match appears good. The high latitude component of the coronal emission coincides spatially with the faculae and the low latitude component with the butterfly diagram.

In the case of the torsional oscillations (TO), the analysis of Howard and LaBonte (1980) that led to the inference of a pole to equator travelling wave pattern with k=2 contained a mathematical artifact and the pattern when freed of the artifact consists of

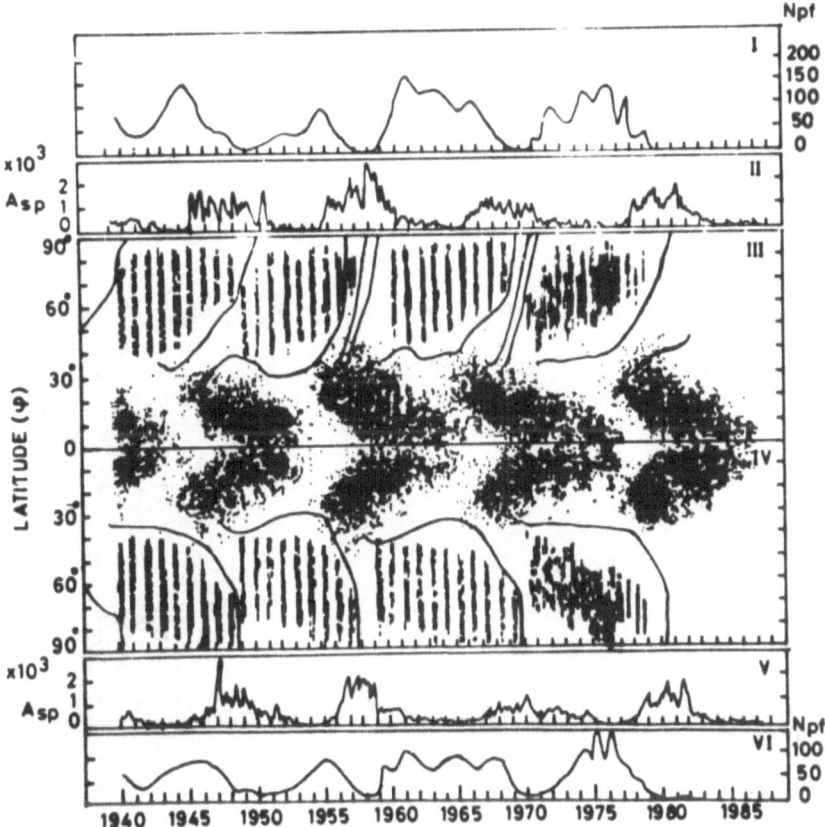

Figure 4. Boxes III and IV show the latitude distribution of polar faculae and sunspots (butterfly diagram) during 1940–1985. The superposed lines are the migration trajectories of filaments reproduced from Figure 1 after smoothing, to show the epochs of polar reversals. Boxes II and V show the sunspot areas A_{sp} as in Figure 1. Boxes I and VI show the counts (N_{pf}) of polar faculae in the north and south hemispheres.

a relative polar spin up near solar maximum and a separate single wave that runs from mid to low latitudes during the rest of the cycle (Snodgrass 1985, 1987). The torsional shear (which is the derivative of the net torsional pattern with respect to the latitude) increase and decrease regions at low latitudes match well with the butterfly diagram of sunspots (Snodgrass 1987). The high latitude shear increase zone which has no counterpart in Snodgrass's (1987) interpretation, is seen to match well with the polar faculae regions (Makarov and Sivaraman, 1989).

4.6. Global Modes in the Magnetic Field Pattern

The discovery of the resonant global wave pattern in the magnetic fields of the sun by Stenflo and coworkers has provided a new area from where encouraging results have come out regarding the global

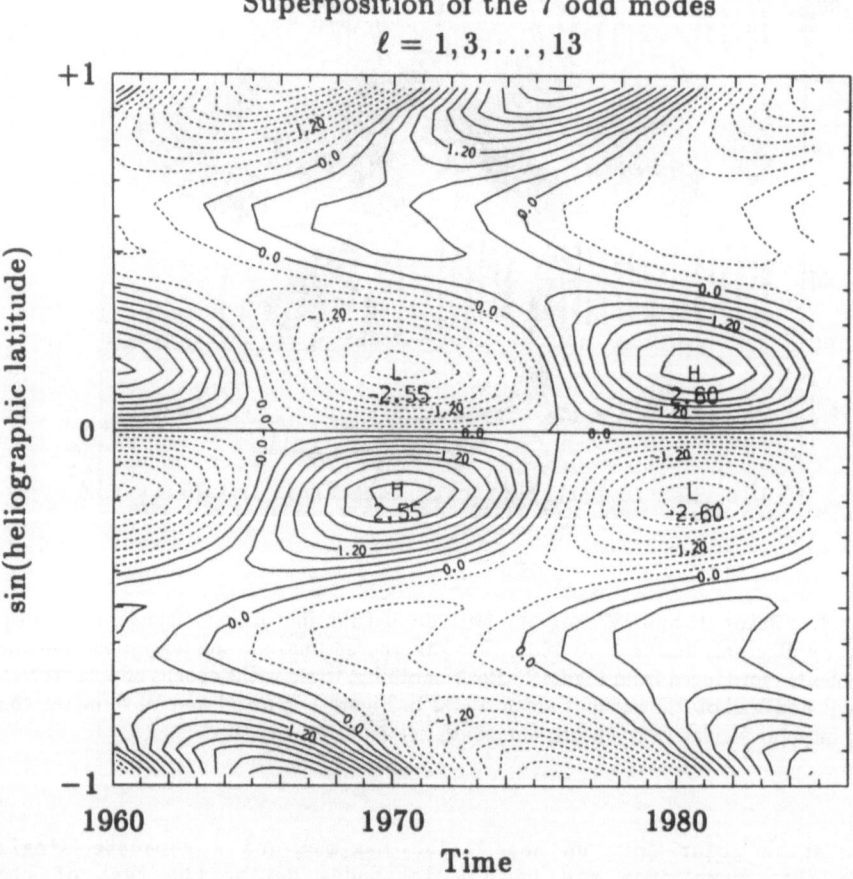

Figure 5. Synthetic evolutionary diagram computed as the superposition of 7 discrete harmonic modes (with purely sinusoidal time variations), with only odd values of $l(l = 1, 3, \ldots, 13)$. (Reproduction of Fig. 4 of Stenflo (1988). By courtesy of Stenflo and Kluwer Academic Publishers.)

evolution of photospheric magnetic fields. The power spectrum of the spherical harmonic coefficients for the zonal modes of the radial magnetic field shows the presence of a set of discrete resonant frequencies (Stenflo and Vogel, 1986; Stenflo and Güdel, 1988; Stenflo, 1988). The harmonic modes can be characterised by their degree ℓ and order m. The rotationally symmetric modes (with m=0) obey a parity selection rule: the modes with odd parity (odd values of ℓ) are dominated by power that correspond to a period of 22 years for all values of ℓ; while the even parity modes show power at higher frequencies that vary with value of ℓ, with no trace of the 22 year cycle. Gokhale et al. (1989) have obtained similar results from their analysis of 80 years of sunspot data. The zonal magnetic field pattern averaged over all longitudes shows the evolution of the field in the solar latitude-time domain (Stenflo, 1988). The main features are:

i. The polarities reverse sign every 11-years both at high and low latitudes.

ii. On the high latitude zones, the magnetic field pattern drift steeply towards the poles and

iii. On the low latitude zones, the field shows drift towards the equator in the course of the 11-year cycle, which is the butterfly diagram.

Stenflo (1988) has constructed synthetic contours by approximating the true power spectrum by δ - functions, one for each value of ℓ. The resulting pattern derived by superposing 14 discrete modes reproduces the observed zonal pattern well. The pattern can be considerably refined by separating out the antisymmetric component (i.e. by a superposition of only 7 odd modes; $\ell = 1, 3, \ldots 13$), which brings out the equatorial and poleward drifts of the magnetic patterns and the 22-year periodicity most conspicuously (Fig.5). This picture of the evolution of the magnetic fields totally agrees with the composite global field pattern with the polar faculae and the sunspot butterfly diagram shown in Fig.4. The similarity of the polarity distribution of the two components and the polarity reversals also match exactly.

5. SCENARIO OF THE GLOBAL MAGNETIC FIELD PATTERN

The good agreement among the observations of different parameters, (coronal emission, polar facule, global modes, geomagnetic activity, ephemeral active regions) all of them connected with the magnetic fields can now be pooled together to evolve a working empirical model. According to this model, the global photospheric magnetic field consists of two components: the high latitude and the low latitude components (Fig.4 and Fig.6). The high latitude component is represented by the coronal emission, the polar faculae and the ephemeral active regions, which makes its appearance immediately after the polar reversal and is characterised by the poleward drift. The second and the more powerful component is represented by the sunspots that appear when the first component is already at its

292

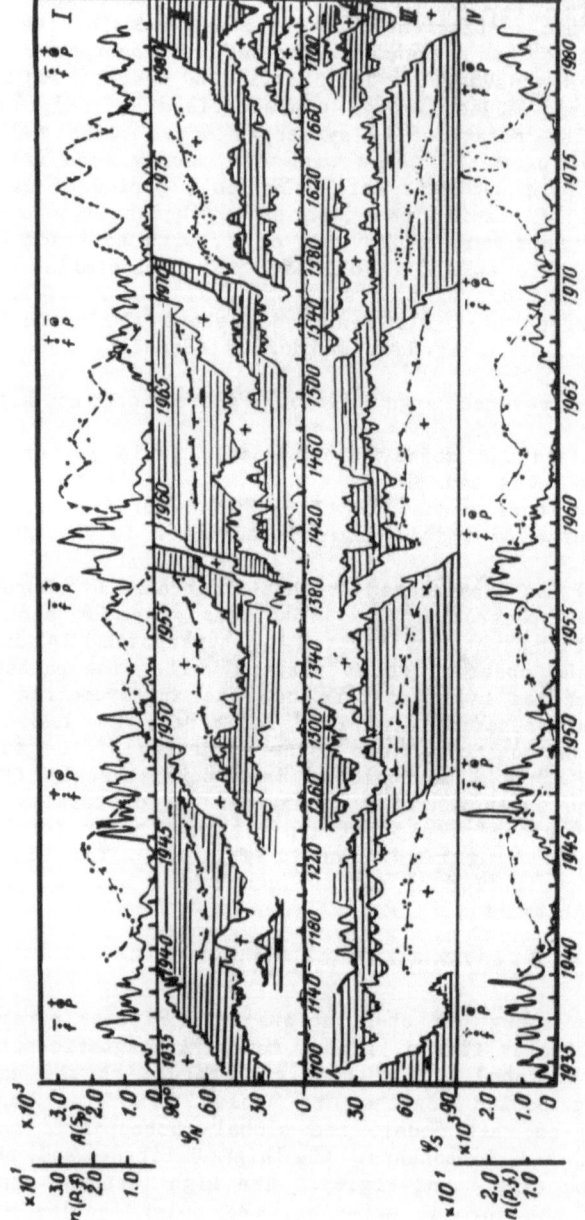

Figure 6. Global solar cycle activity during 1935–1980 as in Fig. 1. Boxes I and IV: The broken curves represent the number of polar faculae n $(p.f.)$. The solid curves represent the areas of spots $A(Sp)$ as in Fig. 1. Boxes II and III: The dots represent the latitudes of the polar faculae for the N and S hemispheres. The broken line is a free hand fit through these points. The rest is the same as in Fig. 1.

peak and has an equatorward drift which is the butterfly diagram. The theoretical support for the magnetic waves of solar activity is provided by the investigations of Makarov et al. (1987c). The two components appear on the sun at different latitude zones and with a shift in time by 5-6 years which gives the notion of the extended duration of the solar cycle. It is not known at this stage whether the two components could be connected by a common causal agency. The agreement or otherwise with the TO pattern is also unclear at this stage, due to paucity of the TO data. With the arrival of more results on TO and ephemeral active regions, the picture should become more meaningful.

Our understanding of the solar magnetism leans heavily on the empirical results derived from observations of the surface fields. Hence, it is important to be able to offer as much information based on observations as possible to provide clues in framing the global picture of the solar magnetic fields. From the theoretical side, it seems quite satisfying to note that the solar magnetic field can be described as a linear superposition of discrete global modes. This approach appears to be promising and may provide clues in our endeavour in understanding the magnetic fields and their variability on the sun and other stars too.

REFERENCES

Altrock, R.C. (1988) 'Variation of Solar Coronal Fe XIV 5303A Emission during solar cycle 21', in Richard C. Altrock (ed), Solar and Stellar Coronal Structure and Dynamics; A Festschrift in Honour of Dr. John Evans; National Solar Observatory, Sunspot, New Mexico, p.414-420.

Babcock, H.W. and Babcock, H.D. (1955) 'The Sun's Magnetic Fields, 1952-1954', Astrophys. J. 121, 349-366.

Babcock, H.W. (1961) 'The Topology of the Sun's Magnetic Field and the 22-year Cycle'. Astrophys. J. 133, 572-587.

Bumba, V. and Howard, R. (1965) 'Large Scale Distribution of Solar Magnetic Fields'. Astrophys. J. 141, 1502-1512.

Bumba, V., Rušin, V. and Rybansky, M. (1989) In this proceedings.

Gokhale, M.H., Javaraiah, J., Hiremath, K.M. (1989) 'Study of Sun's Hydromagnetic Oscillations using Sunspot Data'. (In this proceedings).

Grigoriev, V.M., Peshcherov, V.S., Osak, B.F. (1983) 'The Measurement of the Background Magnetic Field of the Sun at the Sayan Solar Observatory'. Isseledov. PO Geomag. aeron. i fizike Soln. (In Russian), 64, 80-102.

Hale, G.E. (1908) 'Solar Vortices'. Astrophys. J. 28, 100-116.

Howard, R. (1967) 'Magnetic Field on the Sun (observational)'. Ann. REv. Astron. Ap. 5, 1-24.

Howard, R., Bumba, V., Smith, S.F. (1967) 'Atlas of Solar Magnetic Fields' Carneigie Inst. of Washington Publ. No.626, Washington, D.C.

Howard, R. and LaBonte, B.J. (1980) 'The Sun is Observed to be a Torsional Oscillator with a Period of 11 years'. Astrophys. J. 239, L 33-36.

Howard, R. and LaBonte, B.J. (1981) 'Surface Magnetic Fields during Solar Activity Cycle'. Solar Phys. 74, 131-145.

Legrand, J.P. and Simon, P.A. (1981) 'Ten Cycles of Solar and Geo-magnetic Activity'. Solar Phys. 70, 173-195.

Leroy, J.L. and Noens, J.C. (1983) 'Does the Solar Activity Cycle Extend over more than an 11-year period?' Astron. Astrophys. 120, L1-L2.

Makarov, V.I., Fatianov, M.P. and Sivaraman, KR. (1983) 'Poleward Migration of the Magnetic Neutral Line and the Reversal of the Polar Fields on the Sun'. Solar Phys. 85, 215-226.

Makarov, V.I. (1984) 'Do Prominences Migrate Equatorward?' Solar Phys. 93, 393-396.

Makarov, V.I. and Makarova, V.V. (1984) 'On the Structure of Polar Faculae'. Soln. Dann. (In Russian) No.12, 88-94.

Makarov, V.I. and Sivaraman, K.R. (1986) 'On the Latitudinal Migration of Polar Faculae in their Activity Cycle, Period 1940-1968'. Soln. Dann. (In Russian), 9, 64-70.

Makarov, V.I. and Makarova, V.V. (1987) 'On the Relationship between Polar Faculae, X-ray Bright Points and Ephemeral Active Regions on the Sun'. Soln. Dann (In Russian), 3, 62-70.

Makarov, V.I., Leroy, J.L. and Noens, J.C. (1987a) 'Behaviour of the Coronal Intensity in the Line 5303A and Latitude Zonal Structure of the Magnetic Field: Period 1944-1974' Astron. J. (In Russian), 64, 1072-1078.

Makarov, V.I. and Makarova, V.V. and Sivaraman, K.R., (1987b) 'Butter-fly Digram for Polar Faculae and Sunspots During 1940-1985'. Soln. Dann. (In Russian), No.4, 62-64.

Makarov, V.I.and Ruzmaikin, A.A. and Starchenko, S.V. (1987c) 'Magnetic Waves of Solar Activity'. Solar Phys. 111, 267-277.

Makarov, V.I. and Sivaraman, K.R. (1989) 'New Results Concerning the Global Solar Cycle'. Solar Phys. (In press).

Martin, S.F. and Harvey, K.L. (1979) 'Ephemeral Active Regions During Solar Minimum'. Solar Phys. 64, 93-108.

McIntosh, P.S. (1979) 'Annotated Atlas of H-alpha Synoptic charts' World Data Centre A for Solar Terrestrial Physics, NOAA, Boulder, Colorado.

Severny, A.B. (1966) 'An Investigation of the General Magnetic Field of the Sun'. Izvestia Krymsk. Astrophys. Obs. 35, 97-138.

Sheeley, N.R., (Jr) (1976) 'Polar Faculae During the Interval 1906-1975, J. Geophys. Res. 81, 3462-3464.

Snodgrass, H.B. (1985) 'Solar Torsional Oscillations: A Net Pattern with Wave Number 2 as Artifact'. Astrophys. J. 291. 339-343.

Snodgrass, H.B. (1987) 'Torsional Oscillations and the Solar Cycle'. Solar Phys. 110, 35-49.

Stenflo, J.O. and Vogel, M. (1986) 'Global Resonances in the Evolution of Solar Magnetic Fields' Nature, 319, 285-290.

Stenflo, J.O. (1988) 'Global Wave Patterns in the Sun's Magnetic Field'. Astrophys. Space. Sci. 144, 321-336.

Stenflo, J.O. and Güdel, M. (1988) 'Evolution of Solar Magnetic Fields: Modal Structure' Astron. Astrophys. 191, 137-148.

Topka, K., Moore, R., LaBonte, B.J. and Howard, R. (1982) 'Evidence for a Poleward Meridional Flow on the Sun'. Solar Phys. 79, 231-245.

Trellis, M. (1963) 'Repartition des jets de la couronne en fonction de la latitude an cours du cycle solaire'. C.R.Acad. Sci. 257, 52-53.

Waldmeier, M. (1957) 'Die Polare Protuberanzen Zone'. Z.f. Ap. 42, 34-41.

Wilson, P.R. (1988) 'Solar Cycle Workshop; Second Meeting' (A Review), Solar Phys. 117, 205-215.

ORIGIN OF THE SUN´S DIFFERENTIAL ROTATION

L.L.KICHATINOV
Siberian Institute of Terrestrial Magnetism,
Ionosphere and Radio Wave Propagation (SibIZMIR)
664697 Irkutsk, P.O. Box 4
USSR

ABSTRACT. The present status of the Sun´s differential rotation theories is reviewed. Attention is mainly focused on mechanisms for differential rotation based on the anisotropic viscosity concept and their modern develop ments within the framework of the mean-field hydrodynamics. The models with latitude-dependent heat transport and non-axisymmetric numerical simulations are briefly discussed.

1. INTRODUCTION

Differential rotation of the Sun is a phenomenon known for more than a century (Carrington, 1863). There exists the so-called equatorial acceleration with a decrease in angular velocity of global rotation with increasing latitude. Extensive observational information on details of this latitudinal dependence and its variation with phase of a solar cycle has been accumulated using both direct Doppler measurements and observations of motions of various tracers. Detailed discussions of these data may be found in recent reviews by Howard (1984) and Schröter (1985).

The purpose of this article is to review the theory of solar differential rotation. Theoretical efforts in this area also have a long history and are stimulated by its own interest and by the well-known relation to solar magnetism. Unfortunately, the rapid increase of observational information and theoretical work did not serve to choose between the known mechanisms for differential rotation the most effective one. Relatively independent models of solar rotation centered around different effects do co-exist nowadays.

Nevertheless, there is almost complete agreement between different approaches in appreciation of the fact that a fundamental cause of the Sun´s differential rotation is the interaction between solar convection and rotation.

297

J. O. Stenflo (ed.), Solar Photosphere: Structure, Convection, and Magnetic Fields, 297–307.

The Rossby number for giant solar convection is smaller
than unity (Tyler, 1973; Durney and Latour, 1978), i.e.,
convection is strongly influenced by Coriolis forces. The
back reaction disturbs rotation and makes it differential.
    In what follows we will consider mainly the anisotro-
pic viscosity models of differential rotation and their
modern develop ments made within the framework of the
mean-field hydrodynamics. The models with latitude-dependent heat transport and non-axisymmetric simulations are
discussed more briefly. These approaches concentrate the
main theoretical efforts and dominant number of papers.
Nevertheless, strict volume limitations place some interesting mechanisms for differential rotation beyond the
scope of this treatment. Wholly ignored are variations of
rotation over a solar cycle and torsional oscillations
where theory is only incipient.

2. AXISYMMETRIC THEORIES

The axisymmetric theories consider the steady mean flow
which possesses axial symmetry about the rotation axis.
Velocity V of this flow is a superposition of meridional
circulation $V_m = \left[V_r(r, \theta), V_\theta(r, \theta), 0 \right]$ and rotation $V_R =$
$= \left[0, 0, \Omega(r, \theta) r \sin\theta \right]$. (Here and below the usual spherical coordinates are used.) The full velocity v is the su-
perposition of global flow V and random convective veloci-
ty u with zero mean value: $v = V + u$, $\langle v \rangle = V$, $\langle u \rangle = 0$.
The zonal component of the averaged equation of motion

$$div \left(\rho r \sin\theta \langle u_\varphi u \rangle + \rho r^2 \sin^2\theta\, \Omega\, V_m \right) = 0 \qquad (1)$$

is the basic equation of theories of differential rotation.
Eq.(1) is contributed by Reynolds stress tensor $R_{ij} =$
$= -\rho \langle u_i u_j \rangle$. We shall distinguish in the correlation $T_{ij} =$
$= \langle u_i u_j \rangle$ the dissipative (D_{ij}) and non-dissipative (Λ_{ij})
parts

$$T_{ij} = D_{ij} + \Lambda_{ij} \qquad (2)$$

The tensor D_{ij} is linear in spatial derivatives of angular
velocity and represents the contribution of eddy viscosities. We accept the simple expression for the tensor D_{ij}:

$$D_{ij} = -\nu_T \left(\partial V_i / \partial r_j + \partial V_j / \partial r_i - \frac{2}{3} \delta_{ij}\, div\, V \right) \qquad (3)$$

The simplifications connected with (3) are not essential
for the qualitative discussion to follow.
    The term Λ_{ij} in (2) represents nondissipative cont-
ributions which do not depend on spatial derivatives of
angular velocity. Eq. (1) now reads

$$div(\rho\, V_{\varphi}\, sin^2\theta\; r^2\, \nabla\Omega\,) = \nabla_{j}\; (\rho r\, sin\,\theta\; \Lambda_{\varphi j}\,) +$$
$$+\; div\; (\rho r^2 sin^2\theta\; \Omega\; V_m\,)\,. \tag{4}$$

The left-hand side of Eq. (4) describes viscouos damping of rotational inhomogeneity. If (4) includes only this term, the solution would be a rigid-body rotation, Ω = const, for any reasonable boundary conditions.

However, the difference from zero of the right-hand side of (4) prevents such a rotation from being solution of this equation. In other words, the right-hand side of (4) displays the sources of differential rotation. Two different terms represent two sources of a different nature.

First, convective motions can transport angular momentum and create inhomogeneous rotation. The quantity $\rho\, \Lambda_{\varphi j}$ in (4) is the nondissipative part of the j-th component of convective flux $\rho<uu_{\varphi}>$ of angular momentum. Second, meridional circulation may also serve as a transporter of angular momentum; the quantity $\rho V_{\varphi}V_m = \rho r\; sin\,\theta$ $\Omega\, V_m$ being the corresponding flux.

The steady meridional flow V_m satisfies the continuity equation, $div\,\rho V_m = 0$, and can be characterized by a single stream function. The scalar equation for meridional circulation can be obtained by taking the zonal component of curl of the averaged equation of motion (Kippenhahn, 1963)

$$D\,(V_m) = r\, sin\,\theta\; \partial\Omega^2/\partial z + \frac{1}{\rho^2}\;\left(\nabla\rho \times \nabla P\right)_{\varphi} \tag{5}$$

where $\partial/\partial z = cos\,\theta\; \partial/\partial r - (sin\,\theta/r)\partial/\partial\theta$ is spatial derivative along the axis of rotation, $D(V_m)$ signifies the contribution of effective viscosities.

The right-hand side of (5) represents sources of meridional circulation. There are two different sources again. The first term is the nonpotential part of centrifugal force. The second one is brought about by nonpotential pressure force: $-curl(\nabla P/\rho) = (\nabla\rho \times \nabla P)/\rho^2$.

The full system of equations comprises also equations of state and energy transport. However, they are not needed in the qualitative discussion to follow.

2.1 Models with anisotropic viscosity

Lebedinski (1941) was probably the first to note that anisotropy in the velocity distribution of solar convection should lead to differential rotation. The preferred direction of the anisotropy is singled out by gravity. Hence, it is natural to assume that

$$\langle u_{\theta}^2\rangle = \langle u_{\varphi}^2\rangle = s\,\langle u_{r}^2\rangle\;,\; s \neq 1.$$

Influence of the Coriolis force on the anisotropic convection gives rise to convective fluxes of angular momentum.

Reynolds stresses for slowly rotating turbulent fluids with radial anisotropy were derived by Wasiutinski
(1946) and Bierman (1951):

$$R_{r\varphi} = \rho \chi_r \sin\theta \left[r \, \partial\Omega / \partial r + 2\Omega \, (1-s) \right], \qquad (6)$$

$$R_{\theta\varphi} = \rho \chi_r \sin\theta \, \partial\Omega / \partial\theta$$

where $\chi_r = \tau \langle u_\tau^2 \rangle$ is the eddy viscosity (τ is a turnover
time of convective eddy). Eqs.(6) have been repeatedly
used to model solar differential rotation (Kippenhahn,
1963; Sakurai, 1966; Cocke, 1967; Köhler, 1970). Let us
consider the basic effects involved in these models.
Eqs (6) do not include meridional Λ-effect, i.e., $\Lambda_{\theta\varphi}=0$.
Hence, the neglect of meridional circulation in Eq.(4) to-
gether with usually imposed boundary conditions of vanishing stress, $R_{r\varphi}= 0$, at the upper and lower boundaries of
convection zone would lead to purely radial inhomogeneity
of rotation. Hence, the equatorial acceleration can be obtained only with meridional circulation included. Almost
all anisotropic viscosity models assume the adiabatic stratification of convection zone (a very important assumpti-
on). The term $\nabla\rho \times \nabla P$ in Eq.(5) vanishes in this case.
The only cause of meridional flow which remains is the
nonpotential part of centrifugal force. It is certainly
different from zero for radially-inhomogeneous rotation
caused by radial Λ-effect. Therefore, meridional circulation is unavoidable to occur. This circulation redistribu-
tes angular momentum and creates the latitudinal inhomogeneity of rotation.

The anisotropic viscosity models were capable of reproducing the observed surface distribution of angular ve-

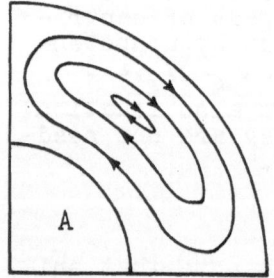
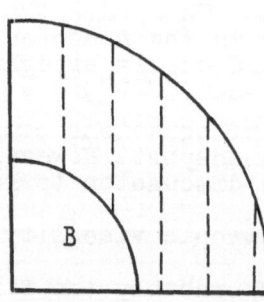

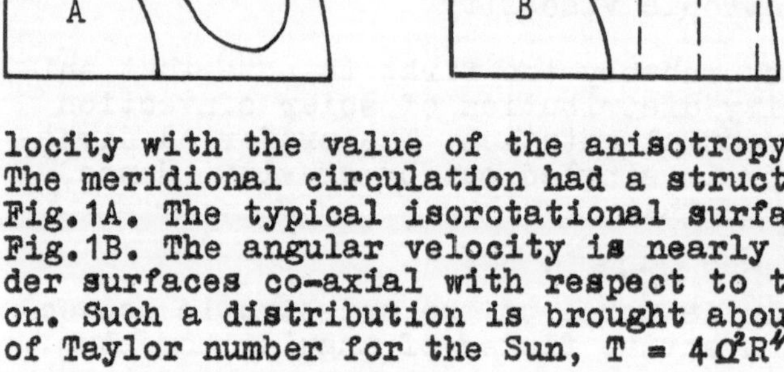

Fig.1. Meridional
circulation (A)
and isorotational
surfaces (B) typi-
cal of the anisot-
ropic viscosity
models.

locity with the value of the anisotropy parameter s ≈ 1.2.
The meridional circulation had a structure shown in
Fig.1A. The typical isorotational surfaces are shown in
Fig.1B. The angular velocity is nearly constant on cylinder surfaces co-axial with respect to the axis of rotation. Such a distribution is brought about by large values
of Taylor number for the Sun, $T = 4\Omega^2 R^4 / \chi_\tau^2 \approx 3 \cdot 10^7$.

In other words, turbulent viscosity is relatively small and viscous drag cannot stabilize meridional circulation on the level where it does not dominate in the process of angular momentum transport. Under these conditions the circulation is stabilized by exhausting their source. The nonpotential part of centrifugal force provides this source in anisotropic viscosity models. The force is proportional to the spatial derivative of angular velocity along the axis of rotation. Hence, the derivative should be small, and we arive at the distribution of Fig.1B.

This result is the consequence of barytropic stratification. The nonpotential part of centrifugal force can be compensated for by viscous force only when $\nabla\rho \times \nabla P = 0$, and the distribution of Fig.1B applies for the solar case of large Taylor numbersindependently of the particular model accepted. Hence, the isorotational surfaces of the type shown in Fig.1B indicate that $\mathrm{curl} < \nabla P/\rho > = 0$ for approximations adopted.

Let us consider whether the assumption $\nabla\rho \times \nabla P = 0$ applies for the Sun. The relative value of deviation from barytropy in a rotating convection zone cannot exceed the relative deviation from adiabaticity of stratification which is extremely small ($\leqslant 10^{-}$) except for the thin (~ 1000 km) surface layer (Baker and Tamesvary, 1966; Spruit, 1974; Gough and Weiss, 1976). However, the centrifugal force for the Sun is about five orders of magnitude smaller than the force of pressure. Hence, the deviation from barytropy can be neglected if only $|\nabla\rho \times \nabla P| \ll 10^{-5} |\nabla\rho||\nabla P|$. It is rather questinable whether this neglect can be justified for the Sun.

It is natural to anticipate that allowance for deviations from barytropy should suppress meridional circulation. Recent numerical simulations by Schmidt (1982) support this point of view. Schmidt treated explicitly the pressure force in his anisotropic viscosity model. The meridional circulation was very slow and latitudinal inhomogeneity of rotation was very weak for Prandtl number unity and reasonable boundary conditions of constant heat flux at the bottom and black-body radiation at the top of convection zone.

The anisotropic viscosity models are capable of reproducing the observed equatorial acceleration but seem to disagree with other observational data. The global equatorward meridional circulation on the surface of the Sun is not observed (Duvall, 1979; LaBonte and Howard, 1982; Howard, 1984; Tuominen et al., 1983). The distribution of Fig.1B seem to disagree with helioseismology data (Deubner et al., 1979; Duvall et al., 1984,1986; Brown, 1985; Brown end Morrow, 1987). On using Eqs (6) we find for the solar surface

$$< u_\theta u_\varphi > \cos\theta = -s\, \nu_r\, \sin\theta\cos\theta\, \partial\Omega/\partial\theta \leqslant 0.$$

This contradicts the positive values of the covariance $<u_\theta u_\varphi>\cos\theta$ infered from sunspot statistics (Ward, 1965; Gilman and Howard, 1984).

Nevertheless, these models provide important insights into the problem at hand, and comparison of their results with observations show some promising directions of theoretical progress.

2.2. Mean-field hydrodynamics and differential rotation theory

A natural way of refining the semi-qualitative anisotropic viscosity models seem to be the derivation of equations for global flows from hydrodynamic equations. Such an approach requires these equations to be averaged over an ensemble of realizations of random convective motions and is termed the mean-field hydrodynamics.

It is necessary to know the properties of convection to perform the averagings required. However, the problem of highly nonlinear rotating solar convection is still impossible to solve. Some properties of convection are again assigned from qualitative consideretions. The interaction between convection and rotation is of primary importance for the differential rotation problem. For this reason attemts have usually been made to describe the rotationally-induced properties of convection in a most consistent way. Other properties, not directly related to rotation, are assumed given. Actually, two different turbulences (convections) are considered. One, to be referred to as "original turbulence", would take place in the presence of real sources of turbulence but with no rotation present. Properties of original turbulence do not explicitly depend on rotation and are assumed to be given. The other one, i.e., a real turbulence perturbed by rotation, will be named "background turbulence". The properties of background turbulence are derived from given properties of original turbulence.

The usually used approximations lead to a linear relation:

$$u_i = B_{ij}(\Omega)\, u_j^o$$

where u and u^o are velocities for background and original turbulences, respectively; rotational influence is involved through the tensor B_{ij} (Rüdiger, 1977,1989).

The main advantage of the approach discussed here as compared those of preceding Section is the possibility of taking account of nonlinearities in the parameter $\omega = 2\tau\Omega$. (Note that $\omega > 1$ holds for giant solar convection.) Various nonlinear derivations (Iroshnikov, 1966; Rüdiger, 1977, 1982; Vandakurov, 1982; Kichatinov, 1986, 1987) being different in details, lead to the same structure of the nondissipative part, $\Lambda_{j\gamma}$, of velocity covariances:

$$\Lambda_{r\varphi} = \Omega \, \mathcal{V}_T \, [V_o(\Omega) + V_1(\Omega)\cos^2\theta] \sin\theta ,$$

$$\Lambda_{\theta\varphi} = \Omega \, \mathcal{V}_T \, H(\Omega)\sin^2\theta\cos\theta \qquad\qquad (7)$$

where V_o, V_1, and H are dimensionless functions.

Comparison of (7) with (6) shows that allowance for nonlinearities in   lead to the appearance of meridional Λ-effect ($\Lambda_{\theta\varphi} \neq 0$). This opens the possibility of establishing agreement of the theories with the Ward profile.

In the rapid rotation limit ($\omega \gg 1$), the Λ-effect induced by convection anisotropy is proportional to ω^{-2} (Rüdiger, 1983; Kichatinov, 1986). Hence, the anisotropy of original convection is efficient in generating differential rotation only when ω is of order unity or smaller. It was found recently that not only anisotropy of convection but also inhomogeneity of convection zone can lead to differential rotation (Kichatinov, 1987, 1988). Note that Λ-effect induced by (density) inhomogeneity tends to a constant value in the rapid rotation limit. Therefore, in the case of rapid rotation ($\omega \gg 1$) the inhomogeneity is more effective in generating differential rotation as compared with convection anisotropy. However, the relation of the roles played by stratification and convection anisotropy in generating differential rotation of the Sun is still uncertain.

The developements of differential rotation theory made within the framework of mean-field hydrodynamics seem to be quite promising. Solutions of the equation for angular velocity using the components (7) of the Λ-tensor were found and requirements imposed by observational data upon the functions V_o, V_1 and H of Eqs (7) were determined (Rüdiger, 1989). However, it is still uncertain whether these requirements can be met with an original turbulence having realistic properties.

2.3. Models with latitude-dependent heat transport

The perturbation of convection by Coriolis forces depends on latitude. For this reason, convective heat flux must also be latitude-dependent. This fact was used by Weiss (1965) and Durney and Roxburg (1971) to explain the differential rotation of the Sun. Inhomogeneous heat flux produces latitudinal temperature inhomogeneity. The meridional circulation arises under these conditions and drives differential rotation.

The dependence of heat flux F_c on latitude was involved through the latitude dependence of the heat transport coefficient K_c:

$$\mathbf{F}_c = - K_c C_p (\nabla T - \nabla T_{ad}), \qquad\qquad (8)$$

$$K = K(r)\left[1 + \varepsilon f(r)P_2(\cos\theta)\right] \tag{9}$$

where $\nabla T - \nabla T_{ad}$ is the superadiabatic temperature gradient, P_2 is the Legendre polynomial and ε is an adjustable parameter.

Early models required equator-to-pole temperature differences of several tens of degrees for the observed equatorial acceleration to be reproduced. However, the attempts to measure the temperature difference between equator and poles (Altrock and Canfield, 1972; Noyes et al., 1973; Falciani et al., 1974) revealed a high degree of homogeneity of global temperature distribution with no differential temperature confidently detected. The upper bound of about 5 K for the pole-equator temperature difference was established.

The latest models (Belvedere and Paterno, 1977, Belvedere et al., 1980) removed this contradiction. However, unreasonably small Prandtl numbers were required to keep the equator-to-pole temperature differences within observational constraints. These models were shown to be highly dependent on the choice of boundary conditions and on whether centrifugal forces are included or not (Moss and Vilhu, 1983).

A combined model was considered by Pidatella et al. (1986) which included both anisotropic viscosities and latitude-dependent heat transport (see also Schmidt, 1982). This permits the comparison of the two mechanisms for differential rotation. Anisotropic viscosities were found to be more effective in generating differential rotation.

The important achievement of the models with latitude-dependent heat transport is the allowance for thermodynamic properties of solar convection, which are mainly ignored by the anisotropic viscosity models. However, the relations (8) and (9) are not satisfactorily substantiated. More rigorous approaches (Durney and Spruit; 1979; Rüdiger, 1982) show that in contrast to (9) the heat transport coefficient is a tensor for rotating convection. Moreover, the heat conductivities are proportional to the intensity of convection and therefore must be dependent on superadiabatic temperature gradient. In other words, the convective heat transport is an essentially nonlinear process. It is rather questionable whether this process can be adequately treated by traditional linear approaches.

3. NONAXISYMMETRIC MODELS

The nonaxisymmetric approaches try to simulate (numerically) the global flows on the Sun as well as the smaller-scale three-dimensional convective motions using the fundamental hydrodynamic equations. In practice, however, some

approximations of these equations are used because of
computers limitations.

Extensive nonaxisymmetric simulations in Boussinesq
approximation were carried out (see, e.g., review by Gilman, 1980). We shall not consider the results of the Boussinesq models but shall proceed with discussion of probably more realistic recent simulations allowing for density
stratification of convection zone (Gilman and Miller, 1986;
Glatzmaier, 1984, 1985 a,b). These simulations start from
particular variant (Gilman and Glatzmaier, 1981) of anelastic approximation (Ogura and Phillips, 1962; Gough, 1969)
of gas dynamics equations; Glatzmaier (1985a,b) take also
magnetic fields into account. Only giant-scale convection

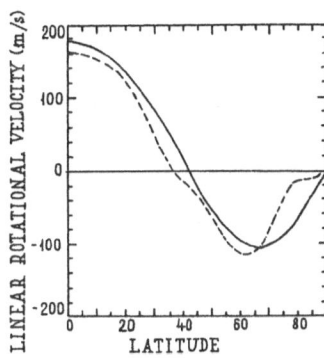

Fig.2. A comparison of observed
(Howard and Harvey, 1970) differential rotation of solar photosphere
(solid) with the results of nonaxisymmetric model by Gilman and Miller
(1986)(broken). Linear velocities of
rotation on the background of solid-
body rotation (Ω = 2.6 rad/s) are
shown.

and global flows have been simulated explicitly. The convective motions of smaller scales were parametrized by effective viscosities and conductivities.

The models by Glatzmaier (1984, 1985 a,b) and Gilman
and Miller (1986) though different in numerical methods
adopted yield essentially the same results. Quite satisfactory agreement with the observed rotation of photosphere has been found (Fig.2), except for high (λ >70°) latitudes. The amplitudes of meridional circulation and equator-to-pole temperature differences were within observational constraints. Covariances of zonal and meridional velocities of simulated convection agreed with the Ward profile.

However, the models discussed did not yield the increase of angular velocity with depth at high latitudes as
suggested by helioseismology. The isorotational surfaces
were nearly cylindrical (Fig.2B). As has been noticed in
Section 2.1, such isorotational surfaces suggest that
$\langle \nabla \rho \times \nabla P \rangle$= 0 under approximations adopted.

The nonaxisymmetric models discussed here are probably the best-developed ones at this moment and agree quite
well with observations. Nevertheless, we may confidently
state that something important is missing in them because

the toroidal magnetic field in Glatzmaier's (1985a) model migrated poleward in contrast to the observed equatorward drift of sunspot activity. It is very temting to suggest that this results from a nearly-cylindrical form of isorotational surfaces.

4. CONCLUDING REMARKS

It seems to follow from the above discussion that the question "whether we know basic mechanisms driving differential rotation of the Sun?" should be answered "yes", but the question "whether a reliable model for the solar rotation exists which reproduces all relevant observational data?" should be answered "no". This is because we are aware of principal physical mechanisms but not of the details of them.

The differential rotation can be produced by meridional circulation and Reynolds stresses. The meridional circulation can be excited by pressure forces and centrifugal forces. The desired nondissipative part of Reynolds stresses can be produced by two causes again: by anisotropy of convective motions and by inhomogeneity of convection zone. Any model of solar rotation faces the choice between one or the other or both of these effects at each step of its develop ment. It may be stated with relatively high confidence that solar differential rotation is the result of the above-mentioned effects, but it is not clear which of them play a dominant role.

REFERENCES

Altrock,R.C. and Canfield,R.C. (1972) Sol. Phys.23,343-354.
Baker,N.and Temesvary,S. (1966) Tables of Convective Stellar Envelope Models, 2nd ed. Goddard Institute for Space Studies, New York.
Belvedere,G. and Paterno,L. (1977) Sol. Phys. 54, 289-312.
Belvedere,G., Paterno,L. and Stix,M. (1980) Geophys. Astrophys. Fluid Dynamics 14, 209-224.
Bierman,L. (1951) Z. Astrophysic 28, 304.
Brown,T.M. (1985) Nature 317, 591-594.
Brown,T.M. and Morrow,C.A. (1987) Astrophys J.314,L21-L26.
Carrington,R.C. (1863) Observations of the Spots on the Sun, Williams & Norgate, London.
Cocke,W.J. (1967) Astrophys. J. 150, 1041-1050.
Deubner,F.-L., Ulrich,R.K. and Rhodes,E.J. (1979) Astron. Astrophys. 72, 177-185.
Durney,B.R. and Roxburg,I.W.(1971) Sol. Phys. 16, 3-20.
Durney,B.R. and Spruit,H.C. (1979) Astrophys.J. 234, 1067.
Durney,B.R. and Latour,J. (1978) Geophys. Astrophys. Fluid Dynamics 9, 241-255.
Duvall,T.L. (1979) Sol. Phys. 63, 3-15.

Duvall,T.L., Dziembowski,W.A., Goode,P.R., Gough,D.O.,
    Harvey,J.W. and Leibacher,J.W. (1984) Nature 310,22-25.
Duvall,T.L., Harvey,J.W. and Pomerantz,M.A. (1986)
    Nature 321, 500-501.
Falciani,R., Rigutti,M. and Roberti,G. (1974) Sol. Phys.
    35,277-280.
Gilman,P.A. (1980) Lecture Notes in Phys. 114, 19-37.
Gilman,P.A. and Glatzmaier,G.A. (1981) Astrophys. J. Suppl.
    45,335-349.
Gilman,P.A. and Howard,R. (1984) Sol. Phys. 93, 171-175.
Gilman,P.A. and Miller,J.(1986) Astrophys.J.Suppl.61,585.
Glatzmaier,G.A. (1984) J. Comput. Phys. 55, 461-484.
Glatzmaier, G.A.(1985a) Astrophys. J. 291, 300-307.
Glatzmaier,G.A. (1985b) Geophys.Astrophys.Fluid Dyn.31,137.
Gough,D.O. (1969) J. Atmosph. Sci. 26, 448-456.
Gough,D.O. and Weiss,N.O. (1976) Mon.Not.Roy.Astr.Soc. 176,
    589-607.
Howard,R. (1984) Ann. Rev. Astron. Astrophys. 22, 131-155.
Howard,R. and Harvey,J. (1970) Sol. Phys. 12, 23-51.
Iroshnikov,R.S. (1969) Astron. J. (SSSR) 46, 97-112.
Kichatinov,L.L.(1986) Geophys.Astrophys.Fluid Dyn.35,93.
Kichatinov,L.L.(1987) Geophys.Astrophys.Fluid Dyn.38,273.
Kichatinov,L.L.(1988) Astron. Nachr. 309, 197-211.
Kippenhahn,R. (1963) Astrophys. J. 314, 664-678.
Köhler,H. (1970) Sol. Phys. 13, 3-18.
LaBonte,B.J. and Howard,R. (1982) Sol. Phys. 80, 361-372.
Lebedinski,A.I. (1941) Astron. Zh. 18, 10-25.
Moss,D. and Vilhu,O. (1983) Astron. Astrophys. 119, 47-53.
Noyes,R.W., Ayres,T.R. and Hall,D.N.(1973) Sol.Phys.28,343.
Ogura,Y. and Phillips,N.A. (1962) J.Atmosph.Sci. 19,173.
Pidatella,R.M., Stix,M., Belvedere,G. and Paterno,L.
    (1986) Astron Astrophys. 156, 22-32.
Rüdiger,G. (1977) Sol. Phys. 51, 257-268.
Rüdiger,G. (1982a) Geophys.Astrophys.Fluid Dyn. 21, 1-25.
Rüdiger,G. (1982b) Astron. Nachr. 303, 293-303.
Rüdiger,G. (1983) Geophys.Astrophys.Fluid Dyn. 25,213-233.
Rüdiger,G. (1989) Differential Rotation and Stellar Con-
    vection, Akademie - Verlag, Berlin.
Sakurai,T.(1966) Publ.Astron.Soc.Japan 18, 174-200.
Schmidt,W. (1982) Geophys.Astrophys.Fluid Dyn. 21, 27-57.
Schröter,E.H. (1985) Sol. Phys. 100, 141-169.
Spruit,H.C. (1974) Sol. Phys. 34, 277-290.
Tayler,R.J. (1973) Mon. Not. Roy. Astron. Soc. 165, 39-52.
Tuominen,J., Tuominen,I. and Kyroläinen,J. (1983) Mon.
    Not. Roy. Astron. Soc. 205, 691-702.
Vandakurov,Yu.V. (1982) Astrophys. Space Sci. 83, 105-116.
Ward,F. (1965) Astrophys. J. 141, 534-547.
Wasiutinski,J. (1946) Astrophys. Norvegica 4, 1.
Weiss,N.O. (1965) Observatory 85, 37-39.

THE SUN'S ROTATION RATE AS INFERRED
FROM MAGNETIC FIELD DATA

J.O. STENFLO
Institute of Astronomy
ETH-Zentrum, CH-8092 Zürich
Switzerland

ABSTRACT. The pattern of solar magnetic fields has been used as a tracer to determine how the sun's rotation rate varies with latitude and time. Two distinctly different rotation laws emerge from such an analysis, one agreeing with the surface Doppler rotation rate, the other corresponding to much more rigid rotation with a small polar spin-up. Detailed analysis shows that this second law cannot be explained in terms of flux redistribution on the solar surface, but that it represents the rotation properties of the sources of magnetic flux, which are likely to be located at the bottom of the convection zone.

The rotational phase velocity of the source pattern is found to be constant with time, which suggests that the depth at which the magnetic flux is stored and amplified inside the sun does not vary with the solar cycle, and that the phase velocity also represents the plasma velocity.

1. The Two Differential Rotation Laws for the Magnetic Field Pattern

We have long been used to the "peaceful coexistence" of various rotation laws on the sun: The determined angular velocity of rotation and its variation with latitude is different for different tracers used, like sunspots, prominences, or coronal holes, and also differs from the rotation rate determined from the Doppler shifts of spectral lines (cf. Van Tend and Zwaan, 1976). It is surprising, however, that the pattern phase velocity of photospheric magnetic fields is also found to be greatly different when different types of correlation analyses are carried out.

This is illustrated in Figure 1. The solid line shows the synodic rotation period determined by Snodgrass (1983) from a cross-correlation analysis of daily Mt Wilson magnetograms. It agrees well with the Doppler rate (dashed-dotted line) of Howard et al. (1983). In striking contrast is the pattern rotation rate determined from an autocorrelation analysis of Mt Wilson – Kitt Peak synoptic magnetic field data over a 26 yr period (Stenflo, 1989). It is given by the four types of symbols in Figure 1, which refer to the four different autocorrelation peaks (with lags from 1 to 4 solar rotation periods) used. While the Snodgrass (1983) rotation law exhibits a steep latitude dependence, with a rotation period of about 38 days near the poles, the autocorrelation results give a rotation period that reaches a maximum of 29-30 days at a latitude of 50-55°, with a tendency for a small polar spin-up.

This huge discrepancy between a quasi-rigid and a steep differential rotation cannot be

J. O. Stenflo (ed.), Solar Photosphere: Structure, Convection, and Magnetic Fields, 309–314.

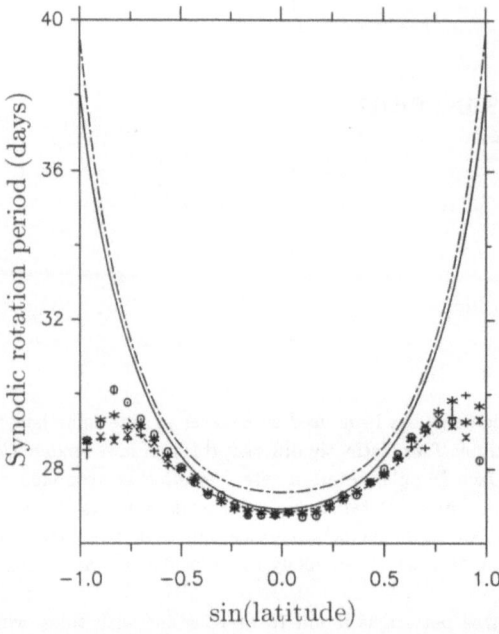

Fig. 1. Differential rotation of the sun's magnetic field pattern. The symbols (circles, stars, pluses, crosses) represent the autocorrelation results of Stenflo (1989), indicating a quasi-rigid rotation law with a small polar spin-up. The solid curve represents the cross-correlation results of Snodgrass (1983), which are in good agreement with the Doppler results (dashed-dotted curve), from Howard et al. (1983).

explained in terms of any statistical errors or artifacts. It also cannot be explained by variations with time of the solar rotation, since a detailed analysis (see below) shows that the quasi-rigid rotation law is time invariant over the 26 yr period to a high degree of accuracy, and Snodgrass (1983) also finds his steep differential law to be time invariant (furthermore the 15 yr period he used is covered by our 26 yr period). Both analyses are based on Mt Wilson daily magnetograms (for the autocorrelation analysis supplemented by Kitt Peak magnetograms from 1976 onwards).

2. Regeneration vs. Redistribution of Magnetic Flux

The coexistence of these two entirely different rotation laws cannot be explained in terms of flux redistribution on the solar surface, which otherwise in principle could cause large differences between the plasma velocity and the pattern *phase* velocity. The most general and elaborate flux redistribution model is due to Sheeley et al. (1987). It shows, both analytically and in particular by numerical simulation, how a quasi-rigid pattern phase velocity can develop from a steeply differential plasma velocity by flux redistribution processes, like turbulent diffusion and meridional circulation. The reason why any such model completely fails to explain the coexistence of the two rotation laws is that if it is used to explain a quasi-rigid rotation law for the autocorrelation analysis, it also predicts practically the same quasi-rigid law if Snodgrass (1983) cross-correlation analysis is applied, in striking contrast with the steep latitude dependence that he found. This conclusion follows both from the numerical simulations and the analytical expressions of such a general redistribution model.

Snodgrass (1983) analyses the longitude variation of the magnetic field in daily magnetograms, and determines by cross-correlation how the field pattern is displaced in longitude through comparison of magnetograms separated in time by 1-4 days. His spatial resolution is about one min of arc. No individual flux features are identified, but the pattern itself is used as a tracer for the rotation. It is therefore surprising that the pattern phase velocity obtained agrees so closely with the plasma velocity obtained from Doppler measurements, in contradiction with redistribution models like that of Sheeley et al. (1987).

The autocorrelation analysis on the other hand has been based on synoptic data sampled at the central meridian in daily magnetograms. This results in a 26 yr time series for each of the 30 latitude zones. As the sun rotates, the pattern recurs at the central meridian after an integer number of rotation periods, resulting in well-defined peaks in the autocorrelation functions at lags of an integer number of rotations. The precise lag at which these peaks occur gives the rotation period for the different latitudes. No significant dependence of the rotation period with peak number is found.

The basic difference between the cross-correlation and autocorrelation methods is thus the time scales involved. In the cross-correlation analysis lags of 1-4 days are used, in the autocorrelation analysis 27 days or more. The reason why flux redistribution models fail is that they implicitly assume that the magnetic flux that we see in magnetograms at any given time is dominated by *old* flux that has been around at the solar surface for a long time (many solar rotations), and is only shuffled around on the surface. In this picture it does not matter if the correlation analysis uses a lag of 27 days or 1 day; approximately the same pattern phase velocity of the old magnetic fluxes will be picked up in both cases.

To explain the *simultaneous* coexistence of the two rotation laws we are therefore forced to introduce the requirement that the magnetic-field pattern is regenerated over a time scale that is shorter than 27 days but longer than 4 days. Such an extremely short pattern turn-over time is indicated by video magnetograph observations of flux emergence and cancellation rates (cf. Martin, 1989). The pattern that recurs at the central meridian after one or more solar rotations then does not consist of the same magnetic fluxes, but of new, recently emerged fluxes. This picture is reminiscent of the old concept of active longitudes. The rotation rate derived from the autocorrelation analysis is then not characteristic of the pattern phase velocity in the photosphere, but of the pattern phase velocity in the source region inside the sun, from which new flux is constantly being "emitted" to the surface.

3. Time Invariance of the Quasi-rigid Rotation Law

Next we want to identify the source region and constrain its properties by determining the possible time variations in the phase velocity of the synoptic magnetic field pattern. Such variations could arise due to the following causes:
- The depth distribution of the sources inside the sun varies with the solar cycle (and the angular velocity varies with depth).
- The phase velocity in the source region may differ from the plasma velocity there, and this difference, being determined by some dynamo wave, may be expected to vary with the solar cycle (the evolution of the dynamo).
- There may be torsional oscillations in the source region.
- The connection between the source and the emergence at the surface may not be strictly in the radial direction, but the path may be curved, connecting different latitudes. In this case the latitude migration of the sources may cause variations in the observed rotation rate.
- We may have overlooked some effect of flux redistribution of old flux at the surface. In this

case, however, since it is the low-latitude active regions that would be the main "source" of the high-latitude flux, and since this "source" migrates in latitude and varies in amplitude with the cycle, time variations in the determined phase velocity would be expected.

As we will see below, however, our analysis shows that the rotation rate is constant, with very tight limits on the possible time variation. This severely constrains or partly rules out the above possibilities, and allows us to make rather far-reaching conclusions concerning the distribution of magnetic fields and angular velocity inside the sun.

To search for possible time variations in the phase velocity of the magnetic-field pattern, the 26 yr time series has been divided into 21 shorter periods, each 16 Carrington periods (16×27.2753 days ≈ 1.2 yr) long. As there are 30 latitude zones, there are $21 \times 30 = 630$ time series to analyse. For each time series a power spectrum has been computed. The sun's rotation shows up in the form of power spectrum peaks at frequencies that are an integer multiple m of the rotation frequency (the inverse of the synodic rotation period). $m = 1$ corresponds to a sine wave with a wavelength equal to the rotation period, $m = 2$ to the second harmonic, etc. The frequency of each peak defines a rotation period of the pattern.

In Figure 2 we illustrate six of the 630 power spectra, for six different latitudes in the southern hemisphere and selected time periods (numbered from 1 to 21). The frequency is

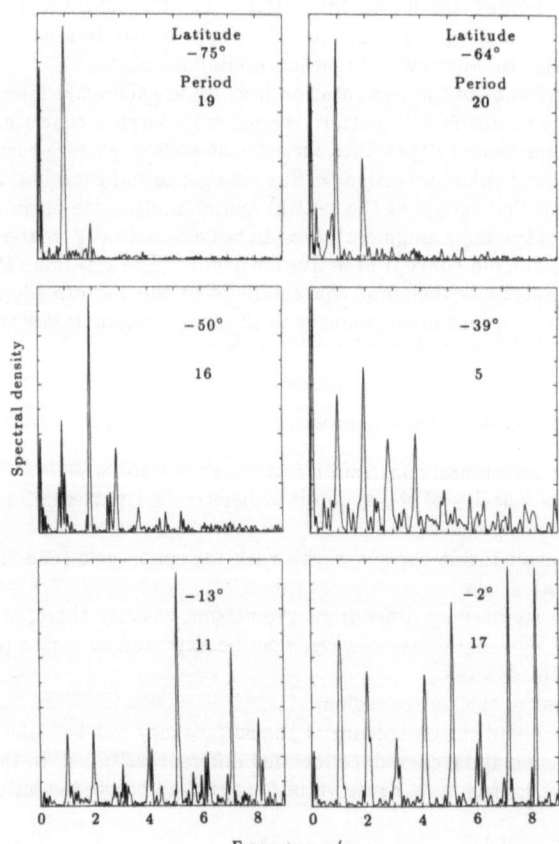

Fig. 2. Power spectra for selected latitude zones and time periods. ν_C is the Carrington frequency.

given in units of the Carrington frequency (1/27.2753 days^{-1}). Near the poles the peaks with the smallest m numbers (representing the large-scale structures) dominate, while at lower latitudes the power is shifted to the higher harmonics (smaller scales).

The outcome of this analysis is that no significant variations of the rotation rate with the solar cycle is found, within the numerical error bars. For each power spectrum from which a well-defined rotation rate can be derived (the majority of the power spectra), the rotation rate is found to agree with the quasi-rigid rotation rate given by the global autocorrelation results. To suppress the apparently random fluctuations around this mean rotation rate to try to bring out any possible solar cycle variations, we have averaged the corresponding latitudes in the north and south hemispheres and applied a 3.7 yr smoothing time window, to filter out the fluctuations with periods much shorter than the cycle period. No systematic, cycle-dependent pattern in the deviations $\Delta\omega$ from the mean sidereal angular velocity ω is found. Also "modal cleaning" (Stenflo, 1988) by harmonic decomposition of the unsmoothed $\Delta\omega$ pattern fails to reveal an underlying solar-cycle pattern.

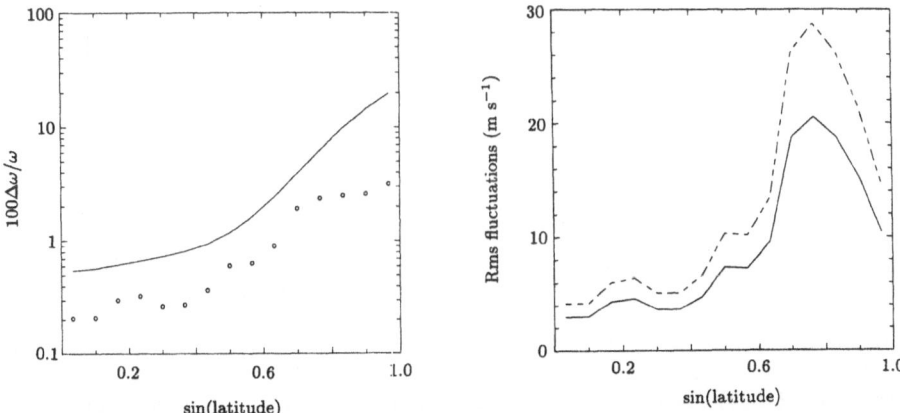

Fig. 3. Open circles in the left diagram: Rms fluctuations of the angular velocity ω around the mean value (the quasi-rigid rotation law), expressed in percent. Right diagram: The rms fluctuations converted to linear velocity (m s^{-1}), assuming that the observed pattern phase velocity refers to the bottom (solid line) or the top (dashed line) of the convection zone. Left diagram, solid line: The difference between the angular velocities of Snodgrass (1983) and the autocorrelation results, expressed in percent.

Figure 3 shows that the rms fluctuations of the smoothed ω values are $\lesssim 0.3\,\%$, or $\lesssim 4$ m s^{-1} at lower latitudes, but increases to 2-3 % (10-20 m s^{-1}) at higher latitudes (where the amount of magnetic flux and thus the signal-to-noise ratio is smaller). While these numbers represent rms fluctuations around a mean level, the *systematic* deviation of the Snodgrass curve reaches as much as 20 % at the poles.

Our values for the rms fluctuations can be regarded as one-sigma upper limits to possible solar-cycle variations of the pattern phase velocity of the sources of magnetic flux in the solar interior. These very low upper limits suggest that the depth distribution of the sources does not vary much with the solar cycle, otherwise the depth variation of the angular velocity would show up as a time variation in the pattern phase velocity. This speaks in favour of flux storage at the bottom of the convection zone with the dynamo operating there, rather than being distributed within the convection zone, in agreement with theoretical arguments by a number of authors (e.g. Spiegel and Weiss, 1980; DeLuca, 1987; Schüssler, 1987).

The absence of significant time variations also seems to favour the view that the pattern phase velocity closely represents the plasma velocity in the source region, since one would expect a difference generated by a dynamo wave to vary with the phase of the solar cycle, which does not seem to be the case. Our upper limits are also similar in magnitude or smaller than the amplitudes of the "torsional oscillations" of the observed surface Doppler velocities (Howard and LaBonte, 1980; LaBonte and Howard, 1982). Torsional oscillations in the source region (at the bottom of the convection zone), if they exist at all, should therefore be smaller than these limits.

There are thus good reasons why the quasi-rigid rotation law should represent the plasma velocity near the base of the convection zone, while the Snodgrass law represents the plasma velocity at the top of the convection zone. This appears to be consistent with the results of helioseismology. Interpolating between the top and bottom of the convection zone we obtain isocontours for the angular velocity which are very similar to the theoretically derived contours in Fig. 2 of Rüdiger and Tuominen (1989).

References

DeLuca, E.E. (1987) 'Dynamo theory for the interface between the convection zone and the radiative interior of a star', NCAR Cooperative Thesis No. 104, Boulder, Colorado.

Howard, R. and LaBonte, B. (1980) 'The Sun is observed to be a torsional oscillator with a period of 11 years', *Astrophys. J.* **239**, L33-L36.

Howard, R., Adkins, J.M., Boyden, J.E., Cragg, T.A., Gregory, T.S., LaBonte, B.J., Padilla, S.P., and Webster, L. (1983) 'Solar rotation results at Mount Wilson', *Solar Phys.* **83**, 321-338.

LaBonte, B.J. and Howard, R. (1982) 'Torsional waves on the sun and the activity cycle', *Solar Phys.* **75**, 161-178.

Martin, S.F. (1989) 'Evolution of small-scale magnetic fields', in these proceedings.

Rüdiger, G. and Tuominen, I. (1989) 'Generators of solar differential rotation and implications of helioseismology', in these proceedings.

Schüssler, M. (1987) 'Magnetic fields and the rotation of the solar convection zone', in B.R. Durney and S. Sofia (eds.), The Internal Solar Angular Velocity, *Astrophys. Space Sci. Library* **137**, 303-320.

Sheeley, N.R., Jr., Nash, A.G., and Wang, Y.-M. (1987) 'The origin of rigidly rotating magnetic field patterns on the sun', *Astrophys. J.* **319**, 481-502.

Snodgrass, H.B. (1983) 'Magnetic rotation of the solar photosphere', *Astrophys. J.* **270**, 288-299.

Spiegel, E.A. and Weiss, N.O. (1980) 'Magnetic activity and variations in solar luminosity', *Nature* **287**, 616-617.

Stenflo, J.O. (1988) 'Global wave patterns in the Sun's magnetic field', *Astrophys. Space Sci.* **144**, 321-336.

Stenflo, J.O. (1989) 'Differential rotation of the Sun's magnetic field pattern', *Astron. Astrophys.* **210**, 403-409.

Van Tend, W. and Zwaan, C. (1976) 'On differences in differential rotation', in V. Bumba and J. Kleczek (eds.), Basic Mechanisms of Solar Activity, *IAU Symp.* **71**, 45-46.

GENERATORS OF SOLAR DIFFERENTIAL ROTATION AND IMPLICATIONS OF HELIOSEISMOLOGY

G. RÜDIGER[1] and I. TUOMINEN[2]

[1]Sternwarte Babelsberg, 1591 Potsdam, German Democratic Republic
[2]Observatory and Astrophysics Laboratory, University of Helsinki
Tähtitorninmäki, SF-00130 Helsinki, Finland

ABSTRACT. We interpret the helioseismological results for the solar rotation law as implying a uniform Ω_0 in depth, if the angular velocity $\Omega(\theta)$ is expanded in series of orthogonal polynomials. One of the possibilities to ensure uniformity of Ω_0 is to exclude any anisotropy from the generating turbulence field, so that only the rotationally originated part of the heat conductivity tensor survive as a generator of meridional flow and consequently equatorial acceleration is produced. In addition to this simplest possible turbulence model, we discuss some possible forms of the nondiffusive parts of the Reynolds stresses which may be compatible with the uniformity of Ω_0.

1. The solar rotation law

The recent observational results of helioseismology have increasingly revealed the rotation law of the solar convection zone. Various inversions of the oscillation data have led to nearly the same results. Apparently the rotation law of the solar convection zone is roughly independent on depth and continues the outer well-known rotation law down to the bottom. For the lower overshooting region the reductions suggest basically rigid rotation (Dziembowski et al., 1988, Christensen-Dalsgaard and Schou, 1988). Roughly speaking the Ω-contours are cylindrical in the close equatorial region, radial at higher latitudes and disk-like at the poles (cf. Fig.2 in Libbrecht, 1988). They are in particular far from the over-all cylindrical structure, $\Omega = \Omega(r\sin\theta)$, envisaged from the Taylor-Proudman theorem and from the conservation law of angular momentum under the presense of a meridional circulation with high Reynolds number. If the often used expression of the surface rotation law

$$\Omega = A + B\cos^2\theta + C\cos^4\theta \tag{1}$$

is transformed to the more physical series

$$\Omega = \Omega_0 + \Omega_2 P_3^1/\sin\theta + \Omega_4 P_5^1/\sin\theta + ..., \tag{2}$$

the numerical values will be

$$\Omega_0 = 2.768, \quad \Omega_2 = -0.087, \quad \Omega_4 = -0.011, \tag{3}$$

J. O. Stenflo (ed.), Solar Photosphere: Structure, Convection, and Magnetic Fields, 315–320.
© *1990 by the IAU.*

(in μrad/s), with a high numerical equality of the surface Ω_0 and the interior Ω_0. The solar rotation law seems thus to be characterized by the properties

$$\Omega_0(r_i) = \Omega_0(r_s) \quad \text{and} \quad \Omega_n(r_i) \approx 0 \quad \text{for} \quad n > 0. \tag{4}$$

Only Ω_0 enters into the angular momentum expression

$$J = \int \rho r^2 \sin^2 \theta \, \Omega \, dv = 4\pi \int \rho r^4 \Omega_0(r) dr, \tag{5}$$

while the remaining terms in (2) do not contribute at all. The property (4) can thus trivially be reformulated with (5) so that the angular momentum in the convection zone should be the same as if it rotates rigidly at the rate of the interior. This statement does not, however, contain more physical information than (4) already expresses (cf. Gilman et al., 1989).

2. Differential rotation for isotropic turbulence

The correlation tensor $\langle u_i' u_j' \rangle$ as well as the turbulent heat flux $\langle T' u_i' \rangle$ contain rotationally induced terms which are nonvanishing for rigid rotation. Due to their existence, rigid rotation can never be a solution of the equations. We must distinguish, however, between two different sets of parameters describing the extra terms due to the basic rotation. The first belongs to the turbulent angular momentum transport (TMT) and exists only for anisotropic turbulences. In particular, its leading term $(V^{(0)})$ which has already been used in the Wasiutinski-Biermann-Kippenhahn formulation of the theory of differential rotation, generally produces radial gradients of the angular velocity. In opposition, the second set of parameters associated to the turbulent heat transport (THT) exists also for *isotropic* turbulences. They produce meridional circulation by pole-equator differences of the mean temperature. Strong radial gradients in Ω are *not* characteristic for their action. From this point of view the helioseismological results could indicate the action of THT-effects alone, working with *isotropic* turbulence. We have then a vanishing Λ-effect and the tensor of eddy conductivity is exceptionally simple. Only the rotation axis, $\boldsymbol{\Omega}$, yields a preferred direction and hence the complete tensorial structure is

$$\chi_{ij} = P(\Omega)\delta_{ij} + Q(\Omega)\Omega_i\Omega_j, \tag{6}$$

so that simply

$$\chi_{rr} = P + Q\cos^2\theta, \quad \chi_{r\theta} = \chi_{\theta r} = -Q\cos\theta\sin\theta, \quad \text{and} \quad \chi_{\theta\theta} = Q\cos^2\theta. \tag{7}$$

Only one parameter, Q, describes the latitudinal dependence of the entire conductivity tensor. It can be expressed with the spectral function of q of the non-rotating turbulence:

$$Q = \frac{8}{5}\Omega^2\chi \int \frac{\omega k^4}{(\omega^2 + \nu^2 k^4)^2} \left[\frac{2\omega k^4(\nu^2 - \chi^2)}{(\omega^2 + \chi^2 k^4)^2} q - \frac{\omega^2 + \nu^2 k^4}{\omega^2 + \chi^2 k^4} \frac{\partial q}{\partial \omega} \right] dk \, d\omega, \tag{8}$$

k and ω being wavenumber and frequency of the turbulence spectrum. The spectral function is always non-negative. The quantity Q has *no* definite sign. It is *positive* for very steep spectral functions, $q \propto \delta(\omega)$, and it is negative for flat spectra of the "white noise"-type, $q \approx$ const and small Prandtl number.

We can numerically fix, on the other hand, the quantity Q by means of the observed surface differential rotation law if the depth-dependence of Q is known. With the plausible idea that already from the dimensional reasons Q depends on the inverse Rossby number

$$Q \approx (t_{\text{corr}}\Omega)^2, \tag{9}$$

we suggest Q to be constant for a giant flow pattern while a relatively strong inward increase may reflect turbulence of mixing length type. We have worked thus with the depth-dependence

$$Q \approx (x_i/x)^\lambda \tag{10}$$

with free λ and x_i the lower fractional radius of the convection zone. As in Rüdiger (1989) and Tuominen and Rüdiger (1989) the Taylor number is 5×10^5, Prandtl number is 0.33 and normalized surface temperature is $T^* = 5 \times 10^7$. Table 1 gives the results. Small values of λ need positive Q for the generation of the observed Ω_2. They also give very strong basal differential rotation, much too strong poleward meridional motion, and large pole-equator temperature difference. When $\lambda = 8$ there appears a reversal of the basal differential rotation, meridional flow slows down and the pole-equator temperature difference becomes reasonably small. Our models favour thus the concept of the mixing-length rather than that of the global convection. Contours of constant angular velocity are given for two models in Fig. 1, with $\lambda = 0$, and for a reasonable one with $\lambda = 12$.

Table 1: Characteristic values for the isotropic turbulence models of the solar differential rotation. Angular velocity values has been here scaled so that at the surface $\Omega_e = 1.06$ and $\Omega_p = 0.76$. The table gives for different λ the surface values of Ω_2, Q at the bottom, polar and equatorial Ω at the bottom, surface meridional flow (m/s), and surface pole-equator temperature difference.

λ	Ω_2/Q_i	Q_i	Ω_p/Ω_e	u_θ	$(T_e - T_p)/T_e$
0	-.214	.19	0.30/1.16	-63	-.027
2	-.015	2.7	-4.16/1.90	-625	-.18
3	.044	-.91	2.12/0.85	+176	+.045
4	.086	-.46	1.40/0.98	+73	-.011
8	.152	-.26	0.96/1.05	+16	+.0033
12	.152	-.26	0.88/1.07	+4	+.0018
15	.141	-.28	0.85/1.07	-0.6	-.0010

3. The horizontal Reynolds stress

Let us interpret sunspot proper motions traditionally, with respect to the mean flow, i.e. with respect to the differential rotation *as well as* to stochastic component of the flow. As we know that the latter is reflected by the horizontal correlation $\langle u'_\theta u'_\phi \rangle$, first derived by means of proper motions of sunspot groups by Ward (1965). In the mean-field approach an isotropic turbulence field causes non-vanishing horizontal correlations via a *positive* eddy viscosity, defined by

$$\langle u'_\theta u'_\phi \rangle = -\nu_T \frac{\partial \Omega}{\partial \theta} \sin \theta \tag{11}$$

so that the correlation is *negative* in the northern hemisphere and positive in the southern. The observed signs are just opposite, so that an extra term must be added:

$$\langle u'_\theta u'_\phi \rangle = -\nu_T \frac{\partial \Omega}{\partial \theta} \sin \theta + \nu_T \Omega H^{(1)} \sin^2 \theta \cos \theta \tag{12}$$

(Rüdiger, 1977). This term is a part of the "Λ-effect" and reflects the rotational influence on the correlation tensor of anisotropic turbulences. Observations require a numerical value of about 1.6 for the quantity $H^{(1)}$. If the radial profile of $d\Omega/d\theta$ is known from helioseismology *and* if the radial profile of $\langle u'_\theta u'_\phi \rangle$ is known from statistics of the sunspot proper motions, then the radial distribution of $H^{(1)}$ can be derived containing information on the inner turbulence regime. In particular this concerns the radial distribution of the correlation time because we have similarly as in (9)

$$H^{(1)} \approx (\tau_{\text{corr}} \Omega)^2 \tag{13}$$

According to this idea, the results from the statistics of sunspot group proper motions can be interpreted in the following way: the rotational velocity decreases with the age of groups (about 5%) and the horizontal correlation decreases as well, being three times larger for the youngest groups than for the old recurrent ones (Tuominen and Virtanen, 1989). Let us locate the information given by the youngest groups at the botom of the convection zone and older ones correspondingly to higher layers. If we assume that the 5% outward decrease of angular velocity occurs in the lower half of the convection zone, we obtain, from the outward decreasing horizontal correlation, for the value of λ about 6-8 in the relation (4). This outward decrease of angular velocity in the lower part of convection zone may not be in conflict with the helioseismological inversions (see Dziembovski et al., 1989).

4. Radial Λ-effect

Ω_0 is also exceptional insofar as the conservation law of angular momentum allows a first integration

$$d\Omega_0/dr = -6A\Omega_0/5\rho r^2 + (V^{(0)} + \frac{4}{5}V^{(1)})\Omega_0/r \tag{14}$$

with A denoting the stream function of the meridional flow and $V^{(0)}$ and $V^{(1)}$ representing the Λ-effect up to the same order as $H^{(1)}$ in (12):

$$\langle u'_r u'_\phi \rangle = \nu_T(-\frac{r}{\Omega} \frac{\partial \Omega}{\partial \theta} + V^{(0)} + V^{(1)} \sin^2 \theta) \sin \theta \Omega. \tag{15}$$

Uniform Ω_0 is thus not only possible for vanishing V's but also if the sum $V^{(0)} + \frac{4}{5}V^{(1)}$ becomes small. Opposite signs of $V^{(0)}$ and $V^{(1)}$ are indeed characteristic for very broad classes of turbulences (Rüdiger, 1989). In Table 2 we present numerical results for such

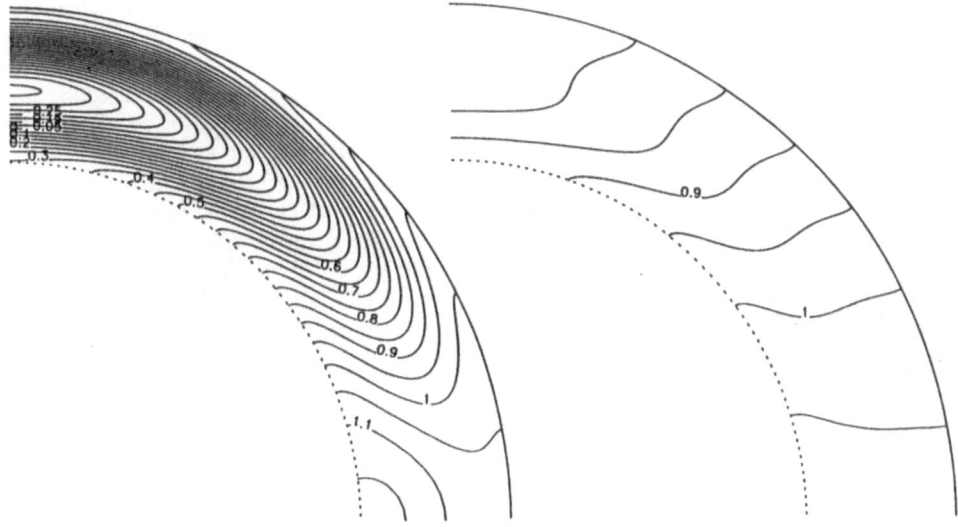

Figure 1. Angular velocity contours for the isotropic turbulence models with $\lambda = 0$ (left) and $\lambda = 12$ (right).

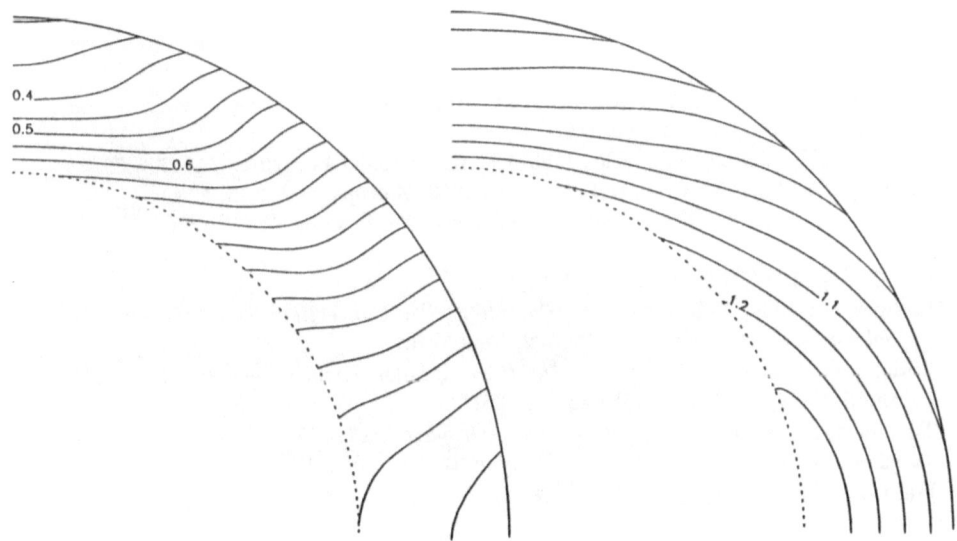

Figure 2. Angular velocity contours for the models with $\lambda = 0$ (left) and $\lambda = 8$ (right) for $V^{(0)} = -1$ and $V^{(1)} = 5/4$.

models, again with small and large λ. $V^{(0)}$ is taken uniform through the whole convection zone. THT-effects are neglected. The numbers clearly demonstrate the existence of some medium where the observed smooth angular velocity profile can be reproduced. A positive $V^{(0)}$ reproduces strong basal differential rotation. The circulation has correct direction and size. On the contrary, the models with a negative $V^{(0)}$ (Fig. 2) have correctly smaller differential rotation at the bottom than at the surface, although for $\lambda=0$ it is everywhere too strong. In this case, however, the Ω-contours have the directions similar to observations.

The results demonstrate the complex nature of the Λ-effect. Furthermore, as we have noted earlier (Rüdiger and Tuominen, 1987; Rüdiger, 1989) with a somewhat larger Taylor number the system becomes unstable (the determinant of the linear system of equations vanishes). It is possible that in this region nonlinear effects in Reynolds stresses become important, limiting the amplitude of Λ-effect by nonlinear feedback, and influence the angular momentum balance in the convection zone.

Table 2: Characteristic numbers for the models with Λ-effect. Scaling of Ω as in Table 1.

λ	$V^{(0)}$	$V^{(1)}$	$H^{(1)}$	Ω_p/Ω_e (sur)	Ω_p/Ω_e (bot)	u_θ
0	1	-5/4	1	0.72/1.07	0.37/1.16	-13
8	1	-5/4	1	0.82/1.04	0.43/0.85	-10
0	-1	5/4	1	0.29/1.18	0.64/1.10	-3
8	-1	5/4	1	0.77/1.06	1.08/1.28	-0.2

References

Christensen-Dalsgaard, J., Schou, J.: 1988, *Proc. Symp. Seismology of the Sun and Sun-like Stars*, ESA SP-286, 149.

Dziembowski, W.A., Goode, P.R., Libbrecht K.G.: 1989, *Astrophys. J.* **337**, L53.

Gilman, P.A., Morrow, C.A., DeLuca, E.E.: 1989, *Astrophys. J.* **338**, 528.

Libbrecht, K.G.: 1988, *Proc. Symp. Seismology of the Sun and Sun-like Stars*, ESA SP-286, 131.

Rüdiger, G.: 1977, *Astron. Nachr.* **298**, 245.

Rüdiger, G.: 1989, *Differential Rotation and Stellar Convection: Sun and Solar-type stars*, Gordon and Breach, New York.

Rüdiger, G., Tuominen, I.: 1987, in *The Internal Solar Angular Velocity*, eds. B.R. Durney and S. Sofia, D. Reidel, Dordrecht, p. 361.

Tuominen, I., Rüdiger, G.: 1989, *Astron. Astrophys.* **217**, 217.

Tuominen, I., Virtanen, H.: 1989, to be submitted to *Solar Phys.*

Ward, F.: 1965, *Astrophys. J.* **141**, 534.

ASYMMETRY OF EMERGING FLUX LOOPS CAUSED BY RADIAL DIFFERENTIAL ROTATION

M. MARIK and K. PETROVAY
Eötvös University, Department of Astronomy
Budapest, Kun Béla tér 2,
H-1083, Hungary

ABSTRACT

Observational and theoretical arguments in favor of a supposed serious asymmetry of magnetic flux loops emerging through the convective zone are briefly summarized. Results from numerical models of flux tubes moving through a differentially rotating upper convective zone are presented (plane parallel geometry, SFT approximation). In most models, especially in the most realistic ones, a remarkable asymmetry of the flux loop is found. It is concluded that in the future observational effects caused by the asymmetry may be used to put quantitative constraints on subphotospheric rotation.

1. BASIC CONCEPTS AND PRELIMINARIES

Recently it was proposed (vanDriel-Gesztelyi and Petrovay, 1989) that the apparently decelerating rotational rate of bipolar sunspot groups (e.g. Tenullo et al., 1981) can be interpreted as a purely geometrical effect arising from the asymmetrical shape and the changing emergence velocity of the magnetic flux loops causing the spots (Fig. 1). The asymmetry would be caused by the aerodynamic drag related to radial differential rotation: the upper layers of the convective zone rotate slower than the flux loops. As the drag, unlike other forces, acts on the surface of the flux tubes, the asymmetry should be stronger for thin (low flux) tubes than for thick (high flux) ones with the same field strength. This can explain the differences in the observed rotational properties of large and small spots (see Howard, 1987 and references therein). As a consequence, it can also be predicted that, on average, the magnetic 0-line will be situated asymmetrically relative to the main spots (Fig. 2); an investigation of Okayama Observatory magnetograms confirmed this result. At the same time, the common "anchoring" or "coupling depth" interpretation of spot proper motions is far more problematic from the observational point of view: after the supposed "decoupling" takes place, there would be no perpendicular forces acting on the loop, so its expansion should stop just when the rotational

321

J. O. Stenflo (ed.), Solar Photosphere: Structure, Convection, and Magnetic Fields, 321–324.
© 1990 by the IAU.

velocity begins to change, in contradiction to what is observed.

While the above observational facts strongly support the asymmetry-hypothesis, it remained to be seen whether the drag arising from differential rotation can actually produce the necessary amount of asymmetry. Petrovay et al. (1989) constructed stationary tube models in order to check this, and they found that the tilts in these crude models ($1°-5°$ for 10^{21} Mx tubes and $10°-30°$ for 10^{18} Mx tubes) produced observational effects that were in order-of-magnitude agreement with observations.

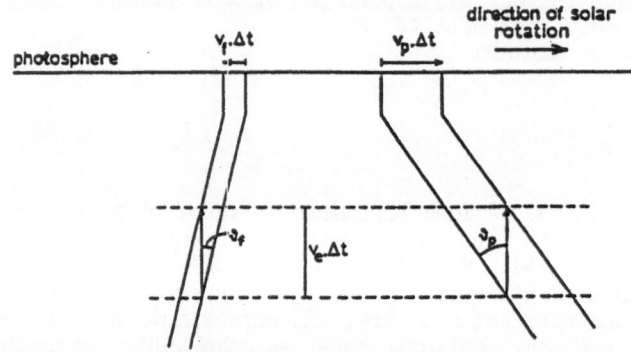

Figure 1: Schematic illustration of the effect of loop asymmetry on spot proper motions. During emergence, the apparent rotational velocity of the sunspot group will be higher than the real one by (v_p-v_f). (This and all the other diagrams are in a reference frame corotating with the loop.)

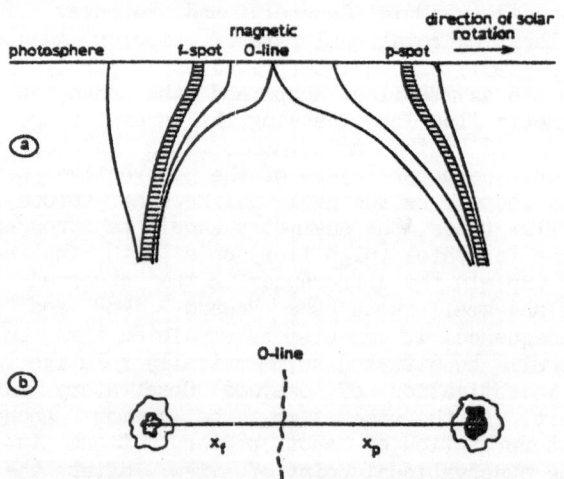

Figure 2: The higher tilt of the thin flux loops leads to an asymmetric position of the magnetic O-line relative to the main spots.

However, as explained in Petrovay et al.(1989), these stationary models are highly unrealistic. So only nonstationary models of emerging flux loops can yield a satisfactory theoretical basis for the asymmetry-hypothesis and make possible more detailed comparisons with observations. Here we present briefly the first results from such nonstationary models; a detailed description of the numerical models will be presented elsewhere (Petrovay, 1989).

2. MODELS AND RESULTS

The models are analogous to those of Meyer et al.(1979): plane parallel geometry, slender flux tube (SFT) approximation and a "half-Lagrangian" integration method are common features of the two families of models. As the SFT approximation breaks down near the photosphere the integration was stopped before the top of the loop reached the surface.

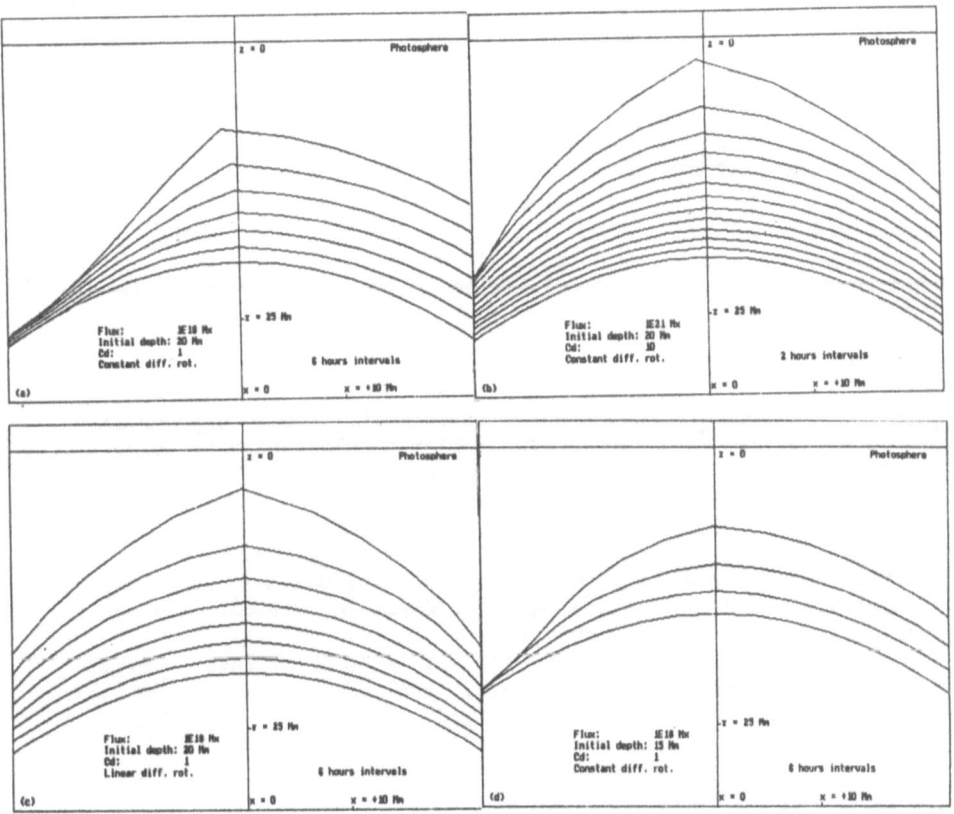

Figure 3: Models of flux loops emerging through a differentially rotating upper convective zone. On each of the four diagrams, the consecutive positions of the tube (considered to be slender) are shown in equal time intervals. Cd is the drag coefficient.

Meyer et al.'s supergranular velocity field was of course replaced by a $v(z)$ horizontal flow with z the depth from photosphere ($v(z)$ is the difference of the rotational speed of the tube and its surroundings). Two extreme cases were examined: $v(z) = 10^4$ cm/s = constant and a linear $v(z) = 4 \cdot 10^2$ cm/s ($z/10^8$ cm $- 25$). The initial shape was a symmetrical, sinusoidal shape in a depth of $z_{top} = 20$ Mm (in other models, 15 Mm; 1 Mm = 10^3 km) with parameters chosen to imitate a typical loop shape in Moreno-Insertis (1986). In the models shown here the curvature was prescribed to vanish at the boundaries. In order to avoid solving the whole nonlinear MHD system of equations the B(z) field strength had to be specified in advance. We used the equipartitional field strength defined by $\varrho v_c^2/2 = B^2(z)/8\pi$. The $\varrho(z)$ density and the $v_c(z)$ convective velocity were taken from mixing-length models (analytic approximations were used). While the initial and boundary conditions used here give lower tilts than real, this form of B(z) yields an upper limit for the tilt, so these models are more or less representative (see Petrovay, 1989 for a more detailed analysis).

The "standard" models are shown on Figures 3(a) and (b). The high tilts obtained earlier in stationary models are confirmed. Figures 3(c) and (d) illustrate the sensitivity of the results to the parameters chosen. This sensitivity offers the possibility to put quantitative constraints on subphotospheric structure and dynamics from photospheric observations. This means that the age-old "tracer method" of the investigation of solar differential rotation might finally be developed into a genuinely exact, quantitative procedure, complementary to the oscillation method. (For details of this envisaged inversion procedure see: Petrovay et al., 1989, Fletcher, Brown and vanDriel-Gesztelyi, 1989).

REFERENCES

Fletcher,L., Brown,J.C., vanDriel-Gesztelyi,L.: 1989, in preparation.
Howard,R.F.: 1987, in: The Internal Solar Angular Velocity, eds. R. Durney and S. Sofia, Reidel, Dordrecht, p.23.
Meyer,F., Schmidt,H.U., Simon,G.W. and Weiss,N.O.: 1979, Astron. Astrophys., 76,35.
Moreno-Insertis,F.: 1986, Astron. Astrophys., 166,291.
Petrovay,K., Brown,J.C., Fletcher,L., Marik,M. and vanDriel-Gesztelyi,L.: 1989, Solar Phys., submitted.
Petrovay,K.: 1989, in preparation.
Tenullo,M., Zappala,R.A., Zuccarello,F.: 1981, Solar Phys., 74,111
vanDriel-Gesztelyi,L. and Petrovay,K.: 1989, Solar Phys., submitted.

CONVECTION AND ITS STABILITY IN THE EQUATORIAL REGIONS OF THE CONVECTION ZONE

A.V. KLYACHKIN
Ioffe Physical-Technical Institute
Academy of Sciences
Ul. Politechnicheskaya 26
194021, Leningrad, USSR

The problem of the existence, evolution, and stability of spatial structures in convection is of considerable importance to astrophysics as well as to geophysical hydrodynamics. The Boussinesq approximation will be used because the considered motions in stars are sufficiently slow. The system of hydrodynamic equations describing convection in a rotating inhomogeneous medium has the form:

$$
\begin{aligned}
D_t \mathbf{U} + 2\mathbf{\Omega} \times \mathbf{U} &= -\rho_m^{-1}\nabla P + (\beta_c C - \beta_T T)\mathbf{g} + \nu_m \Delta \mathbf{U} \\
D_t T + \mathbf{U} \cdot (\nabla T_b - \nabla T_{ad}) &= \chi_m \Delta T \\
D_t C + \mathbf{U} \cdot \nabla C_b &= \mathcal{D}_m \Delta C \\
\mathrm{div}\mathbf{U} &= 0
\end{aligned}
\tag{1}
$$

Here D_t is the total time derivative, $\mathbf{U}$ the velocity, P, T, and C the deviations of the pressure, temperature, and helium abundance (by mass) from the basic equilibrium values, ρ_m, ν_m, χ_m, and $\mathcal{D}_m$ the values averaged over the considered layer of the density, viscosity, thermal and helium diffusivities, β_T and β_c the averaged coefficients of the thermal and helium expansions, $\mathbf{g}$ and $\mathbf{\Omega}$ the gravitational acceleration and angular velocity, ∇T_b and ∇C_b the values of the basic equilibrium temperature and helium gradients, and ∇T_{ad} the adiabatic temperature gradient.

An analysis of finite-amplitude convection in a rotating non-uniform star has been performed by Dolginov and Klyachkin (1987, 1988, 1989) using bifurcation theory methods applied to Eq. (1). It was shown that for a wide range of physical parameters such as rotation rate and degree of chemical inhomogeneity (∇C_b) convection arises when the initial temperature gradient exceeds a critical value (supercritical bifurcation). Furthermore the existence of convective motions for values of the basic temperature gradient less than the critical value (subcritical bifurcation) was found to be possible in the case of slow rotation and a sufficiently large downwards directed helium gradient. In both cases the convection occurs as waves in the form of banana cells elongated along the meridian with an equatorial amplitude maximum. These waves are propagating only azimuthally, both along and opposite to the direction of rotation. The presence of a chemical inhomogeneity in the considered

J. O. Stenflo (ed.), Solar Photosphere: Structure, Convection, and Magnetic Fields, 325–328.
© *1990 by the IAU.*

layer is very important for the physics of the convection processes. It leads not only to a change of the critical temperature gradient, but also to the appearance of slow gyroscopic waves (of Rossby type), with two double-diffusion modes determined by the competition between the thermal and helium diffusions.

It has however not been known whether the spatial structures are stable and how their evolution may be influenced by nonlinear interaction and energy transfer between the different generated modes. For a more detailed analysis of convection in the weakly nonlinear regime we will therefore use the method of amplitude equation, which is based on multiple scalings in space and time. As the propagation of waves is one-dimensional, only the azimuthal coordinate φ is scaled. Let us choose the scaling

$$\tau_1 = \epsilon t, \quad \tau_2 = \epsilon^2 t, \quad \varphi_1 = \epsilon\varphi, \quad \varphi_2 = \epsilon^2\varphi,$$

where ϵ is a measure of supercriticality defined by $R_T = R_0 + \epsilon^2 R_2$. Here R_T is the thermal Rayleigh number, which is proportional to the basic temperature gradient, while R_0 is the critical value of R_T, as obtained with a linear approximation. We seek a solution of Eq. (1) in the form of an expansion in ϵ, which is a measure of the convection amplitude.

$$(V, W, T, C) = \sum_{n=1} \epsilon^n (V_n, W_n, T_n, C_n),$$

where V and W are the poloidal and toroidal velocity potentials ($\mathbf{U} = [\text{curl}^2\mathbf{r}V + \text{curl } \mathbf{r}W]r_0^{-1}$, with $r_0 = R/h$ being the inner radius of the considered spherical shell in units of its depth). According to DiPrima et al. (1971), the multi-scaling method can be used for a Hopf bifurcation in the bounded spatial domain, if the wave number spectrum is sufficiently dense: $\Delta k \lesssim \epsilon^2 < 1$. For a spherical shell, $\Delta k = r_0^{-1}$, and therefore the amplitude equation can correctly describe the convection process in a thin ($r_0 > 1$) spherical layer. Physically it means that in spite of small supercriticality a great number of different spatial harmonics take part in the nonlinear interaction. As a result of the condition that Eq. system (1) has solutions in the $O(\epsilon^3)$ approximation (for details, see Klyachkin, 1989), the amplitude equation follows:

$$\partial A/\partial \tau_2 = \alpha A - \gamma |A|^2 A + \beta \ \partial^2 A/\partial x_1^2 \tag{2}$$

It is a complex ($\beta = \beta_r + i\beta_I$, $\gamma = \gamma_r + i\gamma_I$), one-dimensional Ginzburg-Landau equation. Here $A = A(\tau_2, x_1, x_2)$ is a slowly varying complex amplitude, $x_1 = \epsilon(\varphi - c_g t)$, $x_2 = \epsilon^2(\varphi - c_g t)$, and c_g is the group speed of the most unstable mode. The velocity, temperature, and helium concentration depend both on the slow variables (reflecting the mutual interdependence of the modes) and on the rapid variables (reflecting the spatial and temporal dependence of the most unstable mode). For example

$$T_1 = \text{Re}\hat{T}_1 \ A(\tau_2, x_1, x_2) \ P_\ell^\ell(\cos\theta) \exp\left[i(\ell\varphi + \omega t)\right] \sin\pi(r - r_0),$$

where $\hat{T}_1$ is a constant, chosen from the normalization condition. The coefficients α, β, and γ are uniquely connected with the controlling parameters describing the rotation, degree of chemical inhomogeneity of the medium, and its physical characteristics: viscosity, thermal and helium diffusivities. These expressions for α, β, and γ have been given in Klyachkin (1989). The obvious solutions of Eq. (2) are monochromatic, finite-amplitude rotating waves. It is known that chaotic solutions of Eq. (2) are possible. The condition for such solutions not to appear is also the condition for asymptotically reaching the monochromatic regime (Doering et al., 1988), as well as the condition for modulational stability of the spatially homogeneous solutions $(A = A(\tau_2))$. This condition has the form $\beta_r\gamma_r + \beta_I\gamma_I > 0$. It is satisfied for given values of α, β, and γ when rotation is not very rapid, and in the parameter domain corresponding to supercritical bifurcation in convection. For such parameter values convective motions appearing in the form of slow gyroscopic waves with retrograde propagation represent stable solutions. It is thus not surprising that gyroscopic waves have been found in numerical treatments of convection in a rotating, chemically homogeneous spherical shell in the mean-field approximation (Durney, 1970).

In the domain where subcritical bifurcation takes place, the coefficient $\gamma_r < 0$, and according to Doering et al. (1988) an explosive growth of the solution in a finite time is possible in this case. Phenomenologically this growth can be limited by adding a term with a higher degree of nonlinearity (Landau and Lifshits, 1986):

$$\partial A/\partial \tau_2 = \alpha A - \gamma |A|^2 A - \mu |A|^4 A + \beta \partial^2 A/\partial x_1^2. \tag{3}$$

The evident solutions of Eq. (3), finite-amplitude waves, can possess both small and large amplitudes when the wave number is fixed. The analysis shows that all small-amplitude solutions are modulationally unstable. The spatially homogeneous $(A = A(\tau_2))$ large-amplitude solutions are modulationally stable if

$$(\beta_r\mu_r + \beta_I\mu_I)(\gamma_r^2 + 4\mu_r\alpha)^{1/2} + \beta_I(\gamma_I\mu_r - \gamma_r\mu_I) > 0.$$

For spatially inhomogeneous solutions $A = a\exp[i(kx_1 + \omega\tau_2)]$, the neutral curves of modulational stability relative to perturbations with wave number $\tilde{k}$ have the form (for details, see Klyachkin, 1989)

$$4k^2\beta_r^2 = \left[|\beta|^2\tilde{k}^2 + 2a^2(\beta_r\gamma_r + \beta_I\gamma_I) + 4a^4(\beta_r\mu_r + \beta_I\mu_I)\right]$$
$$\left\{1 + [\beta_I\tilde{k}^2 + \gamma_I a^2 + 2\mu_I a^4]^2[\beta_r\tilde{k}^2 + \gamma_r a^2 + 2\mu_r a^4]^{-2}\right\}^{-1}. \tag{4}$$

Here

$$a^2 = \left[-\gamma_r \pm \sqrt{\gamma_r^2 + 4\mu_r(\alpha - \beta_r k^2)}\right](2\mu_r)^{-1}.$$

Criterion (4) is identical for $\mu = 0$ with the analogous one obtained by Doering et al. (1988).

Let us finally point out that Eqs. (2) and (3) do not describe the processes taking place near the multiple bifurcation line of codimension 2 or the multiple bifurcation point of codimension 3, which occur when a definite relation between rotation, external heating, and degree of chemical inhomogeneity exists.

The author is grateful to Prof. A.Z. Dolginov for fruitful discussions and valuable comments.

References

DiPrima, R.C., Eckhaus, W., and Segel, L.A. (1971) 'Non-linear wave-number interaction in near-critical two-dimensional flows', J. Fluid Mech. 49, 705-744.

Doering, C.R., Gibbon, J.D., Holm D.D., and Nicolaenko, B. (1988) 'Low-dimensional behaviour in the complex Ginzburg-Landau equation', *Nonlinearity* 1, 279-309.

Dolginov, A.Z. and Klyachkin, A.V. (1987) 'The influence of the helium abundance on the
convection threshold in surface layers of rotating stars', *Piz'ma v Astronomicheskii Zhurnal* 13, 764-772.

Dolginov, A.Z. and Klyachkin, A.V. (1988) 'Influence of the chemical composition on the
convection, meridional circulation and differential stellar rotation', in Yu.V. Glagolevsky
and J.M. Kopylov (eds.), *Magnetic Stars*, "Nauka", Leningrad, pp. 264-280.

Dolginov, A.Z. and Klyachkin, A.V. (1989) 'Convection in the equatorial region of a rotating
chemically nonuniform star', *Piz'ma v Astronomicheskii Zhurnal* 15, 33-41.

Durney, B. (1970) 'Nonaxisymmetrical convection in a rotating spherical shell', *Astrophys.
J.* 161, 353-367.

Klyachkin, A.V. (1989) 'The modulation of the autowave convective structures on the rotating chemically nonuniform stars', Preprint No. 1338, Physical-Technical Ioffe Institute,
Leningrad.

Landau, L.D. and Lifshits, E.M. (1986) *Gidrodinamika*, "Nauka", Moscow.

INVERSE CASCADE IN HYDRODYNAMIC TURBULENCE AND ITS ROLE IN
SOLAR GRANULATION

V.KRISHAN
Indian Institute of Astrophysics
Bangalore 560034
India

ABSTRACT.   Self-organization i.e. the formation of large ordered
structures in a turbulent medium is a consequence of inverse cascade
where energy preferentially transfers towards large spatial scales.
It is envisaged that this may be one way of explaining solar granulation at various scales.

1.   Introduction

Radiation and convection are the two main energy transport processes
in the solar interior.   The convective transport becomes operative
where the temperature and density gradients are such that a fluid
element, when displaced from its equilibrium position, keeps moving
away from it.   This stratification, through unstable convection
produces turbulence in the medium.   The fluid eddies of varying
sizes then carry energy as they propagate and dissipate.   The cellular
patterns observed on the solar surface are believed to be the manifestations of convective phenomena occuring in the sub-photospheric
layers.   The cellular velocity fields are seen prominently on two
scales: the granulation and the supergranulation, though mesogranulation and giant cells are also suspected to be present.   The formation of granules with an average size of 1000 km and a life time
of a few minutes can be understood either from the mixing length
(Schwarzschild 1975) or from the linear instability (Bogart, Gierasch
and MacAuslan 1980) description of the convection in the hydrogen
ionization zone of the subphotospheric medium.   The supergranules
with an average size of 30,000 km and a life time of 20 hours do
not have an unambiguous association with a subphotospheric region.
The attempts have been to seek an explanation for the energy concentration at the supergranular scale and to identify the region.
Simon and Leighton (1964) suggested helium ionization to be responsible for accumulation of energy at supergranular scales.   Convective
modes with dominant growth rates at the two scales have been favoured
by Simon and Weiss (1968), Bogart, Gierasch and MacAuslan (1980)
and Antia, Chitre and Narasimha (1984).   Here, a new mechanism
of making large eddies from small eddies is presented.

329

J. O. Stenflo (ed.), Solar Photosphere: Structure, Convection, and Magnetic Fields, 329–332.
© *1990 by the IAU.*

2. Self-organization in Two-Dimensional Turbulence

Formation of ordered structures in a turbulent medium relates to the concept of self-organization which occurs when a system has two or more invariants which suffer selective decay in the presence of dissipation. One invariant has a higher decay rate than the others. The cascading process is such that the slowly decaying quantity transfers towards large spatial scales and thus appears in the form of large organized pattern. The system can be described using a variational principle where the fast decaying quantity is minimised keeping the slowly decaying quantity constant. Kraichnan (1967) found that in a 2-D hydrodynamic turbulence, the energy cascades towards large spatial scales and enstrophy, which is the total squared vorticity ω , towards small spatial scales. It is this property of selective decay and inverse cascade that facilitates the formation of large structures whose dimensions are determined from the ratio of energy and enstrophy. The total energy W and the enstrophy U are defined as

$$W = \int \frac{1}{2} v^2 d^3 r, \quad U = \int \frac{1}{2} \omega^2 d^3 r . \tag{1}$$

Using Kolmogorovic arguments one finds two inertial ranges operating in different spectral regions i.e.

$$W(k) \propto k^{-5/3} \text{ for } k < k_s \tag{2}$$

and

$$W(k) \propto k^{-3} \text{ for } k > k_s \tag{3}$$

where K_s may be the source wave vector. The validity of equations (2) and (3) has been shown by considering inverse cascade through mode-mode coupling (Hasegawa 1985) and by numerically solving Navier-Stokes equations for a 2-D situation.

3. Solar Granulation and 3-D Hydrodynamic Turbulence

But how good is the assumption of 2-D for the solar atmosphere? Levich (1985 and references therein) has shown that the inverse cascade occurs even in 3-D hydrodynamic turbulence. A qualitative description of this phenomenon and a possible answer to the very important question of how a 3-D situation develops into a 2-D or a quasi 2-D are attempted here briefly. The energy injection into the solar atmosphere occurs by the convective upward motions and the energy associated with the latent heat of ionization is released. This begs the question whether the excitation of random small scale motions can lead to large organized structures that are observed in the form of granules, supergranules and giant cells. Here, a picture that emphasizes the role of large helicity fluctuations in the cascading process, as developed by Levich and coworkers is presented. The helicity density, a measure of the knottedness of the vorticity field is given by $\gamma = v \cdot \omega$ and the vorticity $\omega = \nabla \times v$. A turbulent medium exhibits large fluctuations in helicity even though the mean helicity $\langle v \cdot \omega \rangle = 0$. The fluctuating topology of the vorticity field in such a medium is characterized by statistical helicity invariant 'I', represented by conserved mean square helicity per unit volume:

$$I = \lim_{v' \to \infty} \left\langle \left(\int v \cdot \omega \, d^3 r \right)^2 \right\rangle \frac{1}{v'}, \tag{4}$$

and 'I' is a constant for a nondissipative system. Again using Kolmogorov arguments, the inertial range of 'I' and energy W is determined as

$$I(k) \propto k^{-1} \tag{5}$$

$$W(k) \propto k^{-5/3}. \tag{6}$$

Thus, analogous to the 2-D case, 'I' is expected to cascade towards large spatial scales and W(k) to small scales. It is more appropriate to say that the correlation length of helicity fluctuations increases without carrying much energy with it. When this correlation length becomes equal to a fixed vertical scale, as for example restricted by superadiabaticity, the correlation can grow only in the horizontal plane. This gives rise to anisotropy. In the case of highly anisotropic flow characterized by $L_z \ll L_{x,y}$ and $v_z \ll (v_x, v_y)$ it can be shown that the invariant 'I' attains the form $I(K) \propto K^{-5/3}$ which is indistinguishable from that of the energy spectrum in 2-D. Thus, as anisotropy increases, the fraction of energy transferred to larger scales also increases. The growth of large structures in an anisotropic turbulence can again be interrrupted as a result of symmetry breaking, for example cuased by the coriolis force. At the length scale L_c where the nonlinear term of Navier-Stokes equation becomes comparable to the coriolis force, the inverse cascade is inhibited. In the quasi 2-D situation that obtains, coriolis force together with a lack of reflexional symmetry with respect to the horizontal plane favours helical structures with a definite sign of helicity. It is found that in quasi 2-D, the coriolis force favors cyclonic circulation, the sign of helicity corresponding to the updraft cyclonic motion can be fixed. If downward motion is present, it must be anticyclonic to retain the same sign of helicity. There are several related questions of the energetics, the life time, the spatial and temporal structure which need to be investigated keeping in view the available observations of solar granulation and this may give clues as to what more needs to be measured about solar granulation.

References

Antia,H.M., Chitre,S.M. and Narasimha,D. 1984 Convection in the envelope of Red Giants Ap. J. 282: 574-583.

332

Bogart,R.S., Gierasch,P.J., and Mac Auslan,J.M. 1980   Linear modes
    of convection in the solar envelope   Ap. J. 236: 285-293.
Hasegawa,A. 1985   Self-organization processes in continuous media
    Advance in Physics 34; 1-42.
Kraichnan,R.H. 1967, Inertial ranqes in two-dimensional turbulence
    Phys. Fluids $\underline{10}$, 1417-1423.
Levich,E. 1985   Certain problems in the theory of developed hydro-
    dynamic turbulence Phys. Rep. $\underline{151}$, 129-238.
Schwarzschild,M. 1975, On the scale of photospheric convection
    in red giants and supergiants Ap. J. 195: 137-144.
Simon,G.W. and Leighton,R.B. 1964   Velocity fields in the solar
    atmosphere   III. Large-scale motions, the chromospheric network
    and magnetic fields Ap. J. 140:1120-1147.
Simon,G.W. and Weiss,N.O. 1968 Supergranules and the Hydrogen convec-
    tion zone, Zeitschrift fur Astrophysik 69: 435-450.

GENERATION OF TORSIONAL OSCILLATIONS IN THE SUN

Y.V. VANDAKUROV
A.F. Ioffe Physical Technical Institute
Academy of Sciences of the USSR
194021, Leningrad, USSR

ABSTRACT. The hypothesis is considered that the torsional wave observed on the Sun is an eigen-mode oscillation excited in the presence of a weak poloidal magnetic field. We derive asymptotic linear equations for a perturbation with a large number of nodes along the radius, assuming the rotation to be slow and the characteristic perturbation period to be much longer than the rotational period. The results of a preliminary numerical study of the stability of the torsional mode indicate that the superadiabaticity of the solar convection may contribute to the excitation of this mode. In the present work the approximation of harmonic radial dependence of the perturbation has been used.

1. Introduction

Recently Howard and LaBonte (1980) made a detailed analysis of data on the solar horizontal velocity field, and concluded that a solar torsional wave exists, which manifests itself as a modulation of the average rotational velocity. At a fixed latitude the velocity is changed with approximately an 11-yr period having an amplitude of close to 10 m s^{-1} (Howard and LaBonte, 1980; LaBonte and Howard, 1982). This wave is a travelling one showing strong symmetry with respect to the equator. The whole picture seems to be repeated with a 22-yr period.

Many authors consider the torsional wave as a dynamo wave (Yoshimura, 1981; Schüssler, 1981; Kleeorin and Ruzmaikin, 1984). However, the corresponding theory has not been worked out in detail since there have appeared difficulties, which solar dynamo theory itself has not yet overcome.

Wilson (1987), Snodgrass (1987a, 1987b), and Snodgrass and Wilson (1987) have a quite different point of view. They suggest that the torsional wave observed is not in fact an oscillation, but represents a modulation of the mean differential rotation caused by a system of giant azimuthal convective rolls with opposite direction of rotation in any two adjacent rolls of the same hemisphere. They also cite observational evidence that giant-cell convection in the Sun takes the form of equatorward migrating azimuthal rolls. However, it remains unclear whether a rotational velocity distribution exists which is self-maintained, and which satisfies the constraints of the observational data.

A suggestion that the solar torsional wave is excited due to the instability of an appropriate eigen-mode has been considered by Vandakurov (1988). In this case we need to assume that the Sun has a weak, steady, poloidal magnetic field not detectable by current observ-

J. O. Stenflo (ed.), Solar Photosphere: Structure, Convection, and Magnetic Fields, 333–340.

ing techniques. For the oscillations in question, a toroidal magnetic field is generated, the magnetic energy being transformed into kinetic energy and vice versa.

Torsional oscillations have been studied long ago by Walén (1948) and Layzer et al. (1955) in connection with a hypothesis that sunspot fields might represent loops of a toroidal magnetic field gererated by such oscillations and then being pulled up to the surface. However, the characteristic period of the torsional oscillations (for the fundamental mode with a magnetic field strength of around 2 G) turned out to be 25–100 times longer than the solar activity cycle (Layzer et al., 1955).

Nevertheless, the difficulty with the long oscillation period can be eliminated if the torsional mode has numerous nodes along the radius (Vandakurov, 1988). In this case the steady magnetic field is rather weak. An additional restriction of its value follows from the circumstance that in the presence of a steady poloidal magnetic field, an asymmetry should develop between the even- and odd-numbered solar cycles (Boyer and Levy, 1984; Pudovkin and Benevolenskaja, 1984). According to the latter two authors, a maximum value of 0.5 G for the dipole type field gives results consistent with the observations. Such a field can apparently be in accordance with the value 11 yr for the period of the torsional mode (Vandakurov, 1988).

The main question is whether the mode mentioned can be self-excited. This question is considered in the present paper. Asymptotic linear perturbed equations are derived, supposing that the perturbation has a large number of nodes along the radius, and that the torsional oscillation period is much longer than the stellar rotation period. We take into account different types of dissipation. Some results of this study have been discussed briefly in Vandakurov (1988).

2. Asymptotic perturbed equations

Let us assume invariance with respect to φ, the azimuth angle, and consider movements of a viscous, gravitating, compressible medium with finite conductivity in the presence of a magnetic field $\mathbf{B}$. We assume the pressure p and thermal flux $\mathbf{F}$ to be proportional to $\rho T/\mu$ and ∇T, respectively, where ρ is the density, T the temperature, and μ the molecular weight. A different expression for $\mathbf{F}$ will be considered later. In the following r, ϑ, φ are spherical coordinates, and $\mathbf{e}_r$, $\mathbf{e}_\vartheta$, and $\mathbf{e}_\varphi$ are unit vectors.

Let us now write down the φ-component of the equation of motion, as well as the div and $\mathbf{e}_\varphi \cdot$ curl of the same equation:

$$\mathbf{e}_\varphi \cdot (d\mathbf{v}/dt + \mathbf{B} \times \text{curl } \mathbf{B}/4\pi\rho - \nu\nabla^2\mathbf{v}) = 0, \tag{1}$$

$$\text{div}(d\mathbf{v}/dt + \nabla p/\rho + \mathbf{B} \times \text{curl } \mathbf{B}/4\pi\rho - \nu\nabla^2\mathbf{v}) + 4\pi G_0\rho = 0, \tag{2}$$

$$\mathbf{e}_\varphi \cdot \text{curl}(d\mathbf{v}/dt + \mathbf{B} \times \text{curl } \mathbf{B}/4\pi\rho - \nu\nabla^2\mathbf{v}) - \mathbf{e}_\varphi \cdot (\nabla\rho \times \nabla p)/\rho^2 = 0. \tag{3}$$

The other basic equations are

$$\partial\rho/\partial t + \text{div}(\rho\mathbf{v}) = 0, \tag{4}$$

$$\partial\mathbf{B}\partial t - \text{curl}(\mathbf{v} \times \mathbf{B}) - \nu_B\nabla^2 \mathbf{B} = 0, \tag{5}$$

$$dp/dt - (\gamma p/\rho)d\rho/dt - (\gamma - 1)(\rho\epsilon - \text{div } \mathbf{F}) = 0, \tag{6}$$

$$d\mu/dt = 0. \tag{7}$$

Here $d/dt = \partial/\partial t + \mathbf{v} \cdot \nabla$, ν and ν_B are kinematic and magnetic viscosities, and $\mathbf{v}$, G_0, γ, and ϵ denote, respectively, the velocity, gravitational constant, ratio of specific heats, and energy production.

We assume that in equilibrium the toroidal magnetic field is absent, and

$$\mathbf{v} = \mathbf{e}_\varphi r\Omega \sin\vartheta, \qquad B = B_0 \mathbf{b}, \qquad (8)$$

where Ω and B_0 are constant, and $\mathbf{b}$ is the dimensionless meridional vector. The equilibrium conditions follow from Eqs. (1)–(7) if we insert expressions (8). We do not write them down here.

To obtain the perturbed equations we insert, instead of ρ, $\mathbf{v}$, etc., $\rho + \rho^*$, $\mathbf{e}_\varphi r\Omega \sin\vartheta + \mathbf{v}^*$, etc., where $\rho^*(r,\vartheta,t)$ and $\mathbf{v}^*(r,\vartheta,t)$ are Eulerian components of the perturbation. In the linear approximation it follows from Eqs. (3) and (5) that

$$\frac{\partial v_\varphi^*}{\partial t} + 2\Omega(v_r^* \sin\vartheta + v_\vartheta^* \cos\vartheta) - \frac{B_0 \mathbf{b} \cdot \nabla(B_\varphi^* r \sin\vartheta)}{4\pi\rho r \sin\vartheta} - \nu\, L(v_\varphi^*) = 0, \qquad (9a)$$

$$\frac{\partial B_\varphi^*}{\partial t} - B_0 r \sin\vartheta\, \mathbf{b} \cdot \nabla(\frac{v_\varphi^*}{r \sin\vartheta}) - \nu_B\, L\,(B_\varphi^*) = 0, \qquad (9b)$$

where

$$L = \frac{1}{r^2}\frac{\partial}{\partial r} r^2 \frac{\partial}{\partial r} + \frac{1}{r^2 \sin\vartheta}(\frac{\partial}{\partial\vartheta} \sin\vartheta \frac{\partial}{\partial\vartheta} - \frac{1}{\sin\vartheta}).$$

To find the velocities v_r^* and v_ϑ^* in these equations, we need to use the equations of system (1)–(7). Let us write them in the approximations of slow rotation, very slow movements, and small h_*, the radial scale of the perturbation, i.e.,

$$r\Omega^2/g \ll 1, \qquad |\omega^2|/\Omega^2 \ll 1, \qquad h_*/h_p \ll 1, \qquad (10)$$

where $g = -dp/\rho dr$ is the gravitational acceleration, and $h_p = -dr/d\ln p$ is the radial scale of the equilibrium pressure. If $\partial/\partial t \sim i\omega$, then the frequency ω will be of order $\Omega_B r/h_*$ as follows from Eq. (9), with $\Omega_B^2 = B_0^2/4\pi\rho r^2$.

One can see that the main terms in Eq. (3) are the last one and the one that contains the angular velocity Ω. A similar approximation for slow motion (but for the case of large Lorenz force) has been used by Taylor (1963). Note furthermore that from Eq. (2) we obtain the estimate

$$p^*/p \sim (h_*/h_p)(\rho^*/\rho). \qquad (11)$$

Thus, neglecting small terms (but retaining those which are important at small r), we get

$$2\Omega\frac{\partial}{\partial\vartheta}(v_\varphi^* \sin\vartheta) - 2\Omega \cos\vartheta \frac{\partial}{\partial r}(rv_\varphi^*) = \frac{g}{\rho}\frac{\partial\rho^*}{\partial\vartheta}. \qquad (12)$$

In addition, the equations of continuity and energy give

$$\frac{1}{r}\frac{\partial}{\partial r}r^2 v_r^* + \frac{1}{r \sin\vartheta}\frac{\partial}{\partial\vartheta}v_\vartheta^* \sin\vartheta = 0, \qquad (13)$$

$$-\frac{\partial^2}{\partial t^2}\frac{\rho^*}{\rho} + \frac{a}{r}\frac{\partial v_r^*}{\partial t} - \Omega_F\left(\frac{\partial}{\partial r}r^2\frac{\partial}{\partial r} + \frac{1}{\sin\vartheta}\frac{\partial}{\partial\vartheta}\sin\vartheta\frac{\partial}{\partial\vartheta}\right)\left(-\frac{\partial}{\partial t}\frac{\rho^*}{\rho} + \frac{sv_r^*}{r}\right) = 0, \qquad (14)$$

where

$$a = \frac{1}{\gamma}\frac{d\ln p}{d\ln r} - \frac{d\ln \rho}{d\ln r}, \quad s = -\frac{d\ln \mu}{d\ln r}, \quad \Omega_F = -\frac{(\gamma - 1)F_r\, T}{\gamma p r^2 (dT/dr)}.$$

Here we have taken into account that $\partial\mu^*/\partial t = s\mu v_r^*/r$, $T^*/T = -\rho^*/\rho + \mu^*/\mu$. Actually, all terms having an extra factor of order ω^2/Ω^2 or h_*/h_p have been omitted in Eqs. (12)–(14). However, we retain terms with a factor of order $r^2 h_p \Omega^2/gh_*^2$.

Eqs. (9), (12)–(14) constitute a system for the perturbation components $\mathbf{v}$, B_φ, and ρ^*. The boundary conditions are the following. Near the centre (under the condition that $b_r \neq 0$), the term with Ω in Eq. (9a) is small for the perturbation in question. Near the surface where ρ is small, this term is also small. Thus the boundary conditions are the same as those in the absence of rotation, i.e., at the boundaries $\partial(v_\varphi^*/r\sin\vartheta)/\partial r = 0$.

When deriving Eq. (14) we assumed that the heat flux is proportional to ∇T. In the convection zone where $a < 0$, this flux depends mainly on the entropy gradient. Let, for instance, F_r be proportional to $(-a)^\lambda$, where $a < 0$, and λ is a positive constant. Then, with our approximation,

$$F_r^* = (-\lambda r F_r/a)\nabla(\rho^*/\rho - p^*/\gamma p),$$

where the term $p^*/\gamma p$ is negligibly small. We see that Eq. (14) holds true if we put $s = 0$, and replace Ω_F by Ω_{FC}, where

$$\Omega_{FC} = -(\gamma - 1)\lambda F_r/\gamma p r a. \tag{15}$$

3. Model with a radial steady magnetic field

A simple solution of the above equations may be found for the case of an idealized magnetic field distribution: $\mathbf{b} = b_r \mathbf{e}_r$, where $r^2 b_r = \text{const}$, and b_r is positive (negative) if $\vartheta < \pi/2$ ($\vartheta > \pi/2$). In this case, the field direction abruptly reverses when the equator is crossed. Besides, we retain in the equations only terms with the highest radial derivative of the perturbed quantities, assuming the latitudinal derivative not to be large. For example, we replace the operator L by $L_0 = \partial^2/\partial r^2$. In this approximation the term with $\partial v_\varphi^*/\partial\vartheta$ in Eq.(12) may be omitted. The solution of Eqs. (9), (12), and (13) may be expressed in the following form:

$$v_\varphi^* = [r^2(1 - y^2)^{1/2}/y]\, \Lambda\, (\partial^2 E/\partial r\partial y), \tag{16}$$

$$v_\vartheta^* = [r^2(1 - y^2)^{1/2}/2y^2\Omega]\, K\, (\partial^2 E/\partial r\partial y), \tag{17}$$

$$v_r^* = (r/2\Omega)(\partial/\partial y)\left\{[(1 - y^2)/y^2]\, K\, (\partial E/\partial y)\right\}, \tag{18}$$

$$B_\varphi^* = [r^2 b_r B_0(1 - y^2)^{1/2}/y]\, L_0\, (\partial E/\partial y), \tag{19}$$

$$\rho^*/\rho = (2r^3\Omega/g)\, \Lambda L_0\, (E), \tag{20}$$

where $K = r^2 b_r^2 \Omega_B^2 L_0 - \Lambda(\partial/\partial t - \nu L_0)$, $\Lambda = \partial/\partial t - \nu_B L_0$, $y = \cos\vartheta$, and E is a dimensionless function of r, y, and t. Substitution of these expressions in the energy equation (14) yields

$$\left(a\frac{\partial}{\partial t} - sr^2\Omega_F L_0\right)\frac{\partial}{\partial y}\left[\frac{1 - y^2}{y^2}K\left(\frac{\partial E}{\partial y}\right)\right] - qr^2\left(\frac{\partial}{\partial t} - r^2\Omega_F L_0\right)\Lambda L_0\left(\frac{\partial E}{\partial t}\right) = 0. \tag{21}$$

Here $q = 4r\Omega^2/g$.

One can see that the solution of this equation for one mode is

$$E(r, y, t) = Y(y)\exp(ikr + i\omega t). \tag{22}$$

Then

$$\frac{d}{dy}\left[\frac{1 - y^2}{y^2}\frac{dY}{dy}\right] = -j^2 Y, \tag{23}$$

where $j = \text{const.}$, while the frequency ω satisfies a cubic equation studied in Vandakurov (1988). In this paper Eq. (23) has also been investigated.

If ν and ν_B are small, the stability condition is (Vandakurov, 1988)

$$\Omega_B^2 \Omega q(a - s) \geq 0, \tag{24}$$

i.e., thermal dissipation in zones with a superadiabatic temperature gradient serves to self-excite torsional modes with numerous nodes along the radius. In contrast, both ordinary and magnetic viscosity tend to dampen these modes (Vandakurov, 1988).

4. Approximation of harmonic dependence of the perturbation on radius

The equilibrium magnetic field studied in the preceding section had a steplike change near the equator. For a more realistic field distribution, one needs to solve the complicated system of equations in partial derivatives. Since we study perturbations having many radial nodes, the dependence on the boundary conditions becomes of small significance. Then, to form a general concept of the stabilizing or destabilizing contribution of some layer, it seems sufficient to use the approximation of harmonic radial dependence of the perturbation. Thus we assume the perturbation to be proportional to $\exp(ikr + i\omega t)$, where k and ω are constant. Besides, we do not use the approximation that the latitudinal derivative of the perturbation is much smaller than the radial one. Furthermore, we assume the equilibrium magnetic field to be

$$b_r = 2\cos\vartheta, \qquad b_\vartheta = -\beta\sin\vartheta. \tag{25}$$

Here $\beta = \partial\ln(r^2 b_r)/\partial\ln r$, and $|\beta/kr| \ll 1$. Eliminating B_φ^* from Eq. (9), we find

$$\left(1 + \frac{r^2\beta^2\Omega_B^2}{i\omega\nu}\sin^2\vartheta\right)\frac{1}{\sin\vartheta}\frac{\partial}{\partial\vartheta}\sin\vartheta\frac{\partial W}{\partial\vartheta} = \frac{2r^2\Omega}{\nu}(w_r^* + w_\vartheta^*\cot\vartheta) -$$

$$-2\cot\vartheta\frac{\partial W}{\partial\vartheta} + k^2 r^2 W + \frac{2kr^3\Omega_B^2}{\omega\nu}\left[\beta\sin2\vartheta\frac{\partial W}{\partial\vartheta} - (2ikr\cos^2\vartheta + \beta)W\right], \tag{26}$$

$$\frac{\Omega_F}{i\omega\sin\vartheta}\frac{\partial}{\partial\vartheta}\sin\vartheta\frac{\partial\Theta}{\partial\vartheta} = \left(1 + \frac{k^2 r^2\Omega_F}{i\omega}\right)\Theta + (a - s)w_r^*, \tag{27}$$

$$s\frac{\partial w_r^*}{\partial\vartheta} = \frac{\partial\Theta}{\partial\vartheta} + \frac{2i\omega r\Omega\sin\vartheta}{g}\left(\sin\vartheta\frac{\partial W}{\partial\vartheta} - ikr\cos\vartheta\, W\right), \tag{28}$$

$$\frac{1}{\sin\vartheta}\frac{\partial}{\partial\vartheta}\sin\vartheta\, w_v^* = -ikrw_r^*, \tag{29}$$

where $\mathbf{w}^* = \mathbf{v}^*/i\omega r$, $W = w_\varphi^*/\sin\vartheta$, $\Theta = T^*/T$. Here we put $\nu_B \approx 0$.

These equations are equivalent to six first-order differential equations for six variables: W, $\partial W/\partial\vartheta$, w_r^*, w_ϑ^*, Θ, and $\partial\Theta/\partial\vartheta$. In the vicinity of the polar axis ($\vartheta = 0$, or $\vartheta = \pi$), the following expansions are valid:

$$W = W_0 + W_2\sin^2\vartheta + \ldots, \quad w_r^* = V_0 + V_2\sin^2\vartheta + \ldots, \quad \Theta = \Theta_0 + \Theta_2\sin^2\vartheta + \ldots, \quad (30)$$

where

$$W_2 = \frac{1}{8}\left[k^2r^2\left(1 - \frac{4ir^2\Omega_B^2}{\omega\nu}\right)W_0 - \frac{r^2\Omega}{\nu}(ikr - 2)V_0\right], \quad (31)$$

$$\Theta_2 = \frac{i\omega}{4\Omega_F}\left[(a - s)V_0 + \left(1 + \frac{k^2r^2\Omega_F}{i\omega}\right)\Theta_0\right], \quad (32)$$

$$V_2 = \frac{1}{s}\left(\Theta_2 + \frac{kr^2\omega\Omega}{g}W_0\right). \quad (33)$$

Thus the constants W_0, V_0, and Θ_0 remain undetermined. This fact permits us to construct three independent solutions. The whole solution calculated with the initial point at $\vartheta = 0$ coincides with that found with the initial point at $\vartheta = \pi$ if at the equator ($\vartheta = \pi/2$) the quantities $\partial W/\partial\vartheta$, $\partial\Theta/\partial\vartheta$, and w_ϑ^* are zero. These conditions give three linear algebraic equations for W_0, V_0, and Θ_0. The condition that these three equations are solvable,

$$D(\omega) = 0, \quad (34)$$

provides an equation for the complex eigenvalue ω.

5. The case of a chemically homogeneous medium

If $s = 0$, Eqs. (26)–(29) need some modification. Differentiating Eqs. (26) and (27) with respect to ϑ, and excluding (using also Eq. (28)) the derivatives $\partial\Theta/\partial\vartheta$ and $\partial w_r^*/\partial\vartheta$, we find

$$\left[\frac{\nu}{r^2} + \left(\frac{\beta^2\Omega_b^2}{i\omega} + \frac{4r\Omega^2\Omega_F}{ag}\right)\sin^2\vartheta\right]\frac{\partial}{\partial\vartheta}\sin\vartheta\frac{\partial}{\partial\vartheta}\sin\vartheta\frac{\partial W}{\partial\vartheta} =$$

$$= 4kr\left(\frac{\beta\Omega_B^2}{\omega} + \frac{ir\Omega^2\Omega_F}{ag}\right)\sin^2\vartheta\cos\vartheta\frac{\partial}{\partial\vartheta}\sin\vartheta\frac{\partial W}{\partial\vartheta} +$$

$$+ \left\{\nu k^2\sin^2\vartheta - \frac{2kr\Omega_B^2}{\omega}(2ikr\cos^2\vartheta + 3\beta)\sin^2\vartheta + \right.$$

$$+ \left.\frac{4r\Omega^2}{ag}[(i\omega + k^2r^2\Omega_F)\sin^2\vartheta + 2ikr\Omega_F]\sin^2\vartheta\right\}\frac{\partial W}{\partial\vartheta} +$$

$$+ 4ikr\left[\frac{2kr}{\omega}\Omega_B^2 - \frac{r\Omega^2}{ag}(i\omega + k^2r^2\Omega_F)\right]\sin^3\vartheta\cos\vartheta\,W -$$

$$- \Omega\left[ikrw_r^*\sin 2\vartheta - 2w_\vartheta^*(\sin^2\vartheta - 2)\right]. \quad (35)$$

Now Eqs. (35) and (29) reduce to a system of four first-order differential equations for W, $\sin\vartheta\ \partial W/\partial\vartheta$, $\sin\vartheta\ (\partial/\partial\vartheta)(\sin\vartheta\ \partial W/\partial\vartheta)$, and w_ϑ^*. The velocity $i\omega rw_r^*$ in these equations is determined by Eq. (26). The expansions in the vicinity of the polar axis turn out to be given by the first two expressions in Eq. (30), in which the constant W_2 is determined by Eq. (31). One can see that the constants V_0 and W_0 are arbitrary, so by setting the variables $\partial W/\partial \vartheta$ and w_ϑ^* at the equator ($\vartheta = \pi/2$) equal to zero, we find two equations for V_0 and W_0. The determinant $D_*(\omega)$ of these two equations should be equal to zero, i.e., the equation for ω is

$$D_*(\omega) = 0 \tag{36}$$

6. Numerical study of the unstable modes

The excitation of the torsional wave observed on the Sun can occur due to the superadiabaticity of the solar convection zone. In general the approximation that the perturbation has a large number of nodes along the radial direction is not well-founded in the case of the convection zone (Vandakurov, 1988). Nevertheless, some preliminary estimates can be made using Eqs. (35), (29), and (26).

We have carried out a numerical solution of these equations considering the convection zone as a chemically homogeneous medium with a superadiabatic temperature gradient ($a < 0$), with a turbulent viscosity, and, of course, with a convective thermal conductivity. We choose the following values for the parameters: $kr = 10$, $a = -10^{-5}$, $\nu = 1.3 \times 10^{12}\,\text{cm}^2\,\text{s}^{-1}$, $\Omega_{FC} = 5\nu$, $r = 6 \times 10^{10}\,\text{cm}$, $r\Omega = 2 \times 10^5\,\text{cm}\,\text{s}^{-1}$, $r\Omega^2/g = 2 \times 10^{-5}$, $\beta = 0.5$, $\Omega_B^2 = 9.72 \times 10^{-19}\alpha_B$, where α_B is either equal to 1 or to 0.1. If $\rho = 0.001$ g cm^{-3}, these values of α_B correspond to 6.6 G (if $\alpha_B = 1$) and 2.1 G ($\alpha_B = 0.1$). In the case of smaller values of α_B, the computation becomes more time-consuming.

Complex solutions of Eq. (36) were found by the Newton method generalized to cover the case of two-dimensional variables. We searched only for a solution with a positive real part of the quantity $i\omega$. Such solutions imply instability. Note that attempts to find similar solutions for the case of some models with positive values of a did not succeed.

It turns out that the dependence of $D_*(\omega)$ on some trial values of ω is extremely complicated, so the procedure mentioned is convergent only if the trial ω-value is sufficiently close to an eigen-solution. Under the conditions $\vartheta_{st} = 0.006$ and $\alpha_B = 1$, we found the solution $i\omega = (0.5285 - i\,0.0116) \times 10^{-8}\,\text{s}^{-1}$, where ϑ_{st} is the initial value of the angle ϑ. For other small values of ϑ_{st}, the quantity $i\omega$ may differ from the above value by several percent, and fixing $i\omega$ exactly appears to be rather troublesome. However, the latitudinal dependence of the perturbation undergoes only minor changes during the procedure of making the frequency ω more accurate. The dependence of w_ϑ^* on latitude is shown in Figure 1. Here we assume that at the pole ($\vartheta = 0$) W is unity. In the region of $\vartheta \gtrsim 20°$ the perturbation amplitude is very small (if $\vartheta = 90°$, then $w_\varphi^* = 9 \times 10^{-9}(1 + i)$, and the real (imaginary) part of w_φ^* goes through zero at $\vartheta = 74°(80°)$).

The radial velocity v_r^* has many nodes in the vicinity and to the left of the point $\vartheta = \vartheta_*$, where $\vartheta_* = 9°.1$. This is because the coefficient in brackets on the left-hand side of Eq. (35) is small. If $\vartheta < \vartheta_*$, then the closer ϑ is to ϑ_*, the larger is w_r^*, with a maximum value as large as $-5220 + i\,2590$. We do not know whether these large values of w_r^* are consistent with our approximations or not.

In the case that $\alpha_B = 0.1$ we found a solution $i\omega = (0.08145 - i\,0.00538) \times 10^{-8}\,\text{s}^{-1}$ which apparently belongs to the same mode as that considered above. These solutions correspond to nearly exponentially growing modes with a characteristic growth time of the order of several years. Overstable modes are possible if there are zones in which the perturbation is propagating. Thus the study of models having not only convective but also radiative zones is needed.

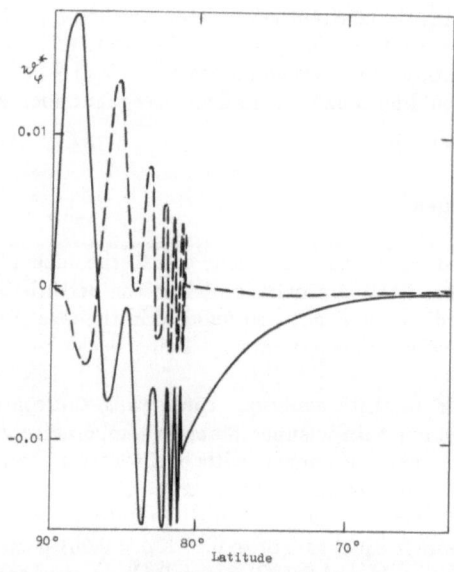

Figure 1. Real (solid curve) and imaginary (dashed curve) parts of w_φ^*, the dimensionless azimuthal displacement, as functions of $\pi/2 - \vartheta$, the latitude.

Note in conlusion that the aforementioned nearly exponentially growing unstable mode can coexist together with the torsional wave found by Howard and LaBonte (1980). We suggest that this mode having large radial velocities in some zones near the poles is the cause of the solar activity observed at high latitudes (Makarov and Sivaraman, 1989). We may relate the existence of such a mode to the weak polar poloidal magnetic fields whose direction reverses periodically. Then the growth of the instability is supposed to begin after the new polar fields have formed. The growth time can be smaller than the cycle duration if the parameter Ω_B is larger than in the previous examples.

7. References

Boyer, D.W., Levy E.H. (1984) *Astrophys. J.* **277**, 848-861.

Howard, R., LaBonte, B.J. (1980) *Astrophys. J.* **239**, L33-L36.

Kleeorin, N.I., Ruzmaikin, A.A. (1984) *Pis'ma Astron. Zh.* **10**, 925-930.

LaBonte, B.J., Howard, R. (1982) *Solar Phys.* **75**, 161-178.

Layzer, D., Krook, M., Menzel, D.H. (1955) *Proc. Roy. Soc.* **A233**, 302-310.

Makarov, V.I., Sivaraman, K.R. (1989) This Symposium

Pudovkin, M.I., Benevolenskaja, E.E. (1984) *Astron. Zh.* **61**, 783-788.

Schüssler, M. (1984) *Astron. Astrophys.* **94**, L17-L18.

Snodgrass, H.B. (1987a), *Solar Phys.* **110**, 35-49.

Snodgrass, H.B. (1987b), *Astrophys. J.* **316**, L91-L94.

Snodgrass, H.B., Wilson P.R. (1987), *Nature* **328**, 696-699.

Taylor, J.B. (1963), *Proc. Roy. Soc.* **A274**, 274-283.

Vandakurov, Yu.V. (1988), *Pis'ma Astron. Zh.* **14**, 334-343.

Walén, C. (1948), *On the vibratory rotation of the Sun*, Henrik Lindståhls Bokhandel, Stockholm.

Wilson, P.R. (1987), *Solar Phys.* **110**, 59-71.

Yoshimura, H. (1981), *Astrophys. J.* **247**, 1102-1112.

VI. GENERATION OF SOLAR MAGNETIC FIELDS

ORDER AND CHAOS IN THE SOLAR CYCLE

A.A. RUZMAIKIN
Institute of Terrestrial Magnetism,
Ionosphere and Radio Wave Propagation
Academy of Sciences USSR
142092, Troitsk, Moscow Region, USSR

ABSTRACT. Solar activity varying with an 11-year cycle is chaotic at large time scales. The evidence comes from an analysis of observations of the sunspot number and of radioactive carbon. Thereby an estimate of the dimension of the solar attractor can be obtained.

The origin of the sunspots can be associated with the interactions of the regular, large-scale, chaotic, and intermittent magnetic fields.

1. Introduction

Before the nineteenth century solar activity was observed by people as random events, mainly through the appearance of sunspots. H. Schwabe was the first who discovered the 11-year cyclicity in the middle of the 19th century. Since that time solar scientists have been looking for the order in the solar activity.

Until the present time the standard point of view has been the following: There is order in the large scales of space and time, e.g. larger than a tenth of the solar radius and one year. For example an annual average of the Maunder butterfly diagram for the sunspot latitude distribution represents order. Then there is disorder in the small scales, e.g. a single spot appears and disappears randomly.

Modern knowledge has cast new light on this issue. It appears that the activity becomes chaotic at large time scales, say, hundreds of years. For example, there were no sunspots during the epochs of the Grand Minima, which occur randomly in time. The concept of global stochasticity of solar activity was introduced eight years ago (Ruzmaikin, 1981; see also Zeldovich and Ruzmaikin, 1983). The origin of chaotic behaviour can be associated with a strange attractor in the phase space of the solar dynamical system.

At the same time there exist some quasi-ordered structures at small scales, for example the network of granulation and mesogranulation.

In that sense the active Sun is a kingdom of physics. Its picture resembles the Arabian makama, a genre of literature that mixes prose and verse, erudition and knavish trickery, sermon and joke. Here I will try to demonstrate this picture by crude strokes with the MHD brush. It is magnetohydrodynamic because convection of the conducting fluid and the magnetic field is responsible for all active phenomena on the Sun.

J. O. Stenflo (ed.), Solar Photosphere: Structure, Convection, and Magnetic Fields, 343–353.

2. Solar Attractor

Let us first consider the processes averaged over a time scale exceeding a year and a spatial scale exceeding the basic scale of solar convection, the diameter of a supergranule. The motions and magnetic fields evolving in the convection zone represent an averaged solar dynamical system. The phase space of the dynamical system can be described, e.g. by the components of the mean (large-scale) magnetic field and the mean velocity field. A path in this phase space corresponds to some solution of the averaged MHD equations. Fortunately it is not necessary to investigate all paths (solutions), because normally there are some manifolds in phase space to which all or almost all paths attract. It is sufficient to find the attractors in order to know the behaviour of the dynamical system. This problem is also very difficult. However, some important characteristics of the attractor can easily be found directly from the observational data. The dimension of the attractor can for instance be determined.

2.1. DIMENSION

The simplest attractors have an integer dimension. The attractor of dimension $d = 0$ corresponds to a point. If such an attractor were present at the origin of the dynamical system under consideration, the Sun would not have any large-scale magnetic field and motions. A closed curve as an attractor of dimension $d = 1$ is the limiting Poincaré cycle. It is exactly the 11-year cycle if it can be stable in time. The attractor of dimension $d = 2$, a so-called 2-torus, may describe the 11-year cycle as modulated by a secular variation. The trajectories of the paths are regular for attractors of integer dimensions.

More typical are however attractors of fractal dimensions. Crudely one can imagine an attractor of fractal dimension, e.g. $d = 3 - \epsilon$, where $\epsilon \ll 1$, as a ball with an infinite number of small holes in it. A path travelling across this manifold will certainly seem very irregular or chaotic. A well known example is the Lorenz attractor (Lorenz, 1963), which has $d = 2.09$ (something more than a surface). The dimension of the attractor is a measure of the regular or chaotic behaviour of the dynamical system.

Is it possible to find the dimension from observational data? At a first glance it appears impossible when one as usual only knows the time behaviour of a single quantity (a projection in phase space). One may, however, hope that a sequence of values sufficiently prolonged in time will reveal the basic properties of the attractor. Grassberger and Procaccia (1983) suggested a method of finding the attractor dimension by using a time series of the observed quantity $x(t), x(t + \tau), \ldots, x[t + (N - 1)\tau]$, with a given time increment τ. The attractor sought for must be embedded in a phase space of a large dimension m, defined by the vectors

$$\xi_i(t) = [x(t), x(t + \tau), \ldots, x(t + (m - 1)\tau)], \qquad i = 1, 2, \ldots, N - 1.$$

The dimension d of the attractor is defined as the exponent of the correlation integral $w(r) \sim r^d$ for small distances r between the points ξ_i.

$$w(r) = N^{-2} \sum_{i,j=1}^{N} \theta(r - |\xi_i - \xi_j|),$$

where N is the total number of experimental points used, and θ is the Heaviside function. Thus, to find the dimension one needs to isolate for small r a linear part of the curve ($\log w$

vs. $\log r$), and to determine its inclination.

Applying this method to the series of relative sunspot numbers, a dimension close to 2 has been obtained (Kürths, 1987; Makarenko and Ajmanova, 1989). The series of sunspot numbers is, however, too short to include such features as the Grand Minima. It is better to use more extended sequences, such as the carbon isotope ^{14}C abundance in the annual tree rings available over 9000 years (Suess, 1965, 1978; Damon and Sonett, 1989; Dergachev, private communication). The rate of ^{14}C formation in the Earth's atmosphere due to the bombardment by cosmic ray particles is well anti-correlated with the number of sunspots, as a result of the solar magnetic field acting as a filter for the cosmic rays. Thus ^{14}C is an index of solar activity.

An estimate of the attractor dimension from Suess's data over 6000 years gives $d \approx 3.3$ (Gizzatulina et al., 1988), i.e., the dimension is close to 3. Because the 3-torus according to the theory of stochastic dynamical systems is unstable and will be transformed into a strange attractor, the above result means that the mean solar attractor is fractal, i.e., the dynamical system is chaotic. The estimate of the dimension can be improved by using more extended and accurate data. The first estimates of the attractor dimension indicate a low dimension. This also suggests that the mean solar dynamical system is very simple. It can be described in terms of only few effective modes. Let us consider the probable origin of these modes (Ruzmaikin, 1985).

2.2. 22-YEAR SOLAR CYCLE AS A MANIFESTATION OF DYNAMO WAVES

One of the possible effective modes of the mean solar dynamical system is the 22-year magnetic cycle. The analysis of the annual mean values of the Wolf sunspot numbers W for 19 of the 11-year cycles made by Gudzenko and Chertoprud (1964) shows the existence of a limiting cycle of near elliptical form in phase space $(W, dW/dt)$. This analysis assumes the existence of the derivative dW/dt and hence excludes the possibility for the attractor to be fractal (Makarenko and Ajmanova, 1989). However, as a smooth approximation over a finite time interval it looks reasonable and gives the possibility of extracting one of the effective modes of the dynamical system.

This two-parameter limiting cycle can be explained by the mean field dynamo theory (cf. Krause and Rädler, 1980). The mean magnetic field is excited and evolves in the solar convective zone due to the differential rotation ($\nabla\Omega$), mean helicity (α), and turbulent diffusion (ν_T):

$$\frac{\partial A}{\partial t} = \alpha B_\phi + \nu_T \nabla^2 A,$$

$$\frac{\partial B_\phi}{\partial t} = r \sin\theta (\nabla\Omega \times \nabla A)_\phi + \nu_T \nabla^2 B_\phi,$$

where A is the azimuth component of the vector potential of the poloidal field (B_r, B_θ).

The solution has the form of waves propagating, when the value of $\nu_T/(\alpha R)$ is small, along the surfaces of constant angular velocity (Yoshimura, 1975). For instance, when the angular velocity only depends on the radial coordinate, the dynamo wave travels along the solar surface:

$$A = A(\omega t - k\theta), \qquad B_\phi = B_\phi(\omega t - k\theta + \delta).$$

The direction of propagation is determined by the sign of the dimensionless dynamo number $D = R_\odot^3 (\sin\theta)\alpha_{max}(d\Omega/dr)_{max}/\nu_T^2$. The wave propagates from the pole to the equator when D is negative, and to the pole when D is positive. The frequency of the wave is $\omega = (k|D|/2)^{1/2}$, the phase speed is $\omega/k = (|D|/2k)^{1/2}$, and the group speed is a factor of two smaller than the phase speed. The phase shift between the poloidal and toroidal components of the magnetic field is δ, which is close to $\pi/4$ in the standard kinematic models of the solar dynamo (cf. Parker, 1979; Krause and Rädler, 1980). In this case the radial component of the field is lagging behind B_ϕ by $3\pi/4$.

In the other idealized case when Ω is only dependent on θ, the dynamo wave propagates radially. It is of importance to know the real distribution of the angular velocity as well as the mean helicity in the solar convection zone. According to mixing length models of stellar convection zones (Spruit, 1974), the turbulent diffusivity can be considered as almost constant in the bulk of the solar convection zone.

Makarov et al. (1987) (cf. also Brandenburg and Tuominen, 1988), have made an attempt to find the dynamo wave distribution on the solar surface by using the depth dependence of the angular velocity obtained from current helioseismology data (Fig.1) and the mean helicity estimated via the mixing-length theory. They found that the source for generating the mean magnetic field, $\alpha r(d\Omega/dr)$, has three main maxima in the convection zone: near the solar surface at $0.9R_\odot$, at $0.8R_\odot$, and close to the bottom at $0.7R_\odot$.

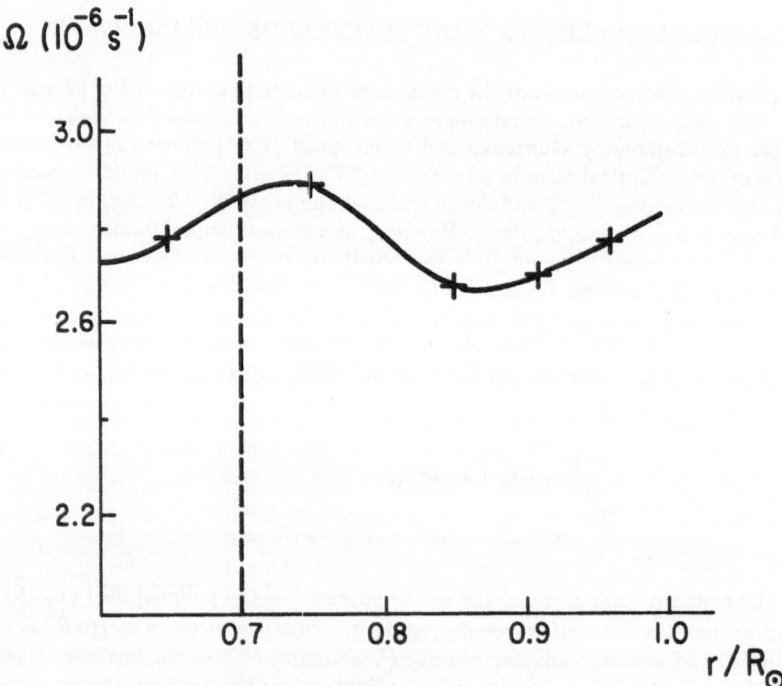

Figure 1. The radial profile of the angular velocity in the solar convection zone appears to have regions with different sign of the gradient.

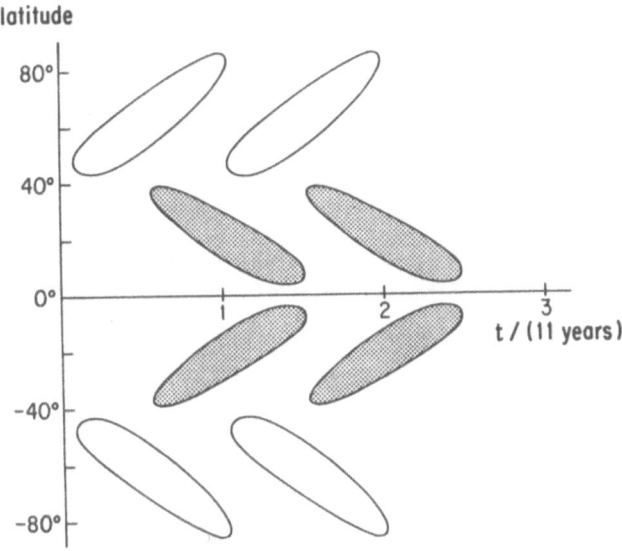

Figure 2. Schematic representation of two magnetic waves on the solar surface (the polar and equatorial dynamo waves, produced by the action of mean helicity and differential rotation of the type given in Fig.1).

In the first region the value of $\alpha(d\Omega/dr)$ is positive, so that the dynamo wave originating there propagates towards the pole in each hemisphere. The maximum of the wave amplitude is reached approximately at a latitude of 65°. This wave probably manifests itself in the activity of polar faculae migrating to the pole. The second maximum, at 0.8 $R_\odot$, has $\alpha(d\Omega/dr) < 0$. It is larger and creates a wave that is ten times larger in amplitude, and which migrates to the equator. This wave is naturally associated with the standard Maunder butterfly diagram of sunspot activity (Fig. 2). The third and deepest maximum of the source also creates a wave propagating towards the pole. This wave is, however, rather weak on the solar surface due to the considerable depth of the source. Makarov et al. (1987) have associated this wave with the apparent extra reversals of the poloidal magnetic field. Let us finally mention an important element of order, which was pointed out by Stenflo (1988). The observed odd rotationally symmetric modes of the solar magnetic field vary with the 22 yr period, while this period is absent for the even modes. This fact agrees with the symmetry and preferable excitation of odd modes in solar dynamo theory.

2.3. ROTATIONAL WAVES AND MERIDIONAL FLOWS

Observational evidence have been found for zones of higher and lower rotational velocities (3–6 m s^{-1} in amplitude), propagating with an 11-year period, superposed on the general differential rotation (LaBonte and Howard, 1982). The study of sunspot motions have shown meridional flows with the same periodicity (Tuominen et al., 1983).

The strong correlation between the motions and the large-scale magnetic field suggests that the motions are created due to the action of dynamo waves. A simple picture of a rotational wave produced by a magnetic dynamo wave has been drawn by Kleeorin and Ruzmaikin (1984 a) as follows:

The rotational velocity variations produced by the magnetic field can be estimated from the balance between the turbulent and magnetic stresses,

$$\rho \nu_T \frac{\partial}{\partial r} \left(\frac{u_\phi}{r} \right) = -\frac{B_r B_\phi}{4\pi}.$$

The magnetic field can be considered as given, because the rotational variations observed are small. In particular, in the case that $\Omega = \Omega(r)$, the field can be written in the form

$$B_r = b_r(r,\theta)\cos(\omega t - k\theta - \pi/2 - \delta),$$
$$B_\phi = b_\phi(r,\theta)\cos(\omega t - k\theta),$$

where the amplitude distribution, frequency, and phase are found by solving the dynamo equations.

For the time-varying part of the rotational velocity one immediately obtains a wave with the double frequency,

$$u_\phi = u_0(r,\theta)\cos(2\omega t - 2k\theta + \pi/2 - \delta).$$

The distribution of the wave amplitude, $u_0 \sim b_r b_\phi$, proves to be more homogeneous as compared with the distribution of the amplitude of the magnetic field. Let the azimuth component of the magnetic field (manifested on the surface in the form of sunspots) be concentrated to mid-latitudes, e.g. as $b_\phi \sim \sin 2\theta$, on the solar surface, and let the radial component increase towards the poles, e.g. as $b_r \sim \sin \theta$. Then $u_0 \sim \sin \theta \cos^2 \theta$ is distributed more homogeneously over the latitudes than the field. The phase of the maximum of the rotational wave is shifted as compared with the phase of the maximum magnetic flux (defined by $|B_\phi|$). The shift is 1/8 of the wavelength of the rotational wave if $\delta = \pi/4$, as the kinematic dynamo theory predicts. This consequence of our simple model is in good agreement with the observations of Howard and LaBonte.

For more details concerning the magnetic action on the rotation and meridional motion, we refer to the papers by Rüdiger et al. (1986) and Kleeorin and Ruzmaikin (1989).

2.4. CAUSES FOR THE STOCHASTIC BEHAVIOUR OF SOLAR ACTIVITY

The second effective mode of the mean solar dynamical system can be identified with the secular modulation of the 22-year solar cycle. Nonlinear beats between odd (dipole type) and even (quadrupole type) magnetic fields may provide a mechanism for the modulation (Kleeorin and Ruzmaikin, 1984 b). Magnetic fields of both symmetries in the form of waves with closely spaced frequencies may be excited by dynamo action. The beats strongly compete with the synchronization that tends to equalize the frequencies. Perhaps this causes the observed irregularity of the secular modulation.

Another reason for this irregular variation is a possible decay of the magnetic field to zero. Actually the existence of a non-magnetic sun does not contradict the equilibrium conditions or other fundamentals. It may even be profitable from some point of view. However, *the non-magnetic state is unstable* due to the dynamo number (the amplitude of the source for the generation of magnetic fields in the solar convection zone) being sufficiently large.

The magnetic field grows when the dynamo number is large. However, the growing field smoothes out the angular velocity gradients and removes the predominance of left-handed vortices over the right-handed ones, i.e., it reduces the mean helicity. As a result the effective dynamo number decreases and the dynamical system can fall down into a non-magnetic state. Some simple models demonstrating such a behaviour have been constructed (Ruzmaikin, 1981; Weiss et al., 1984; Malinetsky et al., 1986). In the first paper based on a Lorenz type model an estimate of the characteristic time to stay in the vicinity of the non-magnetic state is given by $\tau(D - D_{cr})^{-1/2}$, where τ is the decay time of the mean helicity, and D_{cr} is the value of the dynamo number at which excitation of dynamo waves starts. In the more developed model by Malinetsky et al. (1986) it is shown how groups of regular 22-years oscillations are separated by deep minima (Fig.3). The behaviour of the system in phase space is similar to the dynamics of the 3-torus.

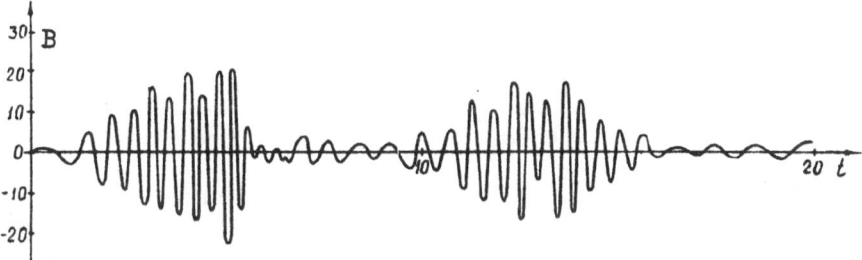

Figure 3.   Time evolution of the mean magnetic field in the model of Malinetsky et al. (1986).

3. Intermittency of solar magnetic fields

Together with the mean magnetic field a fluctuating, small-scale magnetic field is excited. Observations provide evidence that the field is concentrated in thin magnetic flux tubes. The fluctuating field can also be associated with the origin of the sunspots. Let us consider briefly these two problems.

3.1. DISTRIBUTION OF THE FLUCTUATING MAGNETIC FIELD

Let $h_i(r,t)$ represent the deviation of the magnetic field from its mean value, and

$$w(r) = < h_i(\mathbf{r}_1)h_i(\mathbf{r}_2) >, \qquad r = |\mathbf{r}_1 - \mathbf{r}_2|$$

be a space correlation function of the field.

As was shown by Kleeorin et al. (1986) in the framework of a kinematic dynamo model with a velocity field having a short correlation scale, assuming a magnetic Reynolds number of the order of 10^8 typical for the solar convection zone, two modes of the fluctuating magnetic field can be excited (Fig.4).

The first mode corresponds to a simple loop having a diameter comparable to the correlation scale of the turbulent convection $l \sim 10^4$ km (the size of a supergranule), and a

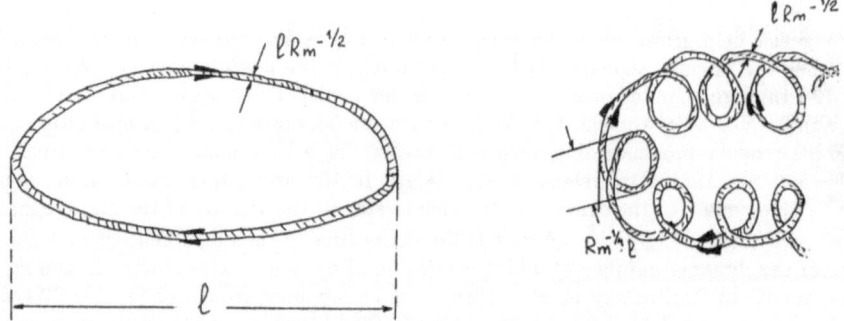

Figure 4. The modes of the fluctuating magnetic field that can be excited by dynamo action in the solar convection zone.

thickness of $lR_m^{-1/2} \sim 1$ km. The second mode is associated with a more complicated loop having a subset of smaller loops of diameter $lR_m^{-1/4} \sim 100$ km. This scale may be compared with the dimensions of observed flux tubes (Muller, 1985).

It is further necessary to take into account the cell structure of the solar convection. It causes advection and concentration of the magnetic field to the cell boundaries (Galloway et al., 1977).

The distribution of the magnetic field in a turbulent flow with a large magnetic Reynolds number should be very inhomogeneous from a fundamental point of view (Molchanov et al., 1985). The magnetic field in such a flow is concentrated into intermittent bundles with large regions in between with a relatively weak field. Thus the appearance of solar magnetic flux tubes can be considered as a natural manifestation of such a behaviour of the magnetic field due to the action of random motions of the highly conducting fluid.

3.2. ON THE ORIGIN OF SUNSPOTS

A sunspot is a region of reduced temperature (by at most 1500 K). It originates randomly in the vicinity of the maximum of the large-scale magnetic field that evolves with the solar cycle. The total area covered by sunspots over the 11-year cycle does not exceed 2% of the solar surface.

Sunspots thus appear as a set of temperature spots. However, they are not like temperature spots in a turbulent flow, where excesses and deficits of the temperature are equally abundant.

The origin of sunspots should be associated with the magnetic field. We recall that solar magnetic fields were discovered by G. Hale in sunspots. A pair of sunspots is normally connected by a loop of the azimuthal magnetic field emerging through the solar surface, probably due to buoyancy (Parker, 1979). It is, however, not clear why separate loops and not the whole latitude belt of the azimuthal field emerge. Let us note that the magnetic field is rather irregular even inside sunspots, although the same field direction is predominant inside the spots (Bray and Loughead, 1969).

It is natural to associate the origin of sunspots with the intermittency of the solar magnetic field, i.e., with random, infrequent but intense concentrations of the fluctuating magnetic field generated by random motions. Deviations from the mean magnetic field occur both to increase and to decrease that field. However, only after an increase the total magnetic field (mean field plus fluctuation) can reach the critical value that makes it float due to buoyancy (Fig.5).

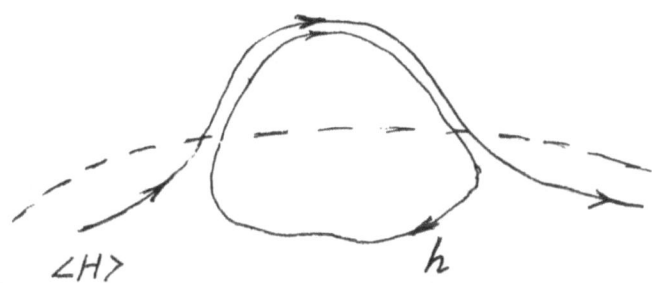

Figure 5. The sum of the mean $< H >$ and the intermittent, fluctuating field h may reach a critical value to allow the field to emerge from the solar convection zone.

The distribution of the intermittent magnetic field is close to a log-normal distribution (Zeldovich et al., 1987):

$$P(\ln H) = \frac{1}{\sqrt{2\pi}\,\sigma} \exp\left[\frac{-(\ln H - a)^2}{2\sigma^2}\right],$$

where $a = < \ln H >$ and $\sigma^2 = < (\ln H - a)^2 >$. The symbol $<>$ means 'mean value'. The successive statistical moments grow with increasing order p:

$$< H > = \exp(a + \sigma^2/2), \quad < H^2 > = \exp(2a + 2\sigma^2), \ldots, \quad < H^p > = \exp(pa + p^2\sigma^2).$$

To eliminate the unknown value of a it is better to use distribution for the normalized magnetic field $H_n = H/ < H >$,

$$P(H_n) = P(\ln H)\frac{d\ln H}{dH_n} = \frac{1}{\sqrt{2\pi}\,\sigma H_n} \exp\left[\frac{-(\ln H_n + \sigma^2/2)^2}{2\sigma^2}\right],$$

determined by the single parameter σ.

The probability for the normalized field to exceed a critical value ζ is

$$Prob\,(H_n > \zeta) = \int_\zeta^\infty \frac{1}{\sqrt{2\pi}\,\sigma H_n} \exp\left[\frac{-(\ln H_n + \sigma^2/2)^2}{2\sigma^2}\right] dH_n =$$

$$= \int_{\ln\zeta}^\infty (\sqrt{2\pi}\sigma x)^{-1} \exp\left[\frac{-(x^2 + \sigma^2/2)}{2\sigma^2}\right] dx = (1/2)\left[1 - F\left(\frac{\ln\zeta + \sigma^2/2}{\sigma}\right)\right].$$

Equating this probability with the relative area covered by sunspots (1% of the total area of the Sun) one can obtain an estimate for the parameter σ. In principle this parameter should be determined from theory.

References

Brandenburg, A. and Tuominen, I. (1988) 'Variation of magnetic field and flows during the solar cycle', COSPAR paper No. 12.4.6, Espoo, Finland.

Bray, R.J. and Loughead, R.E. (1964), *Sunspots*, Chapman and Hall Ltd., London.

Damon, P. and Sonett, C.P. (1989) 'The spectrum of radiocarbon', in Proc. Int. Conf. 'Sun in Time', Arizona Univ. Press, Tucson.

Eddy, J.A. (1976) 'The Maunder Minimum', *Science* **192**, 1189-1202.

Galloway, D.J., Proctor, M.R.E., and Weiss, N.O. (1977) 'Formation of the intense magnetic fields near the surface of the Sun', *Nature* **266**, 686-692.

Gizzatulina, S.M., Ruzmaikin, A.A., Rukavishnikov, V.D., and Tavastsherna, K.S. (1988) 'Radiocarbon evidence of global stochasticity of solar activity', Preprint 40 (794), IZMI-RAN, Moscow (submitted to Solar Phys.).

Grassberger, P. and Procaccia, I. (1983) 'Measuring the strangeness of strange attractors', *Physica* **D9**. 189-208.

Gudzenko, L.L. and Chertoprud, V.E. (1964) 'Some dynamical properties of solar activity', *Soviet Astron.* **41**, 697-706.

Kleeorin, N.I. and Ruzmaikin, A.A. (1984 a) 'On the nature of 11-year torsional oscillations of the Sun', *Pisma Astron.Zh.* (USSR) **10**, 925-935.

Kleeorin, N.I. and Ruzmaikin, A.A. (1984 b) 'Mean-field dynamo with cubic non-linearity', *Astron. Nachr.* **305**, 265-275.

Kleeorin, N.I. and Ruzmaikin, A.A. (1989) 'Large-scale flows excited by magnetic fields in the solar convective zone', Preprint, IZMIRAN, Moscow.

Kleeorin, N.I., Ruzmaikin, A.A., and Sokoloff, D.D. (1986) 'Correlation properties of self-exciting fluctuating magnetic fields', Proc.Varenna-Abastumani Int.School 'Plasma Astrophysics', ESA SP-251-ISSN 0379-6566; (1988) *Kinematics and Physics of Celestial Bodies* **4**, 28-35.

Krause, F. and Rädler, K.-H. (1980) *Mean-field magnetohydrodynamics and dynamo theory*, Akademie-Verlag, Berlin.

Kürths, J. (1987) 'Estimating parameters of attractors in some astrophysical time series', in M. Forkas (ed.), Proc. Int. Conf. on Nonlinear Oscillations, Budapest, pp.664-667.

LaBonte, B.J. and Howard, R. (1982) 'Torsional waves on the sun and the activity cycle', *Solar Phys.* **75**, 161-178.

Lorenz, E.N. (1963) 'Deterministic nonperiodic flow', *J. Atmosph. Sci.* **20**, 130-141.

Makarenko, N.G. and Ajmanova, G.K. (1989) 'K entropy and dimension of solar attractor', submitted to Pisma Astron. Zh.

Makarov, V.I., Ruzmaikin, A.A., and Starchenko, S.V. (1987) 'Magnetic waves of solar activity', *Solar Phys.* **111**, 267-277.

Malinetsky, G.G., Ruzmaikin, A.A., and Samarsky, A.A. (1986) 'A model of longperiodic variations of solar activity', Preprint No. 170, Keldysh Inst. of Appl. Math., Moscow.

Molchanov, S.A., Ruzmaikin, A.A., and Sokoloff, D.D. (1985) 'Kinematic dynamo in random flow', *Sov. Phys. Uspekhy* **145**, 593-628.

Morfill, G. and Voges, W. (1989) 'The solar cycle: Deterministic or chaotic', in Proc. Int. Conf. 'Sun in Time', Arizona Univ. Press, Tucson.

Muller, R. (1985) 'The fine structure of the quiet Sun', *Solar Phys.* **100**, 237-255.

Parker, E.N. (1979) *Cosmic Magnetic Fields*, Clarendon Press, Oxford.

Rüdiger, G., Tuominen, I., Krause, F., and Virtanen, H. (1986) 'Dynamo-generated flows in the Sun: 1. Foundation and first results', *Astron. Astrophys.* **166**, 306-318.

Ruzmaikin, A.A. (1981) 'Solar cycle as strange attractor', *Comments on Astrophys.* **9**, 85-93.

Ruzmaikin, A.A. (1985) 'The solar dynamo', *Solar Phys.* 100, 125-140.

Spruit, H.C. (1974) 'A model of the solar convection zone', *Solar Phys.* **34**, 277-290.

Stenflo, J.O. (1988) 'Global wave patterns in the Sun's magnetic field', *Astrophys. Space Sci.* **144**, 321-336.

Suess, H.E. (1965) 'Secular variations of cosmic-ray produced carbon 14 in the atmosphere and their interpretations', *J. Geophys. Res.* **70**, 5937-5952.

Suess, H.E. (1978) *Radiocarbon* **20**, 1-18.

Tuominen, J., Tuominen, I., and Kyrolinen, J. (1983) 'Eleven-year cycle in solar rotation and meridional motions as derived from the positions of sunspot groups', *Mon. Not. Royal Astron. Soc.* **205**, 691-704.

Weiss, N.O., Cattaneo, F., and Jones, C.A. (1984) 'Periodic and aperiodic dynamo waves', *Geophys. Astrophys. Fluid. Dyn.* **30**, 305-341.

Yoshimura, H. (1975) 'Solar cycle dynamo wave propagation', *Astrophys. J.* **201**, 740-748.

Zeldovich, Ya.B. and Ruzmaikin, A.A. (1983) 'Dynamo problems in astrophysics', *Astrophys. and Space Phys. Rev.* **2**, 333-383.

Zeldovich, Ya.B., Molchanov, S.A., Ruzmaikin, A.A., and Sokoloff, D.D. (1987) 'Intermittency in random medium', *Sov. Phys. Uspekhi.* **152**, 3-32.

SYMMETRY BREAKING IN THE SOLAR DYNAMO:
NONLINEAR SOLUTIONS

R.L. JENNINGS
Department of Mathematics and Statistics
The University
Newcastle upon Tyne NE1 7RU
UK

N.O. WEISS
Department of Applied Mathematics and Theoretical Physics
University of Cambridge
Cambridge CB3 9EW
UK

ABSTRACT. We examine an idealized $\alpha\omega$-dynamo model in which the magnetic fields depend only on latitude and time. The solutions that bifurcate from the field-free state are either symmetric or antisymmetric about the equator (quadrupolar or dipolar respectively). Nonlinear steady and periodic solutions, whether stable or unstable, can be followed numerically as the dynamo number is varied, revealing a rich bifurcation structure with mixed-mode solutions (lacking symmetry about the equator) appearing at secondary bifurcations. These results show how stable asymmetric fields can occur in the sun and illustrate the formation of complicated spatial structure in more active stars.

We assume that the generation of magnetic fields in a star like the sun occurs in a shell at the base of the convective zone and can be described by an axisymmetric mean-field $\alpha\omega$-dynamo (Parker 1979). In order to investigate latitudinal structure and equatorial symmetries we consider a highly simplified model in which only the essential physics is retained; all results are therefore strictly qualitative.

We neglect curvature and replace spherical co-ordinates (r, θ, ϕ) by cartesians (z, x, y) respectively. Then we may consider a magnetic field $\mathbf{B} = (0, B, \partial A/\partial x)$ that depends only on the colatitude x and on time t. In the weak field limit the toroidal field B is created by the sheared azimuthal velocity $\mathbf{u} = \omega z \sin x \, \hat{\mathbf{y}}$ while generation of the poloidal vector potential A through helicity is represented by the parameter $\alpha = \alpha_0 \cos x$ which is antisymmetric about the equator ($x =$

J. O. Stenflo (ed.), Solar Photosphere: Structure, Convection, and Magnetic Fields, 355–358.

$\pi/2$). In the nonlinear regime we introduce parameters τ, κ and λ to represent quenching of the α-effect, quenching of differential rotation and enhanced losses through magnetic buoyancy respectively (Jones 1983; Weiss et al. 1984). Thus we obtain the nondimensionalised system

$$\partial A/\partial t = DB \cos x (1 + \tau B^2)^{-1} + \partial^2 A/\partial x^2 \ , \tag{1}$$

$$\partial B/\partial t = (\partial A/\partial x) \sin x (1 + \kappa B^2)^{-1} + \partial^2 B/\partial x^2 - \lambda B^3 \ . \tag{2}$$

Here the dynamo number $D = -\alpha_o \omega R^3/\eta^2$, where R is the latitudinal length scale and η is a turbulent diffusivity (cf. Stix 1972). The boundary conditions at the poles are $A = B = 0$ at $x = 0, \pi$.

Equations (1) and (2) possess a trivial field-free solution $A = B = 0$. Branches that bifurcate from this trivial solution involve magnetic fields that are either symmetric (quadrupole) or antisymmetric (dipole) about the equator, with $\partial B/\partial x = A = 0$ or $B = \partial A/\partial x = 0$ respectively at $x = \pi/2$. For $D > 0$ we expect to find antisymmetric dynamo waves migrating towards the equator but as D is increased from zero the first bifurcation, at $D \approx 9$, is to a branch of steady quadrupole solutions. A branch of oscillatory dipole solutions appears at $D \approx 102$, followed by oscillatory quadrupole solutions at $D \approx 264$.

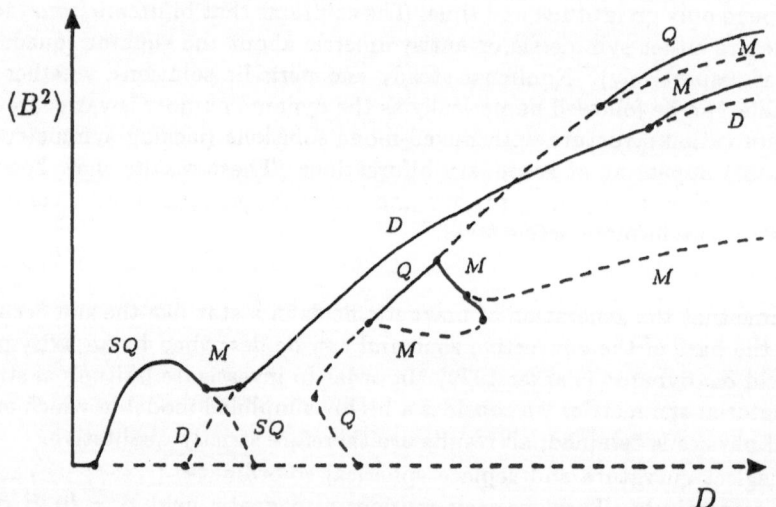

Figure 1. Bifurcation diagram for the case $\kappa = \lambda, \tau = 0$, showing the mean square toroidal field $\langle B^2 \rangle$ as a function of the dynamo number D (not to scale). Solid (broken) lines indicate stable (unstable) solution branches. Steady quadrupole and oscillatory dipole, quadrupole and mixed-mode solutions are denoted by SQ, D, Q and M respectively.

(a)

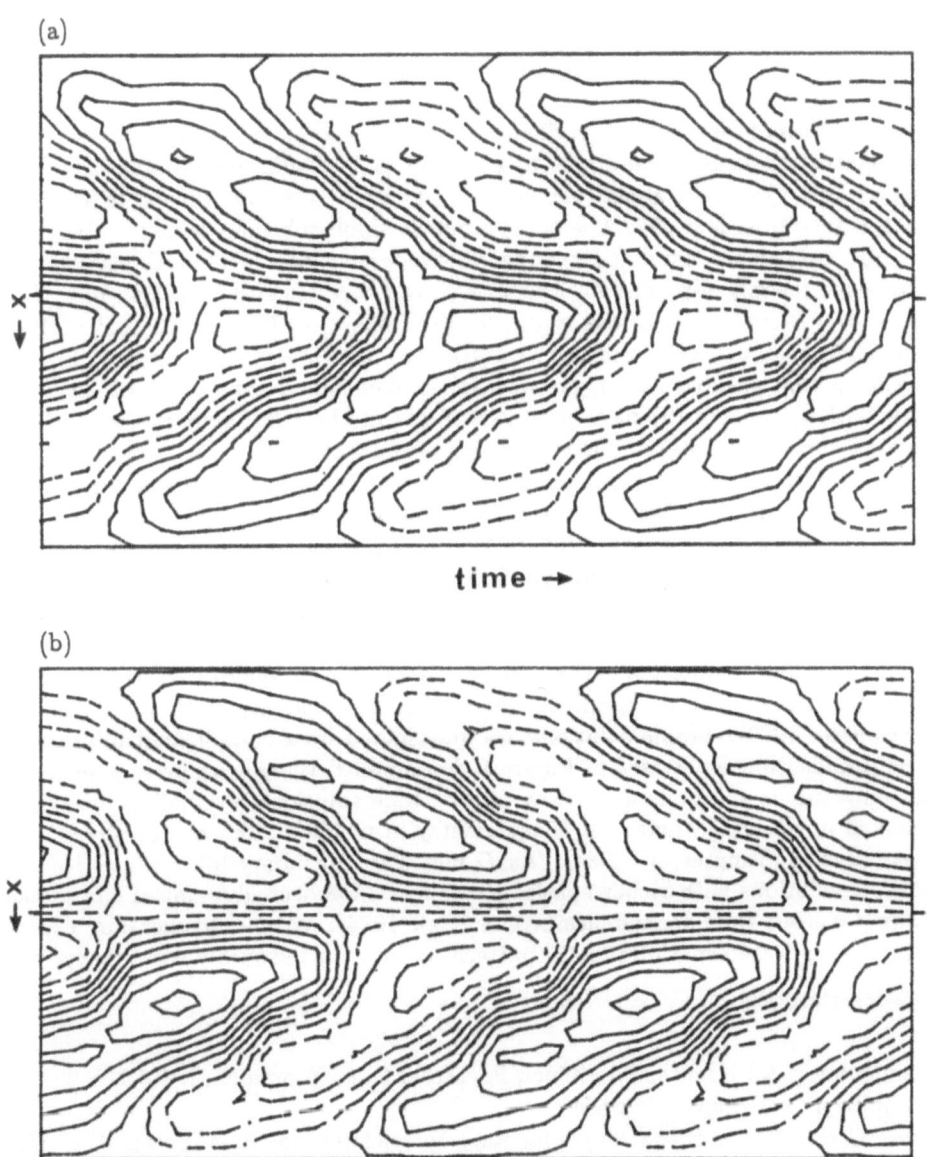

time →

(b)

time →

Figure 2. Butterfly diagrams with contours of the toroidal field $B(x,t)$ for mixed-mode periodic solutions. Full (broken) contours indicate positive (negative) values of B. Both solutions obtained with a total of 14 modes at $D = 2700$, when symmetry (a) is stable and symmetry (b) is unstable.

Nonlinear behaviour can be followed by expanding A and B as truncated Fourier series (with a total of 30 modes). Steady or periodic solutions can then be found numerically both when they are stable and when they are unstable. Figure 1 shows the bifurcation structure when $\kappa = \lambda$ and $\tau = 0$. Stability is transferred from steady quadrupole solutions to a branch of periodic mixed-mode solutions (lacking any symmetry about the equator) at $D \approx 134$ and thence to periodic dipole solutions at $D = 157$. The antisymmetric dipole solutions remain stable until $D \approx 2580$ when they lose stability to mixed-mode solutions. After such a secondary bifurcation the mixed-mode solutions with period P retain either the symmetry (a): $B(x,t) = -B(x, t + \frac{1}{2}P)$ or the symmetry (b): $B(x,t) = B(\pi - x, t + \frac{1}{2}P)$. These symmetries may be lost in a tertiary bifurcation. Solutions with symmetry (a) only have fields that are consistently stronger in one hemisphere than in the other, while those with symmetry (b) have persistent fields of the same sign at the equator. Figure 2 shows examples of butterfly diagrams for both cases.

The branch of oscillatory quadrupole solutions is initially unstable but finally gains stability at $D \approx 2335$. For a range of parameter values there are then two stable periodic solutions with different basins of attraction as well as a number of unstable solutions with different symmetries. At larger values of D there are still more bifurcations whose details depend on the number of terms included in the series.

These results demonstrate that linear theory gives a qualitatively misleading picture of behaviour in the nonlinear regime (cf. Brandenburg et al. 1989; Schmitt & Schüssler 1989). Stable solutions are determined by a complicated bifurcation structure like that illustrated in Figure 1. For small D there exist pure quadrupole and pure dipole solutions as well as mixed-mode solutions resembling behaviour observed in the sun, where a 10% asymmetry in magnetic flux has persisted over several cycles (Tang et al. 1984). As D increases subsidiary bifurcations lead to multiple solutions with richer spatial structure. Moreover, dynamical effects can produce chaotic time dependence (Weiss et al. 1984). We should therefore expect to find complicated spatiotemporal patterns of activity in rapidly rotating stars.

REFERENCES

Brandenburg, A., Krause, F., Meinel, R., Moss, D. & Tuominen, I. 1989. Astron. Astrophys., in press.

Jones, C.A. 1983. In Stellar and Planetary Magnetism (ed. A.M. Soward), p. 159. Gordon & Breach, London.

Parker, E.N. 1979. Cosmical Magnetic Fields. Oxford University Press.

Schmitt, D. & Schüssler, M. 1989. Astron. Astrophys., submitted.

Stix, M. 1972. Astron. Astrophys. 20, 9.

Tang, F., Howard, R. & Adkins, J.M. 1984. Solar Phys. 91, 75.

Weiss, N.O., Cattaneo, F. & Jones, C.A. 1984. Geophys. Astrophys. Fluid Dyn. 30, 305.

EXCITATION OF DYNAMO MODES

P. HOYNG
Laboratory for Space Research
Beneluxlaan 21, 3527 HS Utrecht
The Netherlands

ABSTRACT.   After the very suggestive results of the early days, the theory of the solar dynamo has now entered a period of re-evaluation. It is clear that our initial expectations have been too high. I shall review some of the recent attempts to formulate nonlinear and stochastic mode excitation theoretically. We now have evidence from synoptic observations that the solar dynamo features many periods. Periods both shorter and longer than the fundamental 22 yr cycle have been claimed. The phase stability of any of these periods is uncertain. The phase memory of the 22 yr period may be as short as ~ 10 cycles, but could also be much longer. Linear mean field theories permit only one marginally stable mode; they predict one period with an infinitely long phase memory. Attempts to explain multiperiodicity and finite phase memory effects fall in two categories:
(1). Nonlinear models. These feature a few nonlinearly coupled variables and may exhibit a multiperiodic or chaotic behaviour; (2). If the number of relevant variables is very high, then the dynamo behaves stochastically. It has been argued that this takes the form of stochastic excitation of many dynamo modes (overtones).

1. Introduction

Dynamo theory of the solar cycle has made a very rapid progress in the period, say, 1965 - 1980. Many mean field dynamo models have been constructed during that time which reproduced a number of the basic features of the large-scale solar magnetic field. A number of problems had been recognised at an early stage, too (magnitude α coefficient, role of nonlinearities, E-W orientation of bipolar sunspot groups, etc.), but there appeared to be a genuine optimism in those days that these would be solved in due course. In the meantime, (linear) mean field model building is out of fashion. And in retrospect it is clear that much of the successes of mean field theory, if very suggestive, were only apparent, and that our initial expectations had been pitched too high. As a result, dynamo theory of the solar magnetic field finds itself in a period of reappraisal. Two causes can be pinpointed for the crisis:
(1). The required increase of the angular speed Ω with decreasing r is neither found in selfconsistent simulations (Gilman and Miller 1981; Gilman 1983; Glatzmaier 1985a), nor measured by helioseismology (for a review see Harvey 1988).
(2). The validity of mean field theory is subject to a number of restrictions which have always been somewhat obscure. As a result, mean field modeling is often

J. O. Stenflo (ed.), Solar Photosphere: Structure, Convection, and Magnetic Fields, 359–374.
© *1990 by the IAU.*

handled as an exercice in applied mathematics, with little regard for the physical restrictions and selfconsistency (the 'cookbook approach').

In recent years, attention has shifted from the convection zone to the boundary layer between convection zone and radiative interior as a more likely location for the solar dynamo (Spiegel and Weiss 1980; Galloway and Weiss 1981; Spruit and Van Ballegooijen 1982). A number of problems that beset a dynamo operating in the convection zone, among which in particular the wrong rotation curve mentioned under (1), may be defused in this way, see e.g. Glatzmaier (1985b). The arguments have been reviewed by Schüssler (1984), Stix (1987) and by Gilman et al. (1989). Dziembowski et al. (1989) report an inward decrease in Ω of about 20 nHz in a 50000 km layer at $r/R = 0.73$. If confirmed, this would provide a strong support for the idea of a boundary layer dynamo.

It is not my intention to discuss in detail the relative merits of specific dynamo models (see e.g. Parker 1987). The topic of dynamo mode excitation is more closely connected with the problems indicated under the heading (2) above. Kinematic $\alpha\Omega$ dynamos are strictly periodic in time. The magnetic field behaves as the Phoenix, arising from the ashes of the previous cycle, continually rehearsing the very same act as before. Mode excitation in such dynamos is trivial. There is one eigenmode with an infinite phase memory. Many authors have remarked that this is unsatisfactory, and one way out is to consider nonlinear theory. Another possibility is to no longer neglect the 'rest terms' in the dynamo equation, which act as random forcing terms. In this way, too, other modes may be excited. This topic is closely related to the question whether $<\mathbf{B}>$ is a *two-scale* average or an *ensemble* average.

Let me give one example in support of my allegation that the restrictions of mean field theory are not always appreciated, and then move on to mode excitation proper. For locally isotropic turbulence $\mathbf{v}$, the dynamo coefficients α and β are

$$\alpha = -\tfrac{1}{3}\epsilon_{ijk} \int_0^\infty d\tau \, <v_i(t)\nabla_j v_k(t-\tau)> \simeq - \ <\mathbf{v} \cdot \nabla \times \mathbf{v}> \, \tau_c/3 \ ; \qquad (1.1)$$

$$\beta = \tfrac{1}{3} \int_0^\infty d\tau \, <v_i(t) v_i(t-\tau)> \simeq <v^2> \, \tau_c/3 \ , \qquad (1.2)$$

where τ_c is the correlation time of $\mathbf{v}$. The gradient $\nabla_j \beta$ contains two correlation functions of the type $<a_i \nabla_j b_i>$ which vanish, being isotropic tensors of rank 1. It follows that $\nabla\beta = 0$ and that a position-dependent β must necessarily be tensorial. A similar observation goes for α. The physical reason is that isotropic, inhomogeneous turbulence does not exist. The role of anisotropies in the turbulence has been emphasised by Rädler (1980; 1983); see also Moffatt (1983) and Schüssler (1984). An analysis of what errors ensue if we nevertheless employ a position-dependent scalar α and β is to my knowledge not available. A better known difficulty is that the validity of (1.1), (1.2) and the dynamo equation (2.1) below requires $v\tau_c/\lambda_c \ll 1$ (First Order Smoothing Approximation or *FOSA*; λ_c = correlation length of $\mathbf{v}$), while in reality $v\tau_c/\lambda_c \sim 1$. I shall have no opportunity to discuss attempts to go beyond *FOSA*, and refer to Stix (1987) for a review. It is conceivable that problems such as these have added to the failure of mean field models, in the sense that there may exist as yet totally unknown and unexplored corners in parameter space.

2. Mode Excitation in Mean Field Theory

The transport properties of the mean field $<B>$ are determined by the dynamo equation. Its basic form is (Moffatt 1978; Krause and Rädler 1980):

$$\partial_t <B> = D \cdot <B> + F ; \qquad D = \nabla \times (v_0 \times + \alpha - \beta \nabla \times) . \qquad (2.1)$$

Here, v_0 is the mean flow, and α and β are related to the statistical properties of the turbulent flow v superposed on v_0 as in (1.1) and (1.2). The meaning of $<B>$ is not clear. In the traditional *two-scale approach* (Moffatt 1978, 1983; Krause and Rädler 1980) $<B>$ is a spatial or time average over an unspecified scale in between the largest (size or period dynamo) and the smallest one (eddy size or turn-over time). The term F collects all unwanted terms due to the fact that the averaging operator $< \cdot >$ usually does not commute with ∇, ∂_t, etc. F is generally neglected on the ground that it is formally of order $\lambda_c / R \ll 1$ (R = radius of the dynamo), but there are now indications that F can be large, see sections 5.1 and 6.

Separating the time, $<B> = b \exp(\lambda t)$, we find from (2.1):

$$D \cdot b = \lambda b . \qquad (2.2)$$

Eq. (2.2) has a discrete spectrum of eigenvalues λ_n and eigenfunctions or (eigen)-modes b_n, determined by the geometry of the system and the boundary conditions via the machinery of classical mathematical physics. The solution of eq. (2.1) is

$$<B> = \sum_n c_n b_n(r) \exp(\lambda_n t) . \qquad (2.3)$$

The constants c_n are determined by the initial condition. The mode concept is somewhat ambiguous as one may always choose another basis set $\{b'_n\}$. The eigenvalue problem (2.2) has been solved for a great variety of different models (Roberts 1972; Stix 1978, 1981; Moffatt 1978; Krause and Rädler 1980; Rädler 1986a; Parker 1979, 1987 and references therein). For an infinite, homogeneous dynamo

$$\text{Re} \lambda \simeq (\alpha k \Delta \Omega / 2)^{1/2} - \beta k^2 , \qquad (2.4)$$

$$\text{Im} \lambda \simeq (\alpha k \Delta \Omega / 2)^{1/2} . \qquad (2.5)$$

The most important change for a finite, inhomogeneous dynamo is that the wave vector k is quantised and approximately equal to $2\pi/$(wavelength eigenmode). Furthermore, α and β are understood to be typical values and $\Delta\Omega = |\nabla v_0| \sim$ difference in angular speed in the convection zone; $\Delta\Omega \gg \alpha k$ is assumed ($\alpha\Omega$ approximation). According to (2.4) differential rotation and α-effect amplify, while turbulent mixing (β) damps the mode. $\text{Re} \lambda$ is a decreasing series, see Fig. 1; its relative position with respect to $\text{Re} \lambda = 0$ is determined by the dynamo number D:

$$D = \alpha \Delta \Omega R_\odot^3 / 2\beta^2 . \qquad (2.6)$$

$\text{Re} \lambda = 0$ implies $(kR_\odot)^3 \simeq D$. Modes with smaller (larger) k are unstable (damped); the smallest k is about $2\pi/R_\odot$. The critical value of D is therefore of

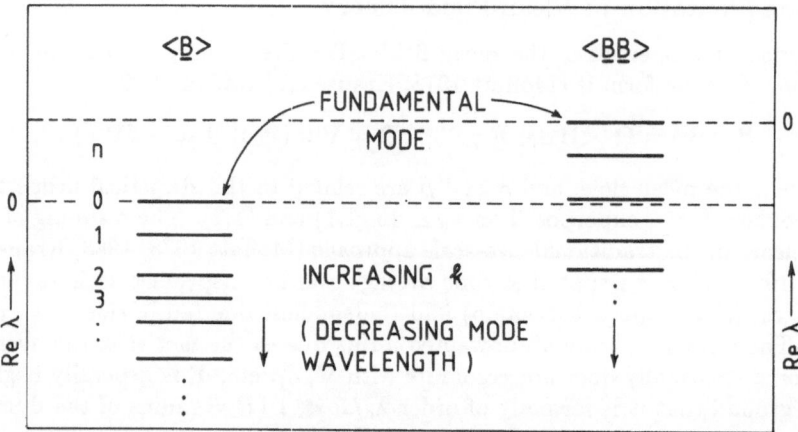

FIG. 1. Growth rates of the eigenmodes b_n of $<B>$. Increasing n means larger k or smaller 'wavelength'. The zero level is determined by D. When $D=D_c$, the largest growth rate of $<B>$ is zero (marginally stable dynamo). However, $<BB>$ still has a linear instability (section 5.1). Right vertical axis: $<BB>$ can be made marginally stable by decreasing D (typically by a factor of order 1) so that $D<D_c$. In practice this means that β is increased, as $\alpha\Delta\Omega$ is fixed by (2.5). (after Hoyng 1988).

the order of $D_c \simeq (2\pi)^3$. For simplicity I ignore the negative branch of D. Just as in any system described by a differential equation, also here two *mode excitation mechanisms* may be distinguished:

(a). Instability $(D \geq D_c)$. One or more modes have $\text{Re }\lambda > 0$. These then grow spontaneously from noise, until the validity of (2.1) breaks down. An idea often invoked in the context of mean field theories is that 'nonlinear effects' change the parameters until $D = D_c$ (e.g. B may reduce the helicity and hence α). The fundamental mode is then *marginally stable* and may be excited at a constant amplitude. The period is $P = 2\pi/\text{Im }\lambda \sim (4\pi R_\odot/\alpha\Delta\Omega)^{1/2}$. Hence, $\alpha\Delta\Omega$ is fixed by requiring $P = 22$ yr, and β follows from $\text{Re }\lambda = 0$. Overtones have shorter periods and are not excited as they have $\text{Re }\lambda < 0$. Such a dynamo would have a single frequency and an infinite quality factor $Q = \omega/\Delta\omega$, as the frequency uncertainty $\Delta\omega$ is zero. However, the fine tuning of D need not be stable. Many authors assume that $D > D_c$, so that overtones are excited. Complicated nonlinear interactions between modes and the velocity field may occur which restrict the amplitudes. In that case the dynamo has several frequencies, each with a finite Q.

(b). External forcing $(D < D_c)$. In that case all $\text{Re }\lambda < 0$ so that according to (2.3), $<B> \to 0$. Subcritical mode excitation may occur if F in (2.1) is sufficiently large. F depends on **v** and thus has in principle the character of a random forcing term. Writing $\mathbf{F} = \Sigma f_n(t)\,\mathbf{b}_n$ and $<B> = \Sigma\,\phi_k(t)\,\mathbf{b}_k$, and supposing for simplicity the eigenfunctions $\mathbf{b}_n$ to be orthonormal, we find from (2.1) and (2.2):

$$\dot{\phi}_k = \lambda_k \phi_k + f_k \quad \rightarrow \quad <\text{B}> = \sum_k \mathbf{b}_k \int_0^\infty d\tau \, \exp(\lambda_k \tau) f_k(t-\tau) \, . \qquad (2.7)$$

Hence those modes for which $f_k \neq 0$ (possibly *all* modes) are excited and each would have a finite Q (finite phase memory). Note that $D < D_c$ provides merely a *lower limit* for β; the numerical value of β is no longer fixed. Note further that $D < D_c$ does not imply that nonlinear interactions are unimportant.

Another external forcing mechanism would be a relic field in the radiative core, which imposes a nonzero boundary condition on eq. (2.1) at the base of the convection zone. This boundary condition could be periodic, in which case the phase stability of the dynamo at that frequency may be very high.

3. Observations

Many solar parameters are now known to vary with the magnetic cycle: luminosity (Hudson 1988; Willson and Hudson 1988), the differential rotation (torsional oscillations; Howard 1984), modulations in the meridional circulation (Ribes and Laclare 1988), the size distribution of surface magnetic flux (Zwaan 1987), and maybe the frequencies of solar oscillations (Gough 1988a; Gelly *et al.* 1988). I shall restrict myself below to observations which pertain directly to dynamo mode excitation.

3.1. VARIATIONS IN THE 22-YEAR PERIOD

The 22 yr magnetic cycle shows fairly large variations in the period lengths. From the epochs of sunspot extrema (Allen 1973) one finds $\delta P_{rms}/P \sim 0.1$, where $P =$ mean half cycle period. Are subsequent period variations δP independent, or is the cycle rather a passive and noisy reflection of a high Q oscillator in the solar core? In the former case there would be a progressive loss of phase memory; in the latter case the phase is locked and never far away from the phase of the core oscillation. Little progress has been made on this old question. Yule (1927) suggested that the phase memory is finite. Dicke (1978) analysed the epochs of sunspot number extrema and concluded that the data indicate phase locking. In a recent study, Dicke (1988) upheld his position. He also found that the transit time for magnetic flux through the convection zone is ~ 12 yr, which is indeed of the order of the turbulent diffusion time d^2/β over the depth d of the convection zone. However, Gough (1978, 1981, 1987) made a very similar analysis of the same data and concluded he could not decide either way. If anything, he found a preference for a random walk in phase. Whitehouse (1985) argues that both models are too simple, as he finds evidence for *systematic* variations in the cycle period.

Barnes *et al.* (1980) have simulated yearly sunspot numbers from narrowband Gaussian noise and obtained a remarkable similarity with the true cycle. On a longer timescale they also see protracted sunspot minima as during the Maunder Minimum (Eddy 1976, 1983). It appears that the data set (AD 1610 - present) is not sufficient to draw a conclusion regarding the phase stability. Unfortunately (for solar physics), the Precambrian Elatina sediment data are no longer believed to contain an extended solar cycle chronology (Williams 1981, 1985; Sonett and

Williams 1987), but are now attributed to *lunar-induced* variations in tidal deposits (Williams 1989; Sonett *et al.* 1988). The phase stability of the 22 yr cycle may ultimately be determined by a careful analysis of the ^{14}C data and of the ^{10}Be record discovered in Greenlandic ice cores (Beer *et al.* 1988).

Advanced techniques exist today to determine whether an irregular time series is stochastic or contains a chaotic attractor of low dimension (Schuster 1988; Atmanspacher *et al.* 1988). For example, Voges *et al.* (1987) have analysed the X-ray variability of *Her X-1* and found an attractor of dimension $2 < D_2 < 3$ in the lightcurve. This suggests that the underlying accretion process may be modeled with only 3 nonlinearly coupled variables (it does not tell which, though). Attractor dimensions have been determined from the sunspot and ^{14}C records (Gissatullina *et al.* 1989; Ostryakov and Usoskin 1989) but these are not yet very reliable. The problem is again the length and quality of the data set (Smith 1988).

3.2. OTHER PERIODICITIES

The sunspot data suggest a ~ 90 year amplitude modulation (Cohen and Lintz 1974). Tree-ring ^{14}C data and ^{10}Be records indicate irregular modulations in the level of solar activity with a typical timescale of a few hundred years (Sonett 1983a; Stuiver and Braziunas 1988; Raisbeck and Yiou 1988). Interesting new results have been reported on short periods. Stenflo and Vogel (1986) and Stenflo and Weisenhorn (1987) have decomposed the radial surface field B_r in spherical harmonics using 25 years of synoptic daily magnetograph data of Mt. Wilson and Kitt Peak. The advantage of this technique is that the results permit a direct comparison with theoretical predictions. The power spectra of the axisymmetric spherical harmonic coefficients ($m = 0$), Fig. 2, show a decoupling between even and odd l. The 22 yr dynamo wave is apparently a linear combination of odd-l components. The absence of higher harmonics indicates that the (surface effect of the) wave is almost sinusoidal in time. Decoupling between even and odd l occurs when α is an odd function of colatitude θ; often $\alpha \propto \cos \theta$ is assumed (Moffatt 1978). The data thus confirm this to be basically correct. Further confirmation might come from comparing the observed amplitudes and phases with theoretical odd-l expansions of the fundamental mode. An unexplained feature is that the power distribution becomes abruptly irregular above $l = 14$. Stenflo and Vogel (1986) suggest that the power at even l in Fig. 2 indicates a resonant modal structure reminiscent of solar p-mode ridges. Hoyng (1987b, 1988) has interpreted this in terms of excitation of dynamo wave overtones. Taking $k \sim l/R_\odot$ one finds with (2.5):

$$\omega \sim (\alpha \Delta\Omega / 2R_\odot)^{1/2} \, l^{1/2} \, , \qquad (3.1)$$

which describes the power ridge in Fig. 2 fairly well for $\alpha \Delta\Omega \sim 1.7 \times 10^{-4}$ cm s^{-2}. This is a rather large value and one might therefore speculate that the power at even l reflects dynamo waves of a *diffuse* dynamo in the convection zone (where α is large), as opposed to dynamo waves of a *boundary layer* dynamo which we would then see at odd l and in the butterfly diagram. Such an idea had been advanced

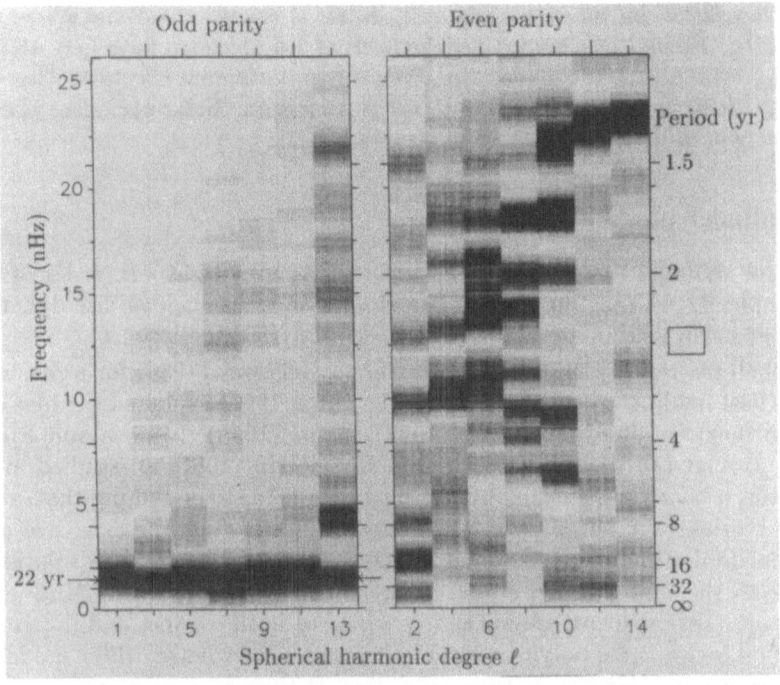

FIG. 2. Power spectrum of the axisymmetric component of the solar radial surface field. Each column is independenty normalised; the average power in even l is about 5 times larger than in odd l. (from Stenflo and Weisenhorn 1987).

earlier by Spruit et al. (1987); see also Golub et al. (1981) and Durney (1988). Stenflo and Güdel (1988) report that there is in fact weak power at *odd l* which interpolates fairly well with the power at even l in Fig. 2, albeit at a systematically lower frequency. Recently, Gokhale and Javaraiah (1989a,b) have confirmed the conclusions of Stenflo and coworkers using magnetic cycle data simulated from the Greenwich sunspot data, which cover a much longer period.

The idea that the magnetic sector structure and coronal holes (Zirker 1977; Hundhausen 1977) may correspond to nonaxisymmetric dynamo modes is relatively old (Stix 1971). These modes, being overtones, are difficult to excite in the linear theory (Stix 1971; Rädler 1986b; Ruzmaikin et al. 1988), but nonlinearities or external forcing can in principle do the job. Other lower main sequence stars are now believed to possess cycles analogous to the solar 11 yr cycle (Vaughan 1983; Baliunas and Vaughan 1985), with periods ranging from 2.6 yr to $\gtrsim$ 20 yr. Very active stars may have more than one period or behave erratically.

In summary, we have several indications that the solar dynamo is multiperiodic. In principle, both nonlinear effects and stochastic excitation may account for that. The phase stability of the 22 yr period is not well known. Its quality factor may be as small as $Q = P/\delta P_{rms} \sim 10$, but could also be much higher. That would be a strong indication for an oscillating core. Little is known about the phase stability at the other frequencies, except near periods of 1.5 yr which have $Q \sim 10$ (Fig. 2). The long-term amplitude modulations might be interference effects of other dynamo modes with periods near 22 yr, but this is uncertain. Solar cycle-like phenomena are clearly seen in other lower main sequence stars.

4. Nonlinear mode excitation

Nonlinear dynamo theory is a large topic in its own right. In order of increasing complexity we may distinguish (a) 'simple' models, such as disc dynamos, (b) nonlinear mean field models, and (c) numerical *MHD* simulations.

I shall restrict myself to (b), and refer to Galloway (1986) for a review of (c). The earliest nonlinear mean field model (Leighton 1969) exploited the idea of losses through magnetic buoyancy. Stix (1972), Jepps (1975), Ivanova and Ruzmaikin (1977), Bräuer (1979) and Kleeorin and Ruzmaikin (1981) all studied models in which the α-effect is quenched by the magnetic field, by assuming that α in (1.1) is a decreasing function of $|<B>|$. These studies showed the existence of stable nonlinear (anharmonic) oscillations for $D > D_c$. At $D = D_c$ the solution bifurcates from the null solution either supercritically or subcritically (Bräuer 1980). Yoshimura (1978a) investigated the quenching of both α-effect and differential rotation and found again only one period. It is not clear whether these models admit indeed only one period since the exploration of their parameter spaces may have been rather incomplete. Yoshimura (1978b) obtained the first *multimodal* mean field model by introducing a time delay in the nonlinear coupling. A single delay time produced a regular modulation of the 22 yr cycle, and with two or more different delay times a more chaotic long-term modulation resulted. The reason for the delay is that the Lorentz force produces an *acceleration* and it will take some time before v and v_0, on which all field regeneration depends, have changed. A time delay therefore effectively increases the order of the equations (= number of independent variables). The early nonlinear work has been reviewed by Stix (1981).

Mean field theory got itself entangled in the net of nonlinear dynamics after Zeldovich and Ruzmaikin had shown how an axisymmetric $\alpha\Omega$ dynamo with nonlinear α-effect quenching can be crudely modeled by the Lorenz equations (Ruzmaikin 1981; Jones 1983; see also Krause and Roberts 1981). These equations are well-known for possessing chaotic solutions (Martens 1984; Schuster 1988). The solution is quasiperiodic but occasionally the trajectory in phase space lingers for a long time near the origin. Accordingly, it was suggested that the Maunder minimum and other 'grand minima' might correspond to a strange attractor (Ruzmaikin 1981), as Yoshimura (1978b) had done implicitly before. This idea was elaborated by Weiss et al. (1984) who included buoyant losses and quenching of differential rotation in addition to α-effect suppression. They derived a 7th order system of equations with two field variables and two variables ω_0 and ω representing the constant part

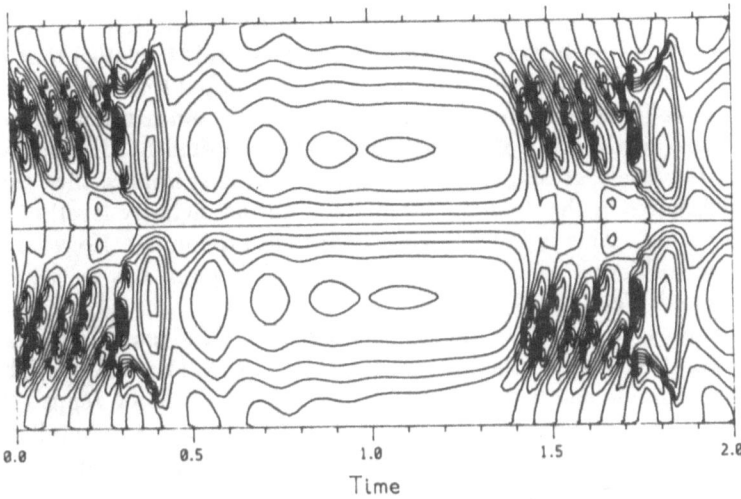

Time

FIG. 3. Butterfly diagram of the nonlinear dynamo model of Belvedere *et al.* (1989). On the horizontal axis time in units of R^2/β (154 year). The magnetic Prandtl number is 0.1 and $D/D_c = 21.3$.

of the differential rotation and the part $\propto \exp(2ikx)$, respectively. Both ω_0 and ω have phenomenological damping coefficients ν_0 and ν. Weiss *et al.* (1984) obtained their most interesting results for the 6th order system with $\nu_0 = \infty$ ($\omega_0 = 0$), and buoyant losses and α-effect quenching switched off. They then observe a series of bifurcations as D increases from D_c, and at each a new frequency appears in the solution. Beyond $D/D_c = 3.84$ the solution is chaotic (for $\nu/\beta k^2 = 0.5$), featuring irregularly modulated cycles, and episodes with almost zero field. For $D/D_c = 4$ the interval between such 'grand minima' is about 11.5 average cycle periods.

Model equations such as those of Ruzmaikin (1981) and Weiss *et al.* (1984) can merely give some indication of the behaviour of real dynamos, since the spatial structure has been severely truncated. As Weiss *et al.* realised, they have only marginally succeeded in this regard. Chaos and 'grand minima' appear only for $\nu \ll \nu_0$ while physically $\nu > \nu_0$ is expected. Moreover, the more complete 7th order system exhibits just aperiodic oscillations without protracted minima. In hindsight it is tempting to regard such studies as a concession to fashion. In those days every respectable field of research had to have its own strange attractor. However, it took almost 5 years before more complete studies were carried out.

Deluca and Gilman (1986, 1988) formulated the first mean field boundary layer dynamo, featuring a selfconsistent mean flow v_0, but phenomenological α-effect quenching and flux losses. They found steady and time-varying solutions (depending very sensitively on the parameter values), but no periodic dynamo waves. They concluded that the magnetic field is unable to generate sufficient shear to drive an

$\alpha\Omega$ dynamo. Schmitt and Schüssler (1989) studied a mean field boundary layer dynamo with either α-effect suppression or flux loss, and concluded that the latter is likely to be the more important of the two. They noted that the mode in which a dynamo finds itself may well depend on its history, since there are sometimes several possible modes for a given D. This is also reported by Brandenburg et al. (1989) and Belvedere et al. (1989), who argue that this implies that there could be large differences in the dynamo parameters (period, activity level) of otherwise very similar stars. This is indeed observed in late type stars. However, it is not clear if these coexisting solutions are all stable. This point has been investigated by Jennings and Weiss (1989).

Brandenburg et al. (1989) analysed a mean field $\alpha\Omega$ dynamo in the convection zone, incorporating only α-effect quenching, $\alpha = \alpha_0 \cos\theta/\{1 + <B>^2\}$. They found periodic single mode solutions, but unlike the earlier works of Jepps (1975) and Yoshimura (1978a) which contain much the same physics, an overall amplitude modulation was observed in certain narrow intervals of α_0. The field structure is then largely dipolar, but changes into a quadrupole when the amplitude is minimal. Belvedere et al. (1989) extended the work of Weiss et al. (1984) and investigated a mean field dynamo in which the only nonlinearity is a selfconsistent variation in the mean flow on top of a given differential rotation. The equations are truncated in r, but full θ dependence is retained. The ensuing gain in computing time enabled a systematic investigation of the transition between the various types of solution. For $D/D_c \lesssim 4.3$ there is a singly periodic solution, for $D/D_c \gtrsim 4.3$ multiperiodic solutions appear on stage, while for $D/D_c \gtrsim 8.5$ the solution features 'relatively long periods of stasis, interrupted by interludes of cyclical behaviour', see Fig. 3.

It is clear that nonlinear mean field theory is still in a very early stage of development. The various studies do not yet show a clear pattern of commonality. It is not known which of the nonlinearities are the more important ones, and the solutions are very sensitive to the functional form of the nonlinear coupling and/or the values of parameters. The main defect is that the nonlinearities are introduced phenomenologically, since dynamically consistent expressions for their dependence on $<B>$ are not available. Investigations to derive these functional forms from first principles are presently of the utmost importance (cf. Malkus and Proctor 1975; Moffatt 1978; Zeldovich et al. 1983). The following illustrates some of the problems lying ahead. All authors considering a selfconsistent mean flow v_0 write the mean Lorentz force as $<f_i> = \{(\nabla \times <B>) \times <B>\}_i/4\pi$, while actually

$$<f_i> = <(\nabla \times B) \times B>_i /4\pi = \nabla_j(2T_{ij} - T_{kk}\delta_{ij})/8\pi . \qquad (4.1)$$

Hence, $<f_i>$ depends on a higher average $T_{ij} = <B_iB_j>$. Both expressions coincide if the fluctuating field $B' = B - <B>$ is small, but that is not the case in the Sun. Only when hurdles such as these have been taken, may we begin to investigate in detail the relations between truncated mode equations and the full nonlinear partial differential equations, as has been done for example by Moore et al. (1983) and Knobloch et al. (1986) in their study of two-dimensional thermosolutal convection.

5. Mode excitation by external forcing

External forcing comes in two varieties: through boundary conditions or fluctuations in the turbulent convection. I shall briefly discuss the former and then deal with the second topic. It has long been speculated that the radiative core of the Sun may contain a relic magnetic field. If such a field exists, it imposes a boundary condition $\mathbf{B} = \mathbf{B}_{core}$ at the base of the convection zone. The field will be smeared out by the turbulent dynamo, but there will be a net polarity and intensity asymmetry in the activity cycle: alternate halves of the full 22 yr cycle have different amplitudes. Sonett (1983b) analysed the sunspot record from this point of view and found a $\sim$ 4% intensity asymmetry. If caused by a fossil dipole field, its strength was estimated to be less than 0.6 G. Levy and Boyer (1982) and Boyer and Levy (1984) analysed kinematic mean field models with a dipolar or quadrupolar field imposed at the base of the convection zone. The dynamo reduces the externally visible magnetic moment by about a factor 3. They also concluded that the poloidal component of the relic field can be no more than a few gauss. Pudovkin and Benevolenska (1985) attribute the occurrence of 'grand minima' to a *periodic* $\mathbf{B}_{core}$ at the bottom of the convection zone ($P = 180$ yr).

Piddington (1971, 1976) and Layzer *et al.* (1979) have sketched oscillator theories for the solar cycle. Piddington visualizes a 22 yr torsional oscillation of the entire core with the poloidal component of the field acting as a spring, while Layzer *et al.* suggest that the oscillation is restricted to the transition layer between core and convection zone. A 22 yr period requires fields of $\sim$ 100 G. Unfortunately, an acceptable model has never been presented (for a detailed critique I refer to Cowling 1981), and it could be worthwhile to try and see if this can be done. It is after all not excluded that the entire Sun performs a 22 yr oscillation of which the magnetic cycle is only one manifestation (Gough 1988b). The dynamo in the convection zone could then be locked in to this oscillation (*e.g.* by a boundary condition) and produce a somewhat noisy surface effect which has the same Q as the driver.

5.1. STOCHASTIC EXCITATION

External forcing by fluctuations in the turbulence has received minimal attention. Parker (1969) and Levy (1972) suggest that a *sudden burst* in the cyclonic convection causes a jump in α, and showed that this may induce a reversal of the (mean) geomagnetic field. This explanation is essentially ad hoc, and the conventional wisdom of today is that such things as reversals in α^2 dynamos and aperiodicity in $\alpha\Omega$ dynamos are caused by nonlinearities. However, even in linear theory the periods of adjacent cycles of an $\alpha\Omega$ dynamo must differ, as the realisation of the turbulence is different. It should be possible to formulate linear mean field theory in the same way as a scalar diffusion process, with a continuous loss of memory.

This problem has been taken up by Hoyng (1987a,b). To be able to evaluate the effect of the fluctuations hidden in $\mathbf{F}$ in eq. (2.1) he interpretes $< \cdot >$ as an ensemble average. In that case $\mathbf{F} = 0$, since ∂_t, ∇, etc., commute *exactly* with $< \cdot >$.† The ensemble average is best understood literally, as an average over many copy systems, e.g. $<q>_{\mathbf{r},t} = \lim_N \{\sum_i q_i(\mathbf{r},t)\}/N$, where q is an arbitrary physical quantity, cf. Krause (1976). Fluctuations now cause *phase mixing*: **B** evolves independently in each system and, in the case of the Sun, the 'copy suns' will distribute themselves evenly over the magnetic cycle (after some time), whence $<\mathbf{B}> \to 0$. Hoyng concludes that (1) only a damped solution of (2.1) is physically meaningful, and (2) the ensemble average $<\mathbf{B}>$ has no relation with observable fields; in particular, it is not equal to the large-scale field. Accepting a damped solution for $<\mathbf{B}>$ implies decreasing D so that $D < D_c$. The freedom in β is used to remove the linear instability of $<\mathbf{BB}>$ (Parker 1979, §17.6), see Fig. 1. The damping time of mode $\mathbf{b}_n$ is interpreted as a measure of its phase stability. For the dipole mode of a simple α^2 dynamo with constant α and β this time is only $\sim 0.15\, R^2/\beta$, less than the timescale for turbulent diffusion through the sphere! It is argued that the dipole component of **B** must rapidly wander over the sphere, contrary to the traditional view which suggests that this α^2 dynamo possesses a constant field (Krause and Rädler 1980, Ch. 14).

Since $<\mathbf{B}>$ is no longer indicative for the field **B** of the dynamo, Hoyng (1988) proposes to expand **B** in terms of the eigenfunctions $\mathbf{b}_n$:

$$\mathbf{B} = \sum_n c_n(t)\,\mathbf{b}_n(\mathbf{r}) \,. \tag{5.1}$$

The coefficients $c_n(t)$ determine the evolution of large scale fields (small n) as well as small scale fields (n large). For an $\alpha\Omega$ dynamo like the Sun, each $c_n(t)$ turns out to be a quasi-periodic random function whose mean period and coherence time are roughly given by the eigenvalue λ_n of $\mathbf{b}_n$. The power spectrum of $c_n(t)$ is:

$$P_n(\omega) \sim \frac{\delta_n}{\pi} <|c_n|^2> \left[\frac{1}{(\omega - \omega'_n)^2 + \delta_n^2} + \frac{1}{(\omega + \omega'_n)^2 + \delta_n^2} \right] \tag{5.2}$$

with

$$\omega'_n \sim \omega_n \equiv \mathrm{Im}\,\lambda_n \;;\quad \delta_n \sim -\,\mathrm{Re}\,\lambda_n \,.$$

Hence, all eigenmodes turn out to be excited, with a finite frequency stability $Q_n = \omega'_n/\Delta\omega'_n \sim \omega_n/\delta_n$, cf. section 2 and Fig. 4. The frequencies ω'_n are shifted from their unperturbed position ω_n because the modes are driven. At present only an estimate of the frequency stability of the 22 yr cycle is available, $Q \sim 1$ (Hoyng 1987b), which is so small that there would be hardly a periodic cycle left.

The theory is currently being applied to the solar dynamo and it seems too early to judge its merits before that has been done. A salient feature is the potentially very drastic influence of the fluctuations. If this turns out to be true then one may have to look for nonlinear effects (yet to be included) as a *stabilising* factor, the reverse of their traditional role. Another feature is that the eigenfunctions $\mathbf{b}_n$ have lost their pretension of representing the dynamo field; they are just a (very handy)

---

† This is also true if $<\cdot>$ is a longitudinal average (Braginskii 1965a,b). I conjecture that in this case the neglect of fluctuations appears in the same way as it does for ensemble averages: a nonzero and finite $<\mathbf{B}>$ *implies* $<\mathbf{BB}> \to \infty$ (Hoyng 1987b).

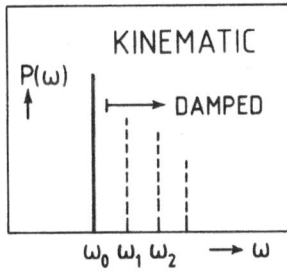 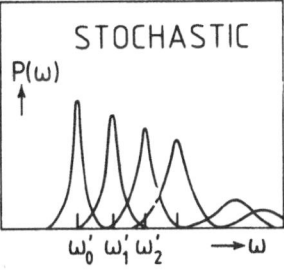 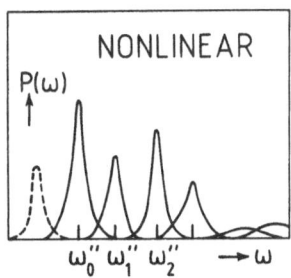

FIG. 4. Left: a marginally stable linear $\alpha\Omega$ dynamo has $D = D_c$ and possesses one frequency and $Q = \infty$. Middle: according to the linear stochastic theory $D < D_c$ and all eigenmodes are excited, each with a finite Q; the frequencies ω'_n are shifted from their unperturbed positions ω_n. Right: inclusion of nonlinear effects further changes the line shapes and positions, and very low frequencies may also appear.

set of basis functions. It remains necessary to identify ensemble averages such as $<c_n(t)>$, $<|c_n(t)|^2>$, with time averages $\overline{c_n(t)}$, $\overline{|c_n(t)|^2}$, but since only scalars are involved depending on time, this is far less controversial than for $\mathbf{B}$.

6. Discussion

In Fig. 4, I have summarized my view on the evolution of mean field dynamo theory. In the early days, the notion of marginal stability ($D = D_c$) was implicitly accepted, for lack of better. Next came nonlinear dynamos operating at $D > D_c$. These feature in principle many nonlinearly interacting magnetic and fluid modes. It has been speculated that their large-scale dynamics can be described adequately with very few modes only and under those circumstances a low dimensional strange attractor might emerge. But if and when this occurs is unknown, also from the point of view of observations. Theoretically, the main obstacle is the lack of selfconsistent nonlinearities. I have pointed out an additional problem, that $<\mathbf{BB}>$ diverges due to the neglect of fluctuations. In the stochastic excitation picture the dynamo would operate subcritically ($D < D_c$), and $<\mathbf{BB}>$ no longer diverges. Many modes are excited and their evolution is coupled. The dimension of the relevant part of phase space (and of attractors) is very high. The situation is analogous to the classical (linear) theory of Brownian motion, and the theory may indeed be regarded as a linear theory for the random walk of an advected vector field. Of course, this idea must still be tested in real dynamo models. Also nonlinear effects remain to be included, and here we face again the problem of selfconsistency.

In moments of complacency one might be inclined to think that dynamo theory has come a long way since the idea of a hydromagnetic dynamo first arose (Larmor 1919) and the pioneering contributions of Cowling (1934), Parker (1955) and Steenbeck, Krause and Rädler (Roberts and Stix 1971). But in fact I don't think it has. I believe we still have a much longer way to go.

Acknowledgements

I am indebted to many colleagues for their help, critical comments and for sending me their recent work: A.A. van Ballegooijen, G. Belvedere, A. Brandenburg, F.H. Busse, S. Childress, A. Gailitis, J.H.G.M. van Geffen, P.A. Gilman, E. Knobloch, M. Meneguzzi, M.R.E. Proctor, K.-H. Rädler, E. Ribes, P.H. Roberts, M. Schüssler, A.M. Soward, J.O. Stenflo, M. Stix and N.O. Weiss.

References

Allen, C.W.: 1973, *Astrophysical Quantities*, Athlone Press (London).
Atmanspacher, H., Scheingraber, H. and Voges, W.: 1988, *Phys. Rev. A37*, 1314.
Baliunas, S.L. and Vaughan, A.H.: 1985, *Ann. Rev. Astron. Astrophys. 23*, 379.
Barnes, J.A., Sargent, H.H. and Tryon, P.V.: 1980, in *The Ancient Sun*, eds. R.O. Pepin, J.A. Eddy and R.B. Merrill, Pergamon Press (New York), p. 159.
Beer, J., Siegenthaler, U. and Blinov, A.: 1988, in *Secular Solar and Geomagnetic Variations in the Last 10,000 Years*, eds. F.R. Stephenson and A.W. Wolfendale, Kluwer Academic Publishers (Dordrecht), p. 297.
Belvedere, G., Pidatella, R.M. and Proctor, R.M.E.: 1989, *Geophys. Astrophys. Fluid Dynamics*, to appear.
Boyer, D.W. and Levy, E.H.: 1984, *Astrophys. J. 277*, 848.
Braginskii, S.I.: 1965, *Sov. Phys. JETP 20*, 726.
Braginskii, S.I.: 1965, *Sov. Phys. JETP 20*, 1462.
Brandenburg, A., Moss, D. and Tuominen, I.: 1989, *Geophys. Astrophys. Fluid Dynamics*, to appear.
Bräuer, H.: 1979, *Astron. Nachr. 300*, 43.
Bräuer, H.: 1980, *Astron. Nachr. 301*, 203.
Cohen, T.J. and Lintz, P.R.: 1974. *Nature 250*, 398.
Cowling, T.G.: 1934, *M.N.R.A.S. 94*, 39.
Cowling, T.G.: 1981, *Ann. Rev. Astron. Astrophys. 19*, 115.
Deluca, E.E. and Gilman, P.A.: 1986, *Geophys. Astrophys. fluid Dynamics 37*, 85.
Deluca, E.E. and Gilman, P.A.: 1988, *Geophys. Astrophys. fluid Dynamics 43*, 119.
Dicke, R.H.: 1978, *Nature 276*, 676.
Dicke, R.H.: 1988, *Solar Phys. 115*, 171.
Durney, B.R.: 1988, *Astron. Astrophys. 191*, 374.
Dziembowski, W.A., Goode, P.R. and Libbrecht, K.G.: 1989, *Astrophys. J. 337*, L53.
Eddy, J.A.: 1976, *Science 192*, 1189.
Eddy, J.A.: 1983, *Solar Phys. 89*, 195.
Galloway, D.J.: 1986, *Adv. Space Res. 6, No. 8*, 19.
Galloway, D.J. and Weiss, N.O.: 1981, *Astrophys. J. 243*, 945.
Gelly, B., Fossat, E. and Grec, G.: 1988, in *Seismology of the Sun & Sun-like Stars*, ed. E.J. Rolfe, ESA SP-286 (Noordwijk), p. 275.
Gilman, P.A.: 1983, *Astrophys. J. Suppl. 53*, 243.
Gilman, P.A. and Miller, J.: 1981, *Astrophys. J. Suppl. 46*, 211.
Gilman, P.A., Morrow, C.A. and Deluca, E.E.: 1989, *Astrophys. J. 338*, 528.
Gissatullina, S.M., Rukavishnikov, V.D., Ruzmaikin, A.A. and Tavastsherna, K.S.: 1989, *Solar Phys.*, submitted.
Glatzmaier, G.A.: 1985a, *Astrophys. J. 291*, 300.
Glatzmaier, G.A.: 1985b, *Geophys. Astrophys. Fluid Dynamics 31*, 137.
Golub, L., Rosner, R., Vaiana, G.S. and Weiss, N.O.: 1981, *Astrophys. J. 243*, 309.
Gokhale, M.H. and Javaraiah, J.: 1989a, *M.N.R.A.S.*, submitted.
Gokhale, M.H. and Javaraiah, J.: 1989b, these proceedings.

Gough, D.O.: 1978, in *Pleins Feux sur la Physique Solaire*, eds. S. Dumont and J. Rösch, Editions du CNRS (Paris), p. 81.

Gough, D.O.: 1981, in *Variations in the Solar Constant*, ed. S. Sofia, US Govt. Printing Office (Washington), p. 185.

Gough, D.O.: 1987, in *Solar-Terrestrial Relationships and the Earth Environment in the Last Millenia*, ed. G. Castalogni-Cini, Soc. Italiana di Fisica (Bologna), p. 95.

Gough, D.O.: 1988a, in *Seismology of the Sun & Sun-like Stars*, ed. E.J. Rolfe, ESA SP-286 (Noordwijk), p. 679.

Gough, D.O.: 1988b, *Nature 336*, 618.

Harvey, J.: 1988, in *Seismology of the Sun & Sun-like Stars*, ed. E.J. Rolfe, ESA SP-286 (Noordwijk), p. 55.

Howard, R.: 1884, *Ann. Rev. Astron. Astrophys. 22*, 131.

Hoyng, P.: 1987a, *Astron. Astrophys. 171*, 348.

Hoyng, P.: 1987b, *Astron. Astrophys. 171*, 357.

Hoyng, P.: 1988, *Astrophys. J. 332*, 857.

Hudson, H.S.: 1988, *Ann. Rev. Astron. Astrophys. 26*, 437.

Hundhausen, A.J.: 1977, in *Coronal Holes and High Speed Wind Streams*, ed. J.B. Zirker, Colorado Associated U.P. (Boulder), p. 225.

Ivanova, T.S. and Ruzmaikin, A.A.: 1977, *Sov. Astron. 21*, 479.

Jennings, R.L. and Weiss, N.O.: 1989, these proceedings.

Jepps, S.A.: 1975, *J. Fluid Mech. 67*, 625.

Jones, C.A.: 1983, in *Stellar and Planetary Magnetism*, ed. A.M. Soward, Gordon and Breach (New York), p. 159.

Kleeorin, N.I. and Ruzmaikin, A.A.: 1981, *Geophys. Astrophys. Fluid Dynamics 17*, 281.

Knobloch, E., Moore, D.R., Toomre, J. and Weiss, N.O.: 1986, *J. Fluid Mech. 166*, 409.

Krause, F.: 1976, in *Basic Mechanisms of Solar Activity*, eds. V. Bumba and J. Kleczec, Reidel (Dordrecht), p. 305.

Krause, F. and Rädler, K.-H.: 1980, *Mean-Field Magnetohydrodynamics and Dynamo Theory*, Pergamon Press (London).

Krause, F. and Roberts, P.H.: 1981, *Adv. Space Res. 1*, 231.

Larmor, J.: 1919, *Rep. Brit. Assoc. Adv. Sci. 1919*, 159.

Layzer, D., Rosner, R. and Doyle, H.T.: 1979, *Astrophys. J. 229*, 1126.

Levy, E.H.: 1972, *Astrophys. J. 171*, 635.

Levy, E.H. and Boyer, D.: 1982, *Astrophys. J. 254*, L19.

Malkus, W.V.R. and Proctor, M.R.E.: 1975, *J. Fluid Mech. 67*, 417.

Martens, P.C.H.: 1984, *Physics Reports 115*, 315.

Moffatt, H.K.: 1978, *Magnetic Field Generation in Electrically Conducting Fluids*, Cambridge U.P. (Cambridge).

Moffatt, H.K.: 1983, *Rep. Prog. Phys. 46*, 621.

Moore, D.R., Toomre, J., Knobloch, E. and Weiss, N.O.: 1983, *Nature 303*, 663.

Ostryakov, V.M. and Usoskin, I.G.: 1989, these proceedings.

Parker, E.N.: 1955, *Astrophys. J. 122*, 293.

Parker, E.N.: 1969, *Astrophys. J. 158*, 815.

Parker, E.N.: 1979, *Cosmical Magnetic Fields*, Clarendon Press (Oxford).

Parker, E.N.: 1987, *Solar Phys. 110*, 11.

Piddington, J.H.: 1971, *Proc. Astron. Soc. Australia 2*, 7.

Piddington, J.H.: 1976, in *Basic Mechanisms of Solar Activity*, eds. V. Bumba and J. Kleczek, Reidel (Dordrecht), p. 389.

Pudovkin, M.I. and Benevolenska, E.E.: 1985, *Solar Phys. 95*, 381.

Rädler, K.-H.: 1980, *Astron. Nachr. 301*, 101.

Rädler, K.-H.: 1983, in *Stellar and Planetary Magnetism*, ed. A.M. Soward, Gordon and Breach (New York), p. 17 and p. 37.

Rädler, K.-H.: 1986a, *Astron. Nachr. 307*, 89.

Rädler, K.-H.: 1986b, in *Plasma Astrophysics*, ESA SP-251 (Noordwijk), p. 569.

Raisbeck, G.M. and Yiou, F.: 1988, in *Secular Solar and Geomagnetic Variations in the Last 10,000 Years*, eds. F.R. Stephenson and A.W. Wolfendale, Kluwer Academic Publishers (Dordrecht), p. 287.

Ribes, E. and Laclare, F.: 1988, *Geophys. Astrophys. Fluid Dynamics 41*, 171.

Roberts, P.H. and Stix, M.: 1971, *NCAR-TN/IA-60*, (NCAR, Boulder, Colorado).

Roberts, P.H.: 1972, *Phil. Trans. Roy. Soc. Lond. A272*, 663.

Ruzmaikin, A.A.: 1981, *Comm. Astrophys. 9*, 85.

Ruzmaikin, A.A., Sokolov, D.D. and Starchenko, S.V.: 1988, *Solar Phys. 115*, 5.

Schmitt, D. and Schüssler, M.: 1989, *Astron. Astrophys.*, to appear.

Schüssler, M.: 1984, in *The Hydromagnetics of the Sun*, ESA SP-220 (Noordwijk), p. 67.

Schuster, H.G.: 1988, *Deterministic Chaos*, VHC Verlagsgesellschaft (Weinheim).

Smith, L.A.: 1988, *Phys. Lett. 133A*, 283.

Sonett, C.P.: 1983a, *Rev. Geophys. Space Phys. 22*, 239.

Sonett, C.P.: 1983b, *Nature 306*, 670.

Sonett, C.P. and Williams, G.E.: 1987, *Solar Phys. 110*, 397.

Sonett, C.P., Finney, S.A. and Williams, C.R.: 1988, *Nature 335*, 806.

Spiegel, E.A. and Weiss, N.O.: 1980, *Nature 287*, 616.

Spruit, H.C. and Van Ballegooijen, A.A.: 1982, *Astron. Astrophys. 106*, 58.

Spruit, H.C., Title, A.M. and Van Ballegooijen, A.A.: 1987, *Solar Phys. 110*, 115.

Stenflo, J.O. and Vogel, M.: 1986, *Nature 319*, 285.

Stenflo, J.O. and Weisenhorn, A.L.: 1987, *Solar Phys. 108*, 205.

Stenflo, J.O. and Güdel, M.: 1988, *Astron. Astrophys. 191*, 137.

Stix, M.: 1971, *Astron. Astrophys. 13*, 203.

Stix, M.: 1972, *Astron. Astrophys. 20*, 9.

Stix, M.: 1978, in *Pleins Feux sur la Physique Solaire*, eds. S. Dumont and J. Rösch, Editions du CNRS (Paris), p. 37.

Stix, M.: 1981, *Solar Phys. 74*, 79.

Stix, M.: 1987, in *Solar and Stellar Physics*, eds. E.H. Schröter and M. Schüssler, Springer Verlag (Berlin), p. 15.

Stuiver, M. and Braziunas, T.F.: 1988, in *Secular Solar and Geomagnetic Variations in the Last 10,000 Years*, eds. F.R. Stephenson and A.W. Wolfendale, Kluwer Academic Publishers (Dordrecht), p. 245.

Vaughan, A.H.: 1983, in *Solar and Stellar Magnetic Fields: Origins and Coronal Effects*, ed. J.O. Stenflo, Reidel (Dordrecht), p. 113.

Voges, W., Atmanspacher, H. and Scheingraber, H.: 1987, *Astrophys. J. 320*, 794.

Weiss, N.O., Cattaneo, F. and Jones, C.A.: 1984, *Geophys. Astrophys. Fluid Dynamics 30*, 305.

Whitehouse, D.R.: 1985, *Astron. Astrophys. 145*, 451.

Williams, G.E.: 1981, *Nature 291*, 624.

Williams, G.E.: 1985, *Aust. J. Phys. 38*, 1027.

Williams, G.E.: 1989, *J. Geol. Soc. London 146*, 97.

Willson, R.C. and Hudson, H.S.: 1988, *Nature 332*, 810.

Yoshimura, H.: 1978a, *Astrophys. J. 220*, 692.

Yoshimura, H.: 1978b, *Astrophys. J. 226*, 706.

Yule, G.U.: 1927, *Phil. Trans. Roy. Soc. London A226*, 267.

Zeldovich, Ya.B., Ruzmaikin, A.A. and Sokoloff, D.D.: 1983, *Magnetic Fields in Astrophysics*, Gordon and Breach (New York).

Zirker, J.B.: 1977, in *Coronal Holes and High Speed Wind Streams*, ed. J.B. Zirker, Colorado Associated U.P. (Boulder), p. 1.

Zwaan, C.: 1987, *Ann. Rev. Astron. Astrophys. 25*, 83.

STUDY OF SUN'S 'HYDROMAGNETIC' OSCILLATIONS USING SUNSPOT DATA

M.H.GOKHALE, J.JAVARAIAH and K.M.HIREMATH
Indian Institute of Astrophysics
Bangalore 560 034
India

ABSTRACT. Spherical-harmonic-fourier (SHF) analysis of sun's 'nominal toroidal magnetic field', computed using sunspot data during 1874-1976, shows that sunspot activity can be considered as possibly originating in interference of sun's axi-symmetric 'hydromagnetic' oscillations of <u>odd</u> degrees up to $\ell \approx 21$ and periods $\approx 22y$. The relative amplitudes and relative phases of these oscillations remain fairly constant, leading to butterfly diagrams with stable latitude-time correlations on all latitude scales $\gtrsim 9°$. The amplitudes and phases do however undergo slow coherent variations. The main modes ($\ell < 11$) represent an approximately standing oscillation.

The mean power spectrum can be fitted excellently to the form $a\ell^3 \exp(-b\ell)$ expected for waves trapped between the poles.

1. Introduction

In an earlier paper (Gokhale and Javaraiah, 1989) we used sunspot data during 1902-1954 to show that SHF analysis of 'sunspot occurrence probability' yields approximately same relative amplitudes and relative phases, during different sunspot cycles, for axisymmetric modes of '11y' periodicity with even degrees up to $\ell \approx 22$, in which most of the SHF power is concentrated. This was shown to be a consequence of a similar stability of relative amplitudes and relative phases of axisymmetric modes, of '22 y' period and odd degrees at least up to $\ell \approx 11$, in a 'nominal toroidal magnetic field' defined by attaching to the sunspot occurrence probability signs (+/-) according to Hale's laws of magnetic polarities. Reasons were given to believe that these modes represent slow MHD oscillations of the sun.

In this paper we present results of the SHF analysis of the 'nominal toroidal field' for the 82 intervals of 22 y length during 1874-1976, each interval being displaced by

J. O. Stenflo (ed.), Solar Photosphere: Structure, Convection, and Magnetic Fields, 375–378.
© *1990 by the IAU.*

one year with respect to the previous one. We also discuss physical meaning of the modes and of the energy spectrum.

2. Data and Method of Analysis

The heliographic colatitudes θ_i, longitudes ϕ_i, and epochs t_i (in days and fractions from 0.0 UT of January 1, 1874), for each sunspot group, on each day of its observation, were determined using relevant information from a magnetic tape of the Greenwich photo-heliographic data which was kindly provided to us by H.Balthassar.

Sunspot occurrence probability 'p' (θ,ϕ,t) during any interval (T_1,T_2) is defined (paper I) as described by a delta function at each data point (μ_i,ϕ_i,τ_i) in (μ,ϕ,τ) space, multiplied by the reciprocal of the total number of data points during (T_1,T_2), where $\mu=\cos\theta$ and $\tau=t_i/(T_2-T_1)$.

The 'nominal toroidal field' $B\varphi$ (θ,ϕ,t) was determined from $p(\theta,\phi,t)$ as in paper I(see sec.1 above).The SHF coefficients of $B\varphi(\theta,\phi,t)$ are also computed using the orthonormality of the SHF functions as described in paper I.

3. Results and Discussion

As in paper I, most of the SHF power is found to be concentrated in axisymmetric modes of odd degrees with '22y' periodicity.

The amplitudes of these modes during the 82 intervals of 22y each, (1874-1895, 1875-1896, etc.) remain in the ranges shown in Fig 1 (a). The relative phases of these modes during the 82 intervals are shown in Fig. 1(b).

These results confirm the conclusion in paper I that $p(\theta,\phi,t)$ is related to $B\varphi(\theta,\phi,t)$ in a nonlinear way as $p(\theta,\phi,t) \approx [B\varphi^2 (\theta,\phi,t)]^{1/2}$ where $B\varphi(\theta,\phi,t)$ can be considered as given by superposition of '22 y' periodic axisymmetric modes of odd degrees at least up to $\ell \approx 21$ whose relative amplitudes and relative phases remain approximately constant so as to yield stable butterfly diagrams in $p(\theta,\phi,t)$. The main modes up to $\ell \approx 9$ describe an approximately stationary oscillation, since their relative phases are near $0°$ or $180°$.

In order to illustrate the significance of the constancy of the relative amplitudes and phases we compare these with those in Figures 2 (a), 2(b) as computed from a simulated data set in which the values of t_i are retained as in the real data but the values of the latitudes $\lambda_i(=90°-\theta_i)$ are simulated as follows: $\lambda_i = (\text{random sign}) \times [\lambda_0 + \Delta\lambda_i]$, in which, during the first seven years of each sunspot cycle, λ_0 reduces linearly from $25°$ to $10°$ and $\Delta\lambda_i$ are random values between $(-9°,+9°)$. During the last four years of each cycle, λ_0 remains constant at $8°$ and $\Delta\lambda_i$ has random values between $(-6°,+6°)$.

From comparison of Figures 1 and 2 it is clear that the constancy and the values of the amplitudes and phases of the modes up to $\ell \approx 13$ can be reproduced by the simulated data. This is because the systematic variation of λ_0 truly reproduces the latitude-time correlation in the butterfly diagram on scales $\gtrsim 16°$. However in Fig.2 the power in the modes $\ell > 13$ is much smaller than that in Fig.1. Also their phase variations become increasingly irregular, till at

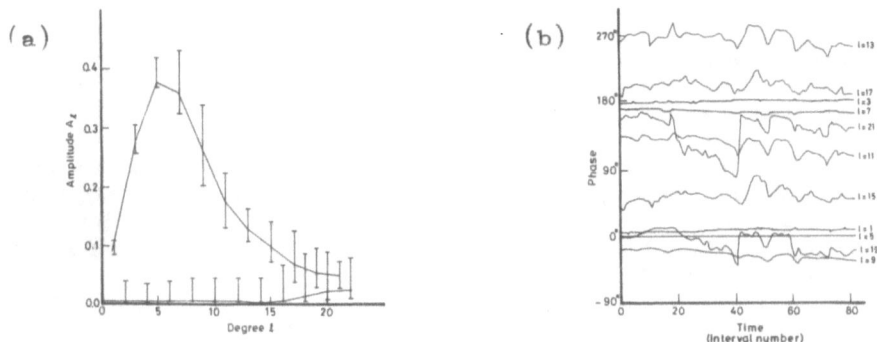

Figure 1. (a) Spectrum of relative amplitudes of the axisymmetric modes of 22y periodicity in 'Bφ' The bars represent total scatter of the 82 values corresponding to the 82 intervals. The continuous curve represents the values obtained from the whole data set of the 103 years. (b) Relative phases $(\varphi_{\bar{\ell}}\varphi_5)$.

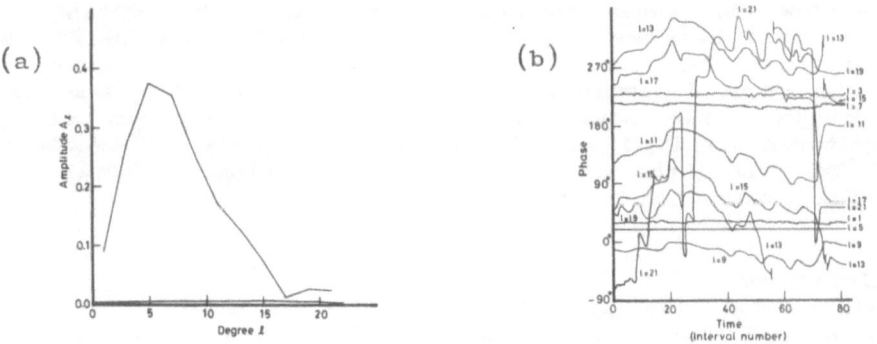

Figure 2. (a) Relative amplitudes and (b) relative phases obtained from the 'simulated' data.

ℓ=21, it becomes extremely irregular. This shows that the value of the amplitudes and the constancy of phases upto $\ell \approx 21$ cannot be obtained by a simulated data set unless it reproduces (θ-t) correlations on __all__ scales down to ~9°. Such a simulation will need ad-hoc 'systematic'assumptions just for reproducing those correlations. It is much __simpler__ and __straightforward__ to conclude that the correlations in the real θ_i and t_i result from interference of the global oscillations of $B\varphi$ with amplitudes and phases as given by the SHF analysis. (We use the word "interference" since p (θ,ϕ,t) is non-linearly related to $B\varphi$ (θ,ϕ,t)).

4. Physical significance

4.1 PHYSICAL NATURE OF THE OSCILLATIONS

As pointed out in paper I, the period indicates slow-MHD nature of the modes which requires these to be __torsional__ if the basic field is poloidal. However these oscillations may have both toroidal and meridional velocity components, constituting essentially an "MHD dynamo".

4.2 'REALITY' OF THE 'NOMINAL' TOROIDAL FIELD

The 'nominal toroidal field' need not represent the real toroidal field. Yet it seems that it does so, since its spatial and temporal frequencies, as well as the amplitude spectrum, are similar to those of the radial field derived from full disc magnetograms (Stenflo, 1988).

4.3 THE FORM OF THE ENERGY SPECTRUM

The slow MHD waves responsible for these oscillations must be trapped between the poles, with small phase shifts and dissipation. The trapping will lead to a power spectrum ~$a\ell^2$ in the small ℓ domain and possibly ~$a\ell^3 \exp(-b\ell)$in the large ℓ domain. The power spectrum given by squares of the amplitudes in Fig.1 can be fitted excellently (χ^2-confidence >99.98%) to the form $a\ell^2\exp(-b\ell)$ with a=0.0142 and b=0.552 $\pm$ 0.015.

References
Gokhale, M.H. and Javaraiah, J (1989) 'Sunspot activity as originating in interference of sun's global MHD oscillations', paper submitted to MNRAS.
Stenflo,J.O. (1988) 'Global wave patterns in the sun's magnetic field', Astrophys & Sp. Sci., 144, 321.

VARIATION OF EVEN AND ODD PARITY
IN THE SOLAR DYNAMO

A. BRANDENBURG[1], R. MEINEL[2], D. MOSS[3], I.TUOMINEN[1]
[1]Observatory and Astrophysics Laboratory, University of Helsinki
Tähtitorninmäki, SF-00130 Helsinki, Finland
[2]Sternwarte Babelsberg, Rosa-Luxemburg-Str. 17a,
1591 Potsdam, German Democratic Republic
[3]Dept. of Mathematics, The University, Manchester M13 9PL, England

ABSTRACT. We have studied axisymmetric nonlinear $\alpha\omega$-dynamo models taking the interaction between even and odd parities fully into account. It turns out that the dominating type of symmetry is not always determined uniquely, but it can vary on a very long time scale compared to the period of the magnetic cycle. In some cases the frequency of this long term variation is close to the beat frequency of the two solutions with purely dipolar and purely quadrupolar parity. The occurrence of a second frequency is typical of solutions whose trajectory describes a torus in the phase space. We argue that this finding is of relevance for understanding secular variations observed in the Sun. For example measurements of sunspots indicate that the spot number on the northern hemisphere at present exceeds the number on the southern hemisphere. The reverse seems to have been the case at the end of last century.

1. Introduction

The magnetic field orientation in sunspot pairs is quite regular. The orientation reverses between the northern and southern hemisheres (odd parity) and alternates with the 22-year solar magnetic cycle (Hale & Nicholson, 1925). Sunspot pairs are usually interpreted as an indicator for an azimuthal magnetic field in the solar convection zone. Flux ropes rise to the surface, because of magnetic buoyancy, where they can produce a pair of sunspots via the Parker (1979) mechanism.

Further, the radial component of the solar mean magnetic field B_r shows a similar systematic variation to the azimuthal field. Stenflo & Vogel (1986) and Stenflo (1988) showed that about 90% of the radial component of the magnetic field is of odd parity and oscillates with the 22-year period. 10% of the B_r-field is of even parity and does not show the 22-year period but there are variations on a much shorter time scale.

Deviation of the solar magnetic field from a pure parity is of importance for understanding the solar dynamo. Traditional spherical mean-field dynamos (Steenbeck & Krause, 1969; Krause & Rädler, 1980) show pure parity magnetic field configurations when the dynamo number is close to the critical dynamo number D_{crit}. Only the solution corresponding to the first D_{crit} is stable and therefore of relevance (Krause & Meinel, 1988).

379

J. O. Stenflo (ed.), Solar Photosphere: Structure, Convection, and Magnetic Fields, 379–382.
© *1990 by the IAU.*

Recently it has been demonstrated that for $\alpha\omega$-dynamos with weakly supercritical dynamo numbers ($D \approx 1.5 D_{crit}$) the first pure solutions are unstable and only stable *mixed parity* solutions with contributions of even and odd parity are found (Brandenburg *et al.* 1989). The parity, measured by a parity parameter

$$P = [E^{(S)} - E^{(A)}]/[E^{(S)} + E^{(A)}], \tag{1}$$

can vary quasi-periodically on a time scale which is typically 3-30 times longer than the magnetic cycle period (Brandenburg, Moss & Tuominen, 1989). $E^{(S)}$ and $E^{(A)}$ denote respectively the energies in the symmetric (even) and antisymmetric (odd) part of the magnetic field.

Oscillatory dynamos with mainly odd parity have been found which show, for approximately one third of the time, strong deviations from the odd parity, see Fig 1b. During such stages considerable contributions of even parity are present and the total magnetic field is weakened. This behavior has been tentatively associated with the phenomenon of Grand Minima (Brandenburg, Krause & Tuominen, 1989). These models would predict an enhanced degree of "parity mixing" during a Grand Minimum.

These results are based on models, in which the α- and ω-effect operate in the entire sphere. However, these conditions are not met in the Sun. The purpose of the present paper is therefore to investigate more realistic models where the α- and ω-effect are confined in a shell.

2. Dynamos in a shell

The $\alpha\omega$-dynamo can operate only in the convection zone and possibly in the thin overshoot layer below. Because of the very long diffusion time the radiative interior can be excluded from the computation of solar-type magnetic cycles. We do this by assuming a perfect conductor boundary condition at a radius $r = 0.7R$, where R is the solar radius. We assume that the field continues as a potential field for $r > R$. Inside the computational domain $0.7R < r < R$ we solve the dynamo equation (cf. Krause and Rädler, 1980)

$$\partial B/\partial t = \mathrm{curl}\,(u \times B + \alpha B - \eta\,\mathrm{curl}\,B) \tag{2}$$

in spherical coordinates (r, θ, ϕ) with $0 \leq \theta \leq \pi$, restricting ourselves, however, to axisymmetry (i.e. $\partial/\partial\phi = 0$). Inside the shell we take $u = \hat{\phi}\Omega\,r\sin\theta$ with $\partial\Omega/\partial r = C_\omega\eta/R^3 = $ const and $C_\omega = -1000$. We assume that α depends on B via $\alpha = C_\alpha\cos\theta\,(\eta/R)(1 + B^2/B_0^2)^{-1}$ with $B_0 = \sqrt{\mu\rho\eta}/R$. We find the critical dynamo numbers for $D_{crit}^{(odd)} = -5620$ and $D_{crit}^{(even)} = -5570$ for even and odd parity, where $D = C_\alpha C_\omega$. Note that, in contrast to dynamos in a complete sphere, the S-type (even) solution bifurcates first from the trivial solution. This feature has been found already by Roberts (1972). (We have investigated here only the case $D < 0$, because only then the field belts migrate toward the equator.)

The minimum and maximum energies and frequencies for both solutions are listed in the Table 1. Note that these values for both parities lie very close together. Near to the first bifurcation ($D = -5570$) the S-type solution is stable. For $D = -8000... - 10\,000$ only the A-type solution is stable. These results were obtained with a numerical resolution of 41×81 mesh points. For a lower resolution of only 21×41 mesh points we find the mixed parity

381

solution still at $D = -10^4$ (this solution is shown in Fig. 1a). Since the bifurcation points are shifted slightly as the resolution is increased we expect that, for a higher resolution, this mixed parity solution corresponds to $D \approx -6000... - 8000$.

Table 1: Energies and frequencies for even and odd parity solutions of the $\alpha\omega$-dynamo in a shell for $C_\omega = -10^3$

	E_{min}			E_{max}			frequency		
C_α	6	8	10	6	8	10	6	8	10
A	0.040	0.279	0.503	0.054	0.339	0.590	138.6	150.3	158.7
S	0.043	0.277	0.499	0.059	0.344	0.601	138.1	149.6	160.3

The temporal variation of E and P for a mixed solution is shown in the figure below (left panel). ($D = -10^4$ with 21 × 41 mesh points.) For comparison we have included in the right panel the solution for the entire sphere for $D = -8500$ and 41 × 81 mesh points (Brandenburg, Krause & Tuominen, 1989).

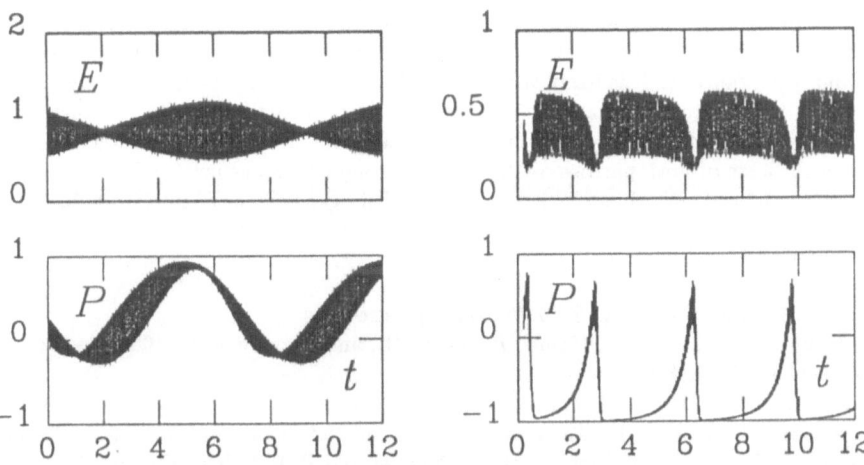

Fig. 1. *Left panel:* Evolution of energy and parity parameter for the shell dynamo and $C_\alpha = 10$. *Right panel:* Same, but for a dynamo operating in the entire sphere

The period of the mixed solutions is very long (about 100 times larger) compared to the period of the basic magnetic cycle. The long term period is in this case close to the beat frequency between the two pure (unstable) solutions. The beat frequency is in this case very low, because the frequencies of the pure solutions lie very close together.

Neglecting the fact that solar Grand Minima occur irregularly (this has been modelled by Belvedere and Proctor (1989) with a highly supercritical one dimensional dynamo) it seems that their recurrence time is not much longer than 10-30 magnetic cycle periods. It would be therefore of interest to look for models with a period ratio smaller then 100, which is the value found for shell dynamos with radially constant α- and ω-profiles.

3. Discussion and conclusions

The shell dynamos discussed in this paper display a somewhat different behaviour to dynamos operating in the entire sphere. The critical dynamo numbers for even and odd parity lie very close together and also the difference in the magnetic cycle periods are so small that the corresponding beat frequency of the mixed mode solution is smaller by a factor 100 than the magnetic cycle frequency.

Previous conclusions concerning the connection to Grand Minima have to be reconsidered in the case where the dynamo is confined to the outer envelope. It is also very important that it is now an oscillatory even parity mode which becomes first excited. This seems somewhat similar to the 1-D model by Jennings and Weiss (1989), which also shows an even mode being excited first. This is however a stationary one.

The shell dynamos have the property that the fields in the northern and southern hemisphere are quite separated, which is not the case for dynamos operating in a full sphere where fields can extend down to the center. This is supported also by the very similar field geometry found for even and odd parity solutions (apart from the opposite sign in either hemisphere).

Of course we must be aware of the high degree of simplifications made in these models. These simplifications may account for some important qualitative discrepancies between our models and the observations. For example, Grand Minima appear irregularly rather than quasi-periodically. It is likely that such features occur once the fields are dynamically coupled to the differential rotation. Also the even parity contributions to the solar field do not show any periodic variations with approximately the 22-year cycle period, whereas in our models fields of both parities oscillate with a common frequency.

Acknowledgements. We thank F. Krause and G. Rüdiger for many interesting discussions.

References

Belvedere, G., Proctor, M. R. E.: 1989, these proceedings

Brandenburg, A., Krause, F., Meinel, R., Moss, D., and Tuominen, I.: 1989, *Astron. Astrophys.* **213**, 411

Brandenburg, A., Moss, D., Tuominen, I., 1989, *Geophys. Astrophys. Fluid Dyn.* (in press)

Brandenburg, A., Krause, F., Tuominen, I.: 1989, in *Turbulence and Nonlinear Dynamics in MHD Flows*, eds. M. Meneguzzi, A. Pouquet and P. L. Sulem, Elsevier Science Publ. B.V. (North-Holland), p. 35-40

Jennings, R., Weiss, N. O.: 1989, these proceedings

Hale, G. E., Nicholson, S. B. : 1925, *Astrophys. J.* **62**, 270

Krause, F., Rädler, K.-H.: 1980, *Mean-Field Magnetohydrodynamics and Dynamo Theory*, Akademie-Verlag, Berlin

Krause, F., Meinel, R. : 1988, *Geophys. Astrophys. Fluid Dyn.* **43**, 95

Parker, E. N.: 1979, *Cosmical magnetic fields*, Clarendon press, Oxford

Roberts, P. H.: 1972, *Phil. Trans. Roy. Soc.* **274**, 663

Steenbeck, M. and Krause, F.: 1969, *Astron. Nachr.* **291**, 49.

Stenflo, J.O., Vogel, M. : 1986, *Nature* **319**, 285

Stenflo, J.O. : 1988, *Astrophys. Spa. Sci.* **144**, 321

SELF-ORDERING OF PHOTOSPHERIC MAGNETIC FIELDS

J. K. LAWRENCE
San Fernando Observatory
Department of Physics and Astronomy
California State University, Northridge
Northridge, California 91330, U. S. A.

ABSTRACT. We model the evolution of photospheric field elements by treating them as mean field structures undergoing a nonlinear self-interaction mediated by much smaller-scale, convectively driven plasma turbulence. Distributed fields can gather into discrete, strong elements of a minimum permitted scale. Also studied are the transport of flux from dissolving elements and to growing elements via weak intermediate fields and the cancellation of adjacent emements of opposite polarity.

1. THE FIELD EVOLUTION EQUATION

The photospheric Reynolds number Re~10^{12} and kinematic viscosity ν~$1\,cm^2/s$ indicate that the convectively driven plasma motions are turbulent down to ~ 1 m. The magnetic Reynolds number Rm ~ 10^6 and plasma magnetic diffusivity η ~$10^7\,cm^2/s$ indicate a strong influence of the turbulence on the magnetic field. Because there are no rigorous derivations of interactions of the magnetic field with the turbulence under these solar conditions, we must combine derivations valid under other conditions, numerical simulations, and observational phenomenology.

Acknowledging that there are unresolved theoretical questions about this procedure, we follow Steenbeck, et al. (Krause and Rädler 1980) and split the magnetic field into a vertical, large-scale, slowly-varying "mean field" $<\vec{B}>=\hat{z}B(x,y,t)$, identified with observed photospheric fields, and a "turbulent" part, carried about by the rapid, small-scale eddies (Stenflo 1988). We assume vanishing helicity and mean velocity, and that statistical properties of the turbulence vary on mean field scales.

We adopt a magnetic "turbulent diffusivity" β~$<u^2>/T$, where u is the turbulent velocity and T the eddy correlation time. See discussions by Parker (1979), Krause and Rädler (1980), Moffatt (1983), and especially numerical simulations by Drummond and Horgan (1986). In the photosphere β ~10^{13} cm^2/s.

A turbulent, conducting fluid behaves diamagnetically with respect to the mean field, leading to an equation, valid for Rm>>1, which

J. O. Stenflo (ed.), Solar Photosphere: Structure, Convection, and Magnetic Fields, 383–386.

reduces to

$$\partial B/\partial t = \vec{\nabla} \cdot [\beta^{\frac{1}{2}} \vec{\nabla}(\beta^{\frac{1}{2}} B)] = \vec{\nabla} \cdot [\beta^{\frac{1}{2}} \vec{\nabla} C] \qquad (1)$$

(Vainshtein and Zel'dovich 1972). See other discussions by Zel'dovich (1956), Rädler (1968), and Moffatt (1983).

Nonlinearity arises from the inhibition of turbulence by strong mean fields. This idea forms the basis of thermal plug models of sunspot cooling. For Rm<<1, Krause and Rädler (1980) show that, assuming the energy source is unaffected, most turbulent modes are suppressed by a strong mean field as $1/B^4$. Peckover and Weiss (1978) found by numerical simulation that overturning convective eddies are unaffected by weak fields, but are suppressed as $1/B^4$ in strong fields, thus reducing the turbulent energy supply. Beckers's (1976) observations of microturbulent velocities in sunspots indicate a turbulent diffusivity about 0.01 that of the photospheric value. We adopt a phenomenological diffusivity to model this behavior: $\beta(B) = \beta_0/(1+|B/B_c|^n)$. B_c is the critical field separating the kinematic and dynamical regimes of eddy suppression, which we estimate at ~100 G.

2. SOLUTIONS AND COMPARISON TO OBSERVATIONS

The functional $C[B] = B[\beta_0/(1+|B/B_c|^n)]^{\frac{1}{2}}$ acts as a potential governing the two-dimensional flow of B. When n>2, B[C] is double valued, so B can vary discontinuously between $B_1<B_m$ and $B_2>B_m$, where $B_m = B_c[2/(n-2)]^{1/n}$, while C remains continuous. When n>2 and $B>B_m$ the functional derivative $\delta C/\delta B<0$, and local maxima and minima of B and C anticoincide. Then fluctuations of B will grow. When $B<B_m$ or n<2 fluctuations are damped.

The smallest-scale fluctuations grow or decay first. A scale cut-off is imposed by requiring B to vary only over distances larger than the turbulent eddies. For numerical solutionsthe minimum scale will be associated with the grid spacing, which we identify with the granular scale ~1 Mm. We present a one-dimensional case using three point spatial derivatives, an explicit, trapezoidal time integration, and modelled random fluctuations. We choose n=4 and closed boundary conditions. An initially uniform field $B_0<B_m=1$ will remain uniform; Figure 1 shows the evolution for the maximally unstable case $B_0=1.3$. $B=B_0$ is metastable for a few hours and then breaks up on the scale of the grid as expected. Separate peaks undergoing a negative fluctuation develop local maxima of C and decay by transporting flux through the intervening weak field to the stronger peaks which, being local minima of C, grow. A stable state is reached after about 200 hours with only one strong peak $B_1>B_m$ and $B_2=B_c^2/B_1$ between; C is uniform. For $B_0=3000$ G, the initial uniform state lasts about 10 days.

The present investigation differs from earlier ones by Kraichnan (1976) and Knobloch (1978). These involve a linear diffusion equation with negative diffusivity. Such a system has no stable static solution.

We find that, under certain conditions, the field gathers itself into isolated, strong elements with weak field between. It evolves to an

ordered state not in a minimum field energy configuration, thus evoking
the "dissipative structures" discussed by Nicolis and Prigogine (1977)
in connection with nonlinear chemical reactions. This process also
suggests the sudden breakup of sunspot umbrae after the appearance of
bright umbral dots (Zwaan 1987). In addition, such solutions include the
growth and decay in place of apparently isolated flux elements, as
observed by Simon and Wilson (1985) and by Topka, et al. (1986).

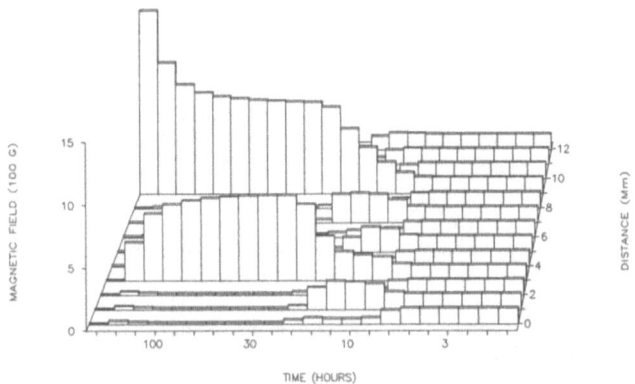

Figure 1.   Magnetic field distribution versus time.

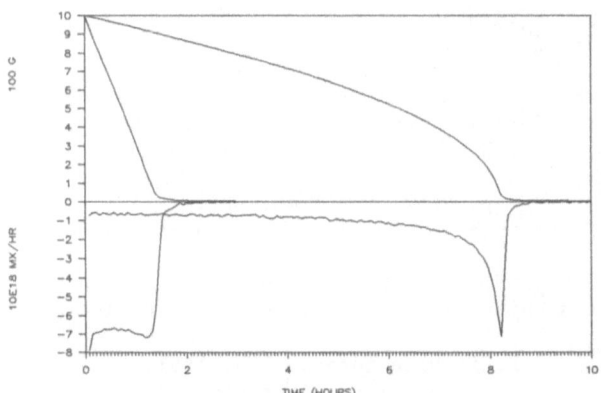

Figure 2.   Peak field strength and rate of flux loss versus time for
n=4 (long lived) and n=2 (short lived) cases.

Martin, et al. (1985) have observed the cancellation in active
regions, over a few hours time, of adjacent but seemingly unconnected
flux elements of opposite polarity. We have carried out a two-dimensional
calculation for such elements of initial strength 1000 G and one
intermediate grid point held at B=0. Figure 2 illustrates the element
peak field value and the flux loss rate in one polarity as a function of

time for n=4 and n=2. The n=4 case extends the lifetime of the peaks to
~8 hr compared to about 1.5 hr for the diffusive n=2 case. For purely
atomic diffusion this would take ~10 years. The flux loss rate is a few
times 10^{18} Mx/hr, in good agreement with observations. A single flux
element with surrounding field held to zero will likewise survive for
several hours; with surrounding field B_c^2/B it will be stable.

3. CONCLUSIONS

Our treatment is far from rigorous. Further, it ignores much of the
physics affecting photospheric fields, such as the balancing of magnetic
and gas pressures which must limit the intensity of flux elements, and
the large-scale velocity flows which can produce regions of $B>B_m$ leading
to fragmentation. All the same, our solutions resemble some field
phenomena which have proven difficult to explain satisfactorily. We
believe we have made a plausible case that a field-turbulence
interaction such as this is relevant to the behavior of photospheric
fields. Such a viewpoint provides a different way to think about
magnetic flux elements: rather than equilibrium structures representing
a balance among various forces, they more resemble shock waves,
maintained by a self-interaction of the mean field which is mediated by
the small-scale turbulence.

4. REFERENCES

Beckers, J. M. (1976) Astrophys. J. 203, 739.
Drummond, I. T. and Horgan, R. R. (1986) J. Fluid Mech. 163, 425.
Knobloch, E. (1978) Astrophys. J. 225, 1050.
Kraichnan, R. H. (1976) J. Fluid Mech. 77, 753.
Krause, F. and Rädler, K. -H. (1980) Mean Field Magnetohydrodynamics and
    Dynamo Theory, Pergamon, London; Akademie-Verlag, Berlin.
Martin, S. F., Livi, S. H. B. and Wang, J. (1985) Aust. J. Phys. 38,
    929.
Moffatt, H. K. (1983) Rep. Prog. Phys. 46, 621.
Nicolis, G. and Prigogine, I. (1977) Self-Organization in Nonequilibrium
    Systems, Wiley, New York.
Parker, E. N. (1979) Cosmical Magnetic Fields, Clarendon Press, Oxford.
Peckover, R. S. and Weiss, N. O. (1978) Mon. Not. R. Astron. Soc. 182,
    189.
Rädler, K. -H. (1968) Z. f. Naturforsch. 23a, 1851.
Simon, G. W. and Wilson, P. R. (1985) Astrophys. J. 295, 241.
Stenflo, J. O. (1988) Solar Phys. 114, 1.
Topka, K. P., Tarbell, T. D. and Title, A. M. (1986) Astrophys. J. 306,
    304.
Vainshtein, S. I. and Zel'dovich, Ya. B. (1972) Usp. Fiz. Nauk. 106,
    431; (1972) Sov. Phys.(Uspekhi) 15, 159.
Zel'dovich, Ya. B. (1956) Zh. Eksp. Teor. Fiz. 31, 154; (1957) Sov.
    Phys. (JETP) 4, 460.
Zwaan, C. (1987) Ann. Rev. Astron. Astrophys. 25, 83.

TORSIONAL OSCILLATIONS AND THE SOLAR DYNAMO REGIME

I. TUOMINEN[1], G. RÜDIGER[2], A. BRANDENBURG[1]

[1]Observatory and Astrophysics Laboratory, University of Helsinki
Tähtitorninmäki, SF-00130 Helsinki, Finland
[2]Sternwarte Babelsberg, Rosa-Luxemburg-Str. 17a,
1591 Potsdam, German Democratic Republic

ABSTRACT. We discuss the observational results of cyclic variations of solar rotation and how these can be used as a means of probing the solar dynamo. We shortly describe two examples of dynamo models where the α-effect has been modified, and compare the resulting flows to the observations.

1. Introduction

The observed temporal dependence of the solar differential rotation can be understood as an empirical constraint to the deep-seated turbulent dynamo. In our opinion the latter forms the physical background for all the various cyclic activity phenomena of the main sequence stars. The construction of the stellar dynamo can only be calibrated by the Sun, which is the best observable star. As only surface features are observed, however, it is not a trivial task to solve the inverse problem of finding either the inner flow or field regime. We continue here to probe the dynamo by means of the cyclic variations of the solar rotational rate, $\Omega = \Omega(\theta, t)$, in addition to the butterfly diagram and the magnetic polar branch. As we have described in much detail (Rüdiger et al., 1986, hereinafter referred to as paper I) all the polynomial modes in the series expansion

$$\Omega = \Omega_0 \sum_{n=1,3,\ldots} \omega_n P_n^1(\cos\theta)/\sin\theta \qquad (1)$$

proved to be periodically time-dependent. The modes with $n > 5$, i.e. the original torsional oscillation (Howard and LaBonte, 1980), form a 4-belt pattern in both hemispheres migrating in 22 years from the poles to the equator. Two belts are faster ("accelerating") and two are slower ("decelerating") than the large-scale modes. The main acceleration belt lies always equatorwards from the centre of the belt of magnetic activity. The pattern starts its migration at the poles during the activity *minimum* — and reaches, after 22 years, the equator also at the minimum. Additionally, also the modes with $n = 3$ and $n = 5$ vary in phase with each other while their *maxima* occurs at *maximal* activity. The full pattern of variable Ω looks somewhat different (Snodgrass and Howard, 1985). There is a speeding up of the rotation at high latitudes during and just after the activity maximum and a wave then emerges at mid-latitudes during the minimum and moves toward the equator. Compared

J. O. Stenflo (ed.), Solar Photosphere: Structure, Convection, and Magnetic Fields, 387–390.

with the temporal variation of the large-scale modes ($n < 7$), the torsional oscillation is the dominating effect, the maximum contribution obtained from the mode with about $n = 9$. A similar power law has recently been derived by Stenflo (1988) for the contribution of the polynomial modes of the radial magnetic field over the cycle. Approximately according to the expectations (the induced flow behaves quadratically with respect to the magnetic field) it was found that the polynomials with $n = 5$ and $n = 7$ provide the largest contribution to the solar radial magnetic field. This finding is also an important empirical constraint for an optimal dynamo model.

These observations in mind we recall the results of Paper I in which the first published solar-type dynamos (Steenbeck and Krause, 1969) were analyzed by means of the solution of the Navier-Stokes equation with mean-field Lorentz force. What we found was that indeed *two* accelerating and *two* decelerating belts of Ω exist for dynamos with a relatively *deep* convection zone. Also the location of the lower fast belt was correct and even the other mentioned properties of $\Omega(\theta, t)$ could be reproduced, besides the behaviour of ω_3 which varied in anti-phase relative to ω_5. As the old dynamo models were only tractable in a very rough way and as new ideas on the inner profile of the α-effect have meanwhile been formulated, we analyze with some recently established dynamo models (Brandenburg and Tuominen, 1988; Brandenburg, 1989) their associated magnetically induced mean flows. We adopt the theoretical formalism of Paper I, also with its shortcomings described there in detail. We mainly change the two most important input parameters of the dynamo theory, i.e. the α-effect and rotation law. With respect to the latter it was the helioseismologically derived solar rotation law which is involved, so that both the radial and the latitudinal gradient of Ω contribute as inducing effects to the dynamo action. The fixing of α is more troublesome. It is certainly positive near the surface due to the dominating density gradient,

$$\alpha \approx \Omega \ell_\rho, \tag{2}$$

where $\ell_\rho = (d\ln\rho/dr)^{-1}$ is the density scale height. But deep in the convection zone the opposite velocity gradient may weaken its value until even the sign is changed,

$$\alpha \approx \Omega \ell_\rho (1 - \ell_\rho/\ell_u), \tag{3}$$

with $\ell_u = (d\ln u_{\text{turb}}/dr)^{-1}$ (Krause, 1967). This consideration may especially hold in the overshoot region below the convectively unstable zone as the density scale length ℓ_ρ becomes larger than ℓ_u (cf. Pidatella and Stix, 1986).

2. Dynamos with negative α

We have thus analyzed first a model with a lower region with a negative α-effect (cf. Fig. 2b in Glatzmaier, 1985). The results for this model are given in Fig. 1. The butterfly diagram indeed exhibits a very solar-like behaviour. The toroidal field belts start at latitude 60° and reach maximum at 15°. The phase relation between B_r and B_ϕ, however, does *not* agree as both components are varying in phase, in opposite to observations (Stix, 1976). Indeed, as a consequence the accelerarion belt is located above the activity zone, contrary to the observations. Furthermore, ω_1 varies much too strongly. That clearly reveals the existence of too large contributions of the lower polynomial modes in the magnetic field representation (cf. Stenflo, 1988). On the other hand, also the other modes in Eq. (1) have "wrong" properties: ω_3 and ω_5 and are *out* of phase. We have thus to state serious contradictions between the model and the real observations, in particular of the time-dependence of the rotational rate (1). Similar difficulties have been described by Glatzmaier (1986) for just this type of dynamo.

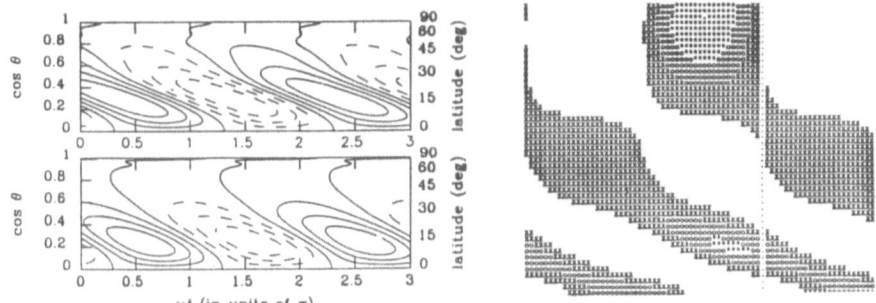

Fig. 1. *Left panel*: Butterfly diagrams for the B_r- and B_ϕ-field for the dynamo model with negative α. *Right panel*: The corresponding torsional wave pattern with $n > 5$. Note that in the right panel the ordinate is linear in θ and the abscissa goes from 0 to 1.5π. The shaded areas define faster rotation.

3. Higher order terms in α

We suggest thus to reject the idea of negative α at the bottom of the convection zone and turn to another modification of the traditional α-profile. As is well known from the theory of differential rotation, it is not enough to consider only the turbulence correlations which are linear in the rotation rate Ω. As the Sun is *not* a slow rotator in the sense of the turbulence theory ("slow" means $\tau_{rot} \ll \tau_{corr}$) also higher order terms must be involved in the calculations (cf. Rüdiger, 1989). The same arguments, of course, hold for the α-effect: Higher order terms also exist with quite another latitude dependence,

$$\alpha = \alpha_0 - \alpha_2 \cos^2 \theta \qquad (4)$$

(Rüdiger, 1980, Schmitt, 1987), so that it even may change its sign at some latitude. For $\alpha_0 = \alpha_2$ the α-effect vanishes at the poles. We want to favour this case rather than that where there is a node at a certain latitude, e.g. at 30° as Schmitt suggests, which seems to be too extreme. Our choice only concentrates the α-effect to lower latitudes instead having maximum at the pole. The hydrodynamical results of this procedure are rather surprising (Fig. 2). The butterfly diagram and polar branch have the right properties and also the phase relation agrees with the observations, i.e. $B_r B_\phi < 0$. In addition the Stenflo constraint is fulfilled, i.e. the lowest modes do not contribute too strongly. The torsional oscillation belts have the correct location in the latitude-time-diagram, compared to the magnetic field, but they start at the activity *maximum* (reaching the equator after a whole 22-year cycle also at the maximum). That is a first contradiction while the second one is the out-of-phase variation of ω_3 and ω_5. But the behaviour of ω_3 itself corresponds to the observations, it has maximum at the activity maximum and minimum at the activity minimum. Obviously, only the term ω_5 is still out of correspondence.

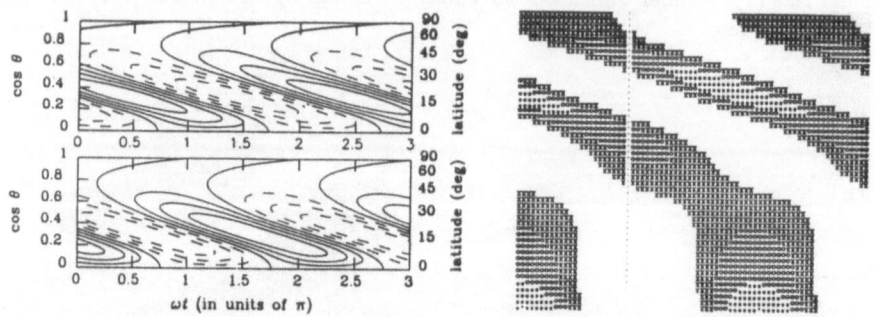

Fig. 2. *Left panel*: Butterfly diagrams for the B_r- and B_ϕ-field for the dynamo model with $\alpha_0 = \alpha_2$. *Right panel*: The corresponding torsional wave pattern.

We understand these examples as an improvement of the preliminary analysis presented in Paper I, where only the depth of the convection zone resulted from the comparison of the calculations and the observations. It seems possible to use the observed flow pattern as a precise tool for analyzing the inner turbulent regime of stellar convection.

References

Brandenburg, A.: 1989, Dissertation, Univ. of Helsinki.

Brandenburg, A., Tuominen, I.: 1988, *Adv. Space Sci.*, **8**, (7)185.

Glatzmaier, G.A.: 1985, *Astrophys. J.* **291**, 300.

Glatzmaier, G.A.: 1986, *Geophys. Astrophys. Fluid Dyn.* **31**, 137.

Howard, R., LaBonte, B.J.: 1980, *Astrophys. J.* **239**, L33.

Krause, F.: 1967, Habilitationsschrift, University of Jena.

Pidatella, R. M., Stix, M.: 1986, *Astron. Astrophys.* **157**, 338.

Rüdiger, G.: 1980, *Geophys. Astrophys. Fluid Dyn.* **16**, 239.

Rüdiger, G.: 1989, *Differential rotation and Stellar Convection:*
     Sun and Solar-type stars, Gordon and Breach, New York.

Rüdiger, G., Tuominen, I., Krause, F., Virtanen, H.: 1986, Astron.
     Astrophys. **166**, 306.

Schmitt, D.: 1987, *Astron. Astrophys.* **174**, 281.

Snodgrass, H. B., Howard, R.: 1985, *Science* **228**, 945.

Steenbeck, M. and Krause, F.: 1969, *Astron. Nachr.* **291**, 49.

Stenflo, J.O. : 1988, *Astrophys. Spa. Sci.* **144**, 321.

Stix, M.: 1976, *Astron. Astrophys.* **47**, 243.

LARGE-SCALE INTERNAL MAGNETIC FIELD OF THE SUN

A.E. DUDOROV[1], V.N. KRIVODUBSKIJ[2], A.A. RUZMAIKIN[3],
and T.V. RUZMAIKINA[4]

[1] *Astronomy Department of Moscow State University*
117899, Moscow, USSR

[2] *Kiev Shevchenko University Astronomical Observatory*
252053, Kiev, USSR

[3] *IZMIRAN, 142092, Troitsk, Moscow region, USSR*

[4] *Institute of Physics of Earth, 123810, Moscow, USSR*

ABSTRACT. The behaviour of the magnetic field during the formation and evolution of the Sun is investigated. It is shown that an internal poloidal magnetic field of the order of $10^4 - 10^5$ G near the core of the Sun may be compatible with differential rotation and with torsional waves, travelling along the magnetic field lines (Dudorov et al., 1989).

1. The fossil magnetic field

Estimates and numerical calculations reveal an intensification of the magnetic field during the contraction of the magnetized protosolar cloud, in spite of some loss of magnetic flux (Ruzmaikina, 1981, 1985; Dudorov, 1986; Dudorov and Sazonov, 1981).

The fossil magnetic field strength of the young Sun has been calculated using a computer code for protostar evolution (Dudorov and Sazonov, 1981). The initial model is a uniform cloud of mass $M = 1.2 M_\odot$, temperature $T_0 = 10$ K, density $n_0 = 2 \times 10^6$ cm^{-3}. The cloud is permeated by a homogeneous magnetic field of strength $B = 4.5 \times 10^{-5}$ G. The evolution of the magnetic flux is investigated in the framework of the kinematic problem for collapse of protostellar clouds. The computer code based on the Lax-Wendroff method solves the gas-dynamic equations together with the induction equation:

$$d\mathbf{B}/dt = \nabla \times [(\mathbf{V} - \mathbf{V}_{AD}) \times \mathbf{B}] + (\mathbf{V} \cdot \nabla)\mathbf{B} - \nabla \times (\nu_m \nabla \times \mathbf{B}), \qquad (1)$$

where $\mathbf{V} = \{V_r, 0, 0\}$ is the gas velocity, $\mathbf{B} = \{B_r \cos\theta, -B_\theta \sin\theta, 0\}$ the strength of the magnetic field, ν_m the magnetic viscosity, and $\mathbf{V}_{AD} = 3.6 \times 10^{31} < \text{curl } \mathbf{B} \times \mathbf{B} > /n^2/x$ (cm s^{-1}) the velocity of ambipolar diffusion. The calculations of ionization degree x take into account the collisional and thermal ionization of elements with sufficiently high cosmic abundances and low ionization potentials. The total ionization rate is determined by the collisional and radiative rates, and by radioactive elements. The recombination processes are radiative recombination and recombination on grains.

J. O. Stenflo (ed.), Solar Photosphere: Structure, Convection, and Magnetic Fields, 391–394.

The calculations show that during the collapse the field strength changes as $B \sim n^k$ with $1/2 < k < 2/3$. When an opaque core in the cloud is formed with $n_i > 10^5 n_0$, the ionization ratio is decreased to $x < 10^{-12} - 10^{-13}$. The coupling between plasma and neutrals weakens, giving rise to ambipolar diffusion of the charged particles with the magnetic field in the neutral gas. An upper limit to the effective flux loss in the dense $(n = 10^9 n_0)$ region is determined by the evaporation of dust grains and the thermal ionization of metals. The degree of magnetic decoupling is strongly influenced by the ionization rates, the grain parameters, and the abundance of metals (Ruzmaikina, 1985; Dudorov, 1986).

The results of the calculations show that the magnetic field strength inside the Sun in the T Tau stage is 10^2 times lower than that of a frozen-in field. For a short period, after the Sun has arrived at the zero-age main sequence, the magnetic field may be of the order of $10^5 - 10^6$ G at the core boundary, and $10^4 - 10^5$ G near the base of the convection zone. The exponent in the relation $B \sim n^k$ is $k = 0.57$ for the core.

2. The dynamo process

The subsequent evolution of the fossil magnetic field depends on hydromagnetic or resistive instabilities, buoyancy, and ohmic and turbulent diffusion. Convection in new-born stars at the T Tauri stage has a dual character: Inside stars with masses $M \approx 0.5 - 1.2 M_\odot$ convection is either absent at the pre-main sequence stage, with the consequence that the strength of the fossil magnetic field is diminished only slightly by ohmic decay, or turbulent convection converts the large-scale magnetic field into a small-scale one. The buoyancy of the magnetic flux tubes may cause the activity of the pre-main sequence stars.

On the other hand turbulent convection with differential rotation drives $\alpha\omega$-dynamos, which begin to operate after sufficient turbulent dissipation of the fossil magnetic field has taken place. The differential rotation is induced by gravitational compression of the protosun. At a nonlinear stabilization level the poloidal (B_p) and toroidal (B_φ) components of the magnetic field are connected with the dynamo number, $N_D = \alpha(\mathrm{d}\omega/\mathrm{d}r)R^4/v_t^2$, and with the turbulent velocity V_t by the following relations (Dudorov et al., 1989):

$$B_p B_\varphi = \pi\omega \, H_p V_t, \quad B_\varphi = N_D^{1/2} B_p, \tag{2}$$

where ω is the angular velocity, H_p the pressure scale height. We can estimate V_t, assuming that the protosun luminosity L is entirely determined by convective energy transport, which gives $V_t = (L/M \cdot H_p)^{1/3}$, where H_p is the pressure scale height. This leads to

$$B_p = (8\pi H_p \rho/3)^{1/3} \, (H_p/R)^{1/2} \, (\omega/R)^{3/4} \, V_t^{3/4}, \tag{3}$$

$$B_\varphi = (24\pi H_p \rho)^{1/2} \, (R/H_p)^{1/2} \, (\omega^3/R)^{1/4} \, V_t^{1/4}. \tag{4}$$

For the standard solar model with a density near the core of $\rho = 10$ g cm^{-3}, $B_p = 10^4 - 10^5$ G, and $B_\varphi = 10^5 - 10^6$ G. At the base of the convection zone $\rho = 0.1$ g cm^{-3}, $B_p = 3 \times 10^3 - 3 \times 10^4$ G, and $B_\varphi = 3 \times 10^4 - 3 \times 10^5$ G. Such a field evolves with time to a state with $B_\varphi = B_p$.

When the Sun arrives at the main sequence, the internal convection ceases. Two scenarios are possible. In the first, the shrinking convection zone rises to the solar surface on the hydrodynamic time scale. In this case the magnetic field is retained inside the Sun. In the second scenario the convection is attenuated slowly on the hydrostatic time scale. In this case the convection, which preserves its turbulent character, may weaken the magnetic field. The magnetic field will then be reduced to $B_p = 10 - 100$ G and $B_\varphi = 100 - 1000$ G for $\rho = 10$ g cm^{-3}, and $B_p = 1 - 10$ G and $B_\varphi = 10 - 100$ G for $\rho = 0.1$ g cm^{-3}. The fossil magnetic field influences neither the internal structure of the Sun nor e.g. the oblateness of the solar core or the solar neutrinos, but it may change the spectrum of acoustic oscillations either directly, or by reducing the gradient of the angular velocity.

3. The differential rotation and magnetic field

It is difficult to estimate the value of the angular velocity gradient in stars at the T Tau stage. Therefore we assume that the differential rotation inside the Sun is represented by the observations of Duvall, Harvey, and Pomerantz (1986). We study the generation of a toroidal magnetic field by the observed differential rotation taking into account magnetic buoyancy, making use of the induction equation (1), where now $\mathbf{V} = \mathbf{\Omega} \times \mathbf{r}$ is the linear rotational velocity. Instead of V_{AD} we need to insert the buoyancy velocity V_p (Parker, 1979). An estimate shows that a stationary toroidal magnetic field establishes itself very quickly, having

$$B_\varphi = \left[G B_r 8\pi P \left(H_p/u_T \right) \left(R'/L' \right)^2 \right]^{1/3}, \tag{5}$$

where $G = r d\omega/dr$, $u_T = 10$ cm s^{-1} (Parker, 1979), H_p and L' are the pressure and temperature scale heights, and R' the tube radius. At the core boundary $B_\varphi = 3 \times 10^7$ G and 10^6 G, respectively, for the two scenarios of convective behaviour. The poloidal fossil magnetic field leads to the substantially larger value of $B_\varphi = 3 \times 10^8$ G. Such a field is compatible with the virial theorem. One may obtain a great reduction of B_φ if one assumes that differential rotation exists during some period.

Hydromagnetic instabilities lead to an upper limit for the toroidal field. The characteristic time scale for magnetic smoothing of the differential rotation, the Alfvénic time scale, is much shorter than the evolutionary time scale, even for the initial magnetic field of the protosolar cloud, $B_p = 4.5 \times 10^{-4}$ G. Thus a significant and stable jump of the angular velocity is possible for a vanishing poloidal magnetic field, for instance on the surface of a magnetically insulating solar core. Small irregularities of the angular momentum may be influenced by torsional hydromagnetic waves between the core and the envelope, where the magnetic field lines should be rigidly fixed. In this case the core and the envelope of the Sun may be azimuthally shifted relative to each other by substantial angles. The characteristic period of oscillation is again the Alfvénic time scale (see also Rosner and Weiss, 1985; Gough, 1986; Moss, 1987; Mestel and Weiss, 1987).

References

Dudorov, A.E. (1986) 'The fossil magnetic field', *Astron. Circular* **1446**, 1-8.
Dudorov, A.E. and Sazonov, Yu.V. (1981) 'Hydrodynamical collapse of interstellar magnetic clouds', *Nauchn. Inf. Astron. Council USSR AN* **49**, 114-136.

Dudorov, A.E., Krivodubskij, V.N., Ruzmaikina, T.V., and Ruzmaikin, A.A. (1989) 'Large-scale internal magnetic field of the Sun', *Astron. Zh.* **66**, No.4.

Duvall, T.L., Harvey, J.W., and Pomerantz, M.A. (1986) 'Latitude and depth variation of solar rotation', *Nature* **321**, 500-501.

Gough, D.O. (1986) 'What causes the solar cycle', *Nature* **319**, 263-264.

Mestel, L. and Weiss, N.O, (1987) 'Magnetic fields and non-uniform rotation in stellar radiative zones', *Mon. Not. Royal Astron. Soc.* **226**, 123-135.

Moss, D. (1987) 'Internal magnetic field of the Sun', *Mon. Not. Royal Astron. Soc.* **224**, 1019-1031.

Parker, E.N. (1979) *Cosmical magnetic fields*, Clarendon Press, Oxford.

Rosner, R. and Weiss, N.O. (1985) 'Differential rotation and magnetic torques in the Sun', *Nature* **317**, 790-792.

Ruzmaikina, T.V. (1981) 'Magnetic field and turbulence in solar nebula' *Adv. Space Res. COSPAR* **1**, 49-54.

Ruzmaikina, T.V. (1985) 'Magnetic field in collapsing presolar clouds', *Astron. Vestnik* **19**, 101-108.

VII.   CONVECTION AND MAGNETIC FIELDS

IN SOLAR-TYPE STARS

OBSERVING, MODELING, AND UNDERSTANDING STELLAR GRANULATION

DAINIS DRAVINS
Lund Observatory
Box 43
S-22100 Lund
Sweden

ABSTRACT. Numerical simulations of the three-dimensional structure and time evolution of stellar surface convection are now possible, at least for *solar-type* stars. Using the output from such simulations as sets of spatially and temporally varying model atmospheres, synthetic granulation images and spectral line profiles are computed, and compared to observations. Thus obtained disk-integrated data agree with observed lineshapes and bisector patterns in different stars, and also permit stellar rotation to be determined. Such simulations represent the first generation of models that are free from the classical *ad hoc* parameters of *'mixing-length'*, *'micro-'* or *'macro-turbulence'*, parameters which in the past have characterized and limited stellar astrophysics.

To infer the surface structure also in more exotic stars, simpler and parametrized models must still be used to interpret observed line asymmetries. Such models suggest that rapidly rising 'granules' cover only a small surface fraction on *early-type* stars, a situation opposite to that in solar-type ones, and one likely to affect magnetic fields and stellar activity. *Theoretical* challenges for the future include detailed modeling also of early-type, giant, and other non-solar type stars of different rotational velocities; the hydrodynamics of entire stellar envelopes (including the interaction with global oscillations); and the interaction with magnetic fields (including their generation). Greatly increased computing power will be needed for such detailed modeling throughout the Herzsprung-Russell diagram, possibly requiring custom-designed computers.

Signatures of stellar granulation are primarily observed as asymmetries and wavelength shifts in photospheric absorption lines. *Observational* challenges include achieving sufficient spectral resolution to fully resolve such asymmetries; identifying granulation signatures throughout the HR-diagram (including the blended spectra of cool stars); observing how line asymmetries for a given spectral type depend on stellar rotational velocity; measuring wavelength shifts between groups of different lines in the same star, and between different stars; monitoring lineshift variations during stellar activity cycles; and ultimately high-resolution spectroscopy of spatially resolved granulation structures across stellar disks. The latter will require active optics on future very large telescopes, or the use of long-baseline optical interferometers.

1. Introduction

Stellar granulation has now joined chromospheres and coronae as another 'gift' from solar to stellar astrophysics. Building upon the solar experience, observable effects of stellar surface structure have been identified in stellar spectra, and a first generation of inhomogeneous stellar models has been developed, initiated from their solar counterparts. The aim of this paper is in particular to discuss the future potential of such studies, and to identify directions in which future work should be headed.

J. O. Stenflo (ed.), Solar Photosphere: Structure, Convection, and Magnetic Fields, 397–415.
© 1990 by the IAU.

2. Theoretical Models of Stellar Granulation

The aims of stellar granulation modeling may be somewhat different from the solar equivalent because of the more fundamental questions that may be posed regarding the surface structure and emission of radiation in widely different types of stars. After all, surface inhomogeneities on the Sun have been studied in great detail for a long time, and with considerable spatial resolution, while the usual description of stellar atmospheres in the past has been by homogeneous models only. Consequently, already rather primitive models might bring interesting new information. On the other hand, the more limited observational data that can be obtained for stars, seriously limits the number of ways in which models can be tested, and makes it practically impossible to obtain unique 'inversions' of observational data into a numerical description of spatially unresolved structures. The degrees of freedom in parametrized models of three-dimensional stellar atmospheres are potentially very many, thus preventing the application of traditional modeling techniques (i.e. the adjustment of successive parameters until the output fits the observed data).

Such limitations can (and should) be avoided by instead using *computational experiments* to numerically simulate stellar surface convection. Although the solar (and, by implication, stellar) surface structure at first sight may appear to be quite complex, the number of *fundamental* physical processes that determine the structure at large, is probably very limited (at least in non-magnetic regions). This makes the problem well suited to numerical simulations: the finite number of physical processes can be described by laws which are mathematically known and physically understood (e.g. those of hydrodynamics and radiative transfer). The availability of supercomputers then makes it possible to perform numerical simulations with considerable detail and realism. In particular, it becomes possible to model nonlinear and complexly intercoupled phenomena, whose correct interpretation from observations and parametrized models only could be next to impossible. For a general introduction to this new role of supercomputer numerical laboratories, see e.g. Winkler *et al.* (1987).

Using the output from such simulations as sets of time- and space-dependent model atmospheres, line profiles may be obtained as spatial and temporal averages, allowing a direct test against observations. This *confrontation between theory and observation* is a most important and crucial step in the development of realistic models, and for the ultimate understanding of stellar atmospheres. During past decades, the development of realistic theories for stellar convection and motions in stellar atmospheres has been severely hampered due to the lack of any sensitive observational verification. Primary among observable parameters are the disk-averaged line profiles, together with their asymmetries and wavelength shifts, and thus theoretical models should be able to confront observations by predicting (at least) these parameters.

2.1. REALISTIC MODELS OF STELLAR GRANULATION

While important insights about the processes in stellar photospheres certainly can be gained already through e.g. parametrized, few-component, two-dimensional and/or stationary models, a fully realistic description ultimately demands a time-variable modeling in three dimensions (corresponding to the real world). In recent years it has become possible to solve numerically the sets of hydrodynamic equations that describe the three-dimensional time evolution of solar granular convection (Nordlund, 1982; 1985). The equations of motion (which describe the time evolution of the velocity field) and the energy equation (which describes the time evolution of the temperature field) are used to step a numerical representation of the velocity and temperature fields forward in time. The use of realistic background physics (equation of state, absorption coefficients, etc.) taken from standard stellar atmosphere code, a detailed treatment of the radiative transfer (non-grey, three-dimensional), and the use of relevant boundary conditions (in particular the absence of any walls or edges that could artificially constrain the flows), leads to a realistic simulation of the granular convective motions.

Such models naturally contain a number of physical, mathematical and numerical approximations. However, they contain no arbitrary or adjustable *physical* parameters: the results depend in principle only on the effective temperature, surface gravity and the chemical abundance. (These determine the heat flux into the simulation volume, the gravity forces stratifying the atmosphere, and the opacity of the gases, respectively.)

Using an improved version of these computer codes (initially developed for solar modeling), simulations have been made for models of four different stars (Nordlund and Dravins, 1989). The temperature dependence of granulation along the main sequence is explored with models hotter and cooler than the Sun. A T_{eff} = 6600 K model has parameters close to those of *Procyon* (F5 IV-V), while a cooler one with T_{eff} = 5200 K corresponds to the K1 dwarf α *Centauri B*. The luminosity dependence is examined by two models at solar temperature (T_{eff} = 5800 K): one at one *half* solar surface gravity corresponds to the slightly evolved main-sequence star α *Centauri A* (G2 V), while another at one *quarter* solar surface gravity has parameters similar to the subgiant β *Hydri* (G2 IV). The simulations were made with a 32 × 32 × 32 spatial grid: 32 × 32 Fourier components horizontally, and 32 knot points defining cubic splines for the vertical dependence of different parameters.

This small grid of models thus allows a first study of the temperature and luminosity dependence of granulation in the vicinity of the Sun in the Hertzsprung-Russell diagram. Using the output from these simulations as sets of spatially and temporally varying model atmospheres, synthetic stellar surface images and spectral line profiles were computed for a series of different parameters. This permits a study of continuum and spectral line formation in inhomogeneous stellar atmospheres, and also allows a comparison with observed line profiles and asymmetries (Dravins and Nordlund, 1989a; 1989b). Samples from this work are used to illustrate the discussion below.

2.2. THE FORMATION AND EVOLUTION OF STELLAR GRANULATION

In the surface regions of all models, hotter and generally rising features are seen, analogous to solar granules. The granulation phenomenon turns out to be closely connected to the temperature sensitivity of the stellar opacity (in particular that of the negative hydrogen ion, H^-), and less dependent on e.g. the stellar surface temperature. When the temperature in the hot and rising elements falls below ≈ 10,000 K, the opacity is rapidly reduced, the radiation detaches from the gas, and the temperature quickly drops to ≈ 6000 K. The granulation patterns are most pronounced in these regions where radiation detaches from matter. In the hotter *Procyon* model (T_{eff} = 6600 K), this region includes the visible stellar surface, which is thus covered by 'naked' granules of very high temperature contrast. In solar-temperature stars, this region is (barely) beneath the surface. Although the granulation pattern is still visible on the surface, the temperature contrast increases with depth. In the cooler (T_{eff} = 5200 K) star, however, this sharp temperature drop remains 'hidden' beneath the visible surface, to which only low-contrast temperature features are able to penetrate. In this case, the radiation is unable to directly escape, but is rather largely absorbed in the upper photosphere, whose temperature structure is strongly influenced by this absorption of radiation from below.

The normal evolution of stellar granules is characterized by a gradual increase in size until they disintegrate due to various causes. Relatively undisturbed granules often grow until they 'collapse under their own weight'. In order to carry the stellar convective heat flux outward, there must be an upflow of hot gases at a few km/s from beneath the photosphere. Upon reaching optically thin layers, these gases cool off and eventually turn around to descend in the cooler intergranular areas. The necessary *horizontal acceleration* of the gases toward the sinking areas is accomplished by local overpressures that develop over granules (and also over intergranular lanes, in order to *decelerate* the incoming horizontal flow there). The larger the granule, the larger the overpressure required to accelerate away the material over the greater horizontal distances. Ultimately, the overpressure on the surface becomes so great that it impedes the arrival of additional hot gases from below. Having been cut off from its energy supply, the granular center begins to cool off, a dark center develops, and the now ring-shaped granule disintegrates, a process apparently similar to solar *exploding granules*. This mechanism limits the characteristic scale of granules on different stars: the granules cannot grow any larger because the required overpressure would block further convective energy supplies from beneath, thus strangling the granule through a lack of input heat. Characteristic scales range between $\approx 10^3$ km in the cool α *Cen B* model to $\approx 10^4$ km on the subgiant β *Hyi*.

Not all granules evolve in this manner. Some granules that are disturbed by velocity flows in neighboring granules may cleave and break apart into smaller fragments, which in turn grow and merge into new features. Some periods may be characterized by especially vigorous granulation with large and well correlated temperature and velocity amplitudes, while other periods (especially following the breakup of some particularly large granule) may be characterized as 'abnormal', with apparently chaotic temperature and velocity fields, as the photosphere is gradually returning back to a more stable situation (Nordlund and Dravins, 1989).

2.3. THE SURFACE STRUCTURE OF VELOCITY AND TEMPERATURE

As an example of stellar surface structure, Fig. 1 shows a 'snapshot' of the *Procyon* model at a representative instant, i.e. at a time when its spatially averaged properties are roughly similar to the average for the entire simulation sequence. Temperature, pressure and velocities are shown at constant geometrical depth at the stellar 'surface'. (This is here defined as the deepest horizontal level in the simulation volume where the radiative flux exceeds 50% of the total energy flux.) The temperature structures are sharply delineated and correlate well with rising velocities. The pressure patterns are smoother and display the overpressures discussed above. The horizontal velocity field is obviously directed away from the upflows into the downflow regions. The occasional lack of a detailed correlation with the superposed 'granular' contours is due to the time history of the different features: old and decaying granules may already begin to be swept away by horizontal flows originating elsewhere. The vertical cuts into the simulation volume (bottom panels) show in particular the asymmetries in the vertical flows. While the upflows are relatively gentle and spread out over extended volumes, the sinking material rapidly converges into concentrated and strong downflows that preserve their identity to large depths. Gases in these downflows often develop a significant rotational motion beneath the surface (by conserving angular momentum from random surface motions), reminiscent of terrestrial tornados. Analogous to the solar case (Nordlund, 1985), a large fraction of the convective energy flux is carried in these cool and concentrated downflows which occupy only a small fraction of the surface area: stellar convection is thus a highly inhomogeneous and intermittent process.

Finally, it can be noted in the vertical cuts in Fig. 1 that the stellar 'surface' (now defined by the T_{eff} = 6600 K level) is highly 'corrugated' with an amplitude of perhaps

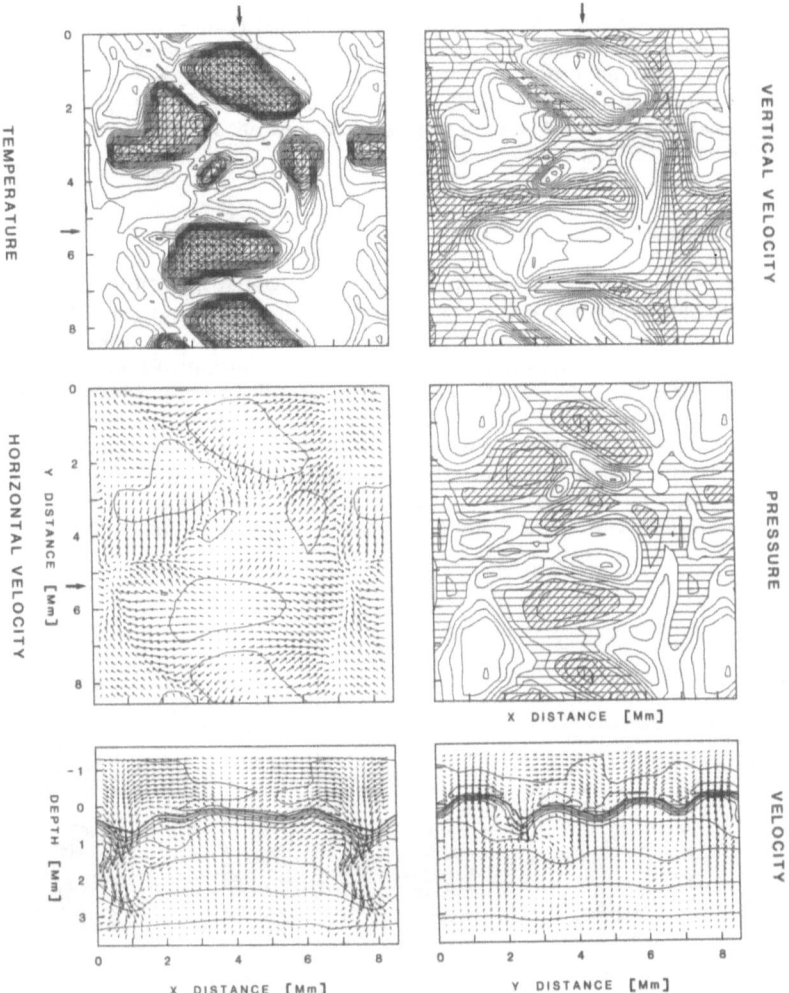

Figure 1. *Overall appearance* of the *Procyon* simulation at a representative time. The top four panels show temperature, pressure and velocities at a fixed geometrical height at the stellar 'surface'. Temperatures are shaded above the T_{eff} value of 6600 K, with contours every 500 K and shading increasing every 1500 K. Downward velocities are shaded with contours every km/s and shading in steps of 3 km/s. Horizontal velocity vectors correspond to 40 seconds of gas motion. The superposed T_{eff} = 6600 K contours delineate the granules. The logarithmic pressure plot is shaded above $10^{3.6}$ Pa, has contours every 0.05 *dex*, and increased shading every 0.1 *dex*. The Fourier representation makes the images periodic in X and Y, with a cycle of $\simeq$ 6840 km (slightly more than $^5/_4$ of one cycle is shown). The two bottom panels show the velocity components in vertical planes. The left cut into the simulation volume (at Y $\simeq$ 5.3 Mm, as marked by arrows) goes across a large growing granule and its associated downflows, while the right one (X $\simeq$ 4.3 Mm) goes across different smaller granules. The bold temperature contour marks T_{eff} = 6600 K, with other contours at 1500 K intervals. (Nordlund and Dravins, 1989)

402

500 km. This is a particular feature of this hotter model (rather less pronounced in the cooler ones), and leads to several consequences in the optical appearance of granulation across the stellar disk, as well as in asymmetries and shifts of the line profiles. Such a 'corrugated' surface on F-stars could be predicted already in a simpler granulation model by Nelson (1980). Another model for granulation on *Procyon* was presented by Atroshchenko *et al.* (1989).

3. The Appearance of Stellar Surfaces

The *output* data from the numerical simulations were used as *input* for radiative transfer calculations to obtain the emerging continuum and line radiation for different spatial points, angles, wavelengths, and spectral line parameters. In Fig. 2, synthetic images are shown for the *Procyon* model, at the same representative instant of time as in Fig. 1. As expected, the geometrical shapes of granules at stellar disk center closely correspond to the temperature features. However, the intensity *contrast* is much lower than could have been naively expected from the temperature contrast at a constant geometrical depth: the temperature-dependent opacities of overlying layers hide the larger temperature contrasts beneath. Nevertheless, by solar standards, the intensity contrast is rather high, and furthermore *increases* toward the stellar limb. This increase originates from the

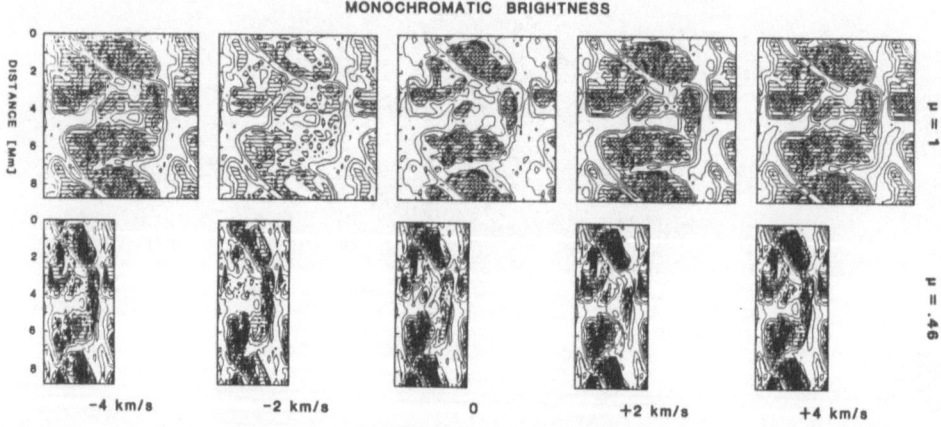

Figure 2. *Synthetic monochromatic images of Procyon* at the same time as Fig. 1. The monochromatic brightness across a weak Fe I line of $\chi = 3$ eV at λ 520 nm is converted to images seen through an ideal narrow-band monochromatic filter, scanning across the line. The region is shown both at disk center ($\mu = \cos \theta = 1$) and as seen (from the left) toward the limb ($\mu = 0.46$). Areas brighter than average are shaded, with contours at every 20% of the average. Noteworthy is the high granulation contrast, which *increases* from disk center toward the limb. This reflects the 'corrugated' surface in this $T_{eff} = 6600$ K model which, under large inclination angles, occasionally makes it possible to view hot bright spots in deeper layers. As on the Sun, the granulation contrast at disk center has a maximum in the red wing of the line, while the changes with wavelength are much smaller near the limb. This reflects the weak brightness-velocity correlation for the *horizontal* velocities seen close to the limb. A further discussion is in Dravins and Nordlund (1989a).

'corrugated' nature of the *Procyon* surface which, at larger inclination angles, occasionally makes it possible to glimpse hot and bright elements in deeper layers. This particular property is *not* shared by the cooler models, and the granulation contrast in e.g. the T_{eff} = 5200 K model is considerably lower and shows no tendency to increase toward the limb.

3.1. LINE FORMATION IN STELLAR PHOTOSPHERES

The simulation data can in particular be used as an input to compute photospheric line profiles at different spatial locations and at different times. The subsequent analysis of such profiles can give considerable insight in the physics of line formation in inhomogeneous stellar photospheres and can predict observable quantities which can verify (or falsify) different models. Further, it may indicate the validity (if any) of classical approximations such as the concepts of '*micro-*' or '*macro-turbulence*'.

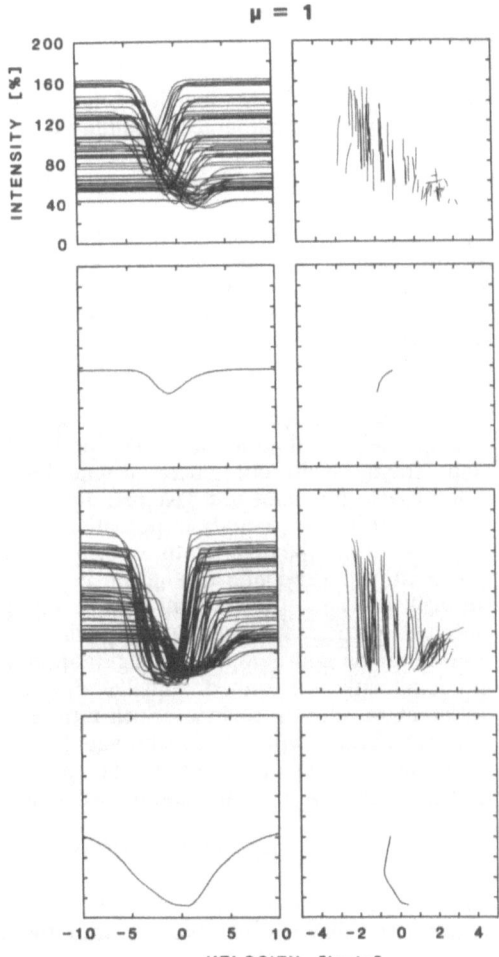

Figure 3. Spatially resolved line profiles at α *Cen A* disk center, and their spatial averages at a representative time. This model differs from the *Sun* only through a halved surface gravity. Line profiles and bisectors are shown for a grid of 8 × 8 = 64 spatial points out of the 32 × 32 = 1024 actually computed at each step in time. The top half of the figure shows data for a weak Fe I line, and the bottom for a very strong one, both with χ = 3 eV at λ 520 nm. 100% intensity corresponds to the average for the entire simulation sequence. The spatially averaged profile and its bisector at this instant in time are also shown. The average profile is *not* representative for spatially resolved points on the star: its width and asymmetry rather reflect the statistical distribution of spatial inhomogeneities. The redshifted profiles of strong lines from intergranular regions are often asymmetric due to vertical velocity gradients over their extended heights of formation. (Dravins and Nordlund, 1989a)

Figure 3 shows a sample of such spatially resolved line profiles (and their averages) at a representative time at the disk center of the α *Cen A* model. An important conclusion from all stellar simulations is that the spatially averaged profile is *not at all* typical for individual points on the stellar surface. The shape, asymmetry and shift of the average profile instead reflect the *statistical distribution functions* of different profiles from different spatial points. The pronounced asymmetry of the spatially averaged profile is not frequent: it only occurs where there happen to be strong velocity gradients in the line-forming layers. Such intrinsic asymmetries are more common for strong lines with extended heights of formation, in particular in intergranular lanes. There the downflow velocities rapidly increase with depth, and this depth gradient can be manifest as asymmetries also in spatially resolved lines. This means that attempts to deduce stellar photospheric structure by interpreting observed line asymmetries as arising in horizontally homogeneous models, will likely lead to fortuitous depth-dependent velocity fields that are not at all present in any real stellar atmosphere.

The correlation between brightness and lineshift is particularly well visible for weaker lines (top panel in Fig. 3). This of course indicates that hot elements generally are rising, and the correlation decreases near the stellar limb, where one mainly sees the effects of horizontal velocities. Closer to the limb, there is a noticeable increase in the velocity spread of the individual lines, reflecting that the horizontal velocities are generally of somewhat larger amplitude than the vertical ones.

3.2. LINE PROFILES IN INTEGRATED STARLIGHT

By a suitable summation of time-averaged data for different center-to-limb positions, disk-integrated line profiles are obtained. These profiles can be compared to observations, and constitute the most important diagnostic tool for stellar granulation studies. For the four models computed, the most pronounced differences in resulting line shapes are between those in different stars. However, for any given star, line asymmetries and wavelength shifts depend systematically on the line's atomic parameters, analogous to the solar case.

The largest differences are seen among lines of different strength, reflecting different average heights of formation. The photospheric structure changes rapidly with height, and already slight differences in line formation conditions may lead to observable differences. For lines of a given strength, potentially observable differences exist among lines of different excitation potentials and/or ionization levels. These differences originate from the temperature sensitivity of the lines: those of a certain excitation potential will predominantly form in regions whose temperatures are sufficient to strongly populate the relevant atomic energy level, yet not so high as to ionize the species. Generally, Fe I lines of higher excitation potential show a larger blueshift, reflecting the more rapidly rising motion of the hotter elements, where these lines predominantly are formed. Smaller differences exist among otherwise similar lines in different wavelength regions. Such differences occur primarily because the granulation contrast changes with wavelength, and also because of changing continuum opacities. It is through such arrays of differential spectral line behavior that also detailed models of stellar granulation structure can be tested against observations.

3.3. COMPARISON TO OBSERVATIONS

The highest-resolution data available on stellar photospheric line profiles and their asymmetries have been obtained with the double-pass scanner of the coudé echelle spectrometer at European Southern Observatory (Dravins, 1987b). The synthetic line profiles for disk-integrated starlight were convoluted with profiles corresponding to the measured instrumental ones, but slightly broadened to include effects of plausible amplitudes of stellar surface oscillations. For each value of the stellar rotational velocity, this produces a grid of line profiles and bisectors for differently strong spectral lines. Observed line profiles and bisectors were then independently compared against these grids, and 'best-fit' values of *V sin i* deduced. Figures 4-5 show the resulting comparison between synthetic and observed data for the well-observed solar near-twin α *Cen A*.

Despite the similarity to the Sun, the strongest lines in α *Cen A* are clearly more asymmetric than corresponding solar ones (Fig.5; compare with solar bisectors plotted

Alpha Centauri A

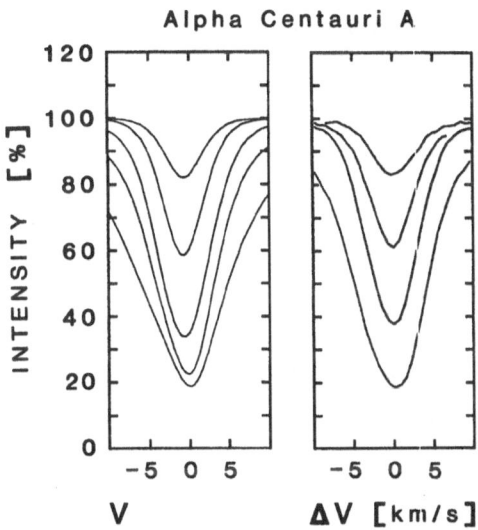

Figure 4. Comparison between *'best-fit'* *synthetic* and *observed line profiles* for the G2 V star α *Cen A*. Synthetic Fe I lines (χ = 3 eV) with zero-rotation depths 20, 45, 69, 80 and 83% of the flux continuum at λ 520 nm are shown at left, and Fe I line profiles, observed with a resolution $\lambda/\Delta\lambda \simeq 200{,}000$ are at right (λ 685.57; 543.63; 633.53; 543.45 nm; for details see Dravins, 1987b). From a grid of rotationally broadened profiles, a 'best-fit' rotation $V \sin i \simeq 1.8 \pm .3$ km/s was obtained. The detailed agreement between observed and synthetic profiles is obtained without involving any classical parameters such as 'mixing-length', 'micro-', nor 'macro-turbulence'. (Dravins and Nordlund, 1989b)

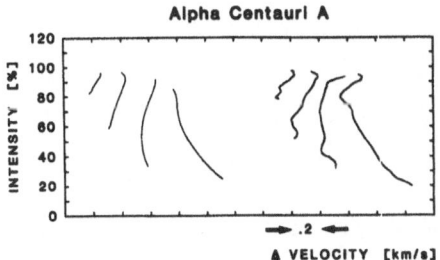

Figure 5. Comparison between *observed* and *'best-fit'* *synthetic bisectors* for α *Cen A*, analogous to Fig.4. A bisector grid gives a best-fit stellar rotation $V \sin i \simeq 1.7 \pm .3$ km/s, independently of profile fits. The observed bisectors are averages for Fe I lines from Dravins (1987b). The detailed agreement shows that also the asymmetric part of the α *Cen A* lines is well modeled. This model differs from the solar one only by slightly lower surface gravity. This leads to a slightly more vigorous convective overshoot, causing a more pronounced asymmetry in the cores of the strongest lines; cf. Fig.3. (Dravins and Nordlund, 1989b)

in the same format in Dravins, 1987b). This difference to the Sun can be traced back to be an effect of the lower surface gravity in α *Cen A*, which permits a more vigorous convective overshoot and slightly larger velocity amplitudes in the high photosphere, where the cores of strong lines often are formed. These velocities give contributions to the absorption not only near the spatially averaged line core, but also further out in the line flanks, in a manner to produce this distinct bisector signature (cf. bottom panels in Fig. 3). It is encouraging that current observations and models permit the study of such subtle differences between the velocity fields in two G2 V stars, revealing the effects on solar granulation of some three billion years of stellar evolution.

4. Understanding Granulation in Different Types of Stars

Powerful as numerical simulations may be, they still have limitations in modeling granulation in situations very different from the solar one. The problems arise mainly from the extensive computing power required, and the multidimensional nature of the problem. This makes it computationally very expensive to e.g. greatly increase the number of spatial resolution points to include features of all significant (but unknown) geometrical scales. It is not yet practical to produce a grid with a large number of granulation models throughout the Herzsprung-Russell diagram, and many theoretical challenges remain.

4.1. THE GEOMETRIC SCALE OF GRANULATION

Numerical simulations confirm that an important factor determining the geometric scale of granulation is the density scale height (at least in solar-type stars). Since that increases greatly from main-sequence stars to giants and supergiants, one may suspect

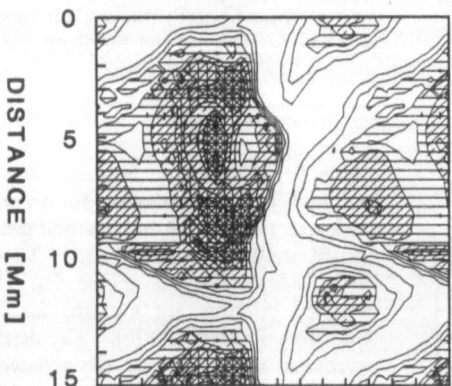
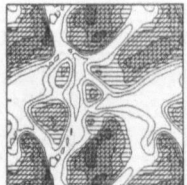

Figure 6. The optical appearance of typical granulation patterns at stellar disk center in different stars, plotted on the same geometric scale. From left to right are models corresponding to β *Hyi* (G2 IV), α *Cen A* (G2 V), and α *Cen B* (K1 V). Areas brighter than average are shaded, with contours at every 20% of the average. These synthetic continuum images at λ 520 nm illustrate the increase in the geometric scale of granulation with increasing density scale height. Adapted from Dravins and Nordlund (1989a).

that granules on giant stars could be very large, perhaps so large that only a small number of them exist on the entire stellar surface (Antia *et al.*, 1984; Schwarzschild, 1975). Figure 6 shows typical 'snapshots' of granulation images from three stellar models, plotted on a constant geometric scale. Clearly, already in this small region of the Herzsprung-Russell diagram, there is a significant change in granular sizes. One can imagine many different processes affecting granules in giant stars, and their detailed modeling remains a challenge for the future.

4.2. 'INVERSE' LINE ASYMMETRIES IN EARLY TYPE STARS

While solar and cool-star bisectors often are similar to the letter 'C', bisectors in early-type stars often show the *inverted* 'Ɔ' shape. Indeed, there seems to be a 'granulation boundary' in the Herzsprung-Russell diagram separating these types of bisectors, running from G supergiants to near F0 on the main sequence (Gray and Nagel, 1989). This boundary largely coincides with that of readily observable chromospheric emission (Dravins, 1981). The generally more rapid rotation in early main-sequence stars makes their spectra more difficult to study, but there are indications of similarly inverted asymmetries in the relatively sharp-lined spectra of the dwarfs *Sirius* (A1 V; Dravins, 1987b) and *τ Sco* (B0 V; Smith and Karp, 1978; 1979).

While one does not expect 'ordinary' surface convection driven by hydrogen ionization in the atmospheres of such hotter stars, there is no reason to believe their atmospheres should be static. Organized motions could well arise due to the He II ionization zone just below the photosphere, or through some other mechanism. Hydrodynamic models for the A0 V star *Vega* by Gigas (1989) indeed show its atmosphere to develop an oscillatory behavior, leading to 'inverse' line asymmetries.

An application of simple few-component models, reproducing the forcefully inverted 'inverse' bisector shapes in F supergiants, suggests that their surfaces may be covered by small 'granules', rising rapidly with perhaps 10–20 km/s (Gray and Toner, 1986; Dravins, 1989). A visualization of the surfaces of such stars, featuring rapidly upwelling currents in small ascending 'geysers', whose output is balanced by a more sluggish downdraft over more extended areas, is certainly quite different from the familiar image of solar granulation. However, to verify whether such a scenario is correct, requires much more detailed modeling.

5. Challenges in Stellar Granulation Modeling

Clearly, detailed granulation modeling should be extended across the Hertzsprung-Russell diagram, and these simulations should be performed without the simplifying assumptions used up to now. In a shorter time perspective, an important improvement will be to include effects of wave propagation and shock formation. Current models strongly suggest that there is material moving at near-sonic velocities in stellar photospheres, at least in the vigorous granulation on hotter and on subgiant stars. Also, a study of the interaction between stellar granulation and magnetic fields should before long clarify the characteristics of small-scale flux concentrations in different types of stars. In a longer time perspective, one very important challenge can be identified:

5.1. MODELING ENTIRE STELLAR ENVELOPES

Current models have only treated very small simulation volumes, corresponding to (at most) a few granules, and not extending very deep into the star. Such simulations clearly are unable to model larger-scale structures, such as stellar *supergranulation* or possible *giant cells*. While they *do* show the convective overshoot in the upper photosphere, and its effects on spectral line profiles, the corresponding effects at the bottom of the convective zone (so important for chemical mixing and for understanding stellar evolution) are *not* modeled. A logical next great step in stellar astrophysics would be the *simulation of entire stellar envelopes*. Such simulations should predict new types of observable parameters such as the temporal power spectrum of global irradiance fluctuations at different wavelengths, or the power in different p-modes of the stellar global oscillations, as well as their typical lifetimes. The capacity of near-future very large ground-based telescopes, as well as planned space instrumentation should permit the accurate observation of such parameters, and thus both demand their modeling, and assure the necessary confrontation between theory and observations.

5.2. NEW COMPUTATIONAL TECHNIQUES

There is a need for *greatly* increased computing power to handle more realistic astro-physical situations, such as the modeling of entire stellar envelopes. Compared to current simulations, the desired increase in stellar area coverage is a factor $\approx 10^6$, a factor $\approx 10 - 100$ in the vertical extent, and ≈ 10 in the spatial (and possibly also temporal) resolution. The latter is needed in particular to handle wave motion and shock formation. Unless radically more efficient program codes can be invented, there is a need for perhaps $\approx 10^9$ more computing power than used at present. At first sight this may appear utterly discouraging since computer performance normally increases by only *one* order of magnitude from one generation of supercomputers to the next. However, such an ambition level may not be entirely unrealistic.

Workers in computer science are currently putting considerable effort into the development of various 'non-classical' concepts, e.g. employing novel computer architecture with many processors working in parallel. (For a general introduction to such concepts, see e.g. Bowler *et al*, 1987.) The greatly improved possibilities for computer-aided design and manufacture of unique electronic circuits at not unreasonable cost promises to make it possible to build a computer with circuits custom-designed for solving the problem of stellar convection. The combination of such circuits with novel computer architecture might well improve computing power by *several* orders of magnitude over today's 'supercomputers'. For example, a dedicated numerical laboratory with 4000 custom-designed CPU's running for three years, will offer more than 100 million CPU-hours, a computing power that may well be adequate to begin realistic simulations of entire stellar atmospheres.

6. Challenges in Stellar Granulation Observations

Effects from stellar granulation can be observed through various subtle signatures in high resolution spectra of integrated starlight, through the time variability of stellar irradiance, and by a few other means. The most accessible observational parameter appears to be that of photospheric absorption line asymmetries. Several authors have now reported observations of such asymmetries, and measurements of relative wavelength shifts between different classes of lines are also becoming available (e.g. Dravins, 1982; 1987a; 1987b; Gray, 1988; and references therein). In the future, we can expect additional data from (a) space-based stellar photometry with micromagnitude precision, allowing the observation of time variability of stellar irradiance due to the time evolution of granular features; (b) space-based astrometric measurements of stellar radial velocities (not involving the stellar spectrum), thus allowing the determination of *absolute* convective wavelength shifts; (c) the use of active optics on the very large telescopes now under construction, which should allow diffraction-limited imaging of surface features on giant stars as well as spatially resolved spectroscopy across stellar disks; (d) high-resolution imaging of stellar surfaces that is now becoming feasible through the development of long baseline optical interferometry and optical aperture synthesis.

6.1. OBSERVING LINE ASYMMETRIES

The faithful observation of the subtle photospheric line asymmetries caused by stellar granulation is a more demanding task than most other stellar spectroscopy applications. In particular one can point to the issues of:

6.1.1. *Finite Spectral Resolution.*
The asymmetries of spectral lines are often represented by their bisectors. In order to 'fully' resolve the bisector shape, some 4 – 5 independent points on the curve may be required. Since the relevant width of most photospheric lines in ordinary stars is on the order of 8 – 10 km/s, and each bisector point is obtained from *two* intensity measurements in opposite flanks of the line, this demands 8 – 10 points across the line, corresponding to a resolution of ≈ 1 km/s ($\lambda/\Delta\lambda \approx 300,000$). Numerical simulations of how bisector shapes are distorted by a lower resolution clearly illustrate this point (Dravins, 1987a; Livingston and Huang, 1986).

Such resolutions are seldom available in stellar spectrometers, where $\lambda/\Delta\lambda \approx 100,000$ is more common. Although such a resolution may be adequate to detect the *presence* of line asymmetries, the study of their detailed shapes or the detection of any differences between different types of lines becomes marginal. The fundamental problem is *not* the lack of spectroscopic instruments of sufficient performance, but the fundamentally poor light efficiency of those that are technically sufficient. For example, *Fourier transform spectrometers* (FTS) offer fully adequate spectral resolution, a good instrumental profile, possibilities for accurate wavelength determination, and other valuable features. The problem is that an FTS, scanning successive Fourier components of absorption-line spectra in the optical, is photon-noise limited, analogous to a wavelength-scanning single-channel spectrometer. (This is because all the light in the instrument falls onto the detector, and contributes photon noise from all wavelengths all the time, while the desired signal is only a slight modulation on top of this.) In order to use FTS's for stellar work one is forced either to accept a lower spectral resolution or to introduce rather narrow bandpass-limiting wavelength filters. Even so, the observation of only a few spectral lines in very bright stars may require several hours (e.g. Wayte and Ring, 1977; Nadeau and Maillard, 1988).

Fortunately, there is reason for optimism due to the current wave of construction of very large telescopes in different parts of the world. Their greatly increased light collecting power will overcome the photon limits in today's 'large' telescopes, and make Fourier transform spectrometers generally available for stellar spectroscopy. An estimate of the likely performance of an FTS on an 8-10 meter class telescope shows that, at spectral resolving power $\approx 10^6$, a signal-to-noise ratio of 100 should be reached for an $m_v = 3$ star in one night's integration over a bandpass of 10 nm (100 Å). Although some astronomers might still think of $m_v = 3$ as representing rather bright objects, the number of stars whose spectra could be studied in great detail will now be counted in the hundreds rather than the handful of the very brightest ones that at present are barely accessible to such studies.

6.1.2. *Problems in Stellar Spectra.* Although spectral lines to be studied obviously are selected to be as undisturbed as possible, they are almost never completely unblended. This makes it awkward to draw conclusions on line asymmetries from measurements of very few lines only: how is one to know if an asymmetry is intrinsic to the stellar photosphere or due to a blend? These problems become especially pronounced for measurements close to the continuum: in weak lines or in the wings of stronger ones. Since one can expect blending lines to be randomly positioned in the wings of different primary lines, one solution is to simply average the asymmetries of several similar lines.

If very high spectral resolution is required, this is not a trivial requirement, since each line to be observed might require hours of observing time. In cooler and/or more rapidly rotating stars it may in practice be impossible to find any significant number of sufficiently undisturbed lines. Solutions could possibly include going to the infrared (where line densities are lower), or face the problem of observing (and modeling!) not individual spectral lines, but rather *line complexes*. For very cool stars this could even be the only alternative. Such complexes of blended lines will carry much the same information as individual lines, but the observational possibilities of averaging bisectors of many similar lines will disappear. Likewise, accurate laboratory wavelengths will be needed for all components making up the blend since average values over similar lines will not suffice.

6.2. DETERMINING WAVELENGTH SHIFTS

Since stellar spectral lines are intrinsically asymmetric, their 'wavelengths' can not simply be defined as single quantities. Ideally, line profiles and their bisectors should be measured on an *absolute* wavelength scale that allows the determination of *convective lineshifts* in different parts of the line, relative to the line's laboratory wavelength (corrected for the stellar radial velocity and its gravitational redshift). Such accurate wavelengths in spectral lines form important constraints on different theoretical models, which sometimes may even predict the same line asymmetry, but different wavelength shifts. This observational task presents several challenges:

6.2.1. *Wavelength Dependence on Spectral Resolution.* The finite spectral resolution in stellar spectrometers causes wavelength shifts of stellar lines. The instrumental convolution of an asymmetric stellar line profile with even a perfectly symmetric instrumental one results in a line profile of different asymmetry, where different parts of the bisector now are at different wavelengths. Since e.g. differently strong lines have different intrinsic asymmetries, they will be differently affected by the same spectral resolution, possibly mimicking the astrophysically expected behavior between different groups of lines. For solar-type asymmetries, such instrumentally induced shifts may reach 100 m/s for resolutions $\lambda/\Delta\lambda \approx 100,000$ (Bray and Loughhead, 1978; Dravins, 1987a; Livingston and Huang, 1986).

6.2.2. *Wavelength Shifts Between Different Lines in the Same Star.* Convective wavelength shifts can be expected to be similar for groups of lines with common properties. Since only *statistical data* for groups of lines are required, it is *not* really necessary to record the full stellar spectrum. Rather, it could suffice with data from radial velocity measuring machines, utilizing the cross-correlation of the stellar spectrum with some spectrum template. The well-known instruments of the Griffin- and CORAVEL-type employ hardware masks and reach precisions on the order of 100 m/s, adequate to begin searches for signatures from stellar granulation.

With different masks, e.g. such preferentially selecting high- or low-excitation lines, one could detect *differential* radial velocities between groups of different lines in the same star. Next-generation radial velocity instruments are likely to avoid mechanical masks in order to improve light efficiency and allow the integration of the full spectrum all the time. Their spectrum templates will then be defined in software by selecting features in the spectrum for cross-correlation. Although there are likely to be many practical hurdles, it should in principle be straightforward to define different templates by selecting interesting groups of lines (from either observed or synthetic; stellar or laboratory spectra) to search for signatures from stellar granulation.

6.2.3. *The Need for Very Accurate Laboratory Wavelengths.* The most useful atomic species for convective lineshift studies appears to be iron. It has high atomic mass (minimizing the thermal broadening of the stellar lines), its hyperfine and isotope splitting has few complications from atomic and isotope structure, and it has a rich and well-studied spectrum. In the Sun and solar-type stars, there are approximately 500 'unblended' Fe I and 50 Fe II lines in the visual spectrum, with another 100 or so lines in the infrared. Reasonable laboratory wavelengths exist, and convective lineshifts have been studied in the visual solar spectrum for Fe I and Fe II (Dravins *et al.*, 1981; 1986), and in the infrared for Fe I (Nadeau, 1988).

Although significant effects are visible for line-group averages, the laboratory wavelength accuracies are *not adequate* to test different granulation models using measurements of *individual* lines only. The noise may be ≈ 100 m/s for stronger Fe I lines, but rapidly gets worse for weaker lines, for high-excitation ones, for Fe II, and for lines in the infrared. Indeed, the lack of sufficiently accurate laboratory data is now the main limiting factor in these studies. Since astronomers do not ordinarily provide such laboratory data, we have to direct our requests to atomic physics groups. The main problem is not the accuracy of measurement, but rather the difficulty of preparing laboratory light sources whose wavelengths are free from systematic effects such as e.g. pressure shifts.

6.2.4. *Different Wavelength Shifts in Different Stars.* As shown in Fig. 7, stars of different temperature and luminosity are predicted to have different amounts of convective lineshift. The vigorous granulation in a $T_{eff} = 6600$ K model of the F5 star Procyon is expected to cause blueshifts in ordinary Fe I lines of ≈ 1000 m/s, while those in the cooler $T_{eff} = 5200$ K model amount to only ≈ 200 m/s (Dravins and Nordlund, 1989b). Observed solar values fall in between, typically around 400 m/s. Understanding such effects is important not only for the study of stellar photospheres *per se*, but is also required for the accurate determination of stellar radial velocities, in particular in systems with small internal velocity dispersions, such as open galactic clusters.

To separate convective shifts from shifts due to stellar motion, requires one to somehow determine the absolute or relative stellar radial velocity without using the spectral lines. (One also needs to correct for different gravitational redshifts in different stars, but stellar models allow this to be done with good accuracy.)

Differential convective lineshifts should be measurable between stars that share the same space velocity (even if the amount of this velocity is not known). The components of *binary stars* must share the same system velocity when averaged over their orbits. Thus, visual binaries with not too long periods could be suitable objects to search for spectral-type dependent differences in orbit-averaged apparent radial velocities.

From galactic dynamics arguments, the velocity dispersion of stars in some young galactic clusters is expected to be only a fraction of one km/s: less than expected differences in convective lineshift between different spectral types. Thus, if one could identify systematic differences in apparent velocities between different classes of cluster members, this would be evidence for different convective lineshifts in different stars.

To determine *absolute* lineshifts is more challenging since it requires the accurate determination of stellar radial velocities without using any spectral lines nor invoking the Doppler principle. This is possible for the Sun, where the solar motion is well determined from planetary system dynamics (rather than from apparent Doppler shifts of photospheric spectral lines), and consequently solar wavelengths can be corrected for the Sun-Earth motion and compared to laboratory values.

In principle, *astrometric* measurements could do the same for stars: if one could accurately determine a star's three-dimensional position in space at different times, the difference in position would yield its space velocity. Although current accuracies are insufficient for such measurements of individual stars, the advent of space astrometry promises to make at least some classes of related measurements possible.

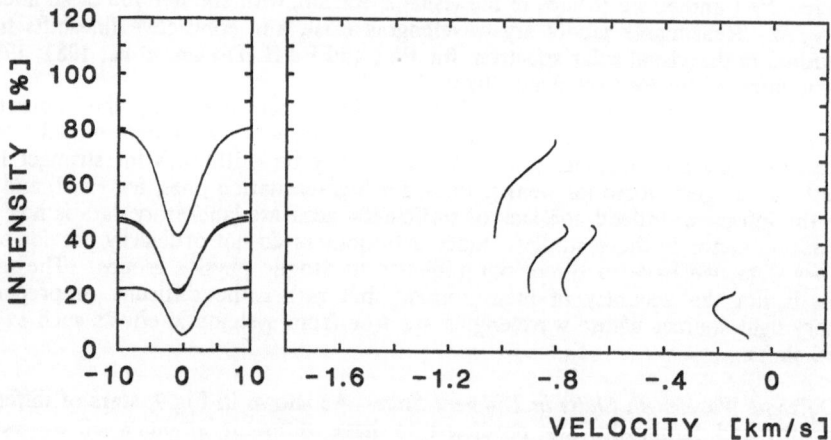

Figure 7. The *same spectral line in different stars*. Temporally and spatially averaged Fe I line profiles and bisectors for integrated starlight from four different models are plotted on the same absolute scale. The intensity is normalized to stellar surface area at *Procyon* disk center, and the wavelength scale is absolute. The oscillator strength for this $\chi = 3$ eV Fe I line at λ 520 nm was chosen such that the disk-center absorption depth in the *Procyon* model is = 60%. From top to bottom, and from left to right, these models represent *Procyon* (F5 IV-V), β *Hyi* (G2 IV), α *Cen A* (G2 V), and α *Cen B* (K1 V). The convective blueshift increases with increasing temperature, and also with increasing luminosity. Observed solar values fall between those of α *Cen A* and α *Cen B*. The corresponding observation of different convective wavelength shifts among different stars remains an observational challenge. A further discussion is in Dravins and Nordlund (1989b).

6.3. CYCLIC CHANGES OF ASYMMETRIES AND SHIFTS

A temporal variation in the granulation pattern may change the convective wavelength shifts, and thus mimic a varying stellar radial velocity. To understand such effects could be very important for one very challenging astronomical problem: the search for possible extrasolar planets.

6.3.1. *The Detection of Planets Around Other Stars.* Among plausible means of detecting such planets, one of the most promising seems to be the long-term monitoring of the radial velocity of the parent star. If the star is moving in conjunction with an unseen planet, one expects a cyclic change in its velocity. For the Sun-Jupiter system, the amplitude of the solar velocity due to its motion around their common center of gravity is 13 m/s, with a period of 12 years. There now exist stellar radial velocity instruments with measuring precision sufficient for this task, but there remains the problem of separating cyclic changes due to stellar motion from those due to changes in convective lineshifts.

Solar granulation structure, bisector curvature and amount of lineshift are observed to vary between solar active regions and quiet ones. Consequently, the different area coverage of active regions during different phases of the solar 11-year cycle must lead to changes in the bisector curvature and wavelength shift also in integrated sunlight. Such changes have been studied by Livingston (1983), Deming *et al.* (1987), Jiménez *et al.* (1988) and Wallace *et al.* (1988).

The exact effect on the apparent radial velocity depends upon precisely how the line is measured, but may correspond to an amplitude of ≈ 30 m/s over the solar cycle. Such a magnitude is consistent with observed differences between active and quiet regions, and their different area coverage in different years of the solar cycle. Since many stars possess activity cycles, one should expect qualitatively similar effects in other stars. Clearly, if the Sun, seen from afar, displays an apparent velocity variation of perhaps 30 m/s with a period of 11 years, that will not be simple to disentangle from the 13 m/s amplitude over 12 years, induced by Jupiter.

To overcome the problem requires a better understanding of stellar granulation properties. Of great importance would be a systematic long-term monitoring of stellar line asymmetries and accurately measured wavelength shifts in different stars. Granulation changes may correlate with active region coverage, and a measure of that is given by the Ca II K chromospheric emission intensity. This should also be monitored in order to identify the phase and period of possible stellar activity cycles. True velocity changes must affect all spectral lines, while granulation changes should affect different lines differently. Variations on the shorter timescales of stellar rotation might identify active region patches of significantly modified granulation (Toner and Gray, 1988), while variations over several years might identify the changing area coverage of such features.

6.4. OBSERVING STRUCTURES ON STELLAR DISKS

Much of the progress in astronomy is based upon improved spatial resolution. Our understanding of planets, nebulae or galaxies would have been very limited indeed, if we only had observed them as point sources. One of the major aims of stellar granulation studies must be to ultimately observe the fine structure of stellar surfaces and to enable the study also of stars as extended objects. Along with this go related aims, such as understanding the physics of stellar line formation in different atmospheric inhomogeneities. The observational problems are of course caused by the small angular extent of stellar disks – no more than a few tens of milliarcseconds even for the largest stars. Nevertheless, with proper techniques, also the fine structure on stellar disks should become accessible for observation.

6.4.1. *Indirect Methods: Deducing Center-to-Limb Line Profile Changes.* Line profiles in disk-integrated starlight are built up by contributions from different disk positions. Since both line asymmetry, convective lineshift and continuum brightness depend on the disk position, the integrated line profile incorporates these effects in a complex manner.

414

What is needed is a tool to disentangle the different quantities. Such a tool could be available in *stellar rotation*.

The line profile contributions from different center-to-limb positions, $\cos \theta = \mu$, are *not* equal for stars of different rotational velocities $V \sin i$. Increased rotation changes the Doppler shift at each μ due to the increased projected velocity, and since the contributions from each μ have a different line asymmetry, these will be differently distorted by different rotational broadening. For more rapid rotation, when the asymmetric line components originating near the stellar limbs begin to affect the wings of the disk-integrated profile, the bisector patterns change significantly, and the line asymmetries may even become enhanced. This phenomenon was suggested by Gray and Toner (1985; Gray, 1986), and studied in more detail for a simulated rapidly rotating Sun by Smith *et al.* (1987).

Numerical simulations permit the computation of synthetic line profiles for different center-to-limb positions in different stars. From such data, synthetic full-disk line profiles and bisectors are obtained for different rotational velocities $V \sin i$ (Dravins and Nordlund, 1989b). What is still lacking is the observational counterpart: a sequence of line profile, line asymmetry and wavelength shift measurements for groups of stars of the same spectral type, only differing by successively more rapid rotation. Such data could effectively constrain granulation models, and permit the analysis of line profile changes across stellar surfaces.

6.4.2. *Direct Methods: Interferometric Imaging of Stellar Surfaces.* Although indirect methods might be quite powerful, they do not replace the ultimate need for direct methods to obtain *images and spatially resolved spectra* across stellar surfaces. The angular (diffraction-limited) resolution of present telescopes in the 4-6 meter class is around 20 milliarcseconds in the visual. Using e.g. speckle interferometry, this allows one to resolve perhaps 10 surface elements on the largest red giants. The forthcoming very large telescopes in the 8 – 10 m class will significantly improve the situation. Of particular promise is the potential of active optics, which could allow the stable imaging of a stellar disk onto a Fourier transform spectrometer for two-dimensional high-resolution spectral observations with good wavelength calibration. It is precisely such data of the center-to-limb changes of stellar line profiles and wavelength shifts that are required to analyze the physics of stellar line formation.

Optical interferometers with baselines on the order of 100 m and more are now becoming operational, and will offer resolutions around 1 milliarcsecond. That is sufficient for thousands of resolution elements on the disks of red giants, and for resolving nearby main-sequence stars. (The baseline requirements for resolving different classes of stars are discussed by e.g. Dupree *et al.*, 1984.) Since granules on some giant stars might subtend a significant fraction of a stellar diameter, their appearance and spectral features might soon become detectable through e.g. speckle spectroscopy. Solar-type granules, however, have sizes only about one thousandth of the stellar diameter, and their imaging requires baselines a thousand times longer than those required to resolve the stellar disk. To achieve this may require kilometric arrays of space-based optical phase interferometers (or possibly ground-based intensity interferometers), and their feasibility is now under active study by different groups in the world.

ACKNOWLEDGEMENT

This work is supported by the Swedish Natural Science Research Council

References

Antia, H.M., Chitre, S.M., Narashima, D.: 1984, *Astrophys. J.* **282**, 574

Atroshchenko, I.N., Gadun, A.S., Kostik, R.I.: 1989, in R.F.Rutten, G.Severino, eds. *Solar and Stellar Granulation*, Kluwer, p. 521

Bowler, K.C., Bruce, A.D., Kenway. R.D., Pawley, G.S., Wallace, D.J.: 1987, *Physics Today* **40**, No. 10, p. 40

Bray, R.J., Loughhead, R.E.: 1978, *Astrophys. J.* **224**, 276

Deming, D., Espenak, F., Jennings, D.E., Brault, J.W., Wagner, J.: 1987, *Astrophys. J.* **316**, 771

Dravins, D.: 1981, *Astron. Astrophys.* **98**, 367

Dravins, D.: 1982, *Ann. Rev. Astron. Astrophys.* **20**, 61

Dravins, D.: 1987a, *Astron. Astrophys.* **172**, 200

Dravins, D.: 1987b, *Astron. Astrophys.* **172**, 211

Dravins, D.: 1989, in R.J.Rutten, G.Severino, eds. *Solar and Stellar Granulation*, Kluwer, p. 493

Dravins, D., Larsson, B., Nordlund, Å.: 1986, *Astron. Astrophys.* **158**, 83

Dravins, D., Lindegren, L., Nordlund, Å.: 1981, *Astron. Astrophys.* **96**, 345

Dravins, D., Nordlund, Å.: 1989a, *Astron. Astrophys.*, in press

Dravins, D., Nordlund, Å.: 1989b, *Astron. Astrophys.*, in press

Dupree, A.K., Baliunas, S.L., Guinan, E.F.: 1984, *Bull. AAS* **16**, 797

Gigas, D.: 1989, in R.F.Rutten, G.Severino, eds. *Solar and Stellar Granulation*, Kluwer, p. 533

Gray, D.F.: 1986, *Publ. Astron. Soc. Pacific* **98**, 319

Gray, D.F.: 1988, *Lectures on Spectral-Line Analysis: F, G, and K Stars*, The Publisher, Arva (Ontario)

Gray, D.F., Nagel, T.: 1989, preprint

Gray, D.F., Toner, C.G.: 1985, *Publ. Astron. Soc. Pacific* **97**, 543

Gray, D.F., Toner, C.G.: 1986, *Publ. Astron. Soc. Pacific* **98**, 499

Gray, D.F., Toner, C.G.: 1985, *Publ. Astron. Soc. Pacific* **97**, 543

Jiménez, A., Pallé, P.L., Régulo, C., Roca Cortés, T., Elsworth, Y.P., Isaak, G.R., Jefferies, S.M., McLeod, C.P., New, R., van der Raay, H.B.: 1988, in J.Christensen-Dalsgaard, S. Frandsen, eds. *Advances in Helio- and Asteroseismology*, Reidel, IAU symp 123, p. 215

Livingston, W.C.: 1983, in J.O.Stenflo, ed. *Solar and Stellar Magnetic Fields: Origins and Coronal Effects*, Reidel, IAU symp. **102**, p. 149

Livingston, W., Huang, Y.R.: 1986, in M.S.Giampapa, ed. *The SHIRSHOG Workshop*, National Solar Observatory, Tucson, p. 1

Nadeau, D.: 1988, *Astrophys. J.* **325**, 480

Nadeau, D., Maillard, J.P.: 1988, *Astrophys. J.* **327**, 321

Nelson, G.D.: 1980, *Astrophys. J.* **238**, 659

Nordlund, Å.: 1982, *Astron. Astrophys.* **107**, 1

Nordlund, Å.: 1985, *Solar Phys.* **100**, 209

Nordlund, Å., Dravins, D.: 1989, *Astron. Astrophys.*, in press

Schwarzschild, M.: 1975, *Astrophys. J.* **195**, 137

Smith, M.A., Huang, Y.R., Livingston, W.: 1987, *Publ. Astron. Soc. Pacific* **99**, 297

Smith, M.A., Karp, A.H.: 1978, *Astrophys. J.* **219**, 522

Smith, M.A., Karp, A.H.: 1979, *Astrophys. J.* **230**, 156

Toner, C.G., Gray, D.F.: 1988, *Astrophys. J.*, **334**, 1008

Wallace, L., Huang, Y.R., Livingston, W.: 1988, *Astrophys. J.* **327**, 399

Wayte, R.C., Ring.J.: 1977, *Monthly Notices Roy. Astron. Soc.* **181**, 131

Winkler, K.H.A., Chalmers, J.W., Hodson, S.W., Woodward, P.R., Zabusky, N.J.: 1987, *Physics Today* **40**, No. 10, p. 28

A MODEL FOR STELLAR CONVECTION AND SPECTRAL LINE ASYMMETRIES

H. M. Antia
Tata Institute of Fundamental Research
Homi Bhabha Road, Bombay 400 005, INDIA

ABSTRACT. A model for stellar convection zones based on linear convective modes using a nonlocal mixing length theory is developed to study the spectral line asymmetries and the line shifts resulting from convective motions in the stellar photospheric region. The amplitudes of these linear convective modes is estimated by requiring the convective flux due to a linear superposition of such modes to reproduce the convective flux in the mixing length model. To study the spectral line asymmetries the convective mode with the largest amplitude in the photospheric line formation region is chosen to represent the stellar velocity field and the accompanying intensity fluctuations. Synthetic spectral line profiles are obtained by summing locally symmetric profiles over the stellar disk according to the local Doppler velocity and intensity fluctuations. The resulting line bisector shapes and the line shifts are compared with observations for α-Cen B. It is found that while the simple model proposed here can explain either the line shifts or the line bisector shape reasonably well, it fails to explain both these characteristics simultaneously.

1. Introduction

The asymmetry and the accompanying line shift in the solar photospheric lines has been known for quite some time. These shifts and asymmetries have been ascribed to the correlated velocity and brightness patterns due to solar convection. Several studies have been made to detect asymmetries in other stars and some results have been obtained (Gray and Toner 1985, Dravins 1987). The asymmetry can be most conveniently characterized in terms of the line bisector which is the loci of points midway between equal intensity points on either side of the line.

A number of attempts have been made to model the convective motions in stars to reproduce the observed asymmetries. These models can be classified into two categories. Firstly there are simple two to four-stream models where the stellar surface is divided into two to four components of different characteristic brightness and velocity. The synthetic line profile is obtained as a summation of line profiles from these components appropriately weighted for their area coverage. By choosing the appropriate combinations of velocity and intensity over each of these components it is possible to model a wide variety of line bisector shapes. Such models are essentially ad hoc and the parameters are chosen to get the "best" fit to observations.

On the other hand Dravins, Lindgren, and Nordlund (1981) and Dravins (1988) have constructed a more sophisticated model of convection zone by actually solving the set of hydrodynamic equations for solar granulation. Because of constraints imposed by computer time such models have to be restricted in extent and hence, all scales of convection can not be included. However, such models give detailed convective velocity field and hence can be used to construct synthetic line profiles using the equation of radiative transfer. These

J. O. Stenflo (ed.), Solar Photosphere: Structure, Convection, and Magnetic Fields, 417–420.

models have essentially no free parameter and can provide a good test for the hypothesis of convective origin of spectral line asymmetries.

Antia and Pandey(1989, hereinafter paper I) considered a model of stellar convection zone in terms of linear convective modes. This model has the advantage of being simpler than the full simulation models and can provide velocity field over the entire stellar volume. Such stellar convection models can give useful insight about the scale of convection in various stars which could be used to construct the three dimensional simulation models.

The linear convective modes in the solar convection zone have been studied by Antia, Chitre, and Narasimha (1983) and it was found that there are two peaks in the growth rate versus horizontal wavelength plot which are in a reasonable agreement with the observed length and time scales of granulation and supergranulation. The linear stability theory used in this approach does not give the amplitudes of these modes which can only be determined by nonlinear processes. However, it is reasonable to assume that the convective flux will be transported by a combination of these modes. Narasimha and Antia (1982) have demonstrated that it is possible to construct a linear superposition of statistically independent unstable convective modes which reproduces convective flux as required by the mixing length theory over the entire convection zone. The resultant vertical velocities were found to be in a reasonable agreement with the observed granular velocities in the solar atmosphere. This may not provide a fully realistic model of stellar convection zone since it is based on the mixing length theory which itself is uncertain to a large extent. However, it can provide a reasonable first approximation to stellar convection zone.

Since it is not possible to estimate the relative phase of each of these convective modes which in any case cannot be constant. Hence, for simplicity we approximate the velocity and temperature fluctuation on the stellar surface as being due to a single convective mode. For this purpose we select the mode with largest amplitude of velocity and intensity fluctuations in the photospheric line formation region. Using the velocity field provided by the convection zone model we construct synthetic line profiles by integrating over the entire stellar surface. At each point on the stellar surface the line profile is assumed to be symmetric, the actual form being given by the usual Voigt function.

Following this approach Antia and Pandey(1989) constructed stellar envelope models for four stars i.e., the Sun, α-Cen A, Arcturus and Procyon. The resulting line bisectors were found to be in a reasonable agreement with observed shapes but the line shifts were suppressed. In this paper the study has been extended to α-Cen B and moreover the line shifts are also included.

2. Numerical Results

Following paper I, a convection zone model for α-Cen B is constructed using a mixing length of $z + 409$ km. The resulting model has a convection zone depth of $\approx 2 \times 10^5$ km. The linear stability of this model is studied to find the convective modes. The turbulent prandtl number is assumed to be 0.33 in these calculations. The growth rate of these mode as a function of the degree of spherical harmonic l is displayed in Fig. 1. It can be seen that there are two peaks in the curve, the dominant peak around $l = 400$ corresponds to a horizontal scale of $\approx 10^4$ km while the secondary peak around $l = 90$ corresponds to a length scale of $\approx 4.5 \times 10^4$ km. The corresponding time scales are ≈ 16 and 130 hrs respectively. Hence for α-Cen B the granules should have a typical length scale of 10^4 km and life-time of several hours.

TABLE 1: Amplitude of convective
modes at optical depth=0.4

ℓ	V_r (ms^{-1})	V_h (ms^{-1})	F_1/F
1	.01	−20	2×10^{-6}
2	.04	−69	8×10^{-6}
4	.09	−97	2×10^{-5}
8	.34	−166	2×10^{-5}
16	.73	−223	2×10^{-4}
24	.94	−216	3×10^{-4}
36	1.76	−270	5×10^{-4}
54	.32	−37	1×10^{-4}
80	2.46	−283	1×10^{-3}
120	19	−829	4×10^{-3}
180	74	−118	2×10^{-3}
270	1600	−2510	4×10^{-2}
400	1800	−4600	8×10^{-2}
450	61	−217	4×10^{-3}

To determine the amplitudes of these modes we obtain a linear superposition which fits the convective flux assumed in the mixing length model. The results are shown in Table 1 which gives the amplitude of velocity and flux perturbation at the level of optical depth 0.4. It can be seen that the perturbations are dominated by $\ell = 400$ convective mode. Unlike the case of the Sun or α-Cen A there is no pronounced secondary peak in these amplitudes. Fig. 2 displays the convective flux profile obtained by superposing 14 modes listed in Table 1. This figure also shows the convective flux profiles of some of the individual modes. It can be seen that the modes with different values of ℓ have peaks at different depths. Further, as explained in paper I we can identify the mixing length at a given depth with the width of flux profile which has peak at that depth. Fig. 3 displays the mixing length as a function of $\log P$, which can be compared with the width of convective flux profiles of individual modes marked by circles in the figure.

Using the velocity and flux perturbations due to the dominant convective mode the synthetic line profiles and bisectors are constructed as explained in paper I. The result is shown in Fig. 4 where the bisector is shown on a scale which is ten times the scale for line profile. For clarity Fig. 5 displays the line bisectors on an expanded scale where the curves have been shifted horizontally so that they do not overlap with each other. These line bisectors can be compared with the observed bisectors in Dravins(1987). It can be seen that the synthetic bisectors shape is in general agreement with the observed shapes.

While for other stars the line shift is not yet measured, but for the Sun the line shift is known to be of order of 400 m s^{-1}. For the five typical line bisectors shown for the Sun from left to right in Fig. 6(a) of paper I the calculated line shifts are found to be respectively $439, 313, 358, 298, 202$ ms^{-1}. While the magnitude of these shifts is comparable to the observed values unfortunately the sign is not correct and all the lines shown in paper I are red shifted rather than the observed blue shifted lines. It is found that if the sign of F_1/F used to construct line profiles is changed then the correct sign and value of line shift is obtained. However, in this case the line bisector essentially gets reflected and hence the shape does not agree with observations. It is not clear to me whether this discrepancy is due to some trivial point that I have missed in the calculations or is due to the inadequacy of the model.

For α-Cen A the red shifts for the typical profiles given in Fig. 6b of paper I are 834, 789, 629, 612, 1050 m s^{-1}. For Arcturus the values are 169, −868, 170, −257, −1066 m s^{-1}. Thus for Arcturus it is possible to get both red shifted and blue shifted lines. This is probably due to the fact that modes with small values of ℓ dominate the fluctuations and for these modes the values may even depend on the orientation of axis assumed in the study. For Procyon all the typical profiles are blue shifted, with blue shifts of 672, 810, 1282, 623 and 1070 m s^{-1} respectively. For α-Cen B the typical profiles shown in Fig. 5 are red shifted by 288, 314, 291, 433, 319 m s^{-1} respectively from left to right.

420

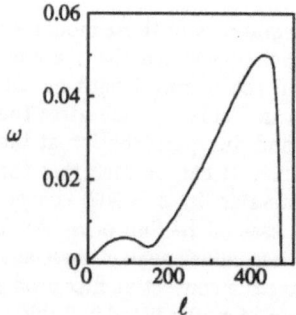

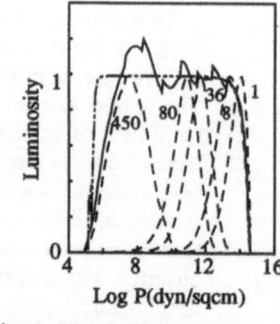

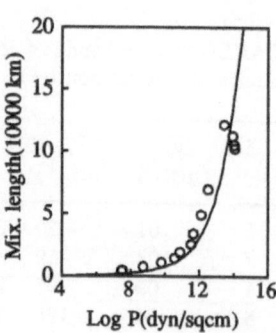

Fig. 1: The growth rate ω (in units of $\sqrt{\frac{3GM}{4\pi R^3}}$) of convective modes as a function of the degree of spherical harmonics ℓ for α-Cen B.

Fig. 2: The dashed curves show the convective luminosity due to individual convective modes for various values of ℓ as a function of $\log P$. The continuous curve shows superposed convective luminosity profile, while dot-dashed curve represents the model convective luminosity.

Fig. 3: The mixing length vs. logarithm of pressure for α-Cen B. The width of the luminosity profile of various convective modes is indicated by circles.

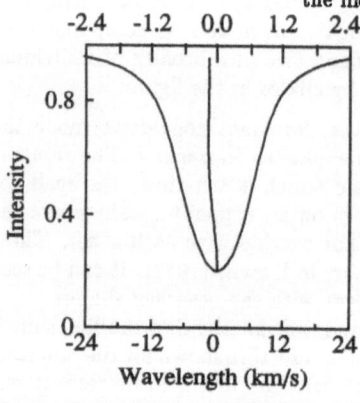

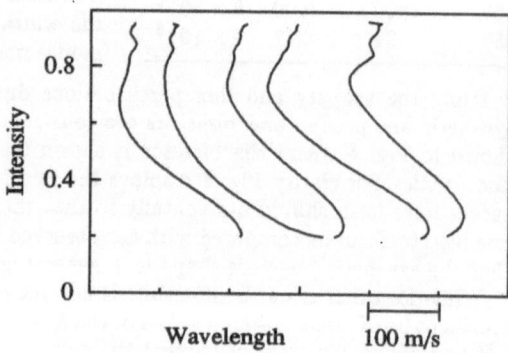

Fig. 4: Representative synthetic line profile and bisector for $\ell = 400$, $V_r = 2.5$ km s^{-1}, $V_h = -7$ km s^{-1}, and $F_1/F = -0.1$. The horizontal wavelength scale (in velocity units) refers to the line profile, while the ten times expanded upper scale refers to the bisector.

Fig. 5: Synthetic line bisectors for some typical values of the amplitudes of convective modes. The curves have been shifted horizontally to avoid overlapping. The values of $(\ell, V_r, V_h, F_1/F)$ for various curves from left to right are: (400, 2.5, −7, −0.1), (400, 2, −8, −0.1), (400, 1.5, −7, −0.2), (400, 2, −7, −0.2), (400, 2, −6, −0.2). Here V_r and V_h are in km s^{-1}.

REFERENCES

Antia, H. M., Chitre, S. M., and Narasimha, D. 1983, *Mon. Not. R. Astr. Soc.*, **204**, 865.
Antia, H. M., and Pandey, S. K. 1989, *Astrophys. J.*, **252** (in press).
Dravins, D. 1987, *Astr. Ap.*, **172**, 211.
Dravins, D. 1988, in R. J. Rutten, G. Severino, eds. *Solar and Stellar Granulation.*
Dravins, D., Lindgren, L., and Nordlund, Å. 1981, *Astr. Ap.*, **96**, 345.
Gray, D. F., and Toner, C. G. 1985, *Publ. Astron. Soc. Pacific*, **97**, 543.
Narasimha, D., and Antia, H. M. 1982, *Ap. J.*, **262**, 358.

COMPARATIVE ANALYSIS OF PHYSICAL CONDITIONS IN THE SOLAR AND PROCYON ATMOSPHERES

I.N. ATROSHCHENKO, A.S. GADUN, R.I. KOSTIK, K.N. PIKALOV
Main Astronomical Observatory
Academy of Sciences of the Ukrainian SSR
252127, Kiev, USSR

ABSTRACT. An analysis is made of the velocity field determined from spectral absorption line profiles as well as inhomogeneous element sizes obtained from stochastic theory. The results of three-dimensional simulations of convective motions in the photospheres of the Sun and Procyon are presented. It is concluded that penetrative convection covers a considerable fraction of the Procyon photosphere.

Velocity fields in the solar and Procyon photospheres. Inhomogeneous element sizes obtained from spectral line profiles

Equivalent widths and central depths of selected lines of neutral iron have been determined from an atlas of the Sun as a star (Kurucz et al., 1984) and the Procyon atlas (Griffin and Griffin, 1979). Spectral line profiles from the Procyon atlas were corrected for instrumental distortion in accordance with data given by Mackle et al. (1975).

The turbulence parameters were determined together with the iron abundance by a method of Gurtovenko and Kostik (1989). The calculations were performed with the SPANSAT program (Gadun and Sheminova, 1988). The Holweger-Müller model (1974) was used for the Sun, the Steffen (1985) model for Procyon with the parameters $T_{eff} = 6500$ and $\log g = 4.04$, and with solar abundances. Rotation was taken into account through direct averaging of the intensity profiles over the disk. Effective line formation depths were calculated using the depression contribution functions.

The sizes of the photospheric inhomogeneities were estimated using stochastic line formation theory (the Uhlenbeck-Ornstein process), which contains two free parameters: L, the element size, and ξ, the RMS velocity, where $\xi^2 = V_{mi}^2 + V_{ma}^2$. Using known chemical abundances, L can then be determined by fitting theoretical equivalent widths and central depths with corresponding observed values.

Fourty-five neutral iron lines with empirical oscillator strengths and excitation potentials of the lower level from 0.05 to 2.61 eV were used. The iron abundance was found to be 7.64 ± 0.3 in the solar and 7.53 ± 1.0 in the Procyon photosphere. The dependence of V_{mi}, V_{ma}, and L on geometrical height are given in Figures 1 and 2. It is seen from these figures that

J. O. Stenflo (ed.), Solar Photosphere: Structure, Convection, and Magnetic Fields, 421–425.

422

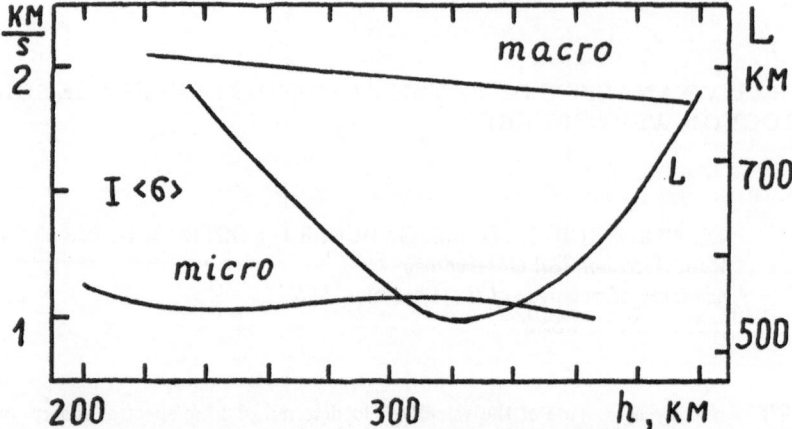

Figure 1. Dependence of V_{mi}, V_{ma}, and L on geometrical height h in the solar photosphere.

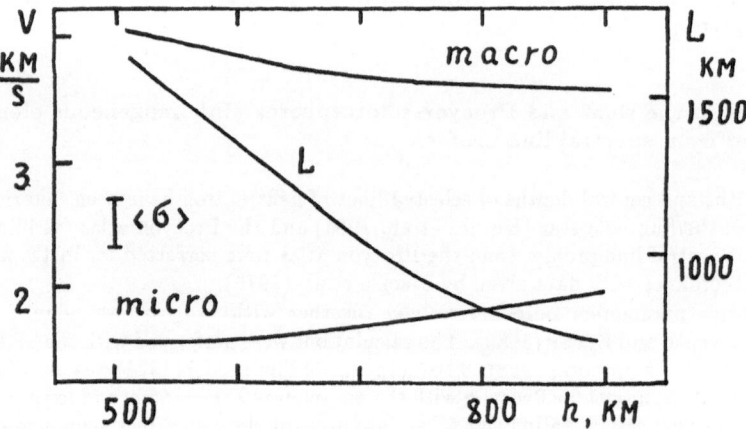

Figure 2. Dependence of V_{mi}, V_{ma}, and L on geometrical height h in the Procyon photosphere.

- the amplitude of the general velocity field in the Procyon photosphere is twice as large as in the solar photosphere;
- the microturbulent velocity in the Procyon photosphere increases with height but with a local decrease at a height of about 300 km. The solar microturbulent velocity decreases with height;
- the behaviour of L in the solar photosphere can be interpreted as follows. Three regions can be identified in the solar photosphere: the penetrative convection region ($h < 310$ km), the intermediate region ($h = 310 - 370$ km), and the oscillation region ($h > 370$ km). Their heights are determined from the observed spectrum of the Sun as a star, since they differ from those for the disk centre (Atroshchenko et al., 1989a). In the case of Procyon penetrative convection is the dominating feature (Figure 2);

– the region where V_{mi} increases coincides with that where the element sizes decrease. Thus an increase of V_{mi} with height may indicate the existence of fragmentation processes of vortices in the Procyon photosphere.

General results of direct numerical simulations of convective motions in the solar and Procyon photospheres

The ideas and computational schemes have been described by Gadun (1986) and Atroshchenko et al. (1989b) in detail. The main results are shown in Figures 3 – 6. The following conclusions may be made:

1. Convective motions penetrate into the stable solar photospheric layers, to heights of 200 – 300 km. In the Procyon case penetrative convection covers virtually the entire computational region (up to 1000 km, from $\tau_5 = 1$).

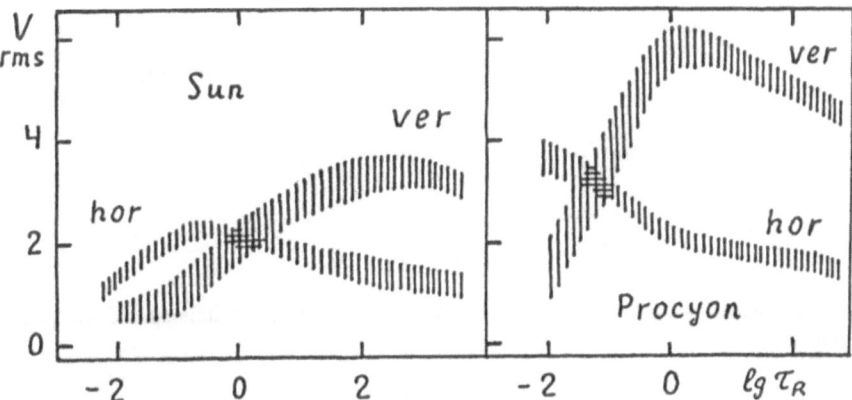

Figure 3. Calculated velocities in the solar and Procyon envelopes.

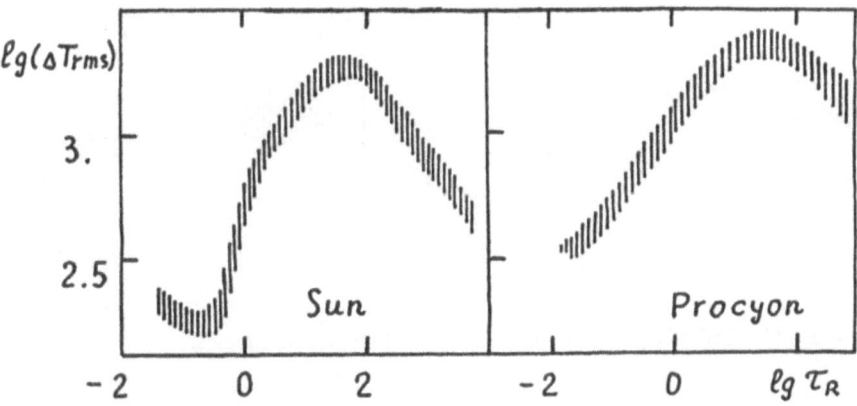

Figure 4. RMS temperature fluctuations from the three-dimensional inhomogeneous solar and Procyon models.

424

2. Calculated convective velocities in the Procyon photosphere are found to be twice as large as in the Sun.
3. The temperature fluctuations for the Sun and Procyon are similar, but they decrease slower with height in the Procyon photosphere.
4. As a consequence, the intensity fluctuations in the Procyon photosphere are higher than for the Sun.
5. The greater penetration of hot elements into the stable Procyon layers is the cause of the inverse limb-effect behaviour as compared with the solar case.

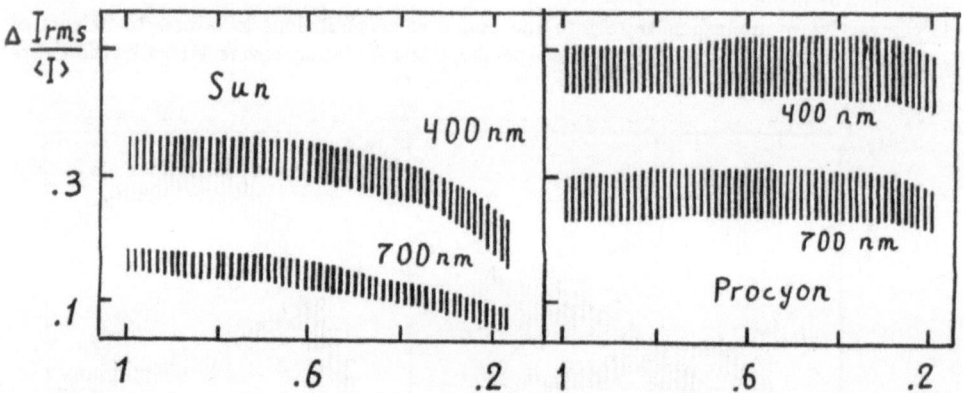

Figure 5. Relative RMS intensity fluctuations from the three-dimensional inhomogeneous models.

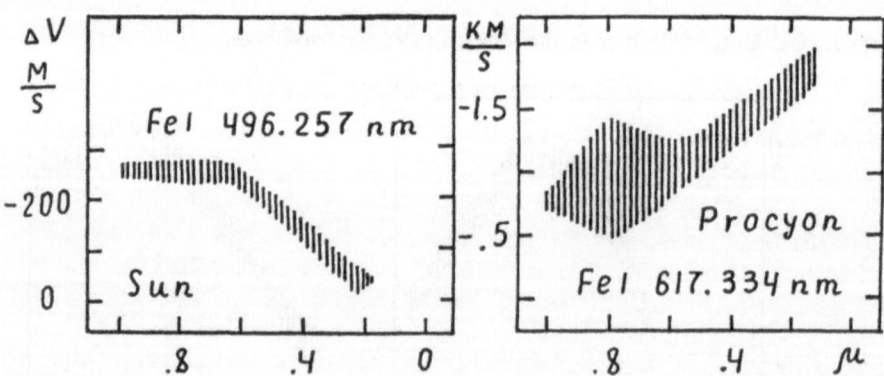

Figure 6. The limb-effect calculated from three-dimensional models.

Accordingly penetrative convection plays a considerably larger role in the photosphere of Procyon than in the Sun.

References

Atroshchenko, I.N., Gadun, A.S., Kostik, R.I. (1989a) *Solar and Stellar Granulation*, NATO ASI Series C **263**, 135.

Atroshchenko, I.N., Gadun, A.S., Kostik, R.I. (1989b) *Solar and Stellar Granulation*, NATO ASI Series C **263**, 521.

Gadun, A.S. (1986) Preprint ITP of Ac. Sc. Ukr. SSR, ITP-86-106R, Kiev.

Gadun, A.S., Sheminova, V.A. (1988) 'SPANSAT: Program for LTE Calculations of Absorbtion Line Profiles in Stellar Atmospheres', Preprint ITP-88-87R, Kiev.

Griffin, R., Griffin, R. (1979) *A Photometric Atlas of the Spectrum of Procyon* 3140 – 7470 Å, Cambridge, UK.

Gurtovenko, E.A., Kostik, R.I. (1989) *Fraunhoferov Spektr i Sistema Sil Oscillatorov* 44, Kiev.

Holweger, H., Müller, E.A. (1974) *Solar Phys.* **39**, 19.

Kurucz, R.L., Furenlid, I., Brault, J., Testerman, L. (1984), *Solar Flux Atlas from 290 to 1300 nm*, Harvard University.

Mackle, R., Griffin, R., Griffin, R., Holweger, H. (1975) *Astron. Astrophys. Suppl. Ser.* **19**, 303.

Steffen, M. (1985) *Astron. Astrophys. Suppl. Ser.* **59**, 403.

MAGNETIC FIELDS ON SOLAR–LIKE STARS: THE FIRST DECADE

S. H. SAAR
Harvard–Smithsonian Center for Astrophysics, Mail Stop 58,
60 Garden Street, Cambridge, MA 02138, USA

ABSTRACT. I review the progress made over the past decade in the measurement of magnetic fields on solar–like stars. I describe the evolution of magnetic analysis techniques, summarize our current understanding of stellar magnetic properties, and outline some future research directions.

1. Introduction

It has now been nearly a decade since Robinson *et al.* (1980) opened a new era of stellar research by making the first measurements of magnetic fields on solar–like stars. In light of this anniversary it is appropriate to review the field's development and its current analysis techniques, summarize the present state of knowledge in this field, and look ahead to future developments. I also encourage the reader to seek out recent reviews by Saar (1987b), Gray (1988), and Linsky (1989).

The reasons for studying magnetic fields on solar–like stars rise directly from the study of magnetic fields on the Sun. The entire family of atmospheric features on the Sun (plages, spots, granules, prominences, flares, etc.) is related to, or significantly affected by, the presence of magnetic flux. The inferred presence of similar features on cool stars suggests comparable, magnetically–related origins (e.g., Linsky 1985). The million degree corona of the Sun requires magnetic fields for both heating (by direct and indirect means) and confinement of the hot plasma. Observations of stellar coronae, with emission often orders of magnitude larger than the Sun, reinforce the concept that magnetic fields are ubiquitous in cool stars (e.g., Golub 1983). These are but two examples. Clearly, data on magnetic fields is vital to understand fully the physical structure, energy balance and evolution of the stellar atmospheres which the fields permeate.

2. Evolution of Analysis Methods

Unfortunately, the solar analogy breaks down when one seeks to employ solar *techniques* for measuring magnetic fields to the stars. Lack of spatial resolution, combined with the (likely) dipolar structure of the magnetic regions, are the the fundamental problems. They conspire to almost completely cancel stellar circular polarization (e.g., Borra *et al.* 1984), and reduce

J. O. Stenflo (ed.), Solar Photosphere: Structure, Convection, and Magnetic Fields, 427–441.

the linear polarized signal from cool stars to a few parts in 10^3 or less, typically (Huovelin *et al.* 1988). Magnetic broadening of the unpolarized line profiles remains detectable, but the splitting (Δv_B) is, in general, considerably less than the intrinsic line width (Δv) at optical wavelengths ($\Delta v \approx 4$ km s^{-1} at 600 nm in the Sun, compared with $\Delta v_B = 1.4 \times 10^{-7} g_{eff} \lambda B = 2$ km s^{-1} for $g_{eff} = 2.5$ and B $= 1$ kG). Furthermore, the Zeeman broadening effect is diluted by the (usually dominant) contributions from non–magnetic regions on the stellar surface. Finally, the lines chosen must be free of blends, and non–magnetic broadening in the lines should be low. As a result of these requirements, data must have high spectral resolution ($\Delta \lambda \leq 2 \Delta \lambda_B$), high S/N ($> 100$–200), and the star should have $\Delta v_B / \Delta v \geq 0.2$ (e.g., $v \sin i \leq 10$ km s^{-1} for a G dwarf at 600 nm). Unfortunately, this last constraint does not allow measurement of some of the most interesting, rapidly rotating stars! Some of the above difficulties can be circumvented by observing in the infrared (pioneered by Giampapa *et al.* 1983) to take advantage of the λ dependence of Δv_B. Ultimately, however, the $v \sin i$ constraint is an erent limitation of the method.

Despite these constraints, by the mid–1970's data of the necessary quality could readily be obtained, thanks to advances in electronic detector technology. New analysis methods were now needed. The breakthrough came when Robinson *et al.* (1980), using the Fourier ratio technique pioneered in solar work by Tarbell and Title (1977), discovered a solar–like field covering a substantial fraction of the active G dwarf, ξ Boo A. Marcy (1982) soon followed with a similar analysis in the wavelength domain. These methods, and all those developed to date, assume that an observed line profile, F_{obs}, can be interpreted with a two–component model, $F_{obs} = fF_m(B) + (1\text{-}f)F_q(B=0)$, where F_m is the profile arising from magnetic regions with a mean field strength B covering a fraction f of the surface, and F_q represents the profile in the field–free (B=0) regions. Methods differ primarily on how F_m and F_q are computed (or derived) and how the resulting magnetic parameters f and B are obtained.

The first analysis methods assumed F_m could be modeled as the appropriately weighted sum (Marcy 1982) or convolution (Robinson 1980) of a triplet of optically thin lines split by a field strength B. The weights reflected the Sears relations for a "disk–averaged" angle (γ) between the magnetic field and the line–of–sight (e.g., Marcy 1982, 1984 used $\overline{\gamma} = 34°$). Lines with low g_{eff} from the same star were used to simulate these magnetic components and F_q. Gray (1984) retained these basic assumptions but extended the Fourier ratio technique to analyze several lines simultaneously by first removing the underlying unsplit profile ($= F_q$) of each line through a radiative transfer calculation.

Some problems began to appear, however. The filling factors on many stars seemed un-realistically large (nearly 90 % for the moderately active K dwarf ϵ Eri, for example). Large magnetic fluxes were recorded for stars with widely different rotation rates and activity levels. Indeed, Gray (1985) showed that the detections made up to that time showed a constant magnetic flux density (i.e., the product, fB) for all stars, with the single exception of the Sun. Taken at face value, this result seemed to suggest two possibilities: Either theories regarding the generation of stellar magnetic fields and their roles in heating chromospheres and coronae were wrong, or something was amiss with magnetic measurements.

It now appears that shortcomings in the analysis played an important role in these problems (Hartmann 1987; Saar 1988a; Basri and Marcy 1988). None of the analyses to that time included treatment of radiative transfer effects *in the Zeeman components themselves*; the construction of F_m was made in the optically thin, weak–line limit. This assumption required large filling factors to duplicate observed "saturation" in optically thick line cores, especially in K dwarfs where the lines employed were generally stronger. When combined with some subtle selection effects (faster rotating, more chromospherically active G dwarfs and slower rotating, less active K dwarfs made up most of the sample; see Stepień 1987), these problems led the early analyses to infer fB = constant.

Clearly, a refined analysis including a radiative transfer model for F_m was needed. Steps in this direction were already taken by Marcy and Bruning (1984), who developed a radiative transfer model to bootstrap comparison between low and high g_{eff} lines of substantially different excitation potential. Full integration of magnetic line transfer effects into the analysis was accomplished by Saar and coworkers (Saar and Linsky 1985; Saar *et al.* 1986a; Saar 1988a), who developed methods which included simple magnetic radiative transfer effects (Unno 1956), the full Zeeman patterns, and some compensation for line blends. These models, however, used convolutions (Gray 1976) to describe velocity broadening, and still required the assumption of an average $\overline{\gamma}$. Bruning (1984) noted that disk integrations are preferred over convolutions for computation of rotational broadening and Saar (1988b) showed use of convolutions could produce errors in f and B values (see also Landolfi *et al.* 1989). The physics of the line transfer was also quite simple, employing a Milne–Eddington atmosphere with a linear source function and all other variables independent of depth.

These shortcomings have also been addressed recently. Basri and Marcy (1988) and Marcy and Basri (1989) have further improved the analysis by numerically integrating the Unno (1956) equations in realistic model atmospheres (rather than using the Milne–Eddington approximation). The latest models also employ full disk integration of intensity profiles to obtain the flux (Saar *et al.* 1989; Marcy and Basri 1989). This step simultaneously eliminates the need to assume an $\overline{\gamma}$, since the angle is automatically accounted for in the disk integration. Known blends can be treated simultaneously by direct line synthesis.

Mathys and Solanki (1989), however, have taken a very different approach. Their technique, based on the solar multi–line regression analysis of Stenflo and Lindegren (1977), correlates line areas measured below the half depth point and the line width at this level with parameters such as excitation potential and a Zeeman broadening term. The Zeeman term, proportional to fB^2, can then be separated into f and B by comparison with similar studies of the areas and widths for different line depths. Typically, large numbers of lines (≈ 40) are used in the analysis. By sorting lines of varying excitation potential and Zeeman pattern, some crude information on the temperature and average orientation ($\overline{\gamma}$), respectively, of the magnetic regions can potentially be determined. The method is also considerably simpler than detailed model calculations for all the lines. The functional form assumed for the regression equation, however, is physically somewhat unclear, and the choice of its terms is not straightforward. Thus, while the technique appears quite promising, additional calibration and tests are needed.

3. A Summary of Current Knowledge

Using the above techniques, our understanding of stellar magnetic properties has grown steadily over the last decade. In this section I summarize the current knowledge of stellar field strengths and filling factors, and present a fresh analysis of the most recent results.

First, Zeeman broadening is definitely present in measurable quantities on many G and K dwarfs (e.g., Robinson et al. 1980; Marcy 1984; Gray 1984), often at levels far exceeding that seen in the Sun. It has not yet been observed in F dwarfs (Gray 1984) or in any bright giants or supergiants. It is also usually not seen in subgiants and normal giants (Marcy and Bruning 1984; Gray and Nagar 1985) unless the stars are active RS CVn variables (e.g., Giampapa et al. 1983). And while the presence of Zeeman broadening is small, often debatable, and always difficult to interpret in optical spectra, it is undeniably present in infrared spectra of M dwarfs, where full splitting patterns are visible (Saar and Linsky 1985; Saar et al. 1987). There is also clear evidence for range in Zeeman broadening at each spectral type, since there are clear non–detections for G (Marcy 1984; Gray 1984; Mathys and Solanki 1989), K (Saar 1987a; Saar et al. 1987; Marcy and Basri 1989) and M (Saar et al. 1987) dwarfs as well. Strong magnetic fields have been detected in stars as cool as M4.5 (EV Lac; Saar et al. 1987), indicating that dynamo generation of magnetic flux is still effective even in stars which are almost fully convective.

To explore trends in magnetic parameters f and B in more detail, I have compiled a critically selected group of recent magnetic measurements in Table 1. The list includes virtually all measurements to date which have been derived using radiative transfer methods. This restriction makes the data set more homogeneous, and also excludes probable systematic effects in the earlier measurements (see section 2). In addition, a few optical M dwarf measurements (Bruning et al. 1987) were excluded due to probable blend problems (e.g., Figure 1 of Saar 1988b), and a few results based on lower S/N optical data (in Saar 1987) were also excluded. I included the infrared detections of λ And (Giampapa et al. 1983; Gondoin et al. 1986) and the results of Mathys and Solanki (1989). In the latter case, B was computed from the mean of the range given (their table 4), and then I assumed $f \equiv \bar{f} = (\sqrt{f}B/\bar{B})^2$. Rotational periods were gathered from Noyes et al. (1984) and Pettersen (1989), and relations in Stepień (1989) and Basri (1987) were used to compute the convective turnover time, τ_c. Two measurements are given for a single star in several cases where different investigators disagreed on the magnetic parameters (e.g., HD 131156A = ξ Boo A). When this occurs, both are plotted and connected with a dotted line.

One of the first things apparent is a general tendency for magnetic field strengths to increase with increasing B–V (i.e., towards later spectral types). This trend, combined with the lack of detections for F dwarfs and most giants and subgiants (likely due to low B; Marcy and Bruning 1984), and the low field strengths seen on the active giant, λ And, strongly suggests a relationship between B and some intrinsic property of the stellar atmosphere. Saar and Linsky (1986a) found a tight correlation between B and the equipartition "pressure–balancing" magnetic field, $B_{eq} \propto P_{gas}^{0.5}$, where P_{gas} is the photospheric gas pressure. I estimated P_{gas} as a function of T_{eff} and gravity by using atmospheres of Kurucz (1979 and unpublished), supplemented with models derived by Mould (1976) for M dwarfs. The gas pressure was determined at optical depth, $\tau_{5000} = 1$, where $P_{gas}(\odot)$ yields $B_{eq} \approx 1.5$ kG (note that the exact choice of τ_{5000} is not critical, since we use P_{gas} only to scale relative

Table 1: A Critical Selection of Recent Magnetic Field Determinations

Star ID	Sp. Type	B–V (R − I)	B (kG)	f (%)	Ref.	adopted B_{eq}	P_{rot} (days)	τ_c (days)
HD 39587	G0V	0.59	1.0	60	1	1.4	5.2	10.0
HD 190406	G1V	0.61	1.8	10	1	1.4	13.5	11.0
HD 1835	G2V	0.66	1.4	32	1	1.5	7.7	13.7
HD 28099	G6V	0.66	1.7	30	2	1.5	8.7	13.7
HD 20630	G5V	0.68	1.5	35	1	1.6	9.4	14.7
HD 10700	G8V	0.72	($\sqrt{f}$B	≤0.2)	12	1.7	31.9	16.9
HD 131156 A	G8V	0.76	1.6	22	3	1.7	6.2	19.0
	G8V	0.76	1.8	35	1	1.7	6.2	19.0
HD 152391	G8V	0.76	1.7	18	1	1.7	11.1	19.0
HD 3651	K0V	0.85	...	...	3	1.9	48	20.5
HD 10476	K1V	0.84	1.0	17	3	2.0	38	20.5
HD 165341	K1V	0.86	1.2	18	3	2.0	19.7	20.5
HD 155885	K1V	0.86	1.5	13	3	2.0	22.9	20.5
HD 22049	K2V	0.88	1.0	30	3	2.2	11.3	20.5
	K2V	0.88	1.9	12	4	2.2	11.3	20.5
HD 115404	K2V	0.93	2.1	20	1	2.2	18.8	20.5
HD 45088 A	K3Ve	0.96	2.4	50	2	2.4	7.4	20.5
HD 209100	K4-5V	1.09	2.6	13	12	2.5	...	20.5
HD 225732	K6V	1.04	1.8	20	7	2.5	...	20.5
HD 131156 B	K4V	1.10	(if 2.6	≤20)	1	2.6	11.5	20.5
EQ Vir	K5Ve	1.18	2.5	80	5	2.7	3.9	20.5
HD 201091	K5V	1.18	(if 1.5	≤5)	1,6	2.7	37.9	20.5
	K5V	1.18	1.2	24	3	2.7	37.9	20.5
BY Dra	K5Ve+	1.19	2.8	60	8	2.7	3.8	20.5
HD 201092	K7V	1.38	(if 1.5	≤10)	8	2.9	48.0	20.5
HD 97101	K9V	1.35	1.8	25	7	2.9	...	20.5
HD 88230	K7V	1.37	(if 1.5	≤10)	8	2.9	...	20.5
	K7V	1.37	0.8	55	7	2.9	...	20.5
GL 205	M1.5V	(0.85)	(if 1.5	≤15)	8	3.4	...	20.5
AU Mic	M1.5Ve	(0.84)	4.0	90	8	3.4	4.8	20.5
=AD Leo	M3.5Ve	(1.12)	3.8	73	6,8	3.8	2.7	20.5
GL 273	M4V	(1.15)	(if 1.5	≤25)	8	4.0	...	20.5
EV Lac	M4.5Ve	(1.15)	5.2	90	8	4.4	4.4	20.5
HD 222107	G8III–IV	1.01	0.6	20	9	0.6	20.5	87.3
	G8III–IV	1.01	1.2	48	10	0.6	20.5	87.3
HD 17433	K1-2IVe	0.96	2.0	60	11	1.3	12.1	89.6

References: [1]Saar (1987); [2]Saar and Linsky (1986); [3]Basri and Marcy (1989); [4]Saar et al. (1986b); [5]Saar et al. (1986a); [6]Saar and Linsky (1985); [7]Bruning et al. (1987); [8]Saar et al. (1987); [9]Gondoin et al. (1986); [10]Giampapa et al. (1983); [11]Bopp et al. (1989); [12]Mathys and Solanki (1989)

to the Sun). The resulting B_{eq} was then determined using $B_{eq} = (P_{gas}/P_{gas}(\odot))^{0.5}B(\odot)$. The results (with $B_{\odot} \equiv 1.5$ kG; Harvey and Hall 1975) are given in Table 1; for G and K stars they are very similar to B_{eq} derived by Zwaan and Cram (1989).

A plot of B and B_{eq} (Figure 1) shows most stars clustering near or below the $B = B_{eq}$ line. I interpret this as evidence that magnetic field strengths in stellar photospheres are largely determined by a horizontal pressure balance in which gas pressure dominates, i.e., B $\leq B_{eq} \propto P_{gas}^{0.5}$. Several theories of magnetic field concentration in stellar photospheres predict this type of relation (e.g., Parker 1978; Spruit and Zweibel 1979; Galloway and Weiss 1981). One RS CVn data point lies considerably above the $B = B_{eq}$ line (Giampapa et al. 1983); Giampapa (1984), however, postulates that the stellar surface had an admixture of umbrae (with much higher B) in the field of view during the observation. In the case of the second RS CVn (HD 17433), which also shows a rather large field strength, enhanced turbulence may play a larger role in the pressure balance (Bopp et al. 1989).

Hartmann's (1987) suggestion that the $B \propto B_{eq}$ relation might be due to line opacity effects is probably not correct, since the analyses used to derive the data in Table 1 explicitly take opacity broadening into account. His comment that stars with low B may be missed due to difficulties in detection (e.g., Marcy 1982; Saar 1988a), however, is certainly valid. An equally important problem is the following: How valid it is to mix results of data

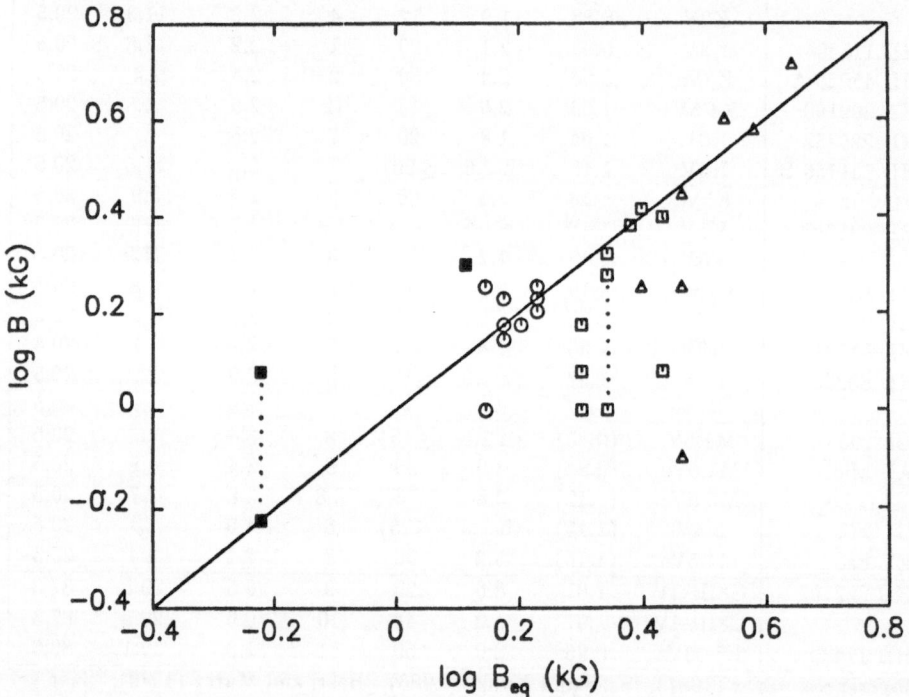

Figure 1. B versus $B_{eq} \propto P_{gas}^{0.5}$. Circles, squares, triangles, and filled squares represent G, K, M dwarfs and RS CVn variables, respectively. The Sun is indicated by $\odot$. The relation $B = B_{eq}$ is shown as a solid line. $B \leq B_{eq}$ is inferred from the data.

sets measured using different lines and wavlengths? Table 1 contains magnetic parameters derived from medium strength Ti I lines at 2200 nm (e.g., Saar and Linsky 1985), Fe I lines at 617 nm (e.g., Saar *et al.* 1986a) and 1600 nm (Gondoin *et al.* 1986) and strong Fe I lines at 846 nm (Marcy and Basri 1989). Continuum opacity (primarily H^-) is similar at 617 and 2200 nm, but is stronger at 846 nm and substantially weaker at 1600 nm (I thank S. Solanki for pointing this out). If all other variables are held constant, fields should appear stronger (with a lower f) at 1600 nm, where the observer sees deep into the atmosphere, and weaker (with higher f) at 846 nm, where the observer sees deep into the atmosphere. Indeed, the B values derived for K dwarfs using the 846 nm line (Marcy and Basri (1989) *are* somewhat smaller (and the f values larger) than those of derived using 617 nm. The considerably greater strength of the 846 nm feature will make some parts of its profile more sensitive to higher atmospheric levels (with lower B and higher f values) than 617 nm. Similarly, optical determinations of B differ from those at 1600 nm by 300–400 G in the Sun (e.g., Stenflo 1989). Thus, it is possible that some of the variation in measured field strengths in K dwarfs is due to the effects of differing line formation heights. Of course, differences in the analysis methods used will also affect the results. A more detailed analysis, using flux tube models (e.g., Steiner *et al.* 1986) to determine the contribution functions (e.g., Grossmann–Doerth *et al.* 1988) for magnetic lines at different wavelengths, will probably be needed to solve this puzzle.

A search for other correlations with B yields little. Plots of B versus f show no correlation (see also Saar and Linsky 1986a), verifying that fB is not a constant. Also, $\sqrt{f}B$ is not constant, as would be expected if errors in the separation of f and B were dominant (Gray 1984; Saar 1988a). No clear relation exists between B and rotation, either. Plots of B with the inverse Rossby number, $\tau_c\Omega$, are basically flat. There appears to be a weak relationship between B and Ω (B $\propto \Omega^{0.3\pm0.2}$; see also Saar 1987a), but this probably represents a selection effect (only stars with large B can be detected at large Ω; Saar 1988a).

The lack of a B–rotation correlation is at first suprising; many simple dynamo models predict B $\propto \Omega$ or $\tau_c\Omega$ (e.g., Stix 1972). However, when it is understood that dynamo theories actually predict the magnetic *flux* generated, rather than field *strength*, the paradox is resolved. Marcy (1984) and Gray (1985)noted the first indications that magnetic flux measurements and stellar rotation were related. Using mostly older data, Stepień (1987) found magnetic flux correlated with $(\tau_c\Omega)^{0.5}$. In the present data set, magnetic flux density (fB) exhibits strong correlations with the stellar rotation: fB $\propto \Omega^{1.3}$ and fB $\propto (\tau_c\Omega)^{1.2}$ (Figure 2; see also Saar 1987a,b). These relations are consistent with the fB $\propto T_{eff}^{-n}\Omega$ relation correlation found by Marcy and Basri (1989), and are broadly consistent with many simple theoretical expectations (e.g., Schatzman 1962). The Durney and Robinson (1982) dynamo model, however, which predicts f $\propto \Omega^{2.5}$, appears to be ruled out.

Since B is not strongly correlated with rotation, the filling factor must be the cause for the fB – rotation relations. A plot of f versus $\tau_c\Omega$ confirms this (Figure 3). A least–squares fit to the Table 1 data set yields f $\propto (\tau_c\Omega)^{0.9}$ (see Linsky and Saar 1987; Saar 1987b; Saar *et al.* 1987; Stepień 1988a). The filling factor must be less than one by definition, so it is inevitable that this simple power law breaks down at some point and f "saturates". Based on the derived power law, stars beyond $\tau_c\Omega \approx 0.8$ may be in this state (see also Saar and Linsky 1986a; Linsky and Saar 1987). Thus, a dynamo model including a saturated state (e.g., Skumanich and MacGregor 1986) fits the data best. Saturated states in chromospheric

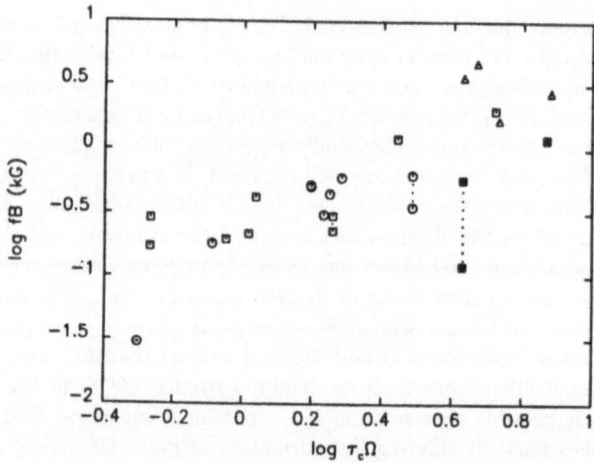

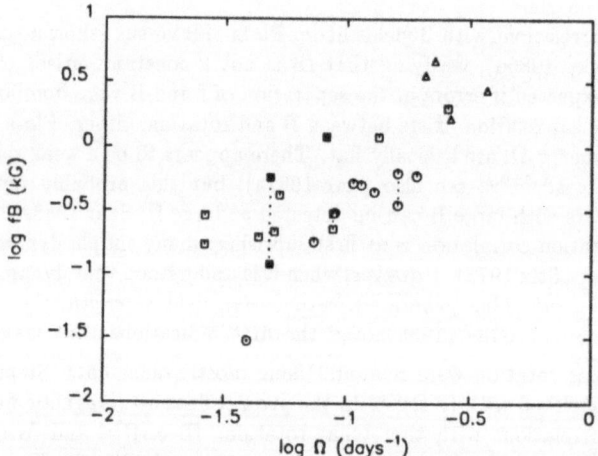

Figure 2. Magnetic flux density (fB) versus rotational parameters. Symbols are same as in Fig. 1. Magnetic flux density correlates with $\tau_c\Omega$ (fB $\propto (\tau_c\Omega)^{1.2\pm0.1}$; top) and Ω (fB $\propto \Omega^{1.3\pm0.1}$; bottom) largely the result of correlations between f and rotation (see Fig. 3).

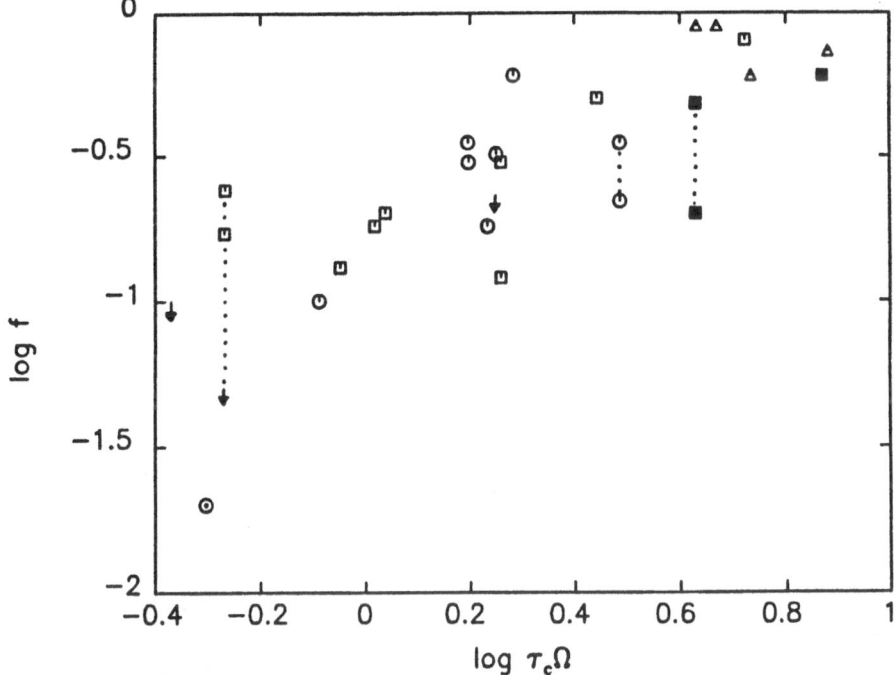

Figure 3. Magnetic filling factor f versus $\tau_c\Omega$. Symbols are as in Fig. 1, with upper limits to f indicated by arrows. A least–squares fit yields $f \propto (\tau_c\Omega)^{0.9\pm0.2}$. The magnetic filling factor appears to be the dominant magnetic parameter controlling rotation–activity and rotation–age relations.

and coronal emission have been recognized for some time (e.g., Vilhu 1984).

As one might anticipate based on the fB – rotation correlations, there are also indications (though the data set is tiny) that fB declines with stellar age (Linsky and Saar 1987; Saar 1989). The magnetic field strength versus age diagram shows only scatter, implying that a reduction in the surface filling factor of active regions is the primary reason for the decline in stellar activity with time (at least for ages ≥ 0.3 Gyr). The decline of fB with time increases with time, but the functional form cannot be determined precisely with the available data. This result, too, is broadly consistent with theoretical models of angular momentum loss (e.g., Weber and Davis 1967; Kawaler 1988; Stepień 1988b).

Unlike the field strength, f exhibits a large range independent of color. Flare stars (listed by variable star name in Table 1) appear to have the highest filling factors. The f values of dM stars suggests that the large Hα filling factors derived by Giampapa (1985) refer to the chromospheric level, where f is indeed much larger than in the photosphere (Saar et al. 1987). Filling factors estimated by Montesinos et al. (1987) are generally much smaller than those observed.

Due to the intimate connection between chromospheric, transition–region, and coronal heating and magnetic fields on the Sun, it is natural to search for similar relations in the stellar data. Marcy (1984), once again, lead the way, noting correlations between magnetic parameters, Ca II, and X-ray emission. Saar and Schrijver (1987) found $F_x \propto (fB)^{0.9}$ and $\Delta F_{CaII} \propto (fB)^{0.6}$ for $fB \leq 300$ G (where ΔF_{CaII} is the residual Ca II flux; Schrijver 1983). Above $fB = 300$ G, they found ΔF_{CaII} was saturated. Schrijver et al. (1989) note that these correlations are consistent with relations derived for the Sun, and with flux-flux relations derived for stars. Saar (1988c) derived power law relations (with ranging from 0.6 and 0.3) between fB and ultraviolet C IV, C II and O I line fluxes. Quillen et al. (1987) found that using X-ray luminosities and the simple coronal loop model of Golub (1983), they could roughly predict stellar magnetic fluxes to within a scaling factor. Estimates of the individual f and B values were rather poor, however. Stepień (1988a) and Jordan et al. (1987) have predicted expected magnetic flux – activity relations with some success (e.g., Jordan et al. estimate $F_X \propto (fB)^{1.8}$).

Groups have also searched for variability in stellar magnetic fields. Already in the first detections, Robinson et al. (1980) noted an apparent change on ξ Boo A. Subsequent observations of the star by Marcy (1981) and Gondoin et al. (1986) yielded no evidence for magnetic line broadening at all, suggesting the star had a large range of magnetic variation (consistent with its Ca II emission). Rotational modulation of chromospheric and transition-region line fluxes with magnetic flux for the active dwarf ξ Boo A support this picture (Saar et al. 1988). The same authors used simultaneous measurements of broadband linear polarization (which measures the net tangential component of the magnetic field) to permit a rough determination of the spatial distribution of active areas on the star. Some of this variability may have been due to changes in turbulence due to a "starpatch", as postulated by Toner and Gray (1988). On the other hand, it is possible (Bruning and Saar 1989) that the line bisector variations seen in both data sets are primarily a magnetic phenomenon (masked in the Toner and Gray analysis by the small g_{eff} leverage: $g_{eff} = 1.0$ versus $g_{eff} = 1.7$). More work on the interaction of magnetic fields and convection, and their role in line shapes, is certainly needed.

Other stars studied in detail so far seem not to vary as much as ξ Boo A. Basri and Marcy (1988) found little change in their ϵ Eri spectra, confirming earlier evidence (Saar et al. 1986b) that its variability range is small. The active subgiant HD 17433 (see Bopp et al. 1989) has also been the subject of a campaign of simultaneous magnetic field, ultraviolet, and optical measurements (Ambruster and Saar 1988). Little variability was observed here either, except during a brief flare, when C IV emission increased significantly and there was weak evidence for a change in the magnetic parameters. Analysis of a similar campaign on the nearly pole–on flare star BD +26 730 is underway (Saar et al. 1989, in preparation).

4. A Comparison of ϵ Eri Measurements

As an example of progress in magnetic measurements and difficulties yet to be resolved, Table 2 shows a compilation of all magnetic measurements to date for the active K2 dwarf, ϵ Eri (HD 22049). All filling factors were adjusted (where possible) to the scale of Marcy (1984; $\overline{\gamma} = 34°$).

Clearly, there has been some evolution of opinion as to the magnetic activity on the star, and the situation is perhaps still not clearly resolved. Early approaches which did not include radiative transfer effects in the Zeeman process (Marcy 1984; Gray 1984) generally

Table 2: Comparison of Magnetic Measurements for the K2 dwarf, ϵ Eridani

Reference	B (kG)	f (%)	fB (kG)	$\sqrt{f}$B (kG)	λ (nm)	# of lines g<1.3/g≥1.3
Marcy (1984)	1.17	67	0.78	0.96	617	1/1
Gray (1984)	1.9	36	0.69	1.14	640	10/6
Saar et al. (1986b)	1.9	13	0.25	0.69	617	3/2
Saar (1988)	(if 2.0	≤5)	≤0.10	≤0.40	2210	1/1
Saar (1988)	3.0:	8:	0.24:	0.85:	617	3/2
Mathys and Solanki (1988)	1.79–2.53	10–20	(0.30)	0.80	580–680	(45 total)
Basri and Marcy (1988)	1.0	35	0.35	0.59	846	1/1
Mathys and Solanki (1989)	2.09–2.96	10–20	(0.35)	0.94	580–680	33/9
Marcy and Basri (1989)	1.0	30	0.30	0.55	846	1/1
Marcy and Basri (1989)	(if 3.0	≤10)	≤0.30	≤0.95	2210	1/0
	(if 1.0	≤30)	≤0.30	≤0.55	2210	1/0

produced higher filling factors. As noted in section 2, this is likely due to the need for higher f to mimic the more "box–shaped" (due to increased line opacity) doppler core of the line, in the absence of a radiative transfer treatment (e.g., Saar 1988a). Introduction of a simple (Unno 1956) radiative transfer model led to smaller filling factors (Saar et al. 1986b; average of 11 measurements). Later, analysis of simultaneous optical and infrared spectra using the same model indicated lower fluxes (Saar 1988a). Indeed the optical results were uncertain enough (due to the low f value) to be deemed a non-detection. The completely different, regression analysis approach of Mathys and Solanki (1988) yields f and B values suprisingly similar to Saar et al. (1986b), though the later recalibration (Mathys and Solanki 1989) produced magnetic field strengths larger than indicated by infrared data (Saar 1988a; Marcy and Basri 1989). The Marcy and Basri (1989) analysis includes improved radiative transfer and explicit disk–integration, and results in lower B and larger f values, though fB is approximately the same. Thus, happily, there is now some convergence in the results for fB and $\sqrt{f}$B, but further work is needed to understand the why the separate f and B values differ. As noted in section 3, part of the disagreement in the derived f and B for ϵ Eri (and other targets) is probably due to the differing wavelengths and lines used, and part is due to the different analysis techniques themseves. Finally, one should remember that even in the best of cases, it is difficult to uniquely separate f and B (Gray 1984).

438

5. Some Future Directions

What questions need to be addressed in stellar magnetic research over the next few years? First, many more magnetic measurements are needed. Determinations for more M dwarfs and RS CVn's as well as first detections of T Tauri's and F dwarfs would be especially useful. Far infrared measurements of the 12 micron lines (Deming *et al.* 1988) and near infrared measurements using cryogenic echelles now being built will substantially increase the accuracy of the f and B determinations. More multiwavelength campaigns would help probe the temporal dependence of magnetic flux and its relationship to atmospheric heating. Observations of broadband linear (Huovelin *et al.* 1988) and circular (Kemp *et al.* 1987) polarization should help unravel some information on the spatial structure of the fields, given suitable theory (Landi Degl'Innocenti 1982; Mürset *et al.* 1988). It may be possible to obtain further spatial information through Stokes V (Donati *et al.* 1989) or Stokes I Doppler imaging (the spot profile "bumps" will be broader in high g_{eff} lines). Multiline observations and analyses (e.g., Mathys and Solanki 1989) should be further explored and refined. Observations over stellar cycle timescales should prove enlightening.

Several important assumptions have been made in stellar Zeeman analyses to date, and it is important to reassess them. The assumption of identical magnetic and non-magnetic atmospheres should eventually be discarded in favor of realistic flux-tube atmospheres for the magnetic component. Infrared data may require a third, umbral component (Sun *et al.* 1987). Turbulent and convective properties in magnetic regions appear to differ from the quiet solar atmosphere (Livingston 1982) and may in stars as well (Toner and Gray 1988). This should be investigated in detail. Magneto-optical effects should also be included in the modeling (Landolfi *et al.* 1989).

In summary, much has been learned in the past ten years about magnetic fields on solar-like stars, and their relationship to both basic stellar properties and to magnetically-related activity. Only the surface of the subject has been really explored, however, and *many* questions remain. Hopefully, ten years hence, many of these will also have been answered, bringing better understanding of the stellar "activity" phenomenon.

Acknowledgements. This work has been supported by the Smithsonian Institution postdoctoral research fellowship program, and NASA grants NAGW-221 and NGL-006-03-057. I am very grateful to NOAO for the generous allocation of telescope time, and to the McMath stellar team in particular for all their help over the years. I would also like to thank G. Basri and G. Marcy for kindly sharing results with me prior to publication, and extend special thanks to many of the people referenced below for a multitude of stimulating discussions!

REFERENCES

Ambruster, C. A., Saar S. H.: 1988, *Bull. A. A. S.*, **20**, 995

Basri, G. S. : 1987 *Astrophys. J.* **316**, 377

Basri, G. S., Marcy, G. W. : 1988 *Astrophys. J.* **330**, 274

Bohigas, J., Carrasco, L., Torres, C. A. O., Quast, G. R. : 1986 *Astron. Astrophys.* **157**, 278

Bopp, B. W., Saar, S. H., Ambruster, C., Feldman, P., Dempsey, R., Allen, M., Barden, S. P. : 1989 *Astrophys. J.* **339**, 1059

Borra, E. F., Edwards, G., Mayor, M. : 1984 *Astrophys. J.* **284**, 211

Bruning, D. H. : 1984 *Astrophys. J.* **281**, 830

Bruning, D. H., Chenoweth, R. E., Marcy G. W.: 1987, in *Cool Stars, Stellar Systems, and the Sun* eds. J. Linsky and R. E. Stencel, New York, p. 36

Bruning, D. H., Saar, S. H.: 1989, in *Solar and Stellar Granulation*, eds. R. J. Rutten and G. Severino, Kluwer, Dordrecht, p. 145

Deming, D., Boyle, R. J., Jennings, D. E., Wiedemann, G. : 1988 *Astrophys. J.* **333**, 978

Donati, J. F., Semel, M., Praderie, F.: 1989, presented at *IAU Colloq. 121, Inside the Sun.*

Durney, B. R., Robinson, R. D. : 1982 *Astrophys. J.* **253**, 290

Galloway, D. J., Weiss, N. O. : 1981 *Astrophys. J.* **243**, 945

Giampapa, M. S. 1984, in *Space Research Prospects in Stellar Activity and Variability*, eds. A. Mangeney and F. Praderie, Obs. de Paris, Meudon, p. 309

Giampapa, M. S. : 1985 *Astrophys. J.* **299**, 781

Giampapa, M. S., Golub, L., Worden, S. P. *Astrophys. J. Letters* 83268121

Golub, L.: 1983, in *Cool Stars, Stellar Systems, and the Sun* eds. P. Byrne and M. Rodonó Reidel, Dordrecht, p. 83

Gondoin, Ph., Giampapa, M. S., Bookbinder, J. A. : 1985 *Astrophys. J.* **297**, 710

Grossmann–Doerth, Larsson, B., ans Solanki, S. K. : 1988 *Astron. Astrophys.* **204**, 266

Gray, D. F.: 1976, *The Observation and Analysis of Stellar Photospheres*, Wiley, New York

Gray, D. F. : 1984 *Astrophys. J.* **277**, 640

Gray, D. F. : 1985 *Publ. Astron. Soc. Pacific* **97**, 719

Gray, D. F.: 1988, *Lectures on Spectral-line Analysis: F, G, and K Stars*, The Publisher, Arva

Gray, D. F., Nagar, P. : 1985 *Astrophys. J.* **298**, 756

Harvey, J. W., Hall, D. S.: *Bull. A. A. S.*, **7**, 459

Huovelin, J., Saar, S. H., Tuominen, I. : 1988 *Astrophys. J.* **329**, 882

Jordan, C., Ayres, T. R. Brown, A., Linsky, J. L., Simon, T. : 1987 *Mon. Not. R. Astr. Soc.* **225**, 903

Kawaler, S. D. : 1988 *Astrophys. J.* **333**, 236

Kemp, J. C. *et al.* : 1987 *Astrophys. J. (Letters)* **317**, L29

Kurucz, R. L. : 1979 *Astrophys. J. Suppl.* **40**, 1

Landi Degl'Innocenti, E. : 1982 *Astron. Astrophys.* **110**, 25

Landolfi, M., Landi Degl'Innocenti, M., Landi Degl'Innocenti, E. : 1989 *Astron. Astro-*

440

phys. **216**, 113

Linsky, J. L. : 1985 *Solar Phys.* **100**, 333

Linsky, J. L.: 1989, *Solar Phys.*, in press.

Linsky, J. L., Saar, S. H.: 1987, in *Cool Stars, Stellar Systems, and the Sun* eds. J. Linsky and R. E. Stencel, Springer, New York, p. 44

Livingston, W. C.: 1982, *Nature*, **297**, 208

Marcy, G. W. : 1981 *Astrophys. J.* **245**, 624

Marcy, G. W. : 1982 *Publ. Astron. Soc. Pacific* **94**, 989

Marcy, G. W. : 1984 *Astrophys. J.* **276**, 286

Marcy, G. W., Basri, G. S.: 1989, *Astrophys. J.* , **345**, in press.

Marcy, G. W., Bruning, D. H. : 1984 *Astrophys. J.* **281**, 286

Mathys, G., Solanki, S. K.: 1988, in *The Impact of Very High S/N Spectroscopy on Stellar Physics*, ed. G. Cayrel de Strobel and M. Spite, Kluwer, Dordrecht, p. 325

Mathys, G., Solanki, S. K. : 1989 *Astron. Astrophys.* **208**, 189

Montesinos, B., Fernandez–Figuerola, M. J., de Castro, E. K. : 1987 *Mon. Not. R. Astr. Soc.* **229**, 627

Mould, J. R. : 1976 *Astron. Astrophys.* **48**, 443

Mürset, U., Solanki, S. K., Stenflo, J. O. : 1988 *Astron. Astrophys.* **204**, 279

Noyes, R. W., Hartmann, L. W., Baliunas, S. L., Duncan, D. K., Vaughan, A. H. : 1984 *Astrophys. J.* **279**, 763

Pettersen, B. R. : 1989 *Astron. Astrophys.* **209**, 279

Parker, E. N.: 1978, *Ap. J.*, **221**, 368

Quillen, A., Golub, L., Harnden, Jr., F. R., Saar, S. H.: 1987, *Bull. A. A. S.*, **19**, 1027

Robinson, R. D. : 1980 *Astrophys. J.* **239**, 961

Robinson, R. D., Worden, S. P., Harvey, J. W. : 1980 *Astrophys. J. (Letters)* **236**, L155

Saar, S. H.: 1987a, *Observations and Analysis of Photospheric Magnetic Fields on G, K, and M Dwarf Stars*, Ph. D. thesis, University of Colorado

Saar, S. H.: 1987b, in *Cool Stars, Stellar Systems, and the Sun* eds. J. Linsky and R. E. Stencel, Springer, New York, p. 10

Saar, S. H. : 1988a *Astrophys. J.* **324**, 441

Saar, S. H.: 1988b, in *The Impact of Very High S/N Spectroscopy on Stellar Physics*, ed. G. Cayrel de Strobel and M. Spite, Kluwer, Dordrecht, p. 295

Saar, S. H.: 1988c, in *Hot Thin Plasmas in Astrophysics*, ed. R. Pallavicini, Kluwer, Dordrecht, p. 139

Saar, S. H.: 1989, in *The Sun in Time*, ed. M. Matthews, Tucson, U. of Arizona, in press

Saar, S.H., Huovelin, J., Giampapa, M.S., Linsky, J.L., Jordan, C.: 1988, in *Activity in Cool Star Envelopes*, Eds. O.Havnes *et al.*, Kluwer, Dordrecht, p.45

Saar, S. H., Linsky, J. L. : 1985 *Astrophys. J. (Letters)* **299**, L47

Saar, S. H., Linsky, J. L.: 1986a, *Advances in Space Physics*, **6**, No. 8, 235

Saar, S. H., Linsky, J. L.: 1986b, in *Cool Stars, Stellar Systems, and the Sun* , eds. M. Zeilik and D. M. Gibson, Springer, New York, p. 278

Saar, S. H., Linsky, J. L., Beckers, J. M. : 1986a *Astrophys. J.* **302**, 777

Saar, S. H., Linsky, J. L., Duncan, D. K.: 1986b, in *Cool Stars, Stellar Systems, and the*

Sun , eds. M. Zeilik and D. M. Gibson, Springer, New York, p. 275

Saar, S. H., Linsky, J. L., Giampapa, M. S.: 1987, in *27th Liegé International Astrophysical Colloquium: Observational Astrophysics With High Precision Data*, eds. L. Delbouille and A. Monfils, Universite de Liège, Liège, p. 103

Saar, S. H., Schrijver, C. J.: 1987, in *Cool Stars, Stellar Systems, and the Sun*, eds. J. Linsky and R. E. Stencel, Springer, New York, p. 38

Schatzman, E.: 1962, *Ann. Astrophys.*, **25**, 18

Schrijver, C., Coté, J., Zwaan, C., Saar, S.H.: 1989, *Astrophys. J.*, **337**, 964

Skumanich, A., MacGregor, K.: 1986, *Advances in Space Physics*, **6**, No. 8, p. 151

Spruit, H. C., Zweibel, H. G. : 1979 *Solar Phys.* **628**, 15

Stenflo, J. O. 1989, *Astron. Astrophys. Rev.*, **1**, 3

Steiner, O., Pneuman, G. W., Stenflo, J. O. : 1986 *Astron. Astrophys.* **170**, 126

Stenflo, J. O., Lindegren, L. : 1977 *Astron. Astrophys.* **59**, 367

Stepień, K. : 1987 *Publ. Astron. Soc. Pacific* **98**, 1292

Stepień, K. : 1988a *Astrophys. J.* **335**, 892

Stepień, K. : 1988b *Astrophys. J.* **335**, 907

Stepień, K. : 1989 *Astron. Astrophys.* **210**, 273

Stix, M.: : 1972 *Astron. Astrophys.* **20**, 9

Sun, W.-H., Giampapa, M. S., Worden, S. P. : 1987 *Astrophys. J.* **312**, 930

Tarbell, T. D., Title, A. M. : 1977 *Solar Phys.* **52**, 13

Toner, C. G., Gray, D. F. : 1988 *Astrophys. J.* **334**, 1008

Unno, W. 1956, *Pub. Ast. Soc. Jap.*, **8**, 108.

Vilhu, O. : 1984 *Astron. Astrophys.* **133**, 117

Weber, E. J., Davis, L.: 1967, *Ap. J.*, **148**, 217

Zwaan, C., Cram, L. E.: 1989, chapter 7 in *The Atmospheres of Cool Stars*, eds. L. E. Cram and L. V. Kuhi, NASA–CNRS Monograph Series, in press

RESULTS OF COORDINATED MULTIWAVELENGTH OBSERVATIONS OF SOLAR-TYPE STARS

J.HUOVELIN
Observatory and Astrophysics Laboratory, University of Helsinki,
Tähtitorninmäki, SF-00130 Helsinki, Finland
S.H.SAAR
Harvard-Smithsonian Center for Astrophysics, Mail Stop 58,
60 Garden Street, Cambridge, MA 02138, USA

ABSTRACT. We present results of nearly simultaneous polarimetric and Ca II(H+K) emission observations taken over a full rotation period for a sample of six solar-type stars. Significant variations are seen for two stars in polarization and four stars in Ca emission, which we interpret as due to rotational modulation of magnetic areas on the stellar surface. The irregularity of the variations suggest changes of the area size and/or distribution in time scales of the rotation period. Preliminary models (including rough surface maps of two stars) and future prospects are discussed.

1. Introduction and Observations

Magnetic activity in late-type stars gives rise to a multitude of effects in their atmospheres, modifying the observed radiation throughout the spectrum. At the photospheric level, the Zeeman effect may broaden the profiles of the magnetically sensitive absorption lines, and recently, several techniques have been developed to analyze this phenomenon using the unpolarized stellar spectra (e.g., Robinson, 1980; Saar, 1988; Mathys and Solanki, 1989). The temperature in magnetic region may be greater or smaller than in the surrounding photosphere, causing both photometric and line profile variations as the region cross the stellar disk. Recently, inversion methods ("Doppler" or "surface" imaging) have been developed to solve the temperature distribution of the stellar surface from a time series of spectroscopic observations (Vogt et al., 1987; Piskunov et al., 1989).

At chromospheric level, the ultraviolet emission lines, Ca II H and K, He D3 (5876Å), Hα, and the Ca IR triplet are indicators of activity. Emission in Ca II (Baliunas et al., 1983) and Mg II (Neff et al., 1989) show convincing evidence of rotational modulation, indicating chromospheric active areas which live longer than one period of rotation.

Chromospherically active late-type stars have also shown indications of variability in broadband linear polarization (Huovelin et al. 1985, 1988, 1989). The variations are likely due to either the effect of variable magnetic field on saturated photospheric absorption lines (i.e. desaturation of perpendicularly polarized Zeeman components), or to Rayleigh scattering in the optically thin layers of the atmosphere, with the former being more probable in solar–like dwarfs (Huovelin et al., 1988). Monitoring several activity indicators simultane-

J. O. Stenflo (ed.), Solar Photosphere: Structure, Convection, and Magnetic Fields, 443–446.

ously over a rotational period is an effective way to obtain information of the stellar surface inhomogeneities. First attempts combining simultaneous polarimetry, magnetic field, and Ca-emission observations are reported by Saar *et al.* (1988) and Huovelin *et al.* (1988).

In October 1987, we made new observations that include simultaneous polarimetry, Ca II H and K, and magnetic field observations for several active solar type dwarfs. The polarimetric observations were made at the Crimean Astrophysical Observatory between October 2 and 14, 1987. The instrument was a five channel (UBVRI) version of the photopolarimeter of the University of Helsinki (Piirola 1973, 1975) connected to the 1.25 meter AZT-11 telescope of the Crimean Observatory. The Ca II H and K line emission observations were obtained contemporaneously with the four-channel photoelectric spectrometer at the Cassegrain focus of the Mt Wilson 1.5 meter telescope, and reduced as described by Vaughan, Preston and Wilson (1978). The S-index values were converted to chromospheric surface fluxes F'_{HK} using the formulae in Huovelin *et al.* (1988). The high resolution spectroscopic observations (617 nm region) for the magnetic field and filling factor determinations were made with the National Solar Observatory McMath echelle/CCD system. We summarize some preliminary results of the analysis of the polarimetric and Ca II observations in this paper.

3. Results and Discussion

For several stars, the overall variations during the observing period were small and fairly irregular. We, therefore, first studied whether statistical fluctuations could be the cause of these changes. As the method we applied a standard χ^2 test (see e.g. Huovelin *et al.*, 1989). The results of this test are shown in Table 1.

Table 1: Programme stars and statistical results. P_{rot} values are from Noyes *et al.* (1984).

HD no.	Name	Sp. type	B-V	P_{rot} (days)	significance	of	variations	(%)
					P(U)	P(B)	P(V)	Ca-em.
1835	9 Cet	G2V	0.66	7.7	≥99.99	96.3	88.8	92.3
17925		K0V	0.87	6.6	85.5	92.8	76.8	≥99.99
20630	κ Cet	G5V	0.68	9.4	99.6	89.4	71.2	≥99.99
22049	ε Eri	K2V	0.88	11.3	99.4	99.98	96.0	≥99.99
39587	χ^1 Ori	G0V	0.59	5.2	84.9	97.9	92.9	≥99.99
206860	HN Peg	G0V	0.59	4.7	52.1	68.4	99.1	87.4

We note that polarization in the ultraviolet may not always be the best indicator of variations, for although the polarization amplitude is generally the largest in U, the observational uncertainties are larger as well, reducing the statistical significance of the detection.

We are developing a procedure to model activity variations, which includes simultaneous calculation of Ca-emission, photometric brightness, linear polarization, and the magnetic field as a function of rotational phase. The surface can be covered with arbitrary number of circular spots of arbitrary size and surface brightness. At present, the Ca-emission is assumed to be confined to the magnetic areas, with a uniform intrinsic intensity, and the surface outside the magnetic regions is assumed to produce no emission. For limb darkening we use a linear law $(1 - \epsilon + \epsilon\mu)$ with $\epsilon = 0.5$ (Schrijver *et al.*, 1989). The current model for the Ca-emission does not yield absolute sizes of the active areas, since the actual levels of emission in the quiet and active areas is unknown (due to unknown and variable contributions from network, spots etc.).

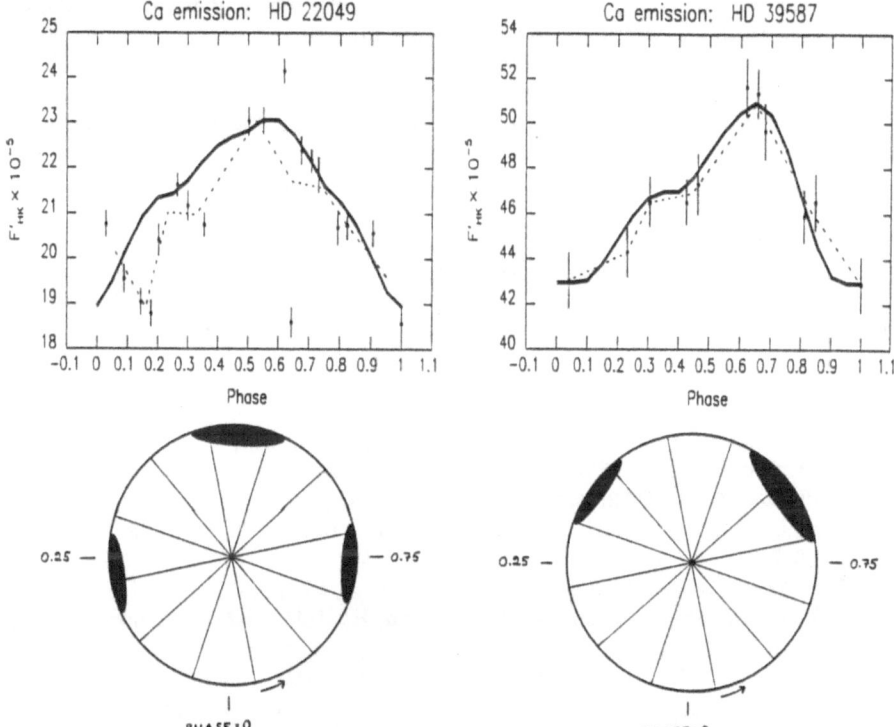

Figure 1. CaII(H+K) flux variations (erg cm^{-2} s^{-1}) vs. rotation phase and corresponding models (thick lines), which are sketched in longitude maps below. Error bars are 1 σ .

Modeling the Ca II data, we obtain rough estimates for the plage longitudes and an upper limit for their relative sizes. In Fig.1 we present the preliminary interpretation of the Ca-emission variations in ϵ Eridani with three and χ^1 Orionis with two equatorial active regions. The plage sizes are exaggerated for visibility. Future work includes the interpretation of simultaneous polarimetry and magnetic field measurements together with Ca-emission, using more detailed models described below.

The polarimetric model is based on differential saturation of magnetically sensitive absorption lines (i.e. magnetic intensification). The basic scheme is an improved version of the Landi Degl'Innocenti (1982) model, employing explicit disk-integration and improved

446

treatment of limb darkening (Saar and Huovelin, 1989 in preparation). For comparison, we can also produce models of the polarization variations assuming Rayleigh scattering as the cause of the polarization. The magnetic field model is the extension of the analysis method of Saar (1988) to inhomogeneous stars, adding the filling factors from the active regions with background fields to form disk-integrated line profiles.

By fitting the combined theoretical model (Ca-emission, linear polarization and magnetic field) to the observations with a suitable inversion method (e.g., Piskunov *et al.*, 1989), we can obtain estimates for the coordinates, sizes and magnetic fields of the active areas, and also for the stellar inclination. From the polarimetry, it is, in principle, also possible to derive the orientation of the stellar rotation axis in the sky. With the present data, however, the surface brightnesses of the active areas remain ambiguous due to lack of simultaneous photometry. Since the current combination of observations should be fairly insensitive to areas considerably cooler than the surrounding photosphere, a reasonable first approximation might be to assume that the active areas and the surrounding photosphere are equally bright. The models can be refined using future simultaneous observations with photometry included.

Acknowledgements.   This work has been supported by a grant from the Aaltonen Foundation (JH) and by the Smithsonian Institution postdoctoral research fellowship program (SS). We are indebted to Dr. S. Baliunas for providing the Ca II data.

References

Baliunas, S., Vaughan, A., Hartmann, L., Middelkoop, F., Mihalas, O., Noyes, R., Preston, G., Frazer, J., Lanning, H.: 1983, *Astrophys. J.*, **275**, 752
Landi Degl'Innocenti, E.: 1982, *Astron. Astrophys.*, **110**, 25
Huovelin, J., Linnaluoto, S., Piirola, V., Tuominen, I., Virtanen, H.: 1985, *Astron. Astrophys.*, **152**, 357
Huovelin, J., Saar, S., Tuominen, I.: 1988, *Astrophys. J.*, **329**, 882
Huovelin, J., Linnaluoto, S., Tuominen, I., Virtanen, H.: 1989, *Astron. Astrophys. Suppl. Ser.*, (in press)
Mathys, G., Solanki, S.K.: 1989, *Astron. Astrophys.*, **208**, 189
Neff, J.E., Walter, F.M., Rodonó, M., Linsky, J.L.: 1989, *Astron. Astrophys.*, **215**, 79
Noyes, R.W., Hartmann, L., Baliunas, S., Duncan, D., Vaughan, A.: 1984, *Astrophys. J.*, **279**, 763
Piirola, V.: 1973, *Astron. Astrophys.*, **27**, 383
Piirola, V.:1975, *Ann. Acad. Sci. Fenn.*, **A VI**, No. 418
Piskunov, N.E., Tuominen, I., Vilhu, O.: 1989, *Astron. Astrophys.*, submitted
Robinson, R.D.: 1980, *Astrophys. J.*, **239**, 961
Saar, S.H.: 1988, *Astrophys. J.*, **324**, 441
Saar, S.H., Huovelin, J., Giampapa, M.S., Linsky, J.L., Jordan, C.: 1988, in *Activity in Cool Star Envelopes*, Eds. O.Havnes *et al.*, Kluwer, Dordrecht, p.45
Schrijver, C., Coté, J., Zwaan, C., Saar, S.H.: 1989, *Astrophys. J.*, **337**, 964
Vaughan, A.H., Preston, G. W., Wilson, O. C.: 1978, *Publ. Astron. Soc. Pacific*, **90**, 267
Vogt, S.S., Penrod, G.D., Hatzes, A.P.: 1987, *Astrophys. J.*, **321**, 496

NONLINEAR DYNAMO MODES AND TIMESCALES
OF STELLAR ACTIVITY

G. Belvedere[*] and M. R. E. Proctor[**]

[*] *Istituto di Astronomia, Università di Catania, Italy.*

[**] *Dept. Applied Mathamtics and Theoretical Physics. University of Cambridge, UK.*

ABSTRACT. A simple mean-field model of a nonlinear stellar α-ω dynamo is considered, in which dynamo action is supposed to occur in a spherical shell, and where the main nonlinearity retained is the influence of the Lorentz force on the zonal flow field. The equations are simplified by truncating in the radial direction, while full latitudinal dependence is retained. The resulting nonlinear p.d.e.'s in latitude and time are solved numerically, and it is found that while regular dynamo wave type solutions are stable when the dynamo number D is sufficiently close to its critical value, there is a wide variety of stable solutions at larger values of D. Furthermore, two different types of dynamo can coexist at the same parameter values. Implications for fields in late-type stars are discussed.

1. INTRODUCTION

The observational evidence of magnetic activity with cyclical and non-cyclical behaviour in stars other than the Sun has suggested that the mechanism giving rise to stellar activity may operate in a variety of different ways. The physics of the interaction between rotation and convective modes is very complex and poorly understood, despite the increasingly detailed models that have been investigated in recent years.

So far, most theoretical work in stellar activity has been done in the framework of the α-ω linear theory and considering the simple theoretical basis, the results are encouraging, giving some useful general principles that may be expected to hold independently of the models.

However, there is no doubt that present and future

447

J. O. Stenflo (ed.), *Solar Photosphere: Structure, Convection, and Magnetic Fields*, 447–453.

theoretical research in stellar activity has to be carried
out in the framework of the more rigorous and self consistent
nonlinear approach, which in principle can describe a large
variety of dynamo operation modes.

Here, a simple mean - field model of a nonlinear stellar
dynamo is considered, in which dynamo action is supposed to
occur in a spherical shell, and where the main nonlinearity
retained is the influence of the Lorentz force on the zonal
flow field.

Weiss, Cattaneo & Jones (1984) have constructed a low
order system of ordinary differential equations describing
stellar cycles, using a severely truncated representation of
the spatial structure, and the simplest nonlinear couplings
between the magnetic field and the mean rotation. They found,
in various parameter ranges, regular cyclical behaviour,
quasi - periodic oscillations and aperiodic cycles with
features similar to those deduced from sunspot number
observations and climatological evidence, including "grand
minima". A drawback of their model is its local nature: the
interaction between active regions in each hemisphere cannot
be represented, and their equations cannot yield steady (non
cyclical) activity although such solutions are known to be
possible for linear α-effect dynamo models.

Our work develops the ideas of the above model. We
retain the α-effect formalism, and the simplest nonlinear
interactions, but use a representation of the spatial
structure only in the radial direction thus allowing full
latitudinal dependence of fields and flows.

While we do not claim to be simulating any particular
stellar dynamo, we do feel that our model can capture many of
the essential physical properties of the nonlinear
interactions.

2. DERIVATION OF THE MODEL EQUATIONS

We begin with the axisymmetric mean field dynamo
equations (see for example Parker 1955, Moffatt 1978) for the
evolution of a magnetic field $B = B(r,\theta)\hat{\phi}+\nabla\times(A(r,\theta)\hat{\phi})$ where
r, θ are spherical polar coordinates and $\hat{\phi}$ is the unit vector
in the azimuthal direction. $B\hat{\phi}$ is the toroidal and $B_p =\nabla\times A\hat{\phi}$
the poloidal part of B. These equations take the form (in the
presumed absence of any mean poloidal flow field)

$$\frac{\partial A}{\partial t} = \alpha F(r,\theta)B+\eta_t \left(\nabla^2 - \frac{1}{r^2 sin^2\theta} \right) A \tag{1}$$

$$\frac{\partial B}{\partial t} = rsin\theta B_p \cdot\nabla \left(\frac{U(r,\theta)}{rsin\theta} \hat{\phi} \right) +\eta_t \left(\nabla^2 - \frac{1}{r^2 sin^2\theta} \right) B \tag{2}$$

Here αF is the usual α-effect, with F representing its spatial structure and α its magnitude. Note also that, consistent with previous models of the solar dynamo, the differential rotation U is considered to be much more potent than the α-effect in producing toroidal field, and the latter term is thus omitted from (2). The quantity η_t is a turbulent diffusivity.

The dynamical influence of the magnetic field enters the model through its effect on the differential rotation $U(r,\vartheta)$. We write $U = u_o + u$, where u_o is a prescribed velocity field and u is a perturbation driven directly by the mean Lorentz force, and subject to viscous damping. The simplest equation that encompasses these features of the evolution of u is

$$\rho \frac{\partial u}{\partial t} = \frac{1}{\mu_o} \langle \nabla \times B \times B \rangle_{\phi} + \rho \nu_t \left(\nabla^2 - \frac{1}{r^2 \sin^2 \vartheta} \right) u \qquad (3)$$

where ν_t is a turbulent viscosity.

The aim, then, is to solve these equations in a spherical shell, representing the convection zone of the star in question. This implicitly assumes that the dynamo is operating throughout the stellar convection zone. However, suggestions have been made in recent years about the possibility that it is instead confined in the thin overshoot layer just beneath the bottom of the C.Z. (see e.g. Spiegel and Weiss 1980; Spruit and van Ballegooijen 1982). However, this is not a crucial point. What we want to show here is that in the non-linear regime a variety of dynamo operation modes arises in a spherical shell. We don't claim to describe real stars, but what *may* occur in real stars. Furthermore, in the context of the radial averaging that we later perform on the equations, the differences between the two scenarios will be manifested only in changes in the coefficients that appear in the truncated equations. We hope in future work to investigate the different consequences of confining the dynamo process to an overshoot layer, and of incorporating the results of the most recent helioseismological data (Brown and Morrow 1987; Brown et al. 1988).

Dimensionless equations are obtained by adopting the following scaling factors: $r_o = 6.96 \times 10^8$ m, $\eta_t = 10^8$ m^{2}s^{-1}, $\tau_o = r_o^2 / \eta_t \simeq 5 \times 10^9$ s $\simeq 160$y, $\Omega^* = 2.57 \times 10^{-6}$ rad s^{-1}, $B^* = \Omega^* r_o \sqrt{(\mu_o \rho^*)}$, $A^* = \eta_t \sqrt{(\mu_o \rho^*)}$, where ρ^* is an appropriate average of density across the c.z.. Thus three basic dimensionless parameters appear, namely r_b (radius of the bottom of the c.z),

$D = \alpha \, \Omega^* r_o^3 / \eta_t^2$ (dynamo number), $P_m = \nu_t / \eta_t$ (magnetic Prandtl number).

We look for solutions of the form: $A = f(r) \chi(\theta, t) / \sin \theta$; $B = g(r) \psi(\theta, t) / \sin \theta$; $u = h(r) \, Q(\theta, t) / \sin \theta$, with the following boundary conditions: $A = 0$ at $r = r_b$ (radiative zone a good conductor); A matches a potential field at $r = 1$; $\partial(rB) / \partial r = 0$ at $r = r_b$ (no tangential current); $B = 0$ at $r = 1$; $\partial(u/r) / \partial r = 0$ at $r = r_b$, $r = 1$ (no stress).

We adopt simple forms of $f(r)$, $g(r)$, $h(r)$, which satisfy the boundary conditions. We also choose the α-effect variation $F(r, \theta) = a(r) \sin \theta \cos \theta$ and the basic zonal field u_o to match the observed latitudinal differential rotation at $r = 1$, while reducing to solid body rotation (with the equatorial value of the surface angular velocity) at the base of the convection zone.

We seek a model problem in which the radial dependence is integrated out (radial truncation) so that A, B, u are functions of θ and t only.

Thus the radially truncated dimensionless equations are given by multiplying equations (1) to (3) by $r^2 f(r)$, $r^2 g(r)$, $r^2 h(r)$ respectively and integrating from r_b to 1. After some algebra, we obtain the equations (in dimensionless variables)

$$C_A \frac{\partial \chi}{\partial t} = C_A D \sin\theta\cos\theta \, \psi + \eta_1 \sin\theta \frac{\partial}{\partial \theta} \left(\frac{1}{\sin\theta} \frac{\partial \chi}{\partial \theta} \right) - \eta_2 \chi \qquad (4)$$

$$C_B \frac{\partial \psi}{\partial t} = C_1 H \sin\theta \frac{\partial \chi}{\partial \theta} - C_2 \frac{\partial H}{\partial \theta} \chi \sin\theta + \frac{C_3 Q}{\sin\theta} \frac{\partial \chi}{\partial \theta} \qquad (5)$$

$$- C_4 \sin\theta \frac{\partial}{\partial \theta} (Q/\sin^2\theta) \, \chi + \eta_3 \sin\theta \frac{\partial}{\partial \theta} \left(\frac{1}{\sin\theta} \frac{\partial \psi}{\partial \theta} \right) - \eta_4 \psi$$

$$C_\Omega \frac{\partial Q}{\partial t} = -(C_3 + C_4) \frac{\psi}{\sin\theta} \frac{\partial \chi}{\partial \theta} - \frac{C_4}{\sin\theta} \chi \frac{\partial \psi}{\partial \theta} \qquad (6)$$

$$+ P_m \left[\nu_1 \sin\theta \frac{\partial}{\partial \theta} \left(\frac{1}{\sin\theta} \frac{\partial Q}{\partial \theta} \right) - \nu_2 Q \right],$$

where C_A, C_B, C_Ω, the C_i, η_i and ν_i are constants that depend only on the functions f, g, h and $H(\theta) = 1 - .189 \, P_2(\cos\theta) - .0394 \, P_4(\cos\theta)$ gives the latitudinal part of the differential rotation (Durney 1974). Note that it is not necessary to define the form a(r); this reinforces our earlier remarks about it not being too necessary to decide the radial dependence of α.

Equations (4) to (6) are to be solved with the boundary conditions, appropriate for axisymmetric fields: $\chi = \psi = Q = 0$, $\theta = 0, \pi$.
In practice, the computations are actually carried out only in the hemisphere $0 \leq \theta \leq \pi/2$, with the symmetry conditions $\psi = \frac{\partial \chi}{\partial \theta} = \frac{\partial Q}{\partial \theta} = 0$ at $\theta = \pi/2$ to simulate a poloidal field of dipole type.

The equations were solved using an explicit time - stepping method of DuFort - Frankel type, for various values of D, P_m, r_b and various initial conditions. The equations were represented on a spatial mesh, with between 20 and 60 mesh intervals. Convergence with respect to temporal and spatial resolution was checked, and found to be quite satisfactory. No problems were experienced at $\theta = 0, \pi$, in spite of the coordinate singularity there. This appears to be due, firstly to a careful treatment of the second order differences so as to give an accurate representation near the poles, and secondly because χ, ψ, Q all vanish quadratically as $\theta \rightarrow 0, \pi$.

3 . RESULTS AND DISCUSSION

It is well known that highly truncated nonlinear dynamo models can exhibit irregular oscillations as shown by Weiss et al.. However these models suffer from the deficiency of having no latitudinal resolution. Not only does this prevent the construction of butterfly diagrams, but it also means that steady solutions of the dynamo equations cannot be found. Our model, on the other hand, is able to describe propagation of dynamo waves and the spherical geometry makes possible a wide range of nonlinear phenomena. In particular, the symmetry between successive cycles may be broken, leading to a preponderance of one polarity over the other.
Furthermore, solutions have been found which spend a long time to an (unstable) steady state and then show a pulsed behaviour. Though these solutions do not closely resemble the modulation of the solar cyle (e.g. the Maunder minimum), they do show that significant variations in activity are possible. Weiss et al.'s calculations showed Maunder minimum - like behaviour but only for the most severely truncated model. The pulsed behaviour occurs for values of the dynamo number ranging (in the particular case $P_m = 0.1$) from D $\simeq$ 625 to beyond D = 2000. Solutions are asym-

452

metric until D ≃ 1800, but symmetric above this value.

In our model dynamo action occurs for D large than about 70. From D ≃ 70 to D ≃ 340 we get periodic symmetric stable solutions. From D ≃ 325 to D ≃ 750 there are stable solutions which are still periodic but asymmetric. However quasi - periodic stable solutions are also allowed in the range D ≃ 350 - 640. The latter resemble quasi - periodic solutions found by Weiss et al. Thus, D ≃ 350 seems to correspond to a point of subcritical bifurcation.

Therefore our results show that two or three different forms of stable solution can coexist in suitable ranges of the dynamo number, depending on the initial conditions. This suggests that there might be large differences in the activity signatures of very similar stars.

Figure 1. shows the time variation of the toroidal field (Psi), the poloidal field (Chi), the differential rotation perturbation (Q) and the related butterfly diagram for four different values of α = 310,750,550,1600 corresponding (top to bottom) to periodic symmetric, periodic asymmetric, quasi periodic and pulsed solutions.

Our calculations were carried out principally for P_m=0.1, but no significant differences appear for other values, provided P_m is not too large. We did find, though, that the bifurcation structure is sensitively affected by the assumed form of the radial dependence of the zonal velocity perturbation h(r) and maybe, by the radial dependences of the poloidal and toroidal fields, but we have not investigated the latter point. Indeed, for some forms of h(r), we did not find asymmetric or pulsed behaviour. Therefore, a further investigation should include a better description of the radial dependencies (ideally by incorporating full radial resolution).

Further developments of the present model will include runs with different values of r_b, to look at dynamos in stars with different depths of the c.z. and, of course, with different rates of rotation (this can be modelled by varying the dynamo number).

REFERENCES

Brown, T.M., and Morrow, C.A.: 1987, Astrophys. J. 314, L21.
Brown, T.M., Christensen - Dalsgaard, J.,Dziembowski, W.A., Goode, P., Gough, D.O. and Morrow, C.A.: 1989, Astrophys, J. in press
Durney, B.R.: 1974, Astrophys. J. 190, 211
Moffatt, H.K.: 1978, *Magnetic Field Generation in Electrically Conducting Fluids*, Cambridge Univ. Press
Parker, E.N.: 1955, Astrophys. J. 122, 293

Spiegel, E.A. and Weiss, N.O.: 1980, Nature 287, 616
Spruit,H.C. and Van Ballegoojen, A.A.: 1982, Astron. Astro-
      phys. 106, 58
Weiss,N.O., Cattaneo, F. and Jones, C.A.: 1984, Geophys.
      Astrophys. Fluid Dyn. 30, 305

FIGURE 1

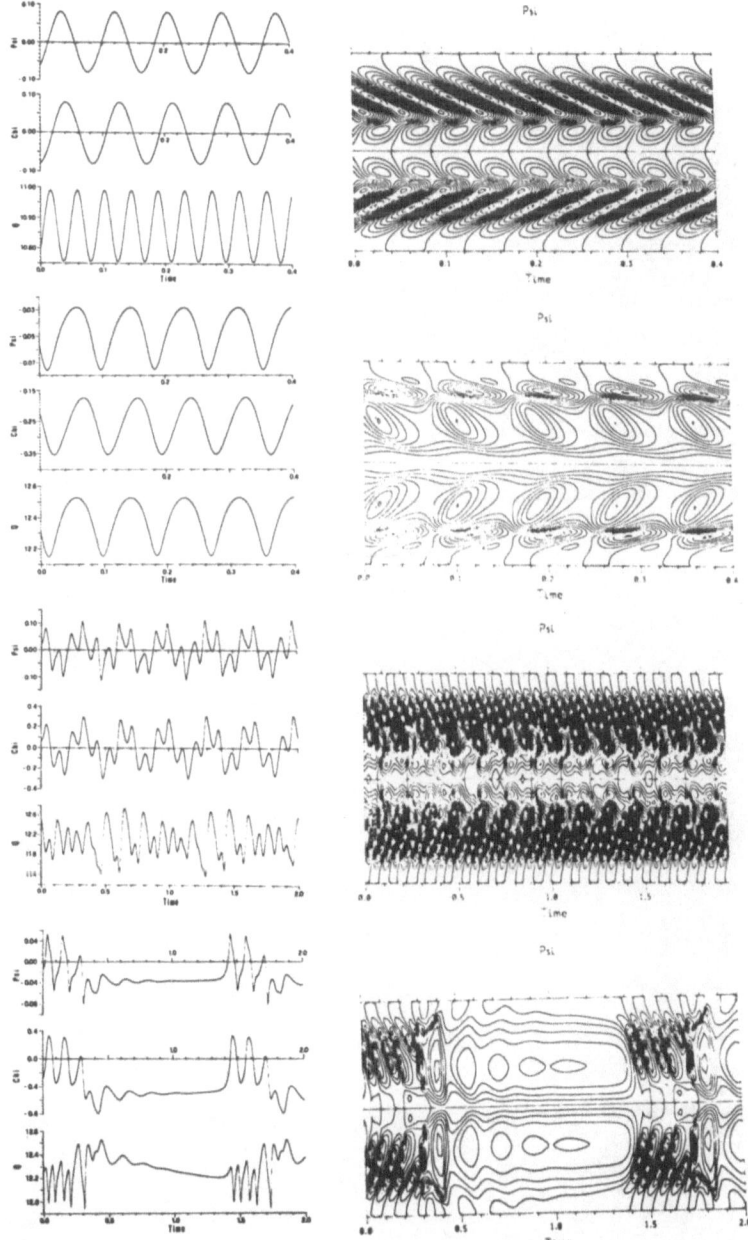

DOES A COMMON DYNAMO MECHANISM EXIST FOR LOWER MAIN SEQUENCE STARS ?

R.B. TEPLITSKAYA and V.G. SKOCHILOV
SibIZMIR
P.O.Box 4, Irkutsk 33
664033, USSR

ABSTRACT. Based on an extended list of lower main sequence stars from Rutten (1987), the relation between chromospheric activity and Rossby number has been revised. The increased statistics changes the shape of the curve as compared with that of Noyes et al. (1984). The saturation at small Rossby numbers has disappeared. The dependence on Rossby number in the range of very large Rossby numbers has weakened. The standard deviation of the activity indices from the mean curve is about 40% . This scatter of individual stars is not due to differences in the spectral type or age of the stars.

It is generally recognized that chromospheric and coronal activity of solar-type stars is closely related to stellar rotation. Since the paper of Noyes et al. (1984) this relation has even been used for evaluating rotation periods and ages of stars for which direct determinations of these parameters are difficult. The 'activity-rotation' relation for the lower part of the main sequence implies that there exists a common curve for stars of all spectral types, from F to M, in the R', $Ro^{(\alpha)}$ diagram. R' is the activity index (the ratio between the emission flux in a chromospheric line F' corrected for the contribution of photospheric emission, and the bolometric flux). $Ro^{(\alpha)}$, the Rossby number, is a measure of the rotation.

$$Ro^{(\alpha)} = P/\tau_c^{(\alpha)},$$

where P is the period of rotation, and $\tau_c^{(\alpha)}$ is the convective turnover time with mixing-length parameter $\alpha = l/H$. The Rossby number is directly related to the dynamo number, so the fact that it is Ro rather than the period or velocity of rotation that controls the activity is convincing evidence for a common mechanism of atmospheric heating by magnetic fields generated by the process of dynamo action.

Since it is not obvious how to choose the optimum index to characterize chromospheric activity, we have tried to introduce the new index

$$A'_{HK} = F'_{HK}/F'^b_{HK} = R'_{HK}/R'^b_{HK}.$$

F'^b_{HK} and R'^b_{HK} are the values of F'_{HK} and R'_{HK} on the basic curve with the same colour index $(B-V)$ as in the star considered. The flux F'_{HK} is measured in the emission reversals

J. O. Stenflo (ed.), Solar Photosphere: Structure, Convection, and Magnetic Fields, 455–459.

of the lines H and K of Ca II. The only possible advantage of A'_{HK} over R'_{HK} is that the former may be colour independent, while this is not the case for the latter. However, the colour independence of A'_{HK} is valid only under special circumstances.

Using data from Rutten (1987), we have considered 182 main-sequence stars of spectral classes F2 to M4.5 and constructed a $\log A_{HK}$ vs. $\log Ro^{(\alpha)}$ diagram (Teplitskaya, 1989). The curve obtained resembles the curve of Noyes et al. (1984) in the sense that it includes all stars, regardless of their spectral type. As concerns its shape it shows interesting differences, however: (1) There are no indications of a saturation of the chromospheric activity at small Rossby numbers, i.e., in the region where because of the large values of $\tau_c^{(\alpha)}$ a large number of late-type stars occur. (2) At large Rossby numbers the index A'_{HK} randomly fluctuates around a constant mean value. In the range of large $Ro^{(\alpha)}$ there are many stars with very thin convective envelopes, because $\tau_c^{(\alpha)}$ is small. (3) The parameter $\alpha = 1.6$ instead of 2.0 in Noyes et al. (4) The scatter of some stars around a mean curve is greater than can be explained by random or systematic errors. The above-mentioned differences occur for stars with very thick and very thin convective envelopes, and may be due to individual features of the dynamo action in some stars.

The first question we have to answer is whether the differences may be attributed to the use of different activity indices, or if they are a consequence of the increased sample of stars (41 stars in Noyes et al. (1984), and 182 stars in Teplitskaya (1989)). To settle this matter we have repeated the investigation of Noyes et al. (1984), but using the 182 stars listed by Rutten (1987). Figure 1 shows plots of $\log R'_{HK}$ vs. $\log \tilde{Ro}^{(2.0)}$ and $\log A'_{HK}$ vs. $\log \tilde{Ro}^{(1.6)}$ ($\tilde{Ro}^{(\alpha)}$ being the Rossby number $Ro^{(\alpha)}$ normalized to its value for $(B - V) = 0.63$). The open circles represent averages in coordinates over 20 or 21 stars.

There are now no fundamental differences in the shapes of the two curves. The indications of saturation are very weak, much less than what was found by Noyes et al. (1984) for 41 stars. On both plots (diagrams to the left) one can observe a branch of constant activity at large Rossby numbers.

Apart from the usual Rossby numbers we have also used the numbers

$$Ro^{(\alpha)\prime} = Ro^{(\alpha)} f(M),$$

where M is the stellar mass, $f(M) = (\alpha H/R_c)^{1/2}$, and R_c and H are the radius and the pressure scale height at the base of the convection zone (Durney and Robinson, 1982). Corresponding 'activity-rotation' curves are shown in the right-hand diagrams of Figure 1. Signs of saturation of the chromospheric activity have become even weaker. Besides, the application of the modified Rossby numbers has almost completely eliminated the branch of constant activity, except for the stars with the largest values of $Ro^{(\alpha)\prime}$. Thus the chromospheric activity of most of the stars under investigation can, on the average, be approximated by the power laws

$$R'_{HK} = 1.89 \times 10^{-4} \left[\tilde{Ro}^{(2.0)\prime} \right]^{-0.683}, \qquad S = \pm 38.7\%$$

$$A'_{HK} = 1.26 \times 10^{-4} \left[\tilde{Ro}^{(1.6)\prime} \right]^{-0.745}, \qquad S = \pm 41.9\% .$$

S is the relative standard deviation of R'_{HK} or A'_{HK} from the mean curves, calculated as $(\text{mod})^{-1} \sigma_y$, where σ_y is the standard deviation along the ordinate axes of Figure 1, and $\text{mod} = \log e \approx 0.43429$.

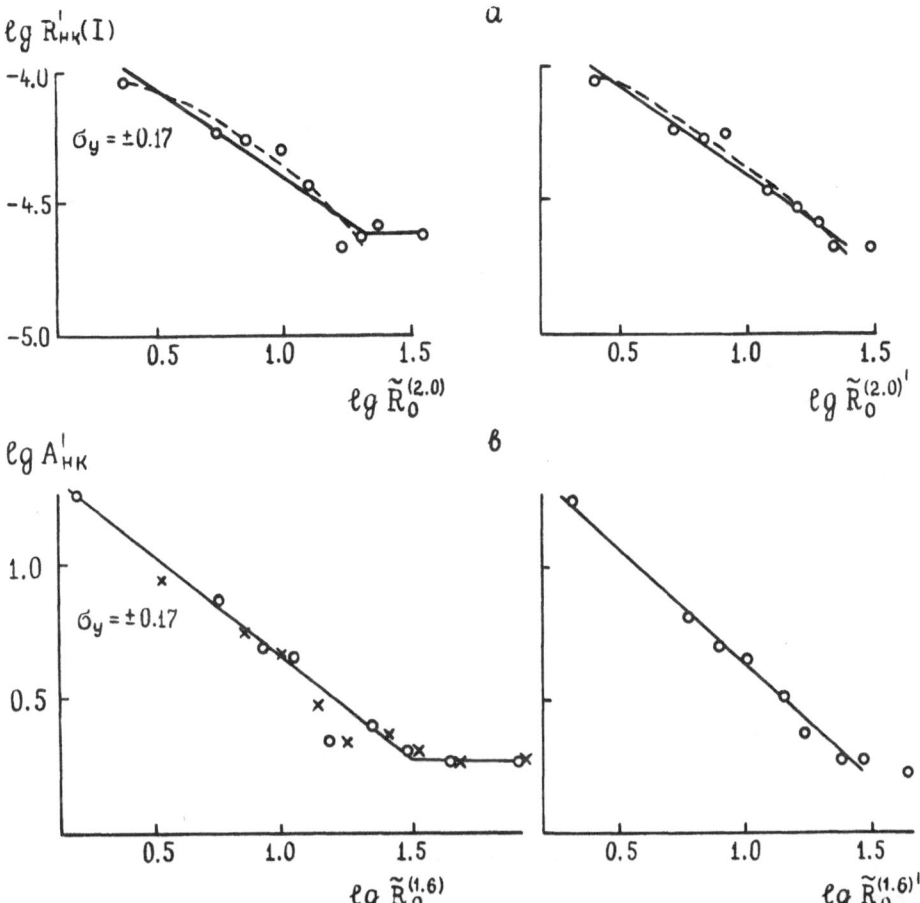

Figure 1. The 'activity-rotation' relation for two indices of chromospheric activity (a: R'_{HK}, b: A'_{HK}) and for two procedures of calculating the Rossby numbers (left and right diagrams).

The considerable scatter of the individual stars is uncorrelated with their colours. There is a slight dependence of the residuals (O-C) on the age of the star as shown in Figure 2. Estimates of the age of the stars and clusters have been taken from Duncan (1981), Catalano and Marilli (1983), and Barry et al. (1987). With the same colour and rotation period young stars tend to be slightly more active than old stars. Part of the scatter may be accounted for by the relative age, i.e., the different rate of evolution. For example, in stars of class F the convective turnover time begins to change already during the main-sequence stage (Gilliland, 1985). Another example of a peculiar behaviour associated with age is that among the stars that have just arrived at the main sequence there are fast rotators, whose rotation velocities are not representative of their rather late spectral class K (Stauffer et al., 1984). However, these factors do not seem to be able to explain the main part of the scatter.

The absence of saturation in late-type stars at a first glance contradicts the concept of

458

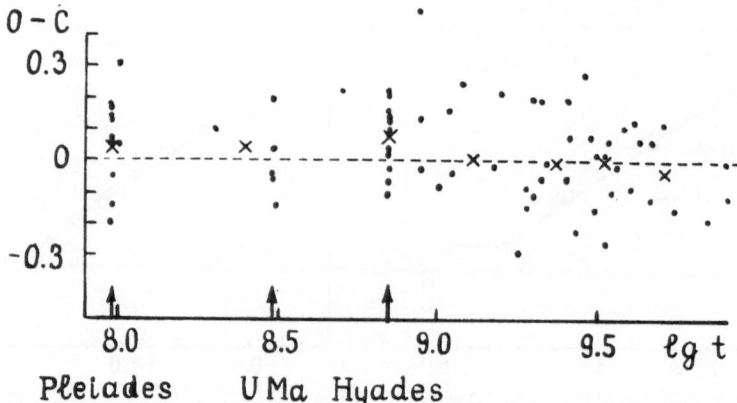

Figure 2. Residuals with respect to the mean curves vs. stellar age. The dots represent individual stars, the crosses averages over 10 stars.

stabilization of the magnetic activity, arrived at from X-ray luminosity observations of M dwarfs. It seems likely that the behaviour of the chromospheric activity indices in low-mass stars is controlled not only by the efficiency of the dynamo mechanism but also, to a significant extent, by geometrical effects, i.e., very large filling factors of active regions, with overlaying magnetic 'canopies' as described by Giovanelli (1980) for the Sun, leading to excess heating of the chromosphere. In the case of very late M stars with $(B-V) \gtrsim 1.50$ Schrijver and Rutten (1987) found the opposite phenomenon, namely a deficiency of chromospheric emission as compared with X-ray emission. This results from a decrease of the role of ionized metal lines in favour of an increased role of the hydrogen lines for the radiative cooling of the chromosphere. In the sample we are investigating there are only two stars with $(B-V) > 1.5$. Accordingly the deficiency caused by radiative transfer effects does not influence the results reported in the present paper.

References

Barry, D.C., Cromwell, R.H., and Hege, E.K. (1987) 'Chromospheric activity and ages of solar-type stars', *Astrophys. J.* **315**, 264-272.

Catalano, S. and Marilli, E. (1983) 'Ca II chromospheric emission and rotation of main sequence stars', *Astron. Astrophys.* **121**, 190-197.

Duncan, D.K. (1981) 'Lithium abundances, K line emission, and ages of nearby solar type stars', *Astrophys. J.* **248**, 651-669.

Durney, B.R. and Robinson, R.D. (1982) 'On an estimate of the dynamo-generated magnetic fields in late-type stars', *Astrophys. J.* **253**, 290-297.

Gilliland, R.L. (1985) 'The relation of chromospheric activity to convection, rotation and evolution of the main sequence', *Astrophys. J.* **299**, 286-294.

Giovanelli, R.G. (1980) 'An exploratory two-dimensional study of the coarse structure of network magnetic fields', *Solar Phys.* **68**, 49-69.

Noyes, R.W., Hartmann, L.W., Baliunas, S.L., Duncan, D.K., and Vaughan, A.H. (1984) 'Rotation, convection, and magnetic activity in lower main-sequence stars', *Astrophys. J.* **279**, 763-777.

Rutten, R.G.M. (1987) 'Magnetic structure in cool stars. XII. Chromospheric activity and rotation of giants and dwarfs', *Astron. Astrophys.* **177**, 131-142.

Schrijver, C.J. and Rutten, R.G.M. (1987) 'Magnetic structure in cool stars. XIV. Deficiency in chromospheric fluxes from M-type dwarfs', *Astron. Astrophys.* **177**, 143-149.

Stauffer, J.R., Hartman, L., Soderblom, D.R., and Burnham, N. (1984) 'Rotational velocities of low-mass stars in the Pleiades', *Astrophys. J.* **280**, 202-212.

Teplitskaya, R.B. (1989) 'On the relation between activity and rotation in the main-sequence stars', *Astron. Nachr.* **310**, in press.

VIII. FUTURE DIRECTIONS

GROUND BASED AND SPACE FUTURE PROSPECTS IN SOLAR INTERFEROMETRY APPLIED TO THE PHOTOSPHERE.

J.P. ROZELOT
Observatoire de la Côte d'Azur
Avenue Copernic 06130 Grasse France

ABSTRACT. This paper sets about a brief review of the current knowledge of ground based interferometry, and then, using the relevant information from this up-to-date technique, its tentative application to the study of solar granular structures. As a matter of fact, most of the fundamental processes in the photosphere take place on very small spatial scales and understanding the formation of the solar granulation, pores as well as tiny photospheric faculae, requires insight into the magnetic network where the thermal instabilities are initiated. The study of fine scale structures, typically of about 15 milliarcsec (11 km) by interferometry, i.e. to a resolution not accessible through classical optical telescopes, is certainly the key to open the door to new scenarios in solar physics. By the way, similar phenomena are likely to be accountered in other type of stars. Particular attention is paid to the role of imaging reconstruction at very high angular resolution. Some of the thus achievable solar programmes are listed, mainly with solar granulation. The new type of instrumentation involved should be in the 90's commissioned. At last, it is pointed out that solar interferometry could be also performed in space, as earlier suggested by J.L. DAME and C. AIME to ESA, for the study of flares, prominences or coronal loops, which are closely linked to the photosphere, and for the 2000's horizon. Some problems inherent to this whole prospect (ground and space) are listed, and discussion is welcomed for determining adequate programmes, and also to highlight promising directions for future investigations.

1. INTRODUCTION

After Michelson's pioneering work of the 20's at Mount Wilson, in the States , interferometry did not make great strides ahead, mainly due to big technical difficulties inherent to this kind of experiment. However, around the 60's, in the visible field, Hanbury-Brown developed the so-called intensity interferometry, the flux performances of which have considerably limited the application field. Thus, the revival of the interferometric technique is essentially due to the painstaking work of A. Labeyrie, which can be roughly divided into four steps:

* the discovery at Mount Palomar Observatory of *speckle interferometry*, with single telescope disclosed of images disturbed by the atmospheric turbulence;

* the initial *fringe observations* using two *small telescopes* of 26 cm in aperture, at the Nice Observatory in 1974. This was followed a few years later, in 1977, by the world premiere measurement using such an interferometer of Capella's angular diameter;

* the first *fringe observations* between *two large telescopes* 1.5 m. in aperture, spaced 13 m. apart. Steps are currently taken to improve its observing efficiency and high resolution astrophysical information will be obtained soon;

463

J. O. Stenflo (ed.), Solar Photosphere: Structure, Convection, and Magnetic Fields, 463–467.
© *1990 by the IAU.*

* The design, studies and construction of a *prototype of an array* of several telescopes, in order to provide a powerful tool for reconstructing images.

Interferometry may be still at teething stage, but its potentials have been well demonstrated by recent measurements using existing facilities, not only in France, but also in the States and in Australia (see for instance Labeyrie, 1988; Davis & Tango, 1985; Shao et al., 1987). The present trend clearly points towards aperture synthesis systems including a large number of subapertures. Moreover, the V.L.T. will be dedicated in an interfero- metric mode in the 1995. With a baseline of about 104 m. between the centres of the most extreme telescopes, resolutions up to approximately 45 milliarcsec at 20 microns wavelength and 0.75 milliarcsec in the blue could be reached.

However despite an intensive work in stellar interferometry, both theoretical and observational, nothing or scarcely anything, has been achieved with solar interferometers. I wish to be clear here: by solar interferometer, I mean a complete instrument with several distinct apertures and a focal laboratory, intended for observations of the Sun. This definition excludes the great amount of theoretical work on solar interferometry carried out by a lot of people, and also solar speckle interferometry practised through a single aperture. With such a concept, a recent proposal was made by Damé (1987), consisting of a "true" interferometer, to be commissioned on the Space Station, and mainly devoted to the ultraviolet field.

Thus, my question remains: why does no such instrument, adapted to the solar case, using two or more telescopes, as yet exist on ground?

III. WHAT IS AN INTERFEROMETER?

Such an instrument is composed of at least two, usually track-movable telescopes, of high mechanical stability, that makes up a North-South baseline. With additional telescopes, it is possible to make up an array, either on the same line, or on a separate one. In this latter case, a-Y or a cross-like pattern can be built, allowing the closure phase. Figure 1 shows what such a device can be like. On the ground, the baseline can be very long (up to 600 m. for stellar interferometers) giving the interfringe angular value (lambda/B). In the focal laboratory, close to the middle of the baseline, the beam recombining optical device lies on a movable carriage. Its displacements are computer-controlled and correct for part the optical path drift induced by the diurnal rotation.

On the image plane, the field is diaphragmed by means of a slit and the light is dispersed by a grating that lets to observe fringes be simultaneously observed over several spectral channels. Then the fringes patterns are recorded on a camera.

For space, it is necessary to have very rigid mechanical structures, and compact configurations in order to avoid pointing and tracking problems. Recently, Damé (1988) suggested a fixed combination of four 20 cm-telescopes fixed on a 2 m. baseline. This gives complete u-v coverage (no phase closure) with imaging allowed when rotating the interferometer through 180°.

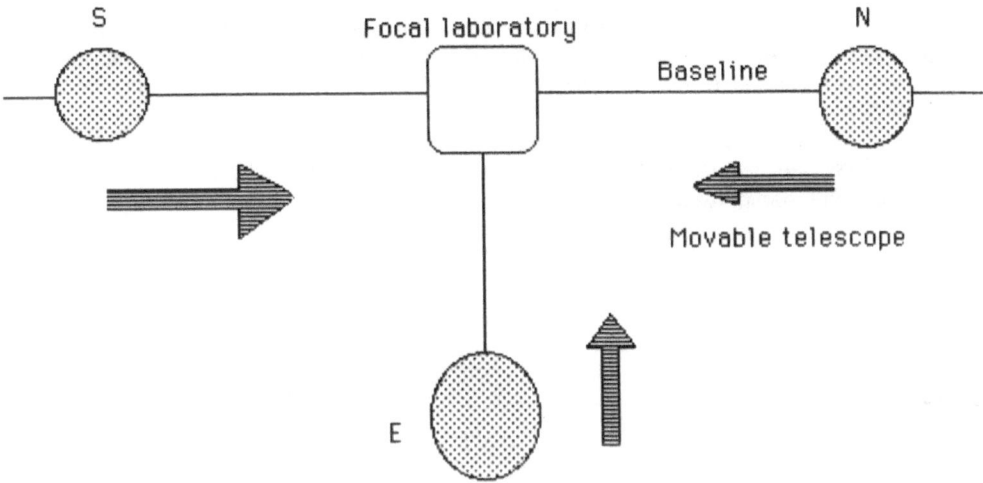

Figure 1.   Schematic view of an interferometer.

IV. THE DELAY LINES

With independently mounted telescopes, the entrance pupils are not coplanar as the telescopes points off the zenith. To maintain the geometrical scaling of the lateral pupil geometry, the exit pupil at the combining optics has to be adjusted (Figure 2). This is made by delay lines, which is one of the two possible solutions (the other one being the use of moving telescopes to ensure equal path length for both beams). Such a prototype of delay lines is currently under construction by Aerospatiale, in France, for the so-called "I3T", a small stellar interferometer, and those delay lines will be put into operation at the "Observatoire de la Côte d'Azur" before late 1989.

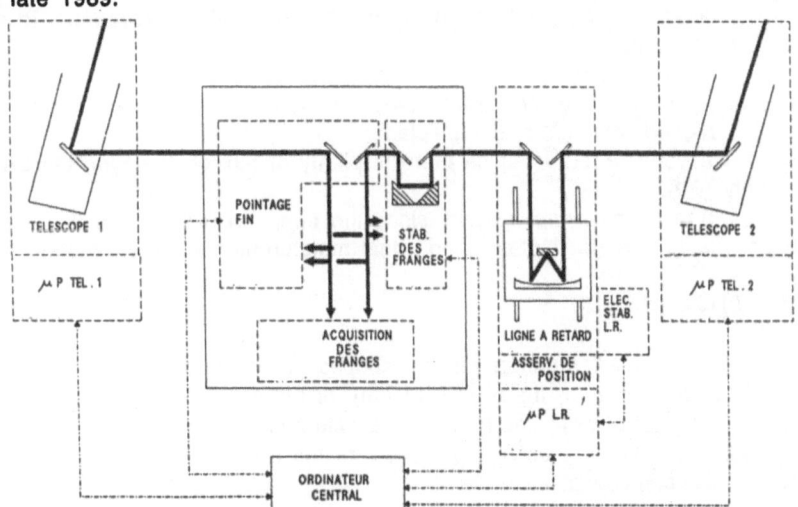

Figure 2.   Principle of the delay lines (after Koehler).

It is proposed that the optical elements which have to be translated during the observation be mounted on a track-movable carriage. The conception of this device is given in Figure 2 (Koehler, 1989). The useful course is about 3 m., with about 50 microns positioning accuracy and approximatively 2 mm/s uniform velocity.

This stage require a smooth motion and a very high mechanical stability. Figure 3 give an example of the so-called "oeil de chat", an optical piece that is movable for path length compensation.

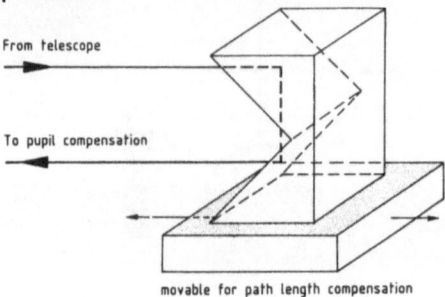

movable for path length compensation

Figure 3.   Schematics of the so-called "oeil de chat", movable prisms.

V. A BRIEF LIST OF CONTRAINTS

Among the numerous problems which must be solved to achieved a solar interferometer, it can be listed (more details can be found in Damé (1988):
 * a very high *pointing system*, to be sure to look at the same features on the Sun, since the observed object is resolved; on the ground, this must be achieved with a greater difficulty than in space, because of the atmospheric turbulence;
 * a very fine accuracy *tracking system*, that means a fraction of the Airy disk or about lambda/10 or lambda/20 (depending on the wavelength under consideration);
 * a good *stabilization* of the whole system, with no vibrations, to access small fields on the Sun;
 * a good *spectral resolution* on lines profiles: high spatial resolution requires analysis of the fringes by a high dipersion spectrograph (about one Å);
 * a good *coverage of the u-v* plane;
 * a good *detector*, since there is plenty of flux on the Sun, the S/N ratio must be very high;
 * a fast *monitoring system*, since the temporal evolution of the structures implies fast electronics (for instance individual measurements every 10 ms).

VI. SOME PROGRAMMES

A solar interferometer, with telescopes of the class 20 to 30 cm diameter can reach a resolution of 15 ms of arc (11 km), or better.
That means that a lot of work can be achieved with such high resolutions. A number of questions still now without a satisfactory answer, and a number of to-day unsuspected questions is still motivating.

It is a commonplace to say that the Sun's atmosphere is highly structured. This is mainly due to the magnetic fields, which strongly interact with the convective motions. But we scarcely know how the fine magnetic elements are associated or linked with bright spots, fibril structures or feet of the loops. As a matter of fact, most of the fundamental processes in the photosphere take place on very small spatial scales and understanding the formation of the solar granulation, pores as well as tiny photospheric faculae, requires insight into the magnetic network where the thermal instabilities are initiated. The study of fine scale structures, typically of about 15 milliarcsec (11 km) by interferometry, a resolution which is not accessible through classical optical telescopes, is certainly the key to open the door for new scenarios in solar physics.

High angular resolution is also of importance to access problems of line diagnostics, loop constitutions, coronal heating, dynamic of flares, filaments and prominences. At last, interferometry will help understand such fundamentally crucial problems, as the nature of granulation, the physics of spicules and the flow of material that goes from the photosphere to the chromosphere and up to the corona.

By the way, similar phenomena are likely to be encountered in other types of stars.

VI. CONCLUSION

Solar interferometers can be defined as instruments of the "second generation", instruments of the first one being coronagraphs, polarimeters... These new instruments, such as heliometers or heliosismology-meters imply ground-based networks or space networks, and in both cases require efficient international cooperation.

The way is quite clear, now, to build such instruments, and I hope that solar physicists will share my faith in these future views.

VIII. REFERENCES

Damé L.: 1987, First IFSUUS meeting on Physics and Astrophysics in the Space station ERA, Venice: "Solar interferometry from the space station",

Damé L.: 1988, Solar Ultraviolet Network, Proposal submitted to the NASA.

Davis J, Tango W.J.: 1985, Proc. Astr. Soc. Australia, "High angular Stellar interferometry", 6, 34-39.

Koehler, B.:1989, Rapport Aeropsatiale.

Labeyrie, A.:1987, "Multiple aperture Interferometry", in Interferometric imaging in astronomy, Oracle, Arizona, USA, p. 97.

Shao M., Colavita M., Staelin D.H., Johnston K.J., Simon R.S., Hughes J.A., Hershey J.L.: 1987, The Astronomical Journal, 93, 1280-1287.

NEW OBSERVATIONAL ASPECTS

ODDBJØRN ENGVOLD
Institute of Theoretical Astrophysics
University of Oslo
P.O.Box 1029, Blindern
N-0315 Oslo 3, Norway

ABSTRACT. The requirements and conditions for high resolution imaging and polarimetry of the Sun are reviewed. Various methods and techniques are discussed for image stabilization and sharpening in solar observations. The new solar facilities in the Canary Islands in particular are frequently reaching diffraction limited resolution and yield new insight in the structure and dynamics of the solar atmosphere. Future ground based telescopes like THEMIS and LEST, as well as planned solar missions in space will trigger a next advance in solar physics.

1   Introduction

The future of solar physics lies in understanding the basic astrophysical processes that can be observed on the Sun. This includes understanding the magnetohydrodynamical processes that occur, and an important, even crucial, tool for understanding these processes is polarimetry, combined with spectroscopy, of small spatial domains over extended periods of time.

It is increasingly apparent to solar astronomers that further progress in understanding the physics of the solar atmosphere relies critically on observations of sub-arcsecond structures. Many fundamental physical processes in the Sun's atmosphere obviously occur at sub-arcsecond scales. Recent analysis of the C IV emission of the solar transition region suggests that emitting structures may be $\leq$ 10 km (Dere 1987). The sizes of the smallest structures in the Sun's atmosphere might possibly be the photon mean free path at $\tau_c \sim 1$ in the photosphere, which is about 50 km in the visible, and $\leq$50 km at the opacity minimum around $1.6\mu m$. In the presence of magnetic fields in the solar atmosphere the oscillatory modes of small flux ropes (Ryutova; this proceedings) may constitute spatial scales that are only a fraction of the assumed flux rope diameters ($\leq$300 km). An even smaller and more relevant scale would seem to be the thickness of boundary layers which are matching local gyro radii and are therefore in the m and cm ranges. The characteristic time scales of the fine structure elements can be tied to the sizes and motions of the structures. Assuming a conservative value for the element size; $d \sim 70km$, and characteristic

J. O. Stenflo (ed.), Solar Photosphere: Structure, Convection, and Magnetic Fields, 469–487.

velocity $\leq 2\ km\ s^{-1}$ and $\leq 10\ km\ s^{-1}$ in respectively the photosphere and chromosphere, we find corresponding time scales $d/v \sim \frac{1}{2}\ min$ and $\leq 5s$.

One will clearly not be able to resolve the smallest of such solar structures in the foreseeable future, but every improvement in spatial and temporal resolution is likely to yield new insight and understanding of the physics of the Sun. Exciting and highly worth while projects for the near future are such as the ground based THEMIS and LEST, and possibly the space borne OSL, and these will presumably offer resolution in the range 0.15 - 0.05 *arcsec*, which is better than the best telescopes today.

2 Observational Requirements

2.1 General Requirements

The requirements for a modern optical solar telescope are: (1) high spatial resolution, (2) high system through-put, and generally also (3) low instrumental polarization. The near infrared wavelength region allows observations to deeper layers of the photosphere and (4) near IR capability is desirable. In addition, the observations done with todays solar telescopes, from ground as well as from space, present a significant challenge in (5) data management capabilities which includes (i) acquistion, (ii) transfer, (iii) calibration, (iv) reduction, (v) analyzing, and (vi) archiving. The problem is not unique to large telescopes only. Given combinations of high spatial, spectral and temporal resolution over even a field of view of few arcsec is generally a major data handling task for any good solar telescope.

The effective spatial resolution attainable with a given telescope system is a function of the optical quality of the system itself and the seeing. The main contribution to the seeing is from the atmosphere, but a non-negligible share comes from the local telescope environment. The careful choice of site is equally important to the optical and mechanical quality of the telescope system. Also a good site may easily be spoiled if the telescope and tower structure generate strong air turbulence.

A solar site is considered to be superb if it offers seeing of 0.1 - 0.3 arcsec a fraction of the time. This fraction can be notably enhanced with the aid of the *real-time* techniques *active* and *adaptive optics* (Beckers 1989c). Furthermore, *solar interferometry* techniques can be applied *after the fact* to recover the high spatial frequency signals down to the diffraction limit of the telescope (von der Lühe and Zirker 1988). The "Knox-Thompson" scheme (Knox and Thompson 1974) shows promise and has been demonstrated by Stachnik *et al.* (1983) and von der Lühe (1988a). The future prospects in solar interferometry is discussed by Rozelot (this proceedings).

Remote operations/observing of stellar telescopes are presently developed and tested for the purpose of enhancing the efficiency in use and lower the cost (Martin and Hartley 1985; Raffi 1988). Remote observing is planned as an option in LEST (Engvold *et al.* 1985).

In the following we shall address some of the points above.

2.2 Spatial Resolution

In a perfect imaging system all energy of a point image is concentrated in an Airy disc corresponding to the diffraction of the circular telescope aperture. The angle corresponding to the radius of the central Airy disc is usually considered as the telescope resolution. The modulation transfer function corresponding to this resolution is 0.0894. This means that, say, a true 15% contrast of solar granulation is reduced to less than 1% in the image produced by a ~10 cm aperture telescope that just "resolves" the granular structure. The effects from telescope aberrations and seeing add to this and serve to reduce the spatial resolution further.

The quality of an imaging optical system is quantified by comparison to the theoretical diffraction image. The ratio of central intensity of the point image theoretical central intensity is the *Strehl Ratio (SR)*. When the *rms* deviations from the perfect surface is expressed in terms of phase error $\Delta\Phi$ the resulting Strehl Ratio becomes (see Dunn 1987):

$$SR = (1 - \Delta\Phi^2) \qquad (1)$$

A common convention is to consider a system diffraction limited when SR is greater than or equal to 0.8, which implies $\Delta\Phi = 0.447$ or the equivalent to $\lambda/14$. An *rms* phase error corresponding to $\lambda/20$ ($\Delta\Phi = 0.314$) yields $SR = 0.90$. The phase errors due to atmospheric seeing will affect the SR in a similar way. A partially correted phase variation of the wavefront due to seeing (see Sections 4.2 and 4.3) will result in a point spread function that consists of a sharp diffraction disc superposed on a residual halo-like image with the shape of the seeing disc (Roddier and Roddier 1986). The resulting image in this case will be a composite of a diffraction limited and a blurred component.

2.3 The Light Flux and Time Resolution

The number of available photons in the detector plane are generally abundant in solar observations at medium resolution (1-2 arcsec) and over medium to broad spectral bands ($\Delta\lambda$ >1 Å). On the other hand, in the case of high resolution, Stokes polarimetry one needs a large aperture solar telescope with high throughput in order to attain the required signal to noise ratio. Besides increasing the collecting aperture area of the telescope one also have to use good detectors. Highly efficient charge-coupled devices ("CCD") detector arrays offer quantum efficiencies up to 50% and more in the visible and near IR, compared to 2 - 25% for photomultipliers, and ~1% for photographic emulsions.

The number of photons per unit time at the detector plane of a telescope with aperture diameter D and focal length f can be expressed by:

$$N = 0.5P(\lambda)\, t(\lambda)\, a\, \Delta\lambda(\frac{D}{f})^2 \qquad (2)$$

$P(\lambda)$ is the flux number of solar photons (the factor 0.5 is atmospheric transmission), $t(\lambda)$ the overall system efficiency, a the detector element area, and $\Delta\lambda$ the spectral element. The system efficiency is the total light loss in reflecting surfaces, transmission optics, beam splitters, monochromators, gratings, and the detectors quantum efficiency. The net through-put is the product of the contribution from the telescope and from the post focus instrument, i.e. $t(\lambda) = t_t(\lambda)\, t_{pf}(\lambda)$.

The number of reflecting surfaces in modern vacuum telescopes are 4 - 5 (the Swedish telescope in La Palma has only three mirrors), and two transmission windows. Assuming that each reflecting surface is freshly coated with aluminum which gives mirror reflectivity will be $\simeq 0.88$ at $\lambda \simeq 5,500$Å, and window transmission 0.92 we get $t_t(5,500$Å$) = 0.45$ to 0.51. The major light loss usually takes place in the focal plane instruments. For example, the estimates of Lites (1987a) for a LEST polarimeter give $t_{pf}(5,500$Å$) \leq 0.016$. The net efficiency will then be $t(5,500$Å$)\sim 0.008$. One may note that a major gain in the overall through-put of a solar telescope is likely to be achieved by clever design of focal plane instruments.

Assuming that photon noise dominates the signal to noise is given by $S = \sqrt{N}$. Let us assume that the angular subtense of the detector pixel elements $\Delta\Theta$ is about one-half of the angular resolution of the system and we get $\Delta\Theta \simeq 2\sqrt{a}/f$. We may then express the relation between the telescope diameter D and S as (cf. MacQueen 1987):

$$D \simeq \frac{2S}{\Delta\Theta} \frac{1}{\sqrt{\pi 0.5 P(\lambda) t(\lambda) \Delta\lambda}} \qquad (3)$$

A signal to noise ratio $S \simeq 300$ will be appropriate for studies of spectral line profiles with $\Delta\lambda = 0.010$ Å. We take $P(5,500$Å$) = 5.5\ 10^{13}\ [cm^{-2}A^{-1}s^{-1}]$ according to Lites (1987a), $\Delta\Theta = 0.1$ arcsec and $t(\lambda) \simeq 0.008$ and find a requisite telescope diameter $D \sim 2.6\ m$ for an integration time of 1s. We may conclude that solar (Stokes) polarimetric work is generally photon starved unless the telescope aperture exceeds 1-2 m.

2.4 Residual Polarization

Observations and interpretations of spectral line profiles recorded in polarized light are invariably complicated by telescopic and instrumental polarization and the effects of yet unknown spatial averaging.

The stringent requirements on instrumental polarization in solar telescopes arises from their inability to provide high spatial resolution, - the polarimetry signal is blurred and therefore substantially weakened. Vector polarimetry will be more practical and simpler when the inherent spatial resolution is ≤ 0.3 arcsec.

The aims of THEMIS and LEST are to reduce the parasitic polarization by using pointed telescopes with rotationally symmetric optics, at least before the position of the polarization analyzer. A recent study by McGuire and Chipman (1988) showed that even rotational symmetric optical systems give a net polarization effect. The effect depends largely on the f-ratio and the reflective coatings of the primary mirrors. In the case of LEST the upper bound of this polarization is .68%. However, since the effect is constant for a given telescope it can in essence be calculated and corrected (see Stenflo 1988; in McGuire and Chipman).

3 Solar Seeing and Sites

3.1 Atmospheric optics

Francois Roddier (1981, 1987) describes the effects of atmospheric turbulence on optical wavefront propagation. The incoming wavefront varies in amplitude $A(x)$ and phase $\Phi(x)$ across the entrance pupil. The image degradation arising from uncorrected amplitude variations ("scintillation") remains small. Image quality is considerably improved by correcting the phase error only.

The variability of wavefront distortion can be represented by a characteristic spatial length (Fried parameter r_o), a disturbance lifetime (atmospheric de-correlation times τ_o), and the angular isoplanatic angel Ω. The Fried parameter (Fried 1966) can be expressed by

$$r_o = 0.185\lambda^{6/5} \cos z^{3/5} \left[\int_o^\infty C_n^2(h)dh \right]^{-3/5} \tag{4}$$

λ is the wavelength, z the zenith distance, and $C_n^2(h)$ is the structure function of the refractive index with height in the atmosphere.

The de-correlation time is due to refractive index inhomogeneities driven by wind across the optical beam and is of the order of

$$\tau_o \simeq r_o/v \tag{5}$$

where v is the average wind velocity.

The angle Ω over which the wavefront perturbation remains approximately the same can be estimated from

$$\Omega \simeq \frac{2}{3}r_o/h \tag{6}$$

where h is an average distance of the turbulent layers. Assuming $r_o = 20\ cm$ at $\lambda = 5,000$Å, and $v \simeq 10\ m\ s^{-1}$ gives $\tau_o \simeq 20\ ms$. For $h = 4,000\ m$ the isoplanatic angle becomes 7 arcsec.

It is essential to be aware of the inherent wavelength dependence of all three parameters, i.e. they all vary as $\lambda^{6/5}$. The effect of this is that the seeing becomes rapidly better with increasing wavelength.

3.2 Meteorological Conditions of Solar Sites

In the past decades both test measurements and actual astronomical observations have shown that high level island and coastal sites in certain latitude belts around the earth show superior performance to inland sites as far as night-time seeing is concerned (Walker 1984). The extensive JOSO testing campaigns performed in the Mediterranean, at the Western coast of the Atlantic ocean and on the Canary Islands (Brandt and Wöhl 1982; Brandt and Righini 1985a) have added some evidence to the validity of this general statement also for solar observations. The location of such good sites is closely connected to the large-scale global circulation pattern, with ascending motion near the equator and descending air masses in subtropical latitudes, forming the so-called trade wind system. High level sites in these latitudes are generally located in semi-permanent high pressure systems above an inversion layer, and are immersed in subsiding, dry and stable air masses (McInnes et al. 1974; Erasmus 1988).

Both the Hawaiian and the Canary Island archipelagos fulfill these conditions and their excellent suitability for astronomical observations is demonstrated by the fact that about half a dozen telescopes had been built and are being operated successfully on each of them. However, their actual performance, especially during daytime, depends critically on the

474

microthermal conditions in the boundary layer (0 to 300 m above the sites), which are strongly influenced by the topography of the sites, heating of the ground, slope winds etc..

The *Canary Islands* are situated at a latitude of approx. 28 N, longitude 17 W in the Azores high pressure system between 350 and 450 km from the African main land. The general climatological situation is well comparable with the one at Hawaii, i.e. a trade wind system with an inversion layer at heights between 1,200 and 1,600 m and subsiding stable air masses above these heights (Brandt and Righini 1985a,b).

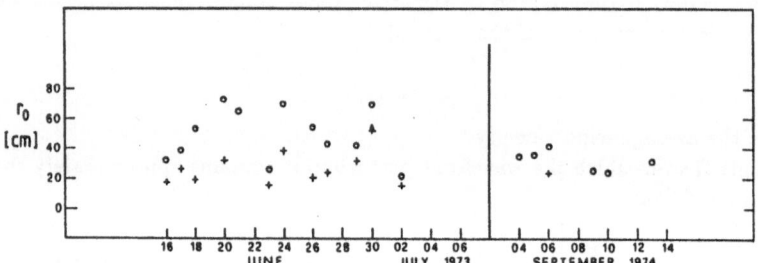

Figure 1: The Fried parameter r_o deduced from radiosonde measurements. Circles include measurements up to 10 km height and crosses up to 17 km. Barletti et al. 1977)

Early in-situ measurements of the temperature fluctuations with radio-sondes could be converted to values of r_o. The Figure 1 from Barletti *et al.* (1977) shows that most often r_o = 20 - 30 cm, and that occasionally r_o exceeds 50 cm. Other measurements have resulted in somewhat smaller values of r_o. Measurements with the 40 cm Newton vacuum telescope at Izaña in 1979 showed that 23% of the hourly measurements taken on 160 observing days yielded an instantaneous image sharpness of $\leq$ 1.2 arcsec FWHM (cf. Brandt and Wöhl 1982). From a 7 day series of solar limb motion measurements in 1986/87 by Kusoffsky (1988) shows a median r_o of 12.4 cm with 10% of all values above 25 cm. Moreover, a number of time sequences of granular evolution and narrow band filtergrams of sub-arcsec resolution have been observed with solar telescopes in the Canary Islands (Brandt *et al.* 1988; Scharmer 1989a; Title *et al.* 1989; Soltau 1989) thus proving the suitability for sub-arcsec solar observations. Programs that permit short exposure observations (less than 1 s) are usually getting diffraction limited images ($\sim \frac{1}{4}$ $arcsec$). One believes that the conditions at these sites may often be good for seeing better than $\frac{1}{4}$ $arcsec$.

4 Image Sharpening Techniques

4.1 Reduction of Telescope Seeing

Turbulence close to the telescope aperture degrades the seeing. Metalic surfaces and other local surfaces heated by the Sun will generate thermal convection and turbulence. A pro-turbing telescope in the wind is a turbulence generator. The latter is largely eliminated if the air can flow smoothly over the structure such as for the domeless and aerodynamic

towers of the vacuum telescope of NSO/Sac Peak (Dunn 1969), the Japanese Hida telescope (Nakai 1980), and the Swedish telescope at La Palma (Wyller and Scharmer 1985,
Scharmer 1989a). Thermal "trouble spots" may be identified by the aid of an IR camera
and be eliminated by means of white paint or cover. Cooling panels works well to suppress
local heating (Nakai and Hattori 1985, Pierce 1987).

Wind shake of exposed structures of heliostats and coelostats often gives rise to disturbing image motions. A *Hammerschlag-Zwaan* type windscreen has been mounted around
the heliostat mirror of the McMath solar telescope and the windshake has been reduced
substantially at minimum cost (Pierce 1989).

Evacuating the telescope light path eliminates internal seeing. Most modern solar
telescopes are therefore vacuum telescopes (Dunn 1969, 1972; Mayfield *et al.* 1969; Zirin
1969; Livingston *et al.* 1976; Nakai and Hattori 1985; Wyller and Scharmer 1985; Soltau
1989). The major drawback of such systems is that thermal and mechanical stresses in
the entrance windows give rise to optical aberration and polarization (Dunn 1984). The
diameter/thickness ratio of a vacuum window is ≥ 10, and vacuum telescope apertures are
therefore in practize limited to less than 1m. Notable interest is invested in the possible
use of helium gas in the telescope light path which may allow the use of a thin (1-3cm)
and larger diameter ($\geq 1m$) entrance window. The idea of filling a telescope with helium
was first put forward by B. Lyot and J. Rösch (Rösch 1965) and later tested by filling the
Kitt Peak vacuum telescope with helium with promising result (Engvold *et al.* 1983). The
advantages in using helium stem from its low refractive index, high thermal conductivity
and relatively high viscosity. Detailed studies are presently carried out for the LEST
project using a full scale mock-up steel tank of the telescope (Engvold *et al.* 1989). The
microthermal fluctuations are quenched by filtering and circulating the gas.

4.2 Correlation Trackers

Image motion originates from instrument shake and guiding errors as well as from random
wavefront tilts averaged over the telescope aperture due to atmospheric turbulence. Correction of wavefront tilt increases the spatial resolution by about a factor of two beyond
the time averaged value, and the use of image stabilizers become rewarding.

In order to compensate for the image jitter caused by atmospheric seeing the system
must be rather fast having a bandwidth of $\sim 10^3 Hz$. This requires a fast and necessarily
small active element in the optical path (cf. von der Lühe 1988b). The error signal that
controls the active mirror is derived from measurements of the image motions via methods
of *pattern recognitions*. For all trackers the tracking area must be $\leq \Omega^2$.

Pattern recognitions can be based on calculation of cross covariance (cross correlation)
function of two images:

$$CC(x,y) = \Sigma_u \Sigma_v \ f(x,y) * g(x+u, y+v) \tag{7}$$

An alternative approach is to calculate the absolute difference of the images (Karud 1989).

$$D(x,y) = \Sigma_u \Sigma_v \mid f(x,y) - g(x+u, y+v) \mid \tag{8}$$

The $CC(x,y)$ function gives weight to each point in proportion to its brightness, which
can lead to false, however small, image shifts. Low frequency intensity variations of the

images must for this reason be removed before the cross correlation function is calculated.

Only areas of non-uniform intensity (intensity gradients) contain information about image displacement. Pixels from uniform areas are "dead" in terms of registering image shifts. Therefore, both algoritms can be optimized by multiplication each cross product, or difference, with a weight factor $w(x, y)$ given by the local intensity contrast. Oskar von der Lühe and collaborators (von der Lühe *et al.* 1989) have built a system for stabilization of image motion that utilizes the contrast of photospheric granulation. The tracker calculates the cross covariance function of a 16 × 16 pixel image by taking the Fourier transform of the two images, multiply one by the complex conjugate of the other, and take the inverse transform. This method is computationally expensive. The tracker developed by the Lockheed group to be used with the SOUP instrument and the OSL coordinated instrument package (Title 1989) uses a fast and simple third scheme that also utilizes the intensity gradients, but which can be used for small shifts ($\sim$ 1 pixel width) and conrasty images. Another simple approach has been developed for the THEMIS telescope using two crossed, one-dimensional resolving detectors. The error signal is derived from cross correlations of successive scans of these two detectors. The system is sensitive to image motions perpendicular to the scanning direction of the detectors and may thus loose track.

Stabilization of image motion i..proves the overall image quality (*Strehl Ratio*) by a factor of 2 and is therefore highly rewarding.

The cross correlation technique is also successfully applied for measurements of horizontal flows in the photosphere (November 1986; Darvann 1988; Brandt *et al.* 1988; Simon *et al.* 1988) and in solar prominences (Darvann and Zirker 1989).

4.3 Adaptive Optics

Adaptive Optics (AO) is the technique in which the optics in a telescope are adjusted continuously with the aim of improving image quality for both short and long exposures. Adaptive optics in this broad definition corrects for telescope aberrations, tracking errors and dome seeing, as well as for atmospheric seeing outside the telescope dome (Beckers 1987). The term *active optics* is adopted for relatively slow adjustments to the telescope optics, aimed primarily at the correction of telescope aberrations.

An AO system consists of (i) a *Wavefront Sensor (WFS)*, (ii) a complex servo loop which applies the error signal to (iii) the *(iii) Adaptive Mirror (AM)*. A concept of an AO system is shown in Figure 2. A number of recent papers have reviewed the techniques of adaptive optics (Title 1985; Beckers 1987, 1989a; Dunn 1987; Hardy 1987; Merkle 1987).

Only the AO system of Lockheed has so far been in use in a solar telescope (Acton 1988). The adaptive mirror of the system consists of 19 hexagonal segments controlled by 57 piezoelectric driven actuators. The actuators have a travel of $5\mu m$, corresponding to $\sim 10\lambda$ in the visible, when operated at 40 volts. The Hartman type *WFS* operates on sunspots and pores. A separate "agile" flat mirror in the system is used to take out the overall wavefront tilt before the beam hits the adaptive, segmented mirror. The Lockheed system was run at the Tower Vacuum Telescope of NSO/Sacramento Peak in July 1988, and achieved occasionally near-diffraction limited images during average seeing conditions. It will be operated at this telescope during the next 2-3 years.

Any reasonable correction of the wavefront will lead to a point spread function consisting of an Airy disc superposed on a broad "halo" (Section 2.2). Current runs and simulations have shown that the residual of light in the "halo" is large when r_o is larger than the mirror segments of the correcting system.

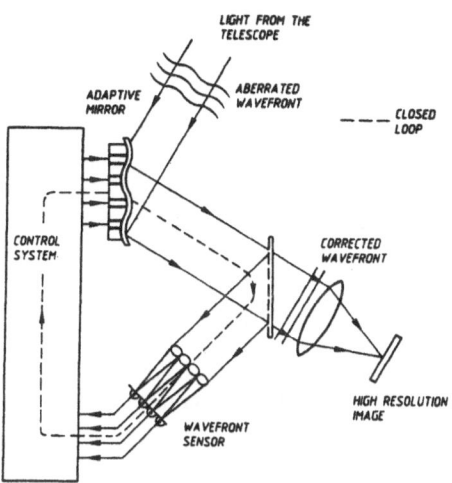

Figure 2: Principle of the application of adaptive optics in an astronomical telescope (Merkle 1987).

There is still a long way from the current successful demonstration of the Lockheed system to a general "user friendly" adaptive optics system for solar observations, but all fundamental and major technical problems appear to be solved. Given appropriate funding a routine-use AO system will be available within a few years.

4.4   Frame Selection

In the cases of "broad" band ($\Delta\lambda \geq 1$ Å) observations one may continuously make short exposures and keep only the sharpest ones. Under the assumption that the atmospheric turbulence obeys the Kolmogorov distribution it is possible to quantify the improvements in spatial resolution from image selection. The degree of improvement is partly a function of the ratio D/r_o. Table 1 from Hecquet and Coupinot (1985) gives the improvement over the long exposure by selection of the listed percentile of best images.

From the Table 1 one finds that even a moderate image selection will result in a notable improvement in image quality. Beckers (1989b) stresses that the formula used for Table 1 applies only when the exposure time is short enough to "freeze" the seeing, and when D represents the aperture diameter corresponding to the actual resolution of the telescope. The reason one hardly ever encounters a factor of four improvement in resolution from frame selection as predicted by Table 1 is that the telescope resolution is usually less than its theoretical limit.

Table 1

D/r_o	76%tile	10%tile	1%tile	0.1%tile
3	2.0	2.7	2.9	3.0
5	1.9	3.1	3.6	4.0
10	1.5	2.4	3.0	3.8
20	1.3	1.9	2.3	3.0

Values of $r_o \geq 10$ cm are not uncommon at sites such as in the Canary Islands. Frame selection has been successfully applied to observations of photospheric granulation with the 50 cm Swedish solar telescope (Scharmer 1989a). This system "freezes" one image every 0.02s and keeps the best frame for every 10s period, corresponding to an image from the best 0.2 percentile. Providing $r_o \sim 15$ cm one gets $D/r_o \sim 3.3$ and one expects to find diffraction limited images already in the 1 percentile. In the case of future large telescopes like $LEST$ ($D=2.4$ m) one would need $D/r_o \sim 8$ ($r_o \sim 30cm$ at $\lambda = 6,000$Å) in order to have 0.10 arcsec resolution images in the 0.1 percentile. At $\lambda = 1.6\mu m$ the same seeing will correspond to $D/r_o \sim 2.5$ and the diffraction limited images are in the 10 percentile.

5 Instrument Development

5.1 Polarimeters

Accurate polarimetry is best achieved through electro optical modulation of the light beam, thereby avoiding the use of moving parts (which inevitably produce noise) as well as eliminating any spurious polarization caused by seeing effects (since the modulation frequency can easily be made much larger than the seeing frequencies). The most simple and stable modulators with the best optical properties are the piezo elastic ones (PEMs), which give a sinusoidal modulation of the retardation at a frequency of typically 50kHz (Kemp 1969; Stenflo 1984a,b).

In order to overcome the much slower fram rates of multichannel detector arrays like CCD's compared to high frequency PEM's, one can introduce an optical demodulation scheme in the form of an electro optical "light chopper" in front of the detector, locked in frequency and phase to the polarization modulation (Stenflo and Povel 1985a,b). Although this optical demodulation scheme provides a solution to the PEM – CCD compatibility problem, it is rather complex and cumbersome to use. A far more elegant solution has been found and is under development by Povel and Stenflo (private communication). With this method the demodulation is done within the CCD itself, by shifting around the charges in synchrony with the polarization modulation. Only certain types of CCDs allow the charges to shifted at the rates required (50 – 100 kHz), and the charges are only shifted by one column in the detector array at such rates. It is therefore necessary to interlace the exposed and unexposed (storage) areas, by optically blocking the light to every second of the array columns, which are used for buffer storage to allow a second image plane to exist. This is illustrated in Figure 3. The charges in the CCD are shifted back and forth between the odd and even columns in synchrony with the PEM modulation.

A prototype of the LEST polarimetry system, including the optical modulation package with two PEMs (which will sit in the LEST secondary focus) and the CCD demodulation scheme (based on synchronous shifting of the charges in the CCD) is being developed for LEST in Zürich and will be tested out in a complete observing system at the Astrophysical Observatory Arosa.

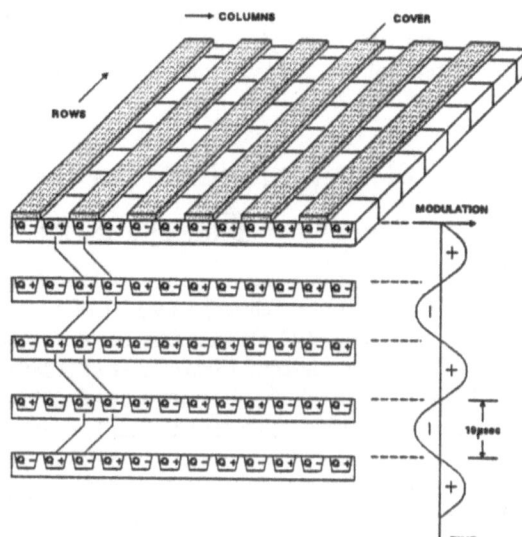

Figure 3: Schematics of the synchronous shift of charges of the CCD detector array of the ETH polarimeter.

Two new polarimeter systems are presently under development in the USA, one at the High Altitude Observatory (HAO) in Boulder, and the other at the Institute for Astronomy, University of Hawaii (UH). In both systems the polarization modulation is performed by the mechanical rotation of a retardation plate (Lites 1987b). Both systems also use CCD arrays for 2-D imaging. The images of the four Stokes parameters are formed through linear combination of consequtive images read out at different position angle setting of the rotating retarder.

The HAO system, called the *Advanced Stokes Polarimeter*, will be set up at the NSO Tower Vacuum Telescope at Sacramento Peak. It will use a spectrometer with moderately high spectral and 1-D spatial resolution, emphasizing high-quality line profile informations for quantitative fluxtube analyzis and Stokes inversion techniques.

The UH system on the other hand, called the *Imaging Vector Magnetograph*, uses a tunable narrow-band filter, thereby sacrificing spectral informations to obtain fast 2-D spatial imaging of the Stokes parameters, with the aim of relating the changing morphology and vector magnetic fields to solar flares.

5.2   Narrowband Filters and Spectrographs

A *Universal Birefringent Filter (UBF)* in tandem with a *Fabry-Perot* interferometer is being developed for 2-D spectroscopy by Bonaccini *et al.* (1989). The filter combination can be operated in the range $\lambda\lambda 4,000$-$7,000$Å with a spectral bandwidth of ~ 20mÅ. The net peak transmission is 4% - 12%, and the positioning accuracy is 1mÅ. A *Triple-Fabry-Perot Universal Filter* is installed and tested on the 30 *cm* refractor of the CSIRO Solar Observatory (Bray 1988). Various options of multichannel universal filters are considered by Ai and Hu (1985) and collaborators.

A versatile Echelle grating spectrograph is being designed and planned for THEMIS and LEST (Mein 1989). The design includes the use of one large mirror ion *Ebert-Fastie* mounting instead of the more commonly used two midle-sized collimator and camera mirrors.

6 New Telescope Facilities in the Canary Islands

6.1 The German Solar Telescope Facilities at Izaña

The German solar telescope installations at Observatorio del Teide, Izaña, consist of the 70 *cm* f/66 Vacuum Tower Telescope (VTT) of Kiepenheuer-Institut, Freiburg, the 45 *cm* f/56 Gregory-Coudé of Göttingen Observatory, Göttingen, and the smaller German-Spanish 40 *cm* Newton telescope (Schröter *et al.* 1985).

The VTT has a classic coelostat configuration and a vertical tower telescope (Mehltretter 1975). The vacuum window is located underneath the coelostat and the window cell has provisions for cooling. The dome is a removable jaw-type similar to the dome of the 60 *cm* vacuum telescope of NSO/Kitt Peak (Livingston *et al.* 1976). The telescope had "first light" in 1987, and the telescope has occasionally approached diffraction limited performance during 1988 (Soltau 1989). The VTT will be equipped with a Multichannel Subtraction Double Pass spectrograph of Meudon (Mein 1977) for 2-D spectroscopic studies. Also the Italian stable UBF will be installed in this telescope.

The Gregory-Coudé telescope was moved from the former German station at Locarno and has been in operation at Izaña since July 1986 (Kneer and Wiehr 1989). The spectrograph is a Czerny-Turner type with collimator and camera mirror focal lenghts 10 *m*. A system for speckle imaging is being prepared.

The spectroscopic studies of the Göttingen group are focused on small scale dynamic of quiescent and active regions, sunspots and prominences.

6.2 The Swedish Solar Station in La Palma

6.2.1 Instrumentation

The Swedish solar telescope at Roque de los Muchachos, La Palma (Scharmer *et al.* 1986) is patterned after the vacuum tower telescope of NSO/Sac Peak (Dunn 1964, 1985). The combination of an optically simple and good system, and a superb site has made this telescope one of the very best in the World for high resolution studies of the Sun.

The domeless turret design is very compact and its aerodynamic shape gives a minimum of local disturbance and eliminates dome seeing completely. Furthermore, the mirrors are located inside vacuum thus eliminating convection near heated surfaces. The 50 *cm* doublet achromatic lens that serves as vacuum window to the telescope. The lens is made from BK7 and F2 glasses, and the focal length is 22.35 *m*. The lens cell is designed to avoid any heating of the lens by the cell. Thermal gradients usually appears in vacuum windows due to local heating which lead to focus changes in the telescope and to spherical aberration (Dunn 1972; Mehltretter 1979). Optical tests have shown that the Swedish telescope has virtually diffraction limited performance (Scharmer 1989a).

The real-time image acquisition and selection system is a valuable asset of the observatory. An important property of the system is that it allows for simultaneous recordings with two synchronized CCD cameras. The two images, which can be a spectrum and its corresponding slit-jaw picture or images at two different wavelengths, "freeze" the atmospheric seeing motion (exposure times 1/30s or 1/60s) and provide unambiguous identifications of corresponding structures. The real-time image selector "grabs" and stores only the best images in a selectable time interval which is commonly set to be 10s. The system has been operated successfully to obtain extended time series of photospheric granulation (Scharmer 1989a). It is expected to work for chromospheric structures and dark filaments as well.

6.2.2 Recent Results

The best time sequence of photospheric granulation that has been analyzed so far covers 79 minutes and was obtained June 16, 1987 (Brandt *et al.* 1988). The spatial resolution was $\approx \frac{1}{4}$ arcsec over 18 arcsec frame size. Local flow velocities were found on the average to be about 1.6 $km\ s^{-1}$. These speeds were larger than similar data recorded earlier with lower spatial resolution (Title *et al.* 1989). The most spectacular discovery of this study was that of a vortex of 5,000 km in diameter and which persisted for the duration of the sequence. Similar recent studies by Müller (1989b) did not find vortex structures.

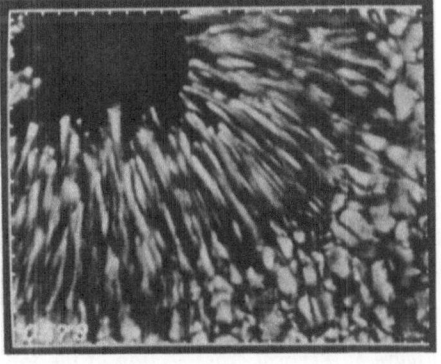

Figure 4: Image of large sunspot observed July 26, 1988, with the Swedish vacuum solar telescope at La Palma, Canary Islands. The observations were made through a 25 Å wide filter centred on 4686Å. The spatial resolution is about $\frac{1}{4}$ arcsec. The image scale is 18 x 25 arcsec². (Courtesy G.B. Scharmer)

A new time series of a medium size sunspot reveals a stunning wealth of dynamic structures (Scharmer 1989b). Figure 4 shows a 18 x 25 $arcsec^2$ view of the penumbra and near by photosphere from this time series. A study of Darvann and Kusoffsky (1989) using a 19 minutes long series of photosperic granulation found that granulation lifetimes decreased from about 12 minutes close to the sunspot pore region to 5 minutes away from the pore.

7   Near Future Solar Telescope Projects

7.1   THEMIS

The French polarization-free solar telescope *"Télescope Héliographique pour l'Etude du Magnetism et des Instabilités Solaires" - (THEMIS)* will be built at Izaña, on the island of Tenerife, Canary Islands. The site is being prepared for construction of the THEMIS building, and the first light in the instrument is expected in 1991.

The pointing Ritchey-Chretien telescope system of 90 cm aperture will be evacuated and sealed by two windows (Mein and Rayrole 1985, 1988). The prime focus (f/17) is the location of the polarization modulator. A high precision correlation tracker and a "tip-tilt" (active) mirror between the prime and secondary foci will be used to stabilize the residual jitter of the solar image from seeing and telescope vibrations. The main instrument, the Multichannel Subtractive Double Pass (MSDP) Echelle spectrograph (Mein 1977, Mein and Rayrole 1988) is mounted behind the secondary focus (f/60).

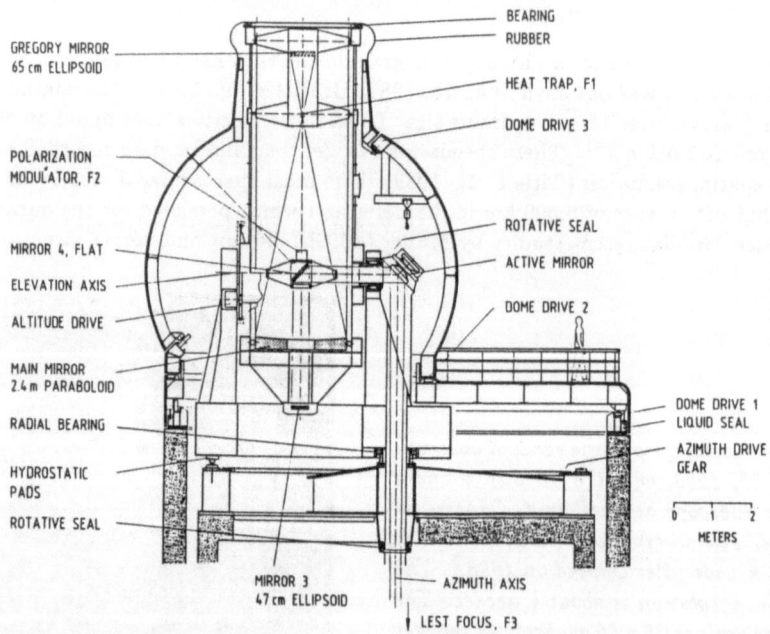

Figure 5: Vertical cross section of the top part of the LEST (Andersen et al. 1985).

7.2   Large Earth-based Solar Telescope - LEST

LEST is the most ambitious ground based solar telescope program today. The current design is shown in Figure 5 (Andersen *et al.* 1984). The LEST aperture is 2.4 *m* and the diffraction limit of the modified Gregorian system is 0.05 *arcsec* at λ5,000 Å. Evacuation becomes impractical for such a large telescope since it will require a rather thick entrance window. It is instead suggested to fill the internal light path with helium gas at nearly ambient air pressure to make possible to use a thin entrance window.

The LEST will be placed either on *La Palma*, Canary Islands (2 360 m a.s.l.) or on the cinder cone *Pu'u Poli'ahu* at Mauna Kea, Hawaii (4 150 m a.s.l.). Besides utilizing a very good site for daytime observations LEST will seek to achieve near diffraction limited resolution within a small field of view by means of image sharpening techniques mentioned above.

The multi-national, non-profit organization behind the project, the LEST Foundation, presently counts 10 member countries (Stenflo 1985; Wyller 1986). The construction of LEST could start in 1992 and the telescope may be ready for first light in 1995.

7.3   Solar Missions in Space

The *Solar and Heliospheric Observatory - SOHO* is a space mission for studies of the solar interior and the outer solar atmosphere. The payload shall include six instruments for studies of structures and dynamics of the Sun's chromosphere and corona (Domingo and Poland 1989).

Two of the SOHO instruments, *Solar Ultraviolet Emitted Radiation* (SUMER) and *Coronal Diagnostic Spectrometer* (CDS), will both build up 2-D images in the order of seconds by scanning their spectroscopic slits across the solar disk with ~2 arcsec resolution. A third instrument, the *Extreme-ultraviolet Imaging Telescope* (EIT) provides high resolution images of the whole Sun at several temperatures. Telemetry of the data will take place via NASA's Deep-Space Network during four designated periods per day.

SOHO is currently scheduled for launch in July 1995 and is being designed for a lifetime of two years, but it will be equipped with sufficient on board consumables for an extra four years.

The *Orbiting Solar Laboratory (OSL)* of NASA is a scaled down version of the former Solar Optical Telescope (SOT) and the High Resolution Solar Observatory (HRSO). OSL will be a free flying, polar orbiting instrument for high spatial and temporal resolution of the Sun over a spectral range from the X-ray to the near IR (Title 1989). The OSL satellite will harbour a 1m aperture telescope optimized for $\lambda\lambda 2,000$-11,000 Å, which feeds a narrow-band filter, a set of fixed broad-band filters, and a visible light Echelle spectrograph. These three instruments are mounted together in a common structure; the *Coordinated Instrument Package (CIP)*. The CIP will have 0.13 arcsec resolution over a 3.9 arcmin field-of-view. An ultraviolet spectrograph and an XUV/X-ray imager are planned co-pointing and co-aligned instruments with overlapping fields of view. The latter is probably an array of normal incidence telescopes with multi layer coatings. The UV instrument will be Naval Research Laboratory's *High Resolution Telescope and Spectrograph (HRTS)*, which covers the wavelength range $\lambda\lambda 1,175$-1,700Å that contains numerous bright emission lines for studies of the chromosphere and chromosphere-corona transition region. The spatial resolution of HRTS is ~1 arcsec. The planned launch dates of OSL are 1995-96, and the design lifetime will be 3 years even though it is reasonable to expect that a full scale operation could be technically possible for 8 - 10 years.

The HRTS instrument will also be flown on rockets during the next few years, presumably as often as about one per year.

8 Prospects for High Resolution Observations of the Sun

New optical telescopes in space (SOHO and possibly OSL) and ground-based (the German and Swedish installations in the Canary Islands, and the future THEMIS and LEST) challenge existing high resolution solar telescopes like Pic-du-Midi (Müller 1989), NSO/Sac Peak (Dunn 1969), Big Bear Solar Observatory (Zirin 1969), and Hida (Nakai and Hattori 1985). The advance of techniques for high resolution imaging and the utilization of excellent sites for daytime seeing will make angular resolution beyond 0.2 *arcsec* in solar observations within reach in a few years.

Acknowledgement

Informations on polarimeters were kindly provided by Jan Olof Stenflo.

References

Acton, D.S.: 1988 *"High Spatial Resolution Solar Observations"*, Sac Peak Workshop Aug. 22-26 (to be published)

Ai Guoxiang and Hu Yuefeng,: 1985 LEST Technical Report No. 14

Andersen, T.E., Dunn, R.B., and Engvold, O.: 1985 LEST Technical Report No. 7

Barletti, R., Ceppatelli, G., Paternò, L., Righini, A., and Speroni, N.: 1977 Astron. Astrophys. **54**, 649

Beckers, J.M.: 1986 SPIE Proceedings **628**, 190

Beckers, J.M.: 1987, LEST Technical Report No. 28, p.55

Beckers, J.M.: 1989a *"Solar and Stellar Granulation"*, Eds.: R.J. Rutten and G. Severino, NATO ASI Series, Kluwer Academic Publishers, p.43

Beckers, J.M.: 1989b   *ibid* p.55

Bonaccini, D., Cavalini, F., Ceppatelli, G., and Righini, A.: 1989 Astron. Astrophys. (in press)

Brandt, P.N. and Wöhl, H.: 1982, Astron. Astrophys. **109**, 77

Brandt, P.N. and Righini, A.: 1985a, LEST Technical Report No. 11

Brandt, P.N. and Righini, A.: 1985b, Vistas in Astronomy **28**, 437

Brandt, P.N. Scharmer, G.B., Ferguson, S., Shine, R.A., Tarbell, T.D. and Title, A.M.: 1988, Nature **335**, 238

Bray, R.J.: 1988 LEST Technical Report No. 35

Darvann, T.A.: 1988 *"High Spatial Resolution Solar Observations"* 10th Sac Peak Summer Workshop Aug. 22-26 (Ed.: O. von der Lühe)

Darvann, T.A. and Kusoffsky, U.: 1989 *"Solar and Stellar Granulation"*, Eds.: R.J. Rutten and G. Severino, NATO ASI Series, Kluwer Academic Publishers, p.313

Darvann, T.A, and Zirker, J.: 1989 IAU Meeting on Solar Prominences Hvar, Yugoslavia September 27-29.

Dere, K.: 1987 Solar Physics **114**, 223

Domingo, V. and Poland, A.I.: 1989 *"THE SOHO MISSION"*, *Scientific and Technical Aspects of the Instruments*, ESA SP-1104, p.7

Dunn, R.B.: 1964 Appl. Optics **3**, 1353

Dunn, R.B.: 1969 Sky & Telescope **38**, 368

Dunn, R.B.: 1972 Space Research **XII**, 1657

Dunn, R.B.: 1984 LEST Technical Report No. 3

Dunn, R.B.: 1985 Solar Physics **100**, 1

Dunn, R.B.: 1987 LEST Technical Report No. 28, p.243

Engvold, O., Dunn, R.B., Livingston, W.C., and Smartt, R.: 1983 Applied Optics **22**, 10

Engvold, O., Andersen, T.E., Carlsson, M., Jensen, J.R., and Klim, K.: 1985 LEST Technical Report No. 15

Engvold, O. *et al.* : 1989 (To be published)

Erasmus, D.A.: 1988, LEST Technical Report No. 31

Fried, D.L.: 1966, J. Opt. Soc. Am. **56**, 1372

Hardy, J.W.: 1987 LEST Technical Report No. 28, p.137

Hecquet, J. and Coupinot, G.: 1985 J. Optics (Paris) **16**, 21

Karud, J.O.: 1989 Master Thesis, Univ. Oslo. (to be published)

Kemp, J.C.: 1969, J. Opt. Soc. Am. **59**, 950

Kneer, F. and Wiehr. E.: 1989 *"Solar and Stellar Granulation"*, Eds.: R.J. Rutten and G. Severino, NATO ASI Series, Kluwer Academic Publishers, p.13

Knox, K. and Thompson, B.: 1974 Astrophys. J. Letters **193**, L45

Kusoffsky, U.: 1988 (unpublished)

Lites, B.W.: 1987a, LEST Technical Report No. 22

Lites, B.W.: 1987b, LEST Technical Report No. 23

Livingston, W.C., Harvey, J., Pierce, A.K., Schrage, D., Gillespie, B., Simmons, J., and Slaughter, C.: 1976 Applied Optics **15**, 33

MacQueen, R.M.: 1987 LEST Technical Report No. 24

Martin, R. and Hartley, K.: 1985 Vistas in Astronomy **28**, 555

Mayfield, E., Vrabec, D., Rogers, E., Janssens, T., and Becker, R.: 1969 Sky & Telescope **37**, 208

McGuire, J.P. and Chipman, R.A.: 1988 LEST Technical Report No. 36

McInnes, B., Hartley, M. and Gough, T.T.: 1974, Observatory **94**, 14

Mein, P.: 1977 Solar Phys. **54**, 45

Mein, P.: 1989 LEST Technical Report No. 37

Mein, P. and Rayrole, J.: 1988 *" High Spatial Resolution Solar Observations"*, 10th Sac Peak Summer Workshop Aug. 22-26 (Ed.: O. von der Lühe)

Mehltretter, J.P.: 1975 JOSO Annual Report p.

Mehltretter, J.P.: 1979 J. Optics **10**, 93

Merkle, F.: 1987, LEST Technical Report No. 28, p.117

Müller, R.: 1989a "Solar and Stellar Granulation", Eds.: R.J. Rutten and G. Severino, NATO ASI Series, Kluwer Academic Publishers, p.9

Müller, R.: 1989b (Private communication)

Nakai, Y.: 1980 Proceedings of The Japan-France Seminar on Solar Physics. Eds.: F.Moriyama, J.C.Henoux, 285

Nakai, Y. and Hattori, A.: 1985 Mem. Fac. Sci., Kyoto University, Ser. Physics, Astrophysics, Geophysics and Chemistry 36, No. 3, 385

November, L.: 1986 Applied Optics 25, 392

Pierce, A.-K.: 1987 Solar Physics 107, 397

Raffi, G.: 1988 "ESO Conference on Very large Telescope and Their Instrumentations". Garching 21-24 March 1988. Ed.:M.-H. Ulrich, p.1061

Roddier, F.: 1981 Progress in Optics 19, 333

Roddier, F.: 1987 LEST Technical Report No. 28, p.7

Roddier, F. and Roddier, C.: 1986 Proceedings SPIE 688, 298

Rösch, J.: 1965 Applied Optics 4, 1672

Scharmer, G.B.: 1989a "Solar and Stellar Granulation", Eds.: R.J. Rutten and G. Severino, NATO ASI Series, Kluwer Academic Publishers, p.161

Scharmer, G.B.:1989b (Private communication)

Schröter, E.H., Soltau, D., and Wiehr, E.: 1985 Vistas in Astronomy 28, 519

Simon, G.W., Title, A.; Topka, K., Tarbell, T., Shine, R., Ferguson, S., Zirin, H., and the SOUP Team: 1988 Astrophys. J. 327, 964

Soltau, D.: 1989 "Solar and Stellar Granulation", Eds.: R.J. Rutten and G. Severino, NATO ASI Series, Kluwer Academic Publishers, p.17

Stachnik, R.V., Nisenson, P., and Noyes, R.W.: 1983 Astrophys. J. 271, L37

Stenflo, J.O.: 1984a, Appl. Optics 23, 1267

Stenflo, J.O.: 1984b, LEST Foundation Technical Report No. 4

Stenflo, J.O.: 1985 Vistas in Astronomy 28, 571

Stenflo, J.O., Povel, H.: 1985a, Appl. Optics 24, 3893

Stenflo, J.O., Povel, H.: 1985b, LEST Foundation Technical Report No. 12

Title, A.M., and Tarbell, T.D. et al.: 1987 in "High Resolution Solar Physics II", NASA Conf. Publ. 2483, 55. Eds.: R.G. Athay and D. Spicer

Title, A.: 1985 "High Resolution in Solar Physics". Lecture Notes in Solar Physics No. 233, 51

Title, A.M.: 1989 "Solar and Stellar Granulation", Eds.: R.J. Rutten and G. Severino, NATO ASI Series, Kluwer Academic Publishers, p.29

Title, A.M. et al.: 1989, publication in preparation

Walker, M.F.: 1984, in "Site Testing for Future Large Telescopes", ESO Conf. and Workshop Proc. No. 18, ed. Ardeberg and Woltjer, 3

von der Lühe, O.: 1988a JOSA A **5**, 721

von der Lühe, O.: 1988b Astron. Astrophys. **205**, 354

von der Lühe, O. and Zirker, J.B.: 1988 *"High Resolution Imagery by Interferometry"*, NOAO/ESO Conf. 15-18 March, Garching, FRG (Ed.: F. Merkle)

von der Lühe, O., Widener, A.L., Rimmele, Th., Spence, G., Dunn, R.B., and Wiborg, P.: 1989 Astron. Astrophys. (submitted)

Wyller, A.A.: 1986 *LEST Large Earth-based Solar Telescope - An Overview*. LEST Foundation, Royal Swedish Academy of Sciences, Stockholm

Wyller, A.A., Scharmer, G.B.: 1985 Vistas in Astronomy **28**, 467

Zirin, H.: 1969 Sky & Telescope **51**, 215

OUTSTANDING THEORETICAL PROBLEMS

C.J. DURRANT
Department of Applied Mathematics
University of Sydney
NSW 2006
Australia

ABSTRACT. The development of model atmospheres from the 'classical' static but deductive models to present-day dynamic but inductive models is sketched. The main problems facing theory are defined in terms of the need to produce a post-classical inductive model. Attention is focused on the most promising tool available today, the computers able to realize simulations of astrophysical systems. The direction of progress in the areas of radiative transfer, convective transport, waves and oscillations, and MHD is reviewed. It is concluded that the major outstanding radiative and hydrodynamic problems are likely to be elucidated in the foreseeable future, especially if there is a suitable commitment by the international community. However, the understanding of the behaviour of magnetic fields and their associated activity will require a longer, but no less urgent, programme.

1. Introduction

It is a curious fact that the first recorded use in the English language of the word 'atmosphere' is a reference in 1638 to the vaporous surroundings of the Moon! But by 1677 the word had gained its commonest connotation, that of the body of air around the Earth. To most people today, the atmosphere means the terrestrial atmosphere, not that of the Sun.

Early last year there was a meeting in Canberra of the IGBP, the International Geosphere-Biosphere Program. The aim of this program is to 'describe and understand the interactive physical, chemical and biological processes that regulate the total Earth system'. It is of course the impact of human society on the terrestrial environment which has given the IGBP its impetus and its urgency. Not much interest was displayed at that meeting in the Sun, although it will have an inexorable effect on human society in the long term if not the short.

Fortunately, the influence of human activity does not yet extend to the Sun, but many of the physical processes operating in the atmosphere of the Earth operate also in the atmosphere of the Sun. Whilst there is already a great body of knowledge about these processes, they 'are so highly interactive [in the terrestrial context] that an adequate quantitative synthesis of them is essential' (Tucker (1988)). The real challenge of the Program is to make the synthesis quantitative.

Exactly the same comments apply to the current situation in solar physics. If we test our understanding by demanding that we should be able to make quantitatively accurate predictions of the behaviour of the atmospheres of other stars, then we would have to admit that we have not advanced very far in the last 70 years.

J. O. Stenflo (ed.), Solar Photosphere: Structure, Convection, and Magnetic Fields, 489–499.
© *1990 by the IAU.*

Theoretical foundations were laid around the turn of the century with the work of Lane (leading up to Emden) on equilibrium stratification, of Schwarzschild on radiative equilibrium and of Milne on local thermodynamic equilibrium. Over the following 60 years the equations governing these three processes allowed the construction of model photospheres which increasingly resembled the empirical one as the frequency distribution of the opacity was treated more accurately (Kurucz (1979)). I shall refer to these as 'classical' models.

As a *simple* model of the photosphere, the classical model is quite impressive. Moreover, the calculations can be extended to any other star (Kurucz (1979)) and so our understanding of the atmosphere at this level meets our practical test.

Unfortunately, this is not quite true. If the assumed equilibria are tested for physical consistency throughout the model, the criterion developed by Schwarzschild himself reveals that hydrostatic and radiative equilibrium cannot be sustained in and below the continuum levels. A new process, that of convection, had to be introduced. To cope with the modifications to the energy and momentum balance resulting from the convective motions, an aerodynamic description of the process was developed by Biermann and Siedentopf in the early 1930s. This changed the nature of the model of the solar atmosphere. Before, it had been built on knowledge and insight about the system of concern; with the adoption of the mixing-length description of turbulent convection, the model had to draw upon observations. It required the empirical determination of parameters and drew heavily on laboratory experiments with turbulence. The models changed from being 'deductive' to 'inductive' (Karplus (1977)).

In practice this has meant that all solar models now contain adjustable parameters of one sort or another. To quote Tucker (1988) again, 'empiricism is required to develop analogs for some component (processes) and to generally "tune" the models, sometimes in an arcane way'. The removal of the adjustable components is an outstanding problem of theory.

We are more fortunate than the IGBP modellers in that we can write down the equations that govern the behaviour of stellar photospheres. The fluid equations and radiative transfer equations can be used (with justification) below the level of the transition region. Kinetic effects can be produced in the electron distribution in the non-LTE regions of the chromosphere (Shoub (1977)) but any dynamical consequences are not of immediate concern in the photosphere. The classical model was a consistent solution of these governing equations in static equilibrium. The convective models were not consistent solutions because the actual terms in the equations were replaced by model terms constructed from empirical knowledge. We have still to understand what the *full* equations will yield under circumstances appropriate to the Sun.

I must stress here that at least as important as the equations themselves are the boundary conditions and on this point I echo the concern expressed by Thomas (1983) about treating model systems as isolated systems. The construction of physically consistent boundary conditions is part of the theoretical task. It does not take us outside current physical theory. I confess to being in Thomas' words a 'speculative theorist'. We should understand what that theory really tells us about stellar atmospheres before appealing to new physics.

2. The way ahead

Let us look first at the tools at our disposal. Since direct experiment is impossible, we have to pursue indirect investigations. Tucker (1988) enumerates the possibilities:

1. mathematical analysis of the governing equations,

2. experiment with miniature analogue systems,

3. experiment with numerical simulation.

Analysis is the classical avenue but it has proved inadequate as far as the full set of equations is concerned. However, analysis has produced an enormous corpus of knowledge about individual processes, gained by isolating and adjusting terms (i.e. the physical interactions) until a mathematical system of sufficient simplicity emerges. Truesdell (1980) claims that this style of investigation originated in the study of thermodynamics in the 19th Century. It results in what can be called 'scenario physics'. If the scenario could be realized, we would understand the physics. But scenarios are generally competitive, what one assumes is contradicted by another. In the Sun, the processes are mixed and modified by their interaction. To understand the Sun, we have simply got to address this problem.

Some of the consequences of the interactions, certainly more than have been elucidated by analysis, can be demonstrated by the second tool, analogue experiments. However, because laboratory circumstances are so different from those in the solar atmosphere these experiments are rarely of direct relevance. Outstanding exceptions are the 'ice-water' experiment of Townsend (1964), which produces convective overshoot, and the Spacelab 3 experiment which produces convection in a rotating, stratified medium (Hart *et al.* (1986)). Although limited in number and scope, such experiments are the only means by which the ideas that guide the development of scenario physics can be tested in a controlled manner. I hope that ingenuity will continue to be expended on devising more of these experiments.

However, I think that the means of breaking out of the straightjacket of the past few decades in provided by the third avenue, the development of the high-speed computer. It is the only tool that we possess that can in principle bridge the gap between the laboratory and the Sun. Since 1940 each decade has seen a phenomenal growth in the power of computers, about half due to improvements in hardware and about half due to improvements in software. Computer manufacturers claim that this trend will continue over the next decade and are confident that massively parallel systems will maintain the momentum into the decades beyond. In 1984 Nordlund wrote that detailed numerical simulations using the full set of magnetohydrodynamic equations were '(at least marginally) feasible with present day computers'. It will clearly not be long before the cautious qualification can be removed.

I should like to stress, though, that numerical simulation is a form of experimentation. The numbers obtained should be regarded as the outcome of an experiment, an experiment designed to demonstrate physical processes that require to be understood. This has two consequences. Firstly, if the result is not interpreted in a manner that increases our physical understanding, the experiment has no value. Here scenario physics finds a legitimate rationale. It provides the building blocks of cause and effect which allow the results to be expressed in terms of physical processes. The simulation sifts out the relevant scenarios from the irrelevant. An admirable example of this approach is set by the work of Nordlund (1984, 1985a, 1985b), in which close attention is given to elucidating the physical processes underlying the model results.

However, there is a step beyond this. As the theoretical jigsaw starts to fall into place, we should begin to recognize the quantitative importance of the interrelationships so that the full-scale model can be broken up into quasi-independent compartments. This is precisely the opposite direction to the analytical path in which the parts are studied before the whole. Only when we know to what extent the compartments influence one another, can we construct simple self-consistent models of the solar atmosphere beyond the classical level. Only then can the resulting theoretical structure be applied with confidence to stars other than the Sun. Only then will discrepancies really point to 'new physics'.

The second consequence of the experimental nature of simulation is the need for appropriate attention to the experimental procedure. Experiments should be reproducible and should be reported in sufficient detail to allow them to be reproduced. This requires specification of

1. the governing equations,

2. the algorithm used to solve them,

3. the physical boundary conditions,

4. the implementation of the boundary conditions.

The major area of uncertainty in astrophysical simulations lies in the boundary conditions at the inner and outer limits of resolution. I shall return to this subject below.

To illustrate these points I want to look briefly at four areas of solar physics in the order in which I believe progress will be made, radiative transfer, convective transport, overshoot and waves, and MHD.

3. Radiative transfer

In this case both the governing equations and appropriate upper and lower boundary conditions are well established. The development of theories of radiative energy transport has differed essentially from that of hydrodynamic transport in that the nonlocal nature of the process was recognized from the outset. The radiation field is obtained by solving self-consistently along all rays, taking explicit account of any inhomogeneity and anisotropy of the system.

Faced with this need, a great deal of effort has been expended on devising efficient algorithms and ones which exploit the ability to do parallel computations along different rays at different frequencies. These are the subject of a recent monograph (Kalkofen (1987)). The need to incorporate a realistic model of the 'microphysics' has also been accepted and calculations can be made using non-LTE where necessary.

The problem of resolution has already been mentioned and it has been overcome, as computing resources have improved, by increasing the number of points at which the frequency is sampled. Indeed, at this point the subject has progressed to the stage of looking for recipes for simpler models that capture the behaviour established in more detailed calculations. The representation of the opacity distribution in frequency by statistical sampling or distribution functions (Kurucz (1979), Carbon (1979)) is a start in this direction. Nordlund (1982) has suggested source function averaging.

I would suggest that there are no outstanding theoretical problems as far as the radiation physics of the photosphere is concerned, though the treatment of non-LTE transfer in rapidly moving structures in the chromosphere may produce significant effects of some subtlety which do not form a part of our intuitive thinking at present. The accuracy of transfer calculations is generally limited though by the desperate shortage of accurate atomic data, particularly of collisional cross sections.

4. Convective transport

Convection is the next most important ingredient in the photospheric system. It governs the mean structure of the lowest levels and modulates that above. It also underlies, in one way or another, the entire horizontal structuring of the photosphere (a classical static atmosphere has no such structure).

Here the governing equations are well known. The anelastic approximation has been extensively used (e.g. Nordlund (1982 *et seq.*)) because it allows an explicit algorithm to be used to study convective motions without inducing sound waves. However, efficient implicit schemes now allow the full compressible equations to be solved (Chan and Sofia (1986)). This becomes essential when the magnetic field is included.

Also well known are the various processes implicit in the equations—buoyancy, pressure gradients, radiative 'conduction', kinetic energy transfer by the nonlinear inertial term and viscous dissipation. But we do not understand how they interact to determine the structure of convective flow in a highly stratified turbulent medium like the outer parts of the solar convection zone. Computer simulations are already proving their value as an experimental tool. Progress is not direct because no computer can (or is likely to) encompass the range of scales from global (10^8 m) down to viscous (less than a metre).

Moreover, the computation cannot be simplified by reducing the number of dimensions. Astrophysical convection is turbulent and is thus an essentially three-dimensional phenomenon. Artificial two-dimensional models have quite different characteristics–a cascade of energy to large scales, for instance. Two-dimensional simulations are no substitute for three-dimensional simulations. With hindsight and a secure knowledge of the three-dimensional system we may at some future stage be able to justify the use of models of lower dimension for some purposes, but not at present.

Since we are interested primarily in scales in the range 10^5–10^7 m our models must account for the fact that the largest and smallest scales are not present. Periodic boundary conditions in the horizontal plane ensure that the model is embedded in a larger system and that there is no special treatment of side walls that might dictate the geometry of the flow. A cartesian geometry then guarantees that there is statistical uniformity and isotropy in the horizontal plane. Simulations in cylindrical geometry (e.g. Steffen (1987)) cannot avoid introducing anisotropy.

The upper boundary condition will be discussed in the next section. The lower boundary conditions are at present problematical. They must be set within the convection zone as there is no hope of extending a surface simulation down to the bottom. The solution adopted by Nordlund and Steffen is to allow 'free' flow into and out of the boundary, i.e. the vertical velocity gradient and the pressure fluctuation are set to zero. In each case the model flows stream in and out of the lower boundary and show no signs of closing. Thus the properties of the system outside the model should control the true behaviour of the flow just as much as those inside. Yet the model flow is determined entirely by the internal properties. The dilemma might be resolved if the vertical flows were an artifact of the 'free' boundary condition having no dynamical significance, in which case a boundary condition enforcing a circulation (e.g. the vertical velocity being set to zero) would have little effect over most of the box. Alternatively, the flow might be localled controlled at the level of the lower boundary, somewhat in the spirit of mixing-length formulations. Or any resemblence between the isolated model results and the behaviour shown by the same model embedded in a much larger system may be fortuitous.

The answer can be provided definitively only by looking at the embedding problem more carefully. Here is there is a long-outstanding theoretical problem. Despite its obvious fundamental importance, even the basic form of convection in a stratified medium is not yet known. To date there are just two three-dimensional simulations—that of Graham (1977) and that of Chan and Sofia (1986). In the former the dominant mode is a quasi-cellular pattern extending the whole depth of the unstable layer whilst in the latter the vertical velocity structure loses its coherence over a pressure scale height. The models differ in many respects. Graham's model had a low Rayleigh number (ten times the critical value) and a Prandtl number around unity so that the flow was just turbulent. The degree of stratification was also small, just over one pressure scale height. The models of Chan and Sofia were more highly turbulent (only an effective Reynolds number of 300 is quoted) and extend over 4–5 scale heights. Because of this, Chan and Sofia had to model the scales below the level of numerical resolution, the so-called subgrid scales.

The standard treatment is one adopted by terrestrial modellers, in which it is assumed that the smallest resolved scale lies within the inertial regime of the turbulence where there is neither significant driving nor dissipation only a transfer of kinetic energy from large scales to small scales via the nonlinear inertial term in the equation of motion. The plausible assumption of isotropy and homogeneity then yields a Kolmogorov spectrum. The contributions of the unresolved motions to the momentum and energy equations can then evaluated as an eddy viscosity and an eddy conductivity (Rogallo and Moin (1984), Eidson (1985)). This formulation has been tested for incompressible turbulent flows for low Reynolds number $Re \leq 40$ by Clark et al. (1979). For compressible flow, most authors (e.g. Hurlburt et al. (1984)) model the subgrid heat diffusion as

$$\langle K\nabla T\rangle_{sg} = \frac{\mu}{Pr_t}\nabla T,$$

where μ is the subgrid eddy viscosity and Pr_t is the turbulent Prandtl number, taken to be about $\frac{1}{3}$. Chan and Sofia (1986) argue that in a stratified fluid it is the entropy gradient which drives the subgrid diffusion

$$\langle K\nabla T\rangle_{sg} = \frac{\mu}{Pr_t}\left[\nabla T - \nabla_{ad}\left(\frac{T}{\rho}\right)\nabla p\right] = \frac{\mu}{Pr_t}\nabla S,$$

the second term arising from the adiabatic gradient ∇_{ad}. Since the entropy varies much less strongly than the temperature across a stratified convective layer, the subgrid heat transport is much smaller than previously estimated. The resolved motions are therefore more vigorous and appear to break up in a statistical sense over a vertical scale height.

Chan and Sofia describe careful tests of their algorithm and the parametrization of their model. Some unanswered questions remain though.

1. Their choice of upper and lower boundary condition reflects acoustic waves. A transmitting boundary would be more appropriate.

2. The rigid upper and lower boundaries also force the heat in their immediate neighbourhood to be carried by the subgrid motions. Yet the basis for their treatment breaks down when isotropy is broken by the influence of the boundary (Eidson (1985)). A narrow stable layer added at the boundaries would remove the possibility that unrealistic boundary effects propagate into the bulk of the system. They would also facilitate the implementation of transmitting boundary conditions.

3. Does the numerical resolution really reach the inertial range so that the subgrid eddy viscosity is a valid estimate? A great deal of progress has been made recently in deriving parameter-free closures in statistical turbulence theory. Yoshikawa (1988a) offers a method that can yield subgrid models that take account of buoyancy. He finds that the turbulent Prandtl number (for passive scalar diffusion) is not constant but varies with the square of the ratio of the velocity to scalar variance dissipation times. Subgrid models may be tested against full resolved simulations of Rayleigh-Bénard convection up to Rayleigh numbers of 4×10^5 (Grötzbach 1982).

4. Finally, the way in which the models are discussed needs to be systematized. The flow visualizations, i.e. instantaneous plots of velocity vectors, tend to focus attention on the coherent motions, whilst spatially averaged cross correlations between the velocities at different vertical heights reveal that much of the motion is disordered. Clearly, convective plumes of limited spatial extent coexist with random motions. We really need to know how much heat is carried in each, and this requires more sophisticated statistical methods. Adrian (1977) gives a full discussion of how higher-order moments and conditionally averaged statistics can be used to interpret coherent structuring in turbulent convective flows and Kerr (1985) describes skewness and kurtosis factors in numerical turbulence simulations.

I think the organization of motions in stellar convection zones is close to being a solved problem. We would then be able to investigate one of the most enigmatic features of the photosphere, the supergranulation. Nordlund (1985) has suggested how the model might be set up; it should decide whether or not the supergranulation is a direct manifestation of convection zone dynamics.

5. Overshoot and waves

Overshoot has been the subject of the most detailed dynamical modelling to date. Nordlund (1982 *et seq.*) has elucidated many of the properties of the granulation by this means. In particular, an understanding of the heat exchange between the flowing gas and the radiation field emerged only from these simulations. On average heat is extracted from the radiation field to heat the gas which cools as it expands into the stably stratified region. The mean temperature of the atmosphere is depressed below the radiative equilibrium values and the model closely matches the empirical structure.

Better knowledge of the convection zone dynamics would allow the lower boundary condition described above to be improved. The upper boundary condition can be formulated in some detail by fitting appropriate wave-like solutions in the stable region. Nordlund (1982) found this unnecessary for anelastic modelling but will be more important for fully compressible models.

A fully compressible treatment is the next step and will yield the first reliable estimates of the acoustic energy flux produced in the upper parts of the solar convection zone. These will provide a crucial test of our ideas concerning the mechanical heating of the solar chromosphere.

Global acoustic modes will be modified or suppressed by finite box models so any coupling between oscillations and convection would be revealed only by an iteration between global scale simulations with parametrized convective scales and convection models with impressed global motions.

Convective motions overshooting into stable layers distort the gravitational equipotential surfaces and generate internal gravity waves in laboratory fluids. However, solar observations show no strong gravity wave field despite considerable convective overshooting in the granulation. Nordlund has explained their absence in terms of his model. Convective pressure disturbances can penetrate far in a stable layer of decreasing mean pressure and drive an extended circulation rather than an oscillation in the photosphere. Gravity waves can only be excited higher in the atmosphere.

The most uncertain aspect of photospheric dynamical modelling must lie in the subgrid scale motion. Nordlund (1982) modelled the subgrid viscosity heuristically. In reality, the transition from turbulent transport processes in the convective layer to wave transport in the stable is likely to be complicated. Fernando (1988) studied the development of turbulence in a stably stratified layer in the laboratory. A locally generated patch of turbulence grows by entrainment of nonturbulent material until the vertical mixing is limited by negative buoyancy forces. Thereafter an interfacial layer develops separating the turbulent region from the overlying nonturbulent material. Turbulent disturbances at the interfacial layer generate internal waves above. As the waves break small regions of turbulence are created above the interface which eventually merge and are incorporated by the turbulent patch, thereby slowly raising the interface.

6. MHD

The incorporation of the magnetic field into the hydrodynamic system is without doubt the most important outstanding theoretical problem. Without it we cannot answer the questions

1. Does the hydrodynamic convection organize the field on the granular and supergranular scales or might the field organize the structure on the supergranular scale or is the structure due to a subtle process of self-organization?

2. What is the structure of the field in the photosphere, why is there strong field and weak field? Where do the electric currents flow and what is their strength?

3. What is the spectrum of dynamic disturbances exhibited by the photospheric field?

These are the questions crucial to understanding the whole atmosphere of the Sun and other stars above the photosphere. The dynamic behaviour of the field and current system in the photosphere and below provides the driver in electrodynamic theories of the outer layers of stars.

Simulations such as those of Nordlund in my opinion now provide more ccurate estimates of the spectrum of granular dynamics, the root-mean-square values being reported as

$$\langle v_z \rangle = 2 \, \text{km s}^{-1}, \quad \langle v_h \rangle = 3 \, \text{km s}^{-1}, \quad \langle \delta p/p \rangle = 0.5.$$

These are much higher than the observationally derived values, which still suffer from lack of spatial resolution and ambiguity of interpretation. Since the study of magnetic field structures requires even higher resolution and more complicated diagnostics, experiment seems unlikely to provide meaningful information of the current system (for which gradients of field strength must be measured) or of dynamical disturbances (for which the distortion of the field structure must be measured). Numerical simulations seem to offer the only hope of substantial progress here.

Nevertheless, the theoretical problems are immense. The grid size required to resolve a magnetic field element adequately is so small that the size of the simulation box must be reduced correspondingly. Subgrid modelling is complicated by the fact that nonlinear effects can operate either from large scales to small–like the hydrodynamic kinetic energy cascade— or from small to large in an inverse cascade. Yoshizawa (1988b) has recently proposed a subgrid model for MHD turbulence based on statistical theory. However, looking at fully resolved hydrodynamic (Kerr (1985)) and MHD (Meneguzzi et al. (1981)) simulations at quite moderate Reynolds number ($Re \leq 100, Rm \leq 100$) for fluids with $Pr \sim 1$, one is struck by the spatial intermittency of the vorticity and magnetic field on the smallest scales. Moreover, Meneguzzi et al. point out that the field concentrations do not coincide with the vorticity concentrations. It is perhaps well to remember that statistical ensembles may not model an *individual* subgrid volume, even when spatially averaged.

The problem of modelling the embedding of the box in physically realistic surroundings is also that much greater. This difficulty is compounded by the fact that the magnetic field cannot be isolated, even conceptually, within the box as can, for example, an eddy of the flow. The field must leave the box at top and bottom so that we require magnetic boundary conditions. These are best thought of in terms of the equivalent boundary conditions on the current system. Where there is field at the boundary, mass flow and current flow are ducted in an out of the box. Clearly these are no locally determined properties of the system. They adjust in response to the need to evolve the whole magnetic structure.

At the upper boundary it may be possible to establish a form of equivalent circuit for the coronal connections using the theory of Ionson (1982). At the lower boundary the circuit cannot be continued. The structure of the field elements below the surface is unknown—do they collect together in larger units or are they uniformly dispersed? An iteration between high-resolution and low-resolution models, as in the case of convection modelling, might yield a heuristic solution but its implementation seems a long way off.

7. Prospects

In summary, I think the prospects for solving the problems of the hydrodynamic behaviour of the solar photosphere are good. We are almost at the stage where the structure and dynamical effects in a stellar atmosphere can be calculated without recourse to empirical scalings. It remains to be seen, though, to what extent the physical insight provided by complex simulations allows the computations to be simplified. On the other hand, the continued increase in computational power might overtake the need to simplify substantially. Within a decade we may all be able to run off the models on cheap massively parallel computers. In either case, present work should have physical insight as its main aim.

Of course, the immediate future lies with big and expensive machines, available only to few. The time seems ripe for an organized international effort dedicated to astrophysical convection simulations, perhaps along the lines of the summer program at the Center for Turbulence Research at NASA Ames Research Center and Stanford University (Hunt (1988)).

But we should be also be planning the attack on the MHD problem. To return once again to the IGBP. Its ambitious program will cover at least a decade starting in 1990. However, little interest was shown at the Canberra meeting in the influence of the Sun on the variability of the Earth system in either the short term or the long term. In the short term, it is solar magnetic activity which affects us on Earth the most. The IGBP will not be complete without an understanding of the MHD behaviour of the Sun. This is controlled in no small part by the processes at the surface and in the immediate subsurface layers. The understanding of these is the outstanding theoretical problem of the next decade.

8. References

Adrian, R.J. (1977) *Journal of Fluid Mechanics*, **69**, 753-781.

Carbon, D.F. (1979) *Annual Review of Astronomy and Astrophysics*, **17**, 513-549.

Chan, K.L. and Sofia, S. (1986) *Astrophysical Journal*, **307**, 222-241.

Clark, R.A., Ferziger, J.H. and Reynolds, W.C. (1979) *Journal of Fluid Mechanics*, **91**, 1-16.

Eidson, T.M. (1985) *Journal of Fluid Mechanics*, **158**, 245-268.

Fernando, H.J.S. (1988) *Journal of Fluid Mechanics*, **190**, 55-70.

Graham, E. (1977) In E.A. Spiegel and J.P. Zahn (eds.), *Problems of Stellar Convection*, Springer-Verlag, Berlin, pp. 151-155.

Grötzbach, G. (1982) *Journal of Fluid Mechanics*, **119**, 27-53.

Hart, J.E., Glatzmeier, G.A. and Toomre, J. (1986) *Journal of Fluid Mechanics*, **173**, 519-544.

Hunt, J.C.R. (1988) *Journal of Fluid Mechanics*, **190**, 375-392.

Hurlburt, N.E., Toomre, J. and Massaguer, J.M. (1984) *Astrophysical Journal*, **282**, 557-573.

Ionson, J.A. (1982) *Astrophysical Journal*, **254**, 318-334.

Kalkofen, W. (ed.) (1987) *Numerical Radiative Transfer*, Cambridge University Press, Cambridge.

Karplus, W.J. (1977) *Mathematics and Computers in Simulation*, **19**, 3-10.

Kerr, R.M. (1985) *Journal of Fluid Mechanics*, **153**, 31-58.

Kurucz, R. (1974) *Astrophysical Journal Supplement*, **40**, 1-340.

Meneguzzi, M., Frisch, U. and Pouquet, A. (1981) *Physical Review Letters*, **47**, 1060-1064.

Nordlund, Å. (1982) *Astronomy and Astrophysics*, **107**, 1-10.

Nordlund, Å. (1984) In S.L. Keil (ed.), *Small-Scale Dynamical Processes in Quiet Stellar Atmospheres*, Sacramento Peak Observatory, Sunspot, pp. 181-221.

Nordlund, Å. (1985a) In H.U. Schmidt (ed.), *Theoretical Problems in High Resolution Solar Physics*, Max-Planck-Institut für Astrophysik, München, pp. 1-24 and pp. 101-119.

Nordlund, Å. (1985b) *Solar Physics*, **100**, 209-235.

Rogallo, R.S. and Moin, P. (1984) *Annual Review of Fluid Dynamics*, **16**, 99-137.

Shoub, E.C. (1977) *Astrophysical Journal Supplement*, **34**, 259-275.

Steffen, M. (1987) In E.-H. Schröter, M. Vázquez and A.A. Wyller (eds.), *The Role of Fine-Scale Magnetic Fields on the Structure of the Solar Atmosphere*, Cambridge University Press, Cambridge, pp. 47-52.

Thomas, R.N. (1983) *Stellar Atmospheric Structural Patterns*, NASA SP-471, CNRS and NASA, Paris and Washington.

Townsend, A.A. (1964) *Quarterly Journal of the Royal Meteorological Society*, **90**, 248-259.

Truesdell, C. (1980) *The Tragicomical History of Thermodynamics 1822–1854*, Springer-Verlag, New York.

Tucker, G.B. (1988) In *Global Change*, Australian Academy of Science, Canberra, pp. 182-191.

Yoshizawa, A. (1988a) *Journal of Fluid Mechanics*, **195**, 541-55

Yoshizawa, A. (1988b) *Physics of Fluids*, **30**, 1089-1095.

SUMMARY LECTURE

Robert J. Rutten

Sterrekundig Instituut, Postbus 80000, 3508 TA Utrecht, The Netherlands

How to do this summary

I have found four models in the literature for doing conference summaries:

1. *The Literal Summary.*
   One summarizes all that has been presented, preferably interspersed with comments as "of particular interest was...".

2. *The Historical Perspective.*
   One places all (or some) presentations within a historical background, preferably implying that science progresses smoothly in well-planned, orderly fashion.

3. *The Future Perspective.*
   One points out the way to go, preferably in overly optimistic vein.

4. *The personal Impressions.*
   One concedes lack of wisdom to forego balanced summarizing, prefering to discuss primarily one's own interests.

Which model to choose here? Literal summarizing seems superfluous for the oral presentations. They are printed in the preceding pages, each is effectively a summary of work published elsewhere, and many have an author's summary already. It won't be useful to summarize them here once more, but some perspective may be worthwhile.

The poster presentations, on the other hand, are not printed in this volume, obeying current IAU (or Kluwer) policy. The policy may be wise since many posters describe work that will eventually be published in regular journals anyhow; nevertheless, it might be better to have one-page abstracts for these and somewhat more space (though refereed) for those that describe new instruments, new techniques and new methods not easily detailed in journal papers. Symposium proceedings would then possess the added flavour of showing who is doing what, where, and how. This is particular useful for PhD studies and students and for meetings with strong East-West overtones: probably, there were surprises here for you as there were for me in discovering research and researchers I was not aware of before.

There were more than 100 posters, many of them excellent; they should be summarized here. However, I have studied only a minority in detail and that holds probably for

J. O. Stenflo (ed.), Solar Photosphere: Structure, Convection, and Magnetic Fields, 501–516.
© *1990 by the IAU.*

the majority of you as well, indicating that the non-reading of posters poses a larger problem than their non-publishing. We had two specific poster sessions plus the breaks; not enough, I fear, for full merit.

And then the video movies. These are neither printed nor posted, only shown; they are virtually impossible to summarize since they must be seen. Nevertheless, they have constituted a prime ingredient of this meeting; their showing showing that solar physics has entered the video clip era, not only in California where the Hollywood heritage is strongest, but also here at Kiev and elsewhere.

Video movies must be seen to be believed. The same holds for equations and diagrams, but the equations and the diagrams in these proceedings can be studied over and over whereas you and I have only a fleeting remembrance of what we thought (or were told) to see in the movies, and other readers of this volume have none. Movies may be a necessary step in the gleaning of useful information out of the complex manifold offered by the Sun, with the very important advantage of utilizing the superb pattern analysis capabilities of human vision, but ultimately, more formal descriptions are needed. Perhaps video storage will replace printed language, math and diagrams in future but until then, moviemakers must face the problem that showing a movie and publishing results are not the same thing at all.

So much for Summary Model 1. It leaves me with the task to summarize all poster and video presentations and to place these and the talks into perspective. Then, there are Summary Models 2–4. These are attractive too. Let me try them all on you.

How to divide the subject

The next question to be answered before I start summarizing is how to divide the subject. This is not obvious either; there are many possibilities:

1. Evolutionary: *past – present – future*.

   This is the standard order for any research article: first review the preceding work, then give the new stuff and end with predictions. The last item lacks too often. There have been classical examples of predictions in solar physics, as Parker's solar wind and Ulrich's p-modes, and we have seen a few gastronomical ones here too such as siphon flows, dynamo rolls and a missing piece of Napolitan cake, but in general solar physics seems a field in which the object produces unpredicted surprises.

2. Geographically: *West – Western Europe – Eastern Europe – East*.

   This division neglects our single participant from the southern hemisphere; permissably, I feel, since he has given his own Conference Summary already. What is wrong in this division is that it is linear whereas international astronomy runs in circles. The most interesting display of that fact were the Big Bear–Huairou movies shown by Sara Martin, the two video magnetographs working in tandem at an 11–hour difference to produce round-the-clock coverage of active regions. They demonstrate that not only helioseismology gains from worldwide observing networks. The LEST Foundation, already the most international of astronomical telescope-building consortia, might solve its location dilemma by building two Large Earth-based Solar Telescopes: Mauna Kea and La Palma are 9 hours apart. A Large Eastern Solar Telescope at (or preferably in) Lake Baikal would then complete the circle.

3. Spectrally: *X-ray – UV – visual – IR – radio.*
The old division in techniques is less evident nowadays. This conference was primarily visual, mainly because that is where $\tau = 1$ in the photosphere and because the $\lambda = 1.6 \ \mu$m promise has not yet been fulfilled. X-ray means flares and radio means coronal instabilities which we have not discussed; the existence of this conference, the first IAU Symposium on the photosphere, signifies a come-back of optical studies. To quote the Conference Rationale:

> "The photosphere is the interface between the solar interior and the outside, and is the layer of the Sun that is best accessible to observations. The photosphere transforms the energy generated in the solar interior and emits it into the corona and the heliosphere. It makes all the radiative, dynamical, and magnetic processes that transfer solar energy into space available to our detailed observations".

Optical astronomy flourishes in general, and the solar and nighttime developments are strikingly similar. While the longest and the shortest waves exhibit the more spectacular phenomena more obviously, optical imaging, spectrometry, photometry and polarimetry often provide the diagnostics that are required to identify the underlying processes. The LEST and OSL projects are direct counterparts to ESO's VLT and to Space Telescope; SOHO's seismometers resemble HIPPARCOS in obtaining very basic information from a mathematical transformation of a year's data gathering.

In general, there is a transition from doing discoveries with newly-opened non-optical eyes to multispectral interpretation for which spatial resolution is an essential requirement. Spatial resolution is the next observational frontier, using satellite VLBI and optical interferometry from space. For solar physics too: the Abstract Book lists an interesting poster by Damé *et XVII al.* (which I couldn't find though, neither poster nor Damé) describing a Space Station proposal called SUN comprising a non-redundant 4-telescope array giving 10 km resolution on the Sun.

4. Height: *core – convection zone – photosphere – T-min region – chromosphere – corona.*
This meeting on the photosphere covered much additional depth by including convection and dynamos. It covered less additional height, presumably because the photosphere suffers more from below than from above, and in keeping with the current inward-looking trend crowned by helioseismology.

This trend does not imply that all things chromospheric and coronal are now fully understood. Although valuable concepts like loop scaling laws, magnetic helicity, electrodynamical circuits, Alfvén wave heating and magnetic reconnection scenarios have been developed, definitive outer-atmosphere success stories are yet lacking. Deeper down, the granulation does constitute a new and important solar physics success. It indicates, as stressed by Durrant in the preceding pages, that numerical simulation is the way to go and that this way may well lead upwards again, progressing to the larger MHD complexity and instability of the outer atmosphere. Here is a see-saw oscillation: from equating solar physics with the photosphere when optical spectrometry was its prime diagnostic, up to the outer atmosphere when radio and space astronomy came in, down to the surface and digging even deeper now with spatial, Fourier and numerical resolution, back up again in future, perhaps eventually down again to get the dynamo. Damped or unstable?

5. Scale: *granulation – mesogranulation – supergranulation – giant cell – torsion wave*
   or: *filigree – intranetwork field – network – active region – activity complex.*
   Different or the same? The most interesting aspect of these scale sequences is their existence—mesogranulation now firmly established from the SOUP cork movies, but giant cells still questionable. Tarbell mentioned that the magnetic structures seen in an active region display cell sizes ranging as a self-similar set, "straight from Mandelbrot's book". One might have expected such behaviour for all of the surface phenomena, the photosphere being made of turbulent gases, and Muller's claim that the smaller scales possess a Kolmogorov spectrum is still in discussion, but it isn't the case in general. Why?

   Nordlund stressed topology as the key item of the hydrodynamical simulations, the granular scales dominant just at the surface (though not for all other stars as shown by Dravins) but finger-like downdrafts repeatedly connecting in larger and larger patterns deeper down. Noticing simulation behaviour which resembles solar behaviour does in itself not explain the latter, but simulation behaviour is, in contrast to solar behaviour, fully understandable—although having a nice simulation is one thing and understanding it is quite another: simulations require extensive interpretation with clever diagnostics just as observations do. But they do permit physical experimentation, and so deliver a vital element to bridge the gap between noticing patterns and understanding them. The solar hydrodynamical scales are now clearly attackable; the magnetohydrodynamical ones should follow when massive parallelism brings the required orders of magnitude improvement in computer power.

   Topology is also a key item in understanding the larger-scale patterns of magnetic activity. Petrovay's suggestion that differential drag causes typical spot group morphologies asks for simulatory confirmation; more in general, the topological nature of the activity cycle remains the major constraint to dynamo theory, not to be lost out of sight while helioseismology delivers the internal rotation.

   Whether the dynamo itself requires full simulation eventually is yet unclear. Hoyng concluded that dynamo theory is now in a stage of reappraisal and renewed reconnoitring, leaving linear mean field theory to try out new ideas and possibilities in order to admit multiple periods and finite phase memories. Numerical experimentation will be worthwhile to study nonlinearities because the fields are dominated by motions and the motions are dominated by nonlinear advection terms. In particular, Ruzmaikin eloquently explained the globally stochastic nature of the solar MHD generator by putting a strange attractor in its phase space. Evaluation of that concept for any but the simplest nonlinear models requires much computation; however, such studies will be interesting even if it turns out in the end that the Sun works differently.

6. Period: *sec – min – hour – day – month – year – cycle.*
   Why are millisecond radio bursts and 22-year cycles harder to grasp then 5-minute oscillations and 10-minute granules? Perhaps because your attention span, listening now to me, is of minute duration too?

   One use of video techniques, for observations and simulations alike, is to transform solar time scales to our physiological ones to obtain better appreciation. We haven't seen the Greenwich sunspot data speeded up to a few minutes yet; it might prove interesting.

7. Observed features: *granule – exploding granule – vortex – BP – FBP – NBP – XRBP – $K_{2v}BP$ – 160 nm BP – jet – grain – bomb – prominence – p-mode – ridge – torsional mode – butterfly – filigree – knot – pore – spot – umbra – umbral dot – penumbra – EFR – EAR – facula – plage – arch – rosette – ribbon – spicule, etc. etc.*
These and a host of others make up solar dermatology, with terminological fashions such as "grain" replacing "mottle" and "mottle" replacing "flocculus". These features are interesting to most of us, but many non-solar astronomers hate them since they wouldn't see them on their object if it has them which they hope it doesn't, regretting Galileo's announcement of blemishes on what should have remained a perfect sphere.

Of course, Dravins' computer granules and Saar's inferred magnetic regions make stars look more like the Sun and may make astronomers like the Sun more; nevertheless, solar morphological detail is not of obvious interest to others. That is quite understandable (who would be oenologist without savouring a vintage wine from time to time?) but leaves us with the need to explain why solar surface detail needs to be explained using expensive telescopes and supercomputers. Such defense is not yet required of galaxy baggers and other morphologists in our feature-prone science, though "clumpiness" being a current buzz word in galactic and extragalactic research implies that fine-scale structuring becomes important elsewhere too.

In the long run, solar physics gains from having to explain now already why studying structural detail is worthwhile, because that pressure forces more emphasis on physical understanding. That should make solar physics a path finder in the transition from phenomenological to process description and from scenario to self-consistent modelling. All fields of astrophysics have to make this transition at some time or other; it goes with the succession of the second observational revolution (the opening up of the electromagnetic spectrum to discover violent nonthermal behaviour) by the third, consisting of getting the resolution necessary to see what is going on. Solar physics is again at an advantage sitting so close to its scene: the physical scales at which many a solar process occurs are in reach.

8. Not-observed features: *fluxtubes – flux sheets – current sheets – magnetic loops – CO clouds – flare kernels – siphon flows – giant cells – circuits – mirror currents – proton beams – g-modes – oblateness, etc. etc.*
Again a host of phenomena, but invented rather than observed. That makes them much more interesting! For example, granules may be a current breakthrough but fluxtubes attract more attention. They are much more atractive to theorize on, presenting an elegant concept with pleasant geometry offering tractability to many a specialist in hydrodynamics, radiative transfer and magnetohydrodynamics, and they are also easily sold to non-solar theoretical astrophysics for use in other objects were they are also not observed. Accretion disks, for example, now produce tube and loop and circuit papers (typically by former solar tube and loop and circuit persons) but no granule or spicule papers.

The reason is, of course, that tubes, loops and circuits are modelling concepts rather than morphological features. Concepts have wider applicability the more abstract they are and the less constrained they are by observations; perhaps it is unfortunate that Solanki and Keller produce such detailed empirical fluxtube models from FTS observations now, and perhaps the attractivity of fluxtubes will wane when the properties of actual magnetic field concentrations will be further constrained by LEST and OSL and SUN. On the other hand, the fluxtube concept does produce firm and detailed predictions open for observational verification, such as Schüssler's illumination heating; it will be nice to find out whether and how the tiny strong-field fluxtubes do all the things they are currently supposed to do, such as heating the corona and, perhaps, the chromosphere.

One thing they appear to do indeed is exist. Tarbell's high-resolution magnetograms dissolve the plage in an active region into unipolar clusters of small grains, arranged in cells of many sizes. The pixels are yet bigger than the modeller's tubes, but the overall graininess is unmistakable.

One thing they appear not to do is to sit in bipolar clusters in quiet-sun cell interiors and parade unresolvedly as weak polarization. I hesitated whether to list *intranetwork fields* under the "observed features" or not, but now Sara Martin's movie shows patches of unipolar intranetwork field steadily travelling to the network boundaries, the coherency of this motion proving their existence at least to me. In the margin of Tarbell's magnetograms there are quiet cell interiors which do not show anything strongly polarized. These observations together with those from Kitt Peak indicate that intranetwork fields do exist and consist primarily of intrinsically weak fields arranged in patches measuring a few arcseconds, not as strong thin tubes. This issue is of obvious importance, as is the question whether there are areas in the photosphere truly without magnetic field.

So now we have already an eightfold way of dividing our subject matter. And there are more: *Sun − solar-like stars − non-solar-like stars − non-stars* for example, or *analytical theory − numerical theory − theoretical interpretation − observational interpretation − observation − instrumentation*, and others.

Which division to follow here? I take the easy way out; realising that the scientific organisers of this meeting have had the same problem already, it seems easiest to copy their solution by just following the order of the Abstract Book[1]. That implies there are 181 presentations to summarize, beginning with the invited review of Avrett and ending with the poster of Bonaccini *et al.* Quite a list, let me begin quickly.

1.1 Avrett's review

Avrett started his review by showing his familar diagram that specifies the height of formation of various spectral features throughout the solar atmosphere. That diagram is often used for openers, but usually only to show where one's diagnostic comes from before one discards one-dimensional modelling to proceed with inhomogeneous explanations of observed or not-observed features. In this era of realistic 3D simulation, 1D standard modelling drops out of fashion. Over half a century of plane-parallel explanations of the solar spectrum is seen as enough of a good thing, spectroscopy not being regarded as a proper science in its own anymore but rather as a necessary tool. Personally, I do not agree to that view at all; in general, we shouldn't forget that here lies a strong link with stellar astrophysics—the oldest and strongest solar-stellar connection.

---

[1] E.A. Gurtovenko (Editor), 1989, *Solar Photosphere: Structure, Convection and Magnetic Fields*, Naukova Dumka, Kiev, 65 kopecks (or 3 abstracts per kopeck; a bargain compared with this volume)

Solar-stellar perspective

Let us digress to solar-stellar connections for a moment. There are more than one:

1. *Stellar abundance determination.*

The oldest one, dating back to the time in which the whole of astrophysics consisted primarily of solar spectrum analysis. Unsöld's 1955 bible[2] is still the basis of what we now call the classical theory of stellar atmospheres. Although Kurucz and Gustafsson's Uppsala group have put this classical edifice on modern computer footing, it still rests on the assumptions of spherical symmetry, hydrostatic equilibrium, radiative or convective equilibrium, and usually LTE notwithstanding Mihalas' book. In the meantime, solar physics has lost interest in the constitution of its matter (some years ago, Zwaan and I terminated a 50-year Utrecht tradition with the ultimate paper on the solar curve of growth), but stellar abundance determination remains a large field, alive and well, in which many astrophysicists use the solar spectrum for guidance. It behooves us to supply them with the information they require; the solar group here in Kiev sets a good example. (Another good example is the Kurucz *et al.* NSO Atlas Nr. 1; so would, if they existed, Nrs. 2 and higher be.)

2. *Stellar activity.*

Cool-star magnetic activity constitutes what is termed "THE solar-stellar connection" at the moment. It started long ago with the work of O.C. Wilson, Bappu and Sivaraman, but it became a hot topic only after EINSTEIN demonstrated that the topic is hot indeed. In the meantime the amazing sharpness of the flux-flux relations has shown that dynamos work in similar fashion in different stars, and pioneers as Schrijver have returned to the Sun to find out how. Some stars deviate, though. There are also stellar flares which differ much from solar flares, but not enough not to have another connection.

3. *Stellar convection.*

Dravins' presentation of the Dravins–Nordlund stellar granulation simulations gave ample evidence of another blossoming connection. It started with Dravins' and Gray's bisector studies, and has progressed very quickly to the desirable stage in which observations and simulations are compared, and that, also desirable, by various groups (including the Kiev one) using different approaches and different numerical methods. Who would have predicted that the smallest surface features on unresolved stars, the star itself smaller than a solar granule on our sky, would be the first to reach this happy state?

4. *Stellar interiors.*

Helioseismology presents another obvious solar-stellar connection in the making. The GONG and SOHO projects are bound to produce results of interest to stellar evolutionaries; asteroseismology does not seem too farfetched. Again—who would have predicted such rich diagnostics of the invisible layers so far below the surface? This is not a yes/no matter of a few missing neutrinos: the oscillation spectrum contains thousands of lines with measurable frequencies, splittings and amplitudes, and hope-

---

[2] unfortunately like Luther's in German

fully *g*-modes as well.

5. *Stellar dynamos.*

Elsewhere I have pessimistically predicted that helioseismology may lead to another solar physics bout of ghettosis by producing too much structural detail again, this time not on the outer but on the inner surface of the convection zone. Let me be optimistic here. The activity connection shows dynamos working in other cool stars. Rotational modulation, circular and linear polarization, asteroseismology, bisector monitoring and other stellar measurement techniques may well deliver the evidence necessary to constrain possible realizations of dynamos to realistic ones. The solar dynamo is perhaps too deep to fathom; knowing more about others will help.

Solar perspective

Returning to Avrett's review, it is clear that all 1D spectrum interpretation is of direct interest to our stellar colleagues. It is important for them to know whether 1D stratification, hydrostatic equilibrium, radiative equilibrium, LTE and opacity distribution functions are acceptable shortcuts, and that can be checked more easily for the Sun than elsewhere.

Amazingly, these shortcuts seem to become more and more acceptable for the solar photosphere. Even NLTE has gone away—the recent change of the upper photosphere in the Harvard models from the cool HSRA dip back to the gentle slope of the classical Holweger-Müller model largely reduces the NLTE departures found before. The change also led Ayres to move his cool CO clouds to larger height, above the temperature minimum where they do not bother anyone anymore. We now have an 1D upper photosphere in hydrostatic equilibrium and nearly in radiative equilibrium, which explains the continuum pretty well from the near-UV to the IR, which reproduces the wings of the Ca II H & K lines and which fits most visual lines in classical manner assuming LTE, as is clear from the Kiev fits to 2000 lines reported by Gurtovenko.

At the same time, Nordlund's simulations indicate that the spatial and temporal variations should be very large throughout the photosphere. Are the simulations wrong in producing too much inhomogeneity? Or is the averaging such that, fortuitously, the spatially and temporally averaged spectrum can well be described with a 1D atmosphere even while large deviations actually occur?

I don't know the answer to this important question, but I conclude that, either way, the photosphere is nice to abundance determiners. This leads me to formulate a new principle here. Let me call it the "Principle of Solar Communicativity" although I won't object if you call it "Rutten's Law" henceforth[3]. Actually, it consists of two principles:

---

[3]although R.G.M. Rutten may object. Yes, there are two of us. He is René and I am Rob.

Detail Is Beautiful

Detail Must Be Optimally Displayed

A thought experiment will clarify its meaning. Imagine a solar terrestrial physicist, sitting on a cool CO cloud above the photosphere. (Sunspots are also suitable locations for organic chemistry, but they live too briefly to produce DNA molecules whereas the CO clouds have been in the literature for years already and don't seem to go away at all.) She has her telescope trained on a structure on the third planet which, in its center, consists of rectangular granules and dark intergranular lanes. The lanes are bordered at regular intervals by the socalled "intergranular features" for which the following model has been derived: a globular infrared-emitting cloud, which shows a half-orbit modulation from optical thick to optical thin, crowns a vertical tube which is present during the full orbit and which seems firmly anchored in the co-rotating black matter at the $\tau = 1$ level. Our physicist now studies a peculiar fine structure in the IR globule. It appears and disappears in a single pulse during a short segment of each orbit. It consists of clumps which seem randomly distributed over the globule. Each clump is small and possesses a high vertical wavenumber signature. She wonders what they are.

We know. They are the blossoms of the chestnut trees that border the streets here in Kiev. The chestnuts make Kiev one of the most beautiful cities on earth. We know that they are beautiful. We appreciate their beauty especially when they show blossomy detail, as they do now. And we know and appreciate that they have been carefully arranged for optimal display. We take it for granted that a plane-parallel city would not have the beauty of a chestnut-lined one; we savour such morphological surface detail without objection; we love adorning simple structures with surface detail to keep ourselves happy and solar terrestrial physicists as well.

The Sun does exactly the same. Clear proof is its singing. Why should the Sun excite tens of thousands of modes in a beautiful harmonic chord if not for the beauty of it, and to keep terrestrial solar physicists happy? Another proof is its magnetic field. To quote Leighton, without magnetic field the Sun would be as boring as the nighttime astronomers believe it is. Configured in strong-field tubes rather than a weak-field dipole, that field is clearly designed to optimize the amount of beautiful detail displayed to terrestrial astrophysicists.

In fact, the Principle of Solar Communicativity underlies all that the Sun displays to us. A short list of solar terrestrial action items illustrates the solar perspective:

- emit far too few neutrino's, just enough to prove that their detector works;
- sing loudly, to show them internal structure;
- show interesting hydrodynamics with minimal interference from magnetic fields;
- show surface structures at different scales that map different depths;
- have beauty spots;
- have a 1.6 μm opacity dip for Koutchmy;
- have tubes for MHD physicists;
- have loops for plasma physicists;
- have prominences for radiative-transfer-in-slabs specialists;

510

The Principle of Solar Communicativity. There is a benign presence in the Sun gracefully offering magnetostatic fluxtubes to terrestrial astrophysicists to grasp and to hold on to for dear life—illustration by M.P. Ryutova.

- have flares big enough to be spectacular but small enough for safety;
- hide a dynamo as a real brain teaser;
- vary everything on time scales from TV-rates to career lengths;
- have a cycle in step with NASA's planning cycle;
- have a supermaximum when funding is poor, to get on Time Magazine;
- obey Parker's wind theory;
- have a moon for eclipses, such that they occur in interesting faraway places;
- attract comets for tail chasing.

In particular, for Avrett the Principle of Solar Communicativity implies that the Sun cooperates by emitting continua that are very well modellable: the Sun is not a box of Pandora; its genes beget continuaty.

1.2 Title's review

The second entry in the Abstract Book is the invited review by Title. How does the Principle of Solar Communicativity apply to him and his LPARL coworkers, measuring Fe I 6303 with SOUP at the superb vacuum refractor of the Swedish Solar Observatory on La Palma? Let us again do a thought experiment. Imagine yourself to be a bunch of iron atoms somewhere in the photosphere, all set to jump the 6303 Å transition. You are aware there is quite a variety of rather quaint characters interested in you; how do you optimally provide beautiful detail to:

- oscillator strengtheners and plane-parallel layer layers;
- NLTE radiative transfreaks and magneto-optical affectionists;
- k-ω plodders and helioseismologists;
- granulation morphologists and bisectarians;
- convective blueshifters and limbshifters;
- compressible and incomprehensible 2D and 3D simulators;
- fluxtube FTS ETH highschoolers and fluxtube MHD PHD students;
- magnetic field pattern recognizers and self-similar setters;
- activity cyclists and torsional surfers?

All these terrestrial solar physicists are going to study the iron line that you are about to cause, and each of them will use that line in his own particular way for his own particular purpose. It is your task to provide all of them with the beautiful detail that each of them requires to write an interesting paper on his subject, not once but over and over again.

That task is not easy. Nevertheless, the Sun accomplishes it. All these people are here and have new results to show and tell. That implies that any solar signal is a mixture containing diagnostics for all of these diverse interests simultaneously. Title said in his introduction:

"The Sun is a very complicated structure—it has turbulent convection, a whole family of wave motions, magnetic structures *etc*. It is easy to fall into the trap of looking only at those aspects that you can model simplifiedly"

which is an essential point to be taken very seriously, by observers and theoreticians alike.

Solar physics may be part of physics but in many respects it resembles biology, being less reductionist than physics. Our object, the solar photosphere, obeys a rather simple set of basic equations and offers nothing special to those who desire to reduce nature to a few fields and particles; its interest rather lies in the beautiful detail that nature is able to generate out of those simple equations, much richer than Eddington's "cloud-bound physicist" might ever have predicted. To study that rich detail, we cannot stick a thermometer in the photosphere and measure the isolated effect of a single controlled parameter change; we have to take our object as it is, holistically, with only limited experimentation possible through simulation.

Thus, Title's warning must be heeded. Interpreting solar surface phenomena is complicated because the Sun tries to satisfy all of us at the same time. Your message is there, but there are many more messages on the same information carrier; clear reception requires sharp listening to pick your message out of the noise made by the others.

The problem becomes larger when the observing is better. The more data, the less *a priori* selection towards a preconceived idea. This is particularly clear in the LPARL observations. In one of last year's runs the SOUP was used to obtain images of 512×512 pixels, cycling through the Fe I 6303 magnetic line in left and right circular polarization, the continuum and the Ni I 6768 velocity line in 4 wavelengths, one cycle per 50 sec for 2.5 hours, over 2 Gigabyte in total. It takes sophisticated processing to transform such data into Dopplergrams, magnetograms, continuum images and line-center images: gain and dark corrections, reordering, derotation, reregistration, destretching *etc.*, 30 hours computer time per sequence. And that is only the beginning. The previous SOUP analyses have shown how cleverly such data sets must be attacked to distill interesting information: 3D Fourier filtering, local correlation tracking and cork sprinklers were required to isolate the flow fields discussed by Title. Future data sets will be even more comprehensive; the poster by Bonaccini *et al.* describes instrumentation for 2D imaging in 41 wavelengths, effectively giving 2D spectrometry.

Clearly, the Principle of Solar Communicativity has a corollary: there is so much beautiful detail optimally displayed by the Sun that much ingenuity is required for full appreciation. We need new ways of analysing data, not only video clips but also new display formats, analysing techniques (as the multivariate approach in the poster by Caccin *et al.*) and cork-like inventions.

A new display format was displayed by Deubner. Not content with having put the ridges in the k-ω diagram already, he now changed that diagram into a 3D one by adding phase, specifying power per spatial and temporal Fourier component per phase shift between intensity and velocity, both per line and between lines formed at different heights. He so produced clear evidence of gravity waves and new k-ω phase ridges; there is much to be learned from such phase diagrams and from corresponding predictions by Marmolino and Severino (including their missing piece of phase cake).

This indicates that the new LPARL data may fruitfully be analysed for intensity and velocity phase behaviour, and that 4D rather than 3D Fourier analysis may be the next trick to try. The same suggestion applies to simulation results. In Deubner's words[4]:

"To our knowledge there are no theoretical predictions available yet about phase relations in that brackish regime between fresh convection and the overshoot layers salted with all kinds of waves. We strongly encourage our

---

[4]F.-L. Deubner, 1989, *Astron. Astrophys.* **216**, 259

colleagues working on three-dimensional simulation of compressible convection to extract the temporal phase information from their models, since we feel that it bears extraordinary diagnostic potential."

There is another lesson in the paper from which this quote was taken: it re-analyzed data taken 17 years ago. Is it not worthwhile to sprinkle corks on digitized older data, such as the beautiful balloon sequences that were taken by Karpinsky *et al.* already before the Spectrostratoscope, Pic du Midi and La Palma high-resolution imaging? That might recover the supergranulation, too large to fit on a CCD.

2. Historical perspective

I am running out of summary spacetime, speaking time and printing space. Further literal summarizing of all the remaining 179 contributions in Abstract Book order is out of the question; let me apologize to all of you whom I won't mention (and also to those whom I did mention), and skip all other presentations by jumping to Summary Model 2.

Our history is summarized in this list of solar IAU Symposia:

Nr	Year, Place & Title
6	1956 Stockholm — *Electromagnetic Phenomena in Cosmical Physics*
9	1958 Paris — *Paris Symposium on Radio Astronomy*
12	1960 Varenna — *Aerodynamic Phenomena in Stellar Atmospheres*
16	1961 Cloudcroft — *The Solar Corona*
22	1963 München — *Stellar and Solar Magnetic Fields*
35	1967 Budapest — *Structure and Development of Solar Active Regions*
43	1970 Paris — *Solar Magnetic Fields*
56	1973 Surfers Paradise — *Chromospheric Fine Structure*
57	1973 Surfers Paradise — *Coronal Disturbances*
71	1975 Praha — *Basic Mechanisms of Solar Activity*
86	1979 Maryland — *Radiophysics of the Sun*
91	1979 Cambridge Mass. — *Solar and Interplanetary Dynamics*
102	1982 Zürich — *Solar and Stellar Magnetic Fields*
123	1986 Aarhus — *Advances in Helio- and Asteroseismology*
138	1989 Kiev — *Solar Photosphere: Structure, Convection and Magnetic Fields*

At first solar physics was quite cosmical, in keeping with the IAU's roots in the International Union for Cooperation in Solar Research. Later, solar physics became more restrictive. Numbers 22, 43 and 102 indicate that we will have a definitive meeting called *Stellar Magnetic Fields* in the year 2000; the last one on that topic because it will exhaust its title possibilities.

But while we discuss solar and/or stellar magnetic fields throughout the years, the world around us changes. Here in Kiev, with this Symposium embedded between the second and the final rounds of the election of a Kiev representative to the new USSR Congress, that change is outspoken. Magnetic-field discussing solar physicists play only a minor role in world politics, even if they write letters to superpower presidents as we have done; nevertheless, they belong to an exceptionally internationally-oriented community in which cooperation across time zones and borders is the rule and not an exception.

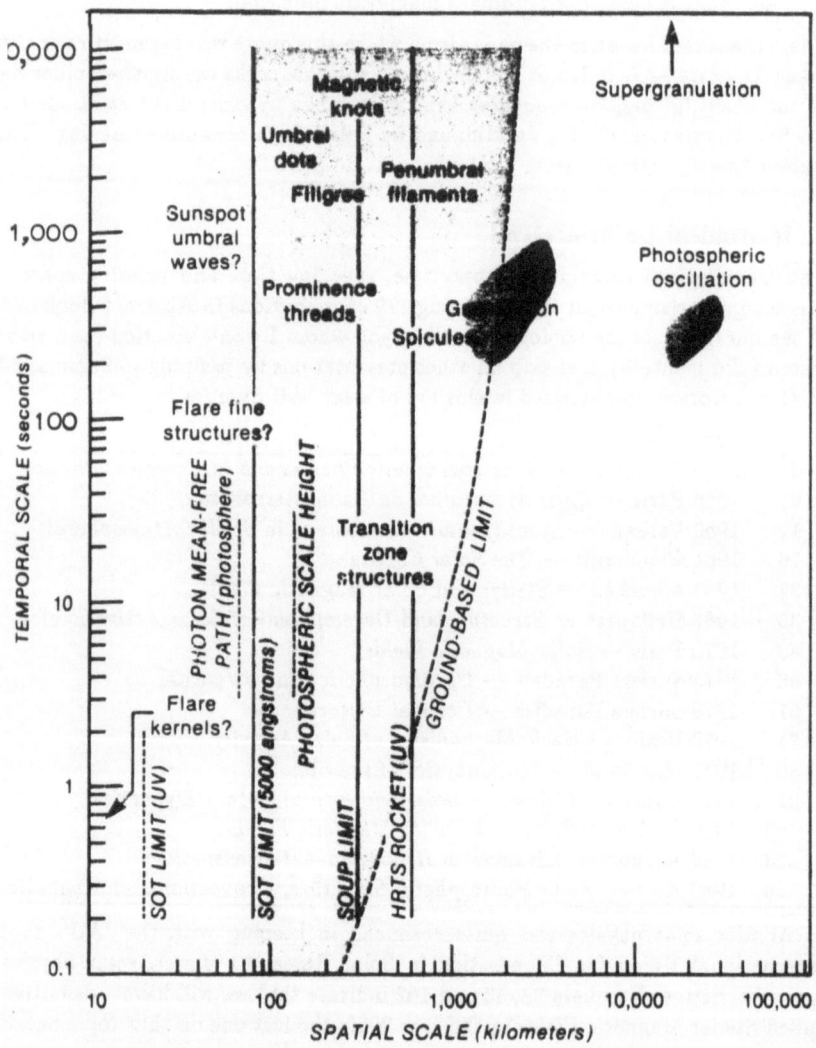

The spirit of frankness, mutual respect and constructive cooperation in which we run our business of understanding the physics which the Sun tries to teach us sets an example which we ourselves do not, perhaps, fully appreciate but which has important value, especially in this city where the awareness of the dangers embedded in physical knowledge is larger than anywhere else. In that historical perspective, the timing and location of this meeting have been significant. I hope that, looking back from the future, we will be happy to mark it as a turning point within and outside solar physics.

3. Future perspective

I base my predictions for the future on a graph made by Dunn, Harvey and Milkey over a decade ago to sell the SOT project which then became the HRSO project which then became the OSL project, and which we will eagerly await for years yet to come. This long delay is very unfortunate. NASA's Orbiting Solar Laboratory is for solar physics what Space Telescope is for nighttime astronomy: not just another space project to be advertised overly loudly but the required next step for nearly all interests in the whole field, a general purpose observatory located where it belongs, above the atmosphere. It should have flown its maidenflight long ago. In keeping with the letter sent from this meeting to Presidents Bush and Gorbachev I note that a very small fraction of the funds misspent on space militarization would have sufficed for space solarization.

However, solar physics does not compete directly with the military for funds but rather with non-solar colleagues who do not regard solar detail as beautiful yet; we must teach that principle by displaying what we do. There is enough to show, but the showing can be better. To quote Beckers[5], who recently had a look at granulation after leaving solar physics a decade ago: "The story of this exciting research should be made very visible so that our other astronomy colleagues can enjoy it as well".

Apart from the SOT-to-OSL name change, there are other changes necessary in the graph that do mark significant progress. The groundbased limit should be shifted and tilted a bit. Its lower part shifts to the left thanks to the good seeing of the Canary Islands. The shift is larger higher up because active mirroring, image grabbing, correlation tracking and destretching result in much longer high-quality time series, connecting periods of good seeing over hours rather than minutes. Hopefully, the realisation of a LEST with adaptive optics will produce a yet larger leftward shift all over. The graph must also be extended upwards since the Big Bear–Huairou magnetographs have already produced uninterrupted movies of over seventy hours. Let us hope there is no turning point now in this cooperation. Finally, there is a new feature to be entered: mesogranulation, at a few thousand kilometers and a few thousand seconds firmly to the right of the groundbased limit.

The area to the right marks the domain where we should see things properly already—and therefore understand them too. That is not altogether true because there are only two resolutions plotted here; temporal Fourier resolution, for example, is missing. Nevertheless, the photospheric oscillation is indeed understood; that was the first success story of modern solar physics. The granulation breakthrough also obeys this graph although

---

[5] J.M. Beckers, 1989, in R.J. Rutten and G. Severino (Eds.): *Solar and Stellar Granulation*, NATO ASI C-263, p. 613, Kluwer, Dordrecht

its essential resolution (simulation computer resources) is also missing.

Any graduate student can now see what subject he should choose for thesis project, depending on whether he prefers quick success or to spend a long career on a single problem. Mesogranulation is the next to go, then follow penumbrae and spicules (amazingly so), after that transition zone structures, and only then come the tiny strong-field concentrations (knots and filigree) of interest to tube and sheet modellers. This graph indicates that it will take some time before these are properly observed, but that may work out well; neutron stars also waited thirty years after their invention before showing themselves, giving time for thought. Perhaps such maturing is part of the Principle too.

4. Personal Impressions

I have found this a very interesting and inspiring conference. Speaking for the other participants as well, I gratefully thank the organisers for taking on and completing so successfully a task that undoubtedly must have brought much more work and problem solving than we can guess and, probably, than they themselves envisaged. Let me assure them that their work was worthwhile.

I sincerely believe that we are at a turning point in solar physics, progressing from riddles to answers. That belief was strengthened here, thanks to the excellent scientific program. We may also be at a turning point in international relationships. This conference, the first solar IAU Symposium in the USSR, exhibited a strong and lively spirit of unrestrained international exchange and cooperation, to which the splendid social program contributed substantially. On behalf of all participants, many thanks for the science and the hospitality presented to us!

LIST OF POSTER PAPERS

I.   Global Properties of the Photosphere

ON THE ADIABATICITY OF THE SOLAR ATMOSPHERE FROM COMPARING
LUMINOSITY AND VELOCITY MEASUREMENTS OF ACOUSTIC MODES
    A. Jimenez[1], M. Alvarez[3], V. Domingo[2], A. Jones[1], E. Ledezma[3], P.L. Palle[1]
    T. Roca Cortes[1]
    [1] Instituto de Astrofisica de Canarias, E–38200 La Laguna, Tenerife, Spain
    [2] Space Science Department of ESA, ESTEC, Noordwijk, The Netherlands
    [3] Inst. de Astronomia, Observ. Astron. Nacional, Ensenada, Baja California, Mexico

NEW ABSOLUTE MEASUREMENTS OF THE SOLAR RADIATION
    K.A. Burlov-Vasiljev, E.A. Gurtovenko, Yu.B. Matvejev, V.I. Trojan
    Main Astronomical Observatory Ac.Sc. Ukr.SSR, Goloseevo, 252127 Kiev, USSR

PRECISION SPECTROPHOTOMETRY OF THE SUN AS A STAR
    S.I. Gandzha, S.N. Osipov
    Main Astronomical Observatory Ac.Sc. Ukr.SSR, Goloseevo, 252127 Kiev, USSR

ON THE DISTINCTION AND PHOTOSPHERIC DEPTHS OF THE PROCESSES
FORMING A FRAUNHOFER LINE
    E.A. Gurtovenko[1], A.P. Sarychev[2], V.A. Sheminova[1]
    [1] Main Astronomical Observatory Ac.Sc. Ukr.SSR, Goloseevo, 252127 Kiev, USSR
    [2] Sternberg State Astronomical Institute
      Universitetsky Prospect, 13, 119899 Moscow, USSR

RESPONSE FUNCTIONS AND THEIR CAPABILITIES FOR DIAGNOSTICS OF
VELOCITIES IN THE PHOTOSPHERE
    V.N. Karpinsky, N.S. Petrova
    Main Astronomical Observatory Ac.Sc. USSR, 196140 Leningrad, USSR

THE IRON IONIZATION EQUILIBRIUM IN THE SOLAR PHOTOSPHERE
    J.H.M.J. Bruls
    Astronomical Institute, NL–3508 TA Utrecht, The Netherlands

SOME ASPECTS OF ABUNDANCES DERIVED FROM THE SOLAR
PHOTOSPHERIC SPECTRUM
    B.T. Baij, M.B. Girnjak, M.M. Kovalchuk, P. Olijnyk, R. Rykaljuk
    Astronomical Observatory of Lvov University, Lomonosov str. 8
    Lvov 290005, USSR

DETERMINATION OF MICROTURBULENT VELOCITIES IN THE SOLAR
PHOTOSPHERE
A.V. Baranov
Solar Station, 692533 Ussurijsk, USSR

WEAK VIOLET CYANOGEN LINES AND THE VELOCITY FIELD IN THE SOLAR
PHOTOSPHERE
G.A. Porfireva
Sternberg State Astronomical Institute, Moscow University, 119899 Moscow, USSR

PHOTOSPHERIC LINE REDSHIFTS FROM QUIET SUN ULTRAVIOLET SPECTRA
D. Samain
Institut d'Astrophys. Spatiale, B.P. 10, F–91371 Verrières-le-Buisson, France

PROFILES OF STRONG FRAUNHOFER LINES IN RESOLVED AND UNRESOLVED
SOLAR SPECTRA
J.M. Kuli-Zade
Azerbaijan State University, P.Lumumba, 23, 370073 Baku, USSR

Ca II EMISSION FROM THE SUN
R.J. Rutten, H. Uitenbroek
Astronomical Institute, NL–3508 TA Utrecht, The Netherlands

ON THE POSSIBILITY OF THE DEFINITION OF THE PHOTOSPHERE BRIGHTNESS
DISTRIBUTION FROM ECLIPSE DATA
I.L. Belkina, N.P. Djatel, M.M. Pospergelis
Astronomical Observatory, Kharkov University, Sumskaya 35, Kharkov, USSR

SOLAR DIAMETER MEASUREMENTS WITH A SOLAR ASTROLABE
F. Laclare[1], S. Debarbat[2], A. Journet[1], G. Merlin[1]
[1] C.E.R.G.A., F–06130 Grasse, France
[2] Observatoire de Paris, F–75014 Paris, France

OBSERVATIONS OF THE SUN AS A STAR AND THE MAPS OF CORONAL
HOLES IN THE He I 10830 Å LINE
E.V. Kondrashev
Main Astronomical Observatory Ac.Sc. USSR, Pulkovo, 196140 Leningrad, USSR

OBSERVATION OF SOLAR BRIGHTNESS FLUCTUATIONS WITH LOW SPATIAL
RESOLUTION
Yu.D. Zhugzhda, N. Lebedev
IZMIRAN, 142092 Troitsk, Moscow region, USSR

GLOBAL SOLAR OSCILLATIONS FROM OBSERVATIONS OF CONTINUUM
BRIGHTNESS FLUCTUATIONS
Yu.D. Zhugzhda, N. Lebedev
IZMIRAN, 142092 Troitsk, Moscow region, USSR

II.   Photospheric Fine Structure

FAST LARGE–SCALE CHANGE IN THE NATURE OF THE SOLAR GRANULATION
STRUCTURE
V.N. Karpinsky
Main Astronomical Observatory Ac.Sc. USSR, Pulkovo, 196140 Leningrad, USSR

STUDIES OF SOLAR GRANULATION USING 3-D FOURIER FILTERING
R.A. Shine, K.P. Topka, T.D. Tarbell, A.M. Title
Solar and Optical Physics Laboratory, O/91-30, B/256
Lockheed Palo Alto Research Lab., 3251 Hanover Street, Palo Alto, CA 94304, USA

SOLAR GRANULATION SPECTROMETRY
R.J. Rutten[1], M. Carlsson[2]
[1] Astronomical Institute, NL–3508 TA Utrecht, The Netherlands
[2] Institute of Theoretical Astrophysics, N–0315 Oslo, Norway

PHOTOMETRY AND SPECTROSCOPY OF THE SOLAR GRANULATION ALONG
THE POLAR AXIS AND EQUATOR
I. Rodriguez Hidalgo, M. Collados, M. Vazquez
Instituto de Astrofisica de Canarias, E–38200 La Laguna, Tenerife, Spain

SOLAR GRANULATION CONTRAST AND COLOR INDEX IN MAGNETIC AND
NON-MAGNETIC REGIONS
C.U. Keller[1], S. Koutchmy[2]
[1] National Solar Observatory, Sunspot, NM 88349, USA and
   Institute of Astronomy, ETH-Zentrum, CH–8092 Zurich, Switzerland
[2] AGFL/NSO–SP, Sunspot, NM 88349, USA and
   Institut d'Astrophysique, CNRS, 98 bis Bd Arago, F–75014 Paris, France

CONCERNING THE HEIGHT DISTRIBUTION OF VELOCITIES IN THE SOLAR
GRANULATION
V.S. Markov
SibIZMIR, 664033 Irkutsk, USSR

THE INTERGRANULAR LANES IN THE QUIET SOLAR PHOTOSPHERE AS
ORGANIZED STRUCTURES
S.I. Amalskaya
Main Astronomical Observatory Ac.Sc. USSR, Pulkovo, 196140 Leningrad, USSR

CONTINUOUS AND LINE SPECTRA OF GRANULES AND INTERGRANULAR
LANES
Z. Suemoto, E. Hiei, Y. Nakagomi
National Astronomical Observatory, Mitaka, Tokyo–81, Japan

MULTIVARIATE ANALYSIS OF BIDIMENSIONAL SPECTROSCOPY OBSERVATIONS
OF QUIET SOLAR REGIONS
B. Caccin[1], G. Cauzzi[2], R. Falciani[3], M. Fofi[4], L. Smaldone[2]
[1] Physics Department, II, University of Rome, Italy
[2] Phys. Sciences Department, University of Naples, Italy
[3] Astronomy Department, University of Florence, Italy
[4] Monte Mario Astronomical Observatory, Rome, Italy

III. Small–scale Magnetic Fields

A THERMODYNAMIC MODEL OF MAGNETIC FIELDS IN THE SOLAR
PHOTOSPHERE
O.V. Chumak
Astrophysical Institute Ac.Sc. Kaz.SSR, 480068 Alma–Ata, USSR

A STATISTICAL STUDY OF THE SIZE AND FLUX SPECTRUM OF SMALL–SCALE
MAGNETIC FIELDS
Wang Jingxiu, Shi Zhongxian, Liu Jianqiang, Han Feng, Liu Guilin
Beijing Astronomical Observatory, Chinese Academy of Sciences

OBSERVATIONS OF PHOTOSPHERIC MAGNETIC FIELDS WITH A NARROW-BAND
TUNABLE FILTER
E.J. Kulagin
Main Astronomical Observatory Ac.Sc. USSR, Pulkovo, 196140 Leningrad, USSR

RING STRUCTURES OF THE TRANSVERSE MAGNETIC FIELD IN THE SOLAR
PHOTOSPHERE
A.A. Pevtsov
SibIZMIR, 664033 Irkutsk, USSR

IV. Magnetohydrodynamics of the Photosphere

SYNERGETICAL ASPECT OF SOLAR MAGNETOHYDRODYNAMICS
E.I. Mogilevskij
Inst. of Terrestrial Magnetism, Ionosphere and Radio Wave Propagation
IZMIRAN, 142092 Troitsk, Moscow region, USSR

ABOUT SOME QUANTITATIVE CHARACTERISTICS OF PHOTOSPHERIC MHD
TURBULENCE
V.I. Dolgopolov
Kiev University Astronomical Observatory, Observatornaya 3, 252053 Kiev, USSR

PHENOMENOLOGY OF LOCAL MAGNETIC STRUCTURES NEAR THE
PHOTOSPHERE: APPROACH IN TERMS OF SCALE SPECTRUM
N.N. Kontor
Inst. of Nuclear Physics, Moscow State University, Moscow 119899, USSR

DYNAMICS AND THERMODYNAMICS OF 3-D CONVECTION
Kwing L. Chan
Applied Research Corporation, Landover, MD, USA

SIMULATION OF SOLAR GRANULATION
U. Uus[1], E.V. Poljakov[2], V.N. Karpinsky[2]
[1] Inst. of Astrophysics and Physics of Atmosphere Ac.Sc. Est.SSR
202444 Tartu, USSR
[2] Main Astronomical Observatory Ac.Sc. USSR, Pulkovo, 196140 Leningrad, USSR

ON THE APPLICABILITY OF A COMPACT DIFFERENCE HIGH ORDER ACCURACY
SCHEME FOR THE NUMERICAL SIMULATION OF TURBULENT CONVECTION IN
STELLAR ATMOSPHERES
I.N. Atroshchenko, A.S. Gadun
Main Astronomical Observatory Ac.Sc. Ukr.SSR, Goloseevo, 252127 Kiev, USSR

THE ROLE OF THE PHOTOSPHERIC OSCILLATIONS AND CONCENTRATED
MAGNETIC FIELDS IN THE HEATING OF THE CHROMOSPHERE

C.E. Alissandrakis[1], H.C. Dara[2], S. Koutchmy[3]

[1] University of Athens, 15783 Athens, Greece

[2] Center for Astronomy and Applied Mathematics, 10673 Athens, Greece

[3] Institut d'Astrophysique de Paris, CNRS, 98 bis, Bd. Arago, F-75014 Paris, France

LINEAR WAVE TRANSFORMATION AND THE EXCITATION OF WAVES WITH
SHORT WAVELENGTH

Yu.D. Zhugzhda, N.S. Dzahlilov

Solar Electrodynamic Sector, IZMIRAN, 142092 Troitsk, Moscow region, USSR

WAVE PROPAGATION IN MAGNETIC FLUX TUBES

S. Nasiri, Y. Sobouti

Department of Physics and Biruni Observatory, Shiraz, Iran

LONG NONLINEAR NON-AXISYMMETRIC SURFACE WAVE PROPAGATION IN
A MAGNETIC TUBE

M.S. Ruderman

Inst. for Problems in Mechanics, Prospect Vernadskogo 101, 117526 Moscow, USSR

KINETIC ALFVÉN WAVES IN THE SOLAR ATMOSPHERE

Yu.M. Voytenko, A.N. Kryshtal, A.K. Yukhimuck

Main Astronomical Observatory Ac.Sc. Ukr.S.S.R., Goloseevo, 252127 Kiev, USSR

EXCITATION OF MAGNETIC FLUXTUBE OSCILLATIONS DURING SOLAR FLARES

V.V. Zaitsev

Inst. of Applied Physics Ac.Sc. USSR, Gorky, USSR

DIRECT MEASUREMENTS OF SHORT PERIOD TORSIONAL OSCILLATIONS
OF SUNSPOTS

S.A. Druzhinin, A.A. Pevtsov, A.A. Levkovsky, M.V. Nikonova

SibIZMIR, 664033 Irkutsk, USSR

MAGNETOHYDRODYNAMIC DISTURBANCES IN SUNSPOTS

S.V. Kootz, P.P. Malovichko, A.K. Yukchimuk

Main Astronomical Observatory Ac.Sc. Ukr.SSR, Goloseevo, 252127 Kiev, USSR

INVESTIGATION OF THE DYNAMICS OF THE MAGNETIC FIELD AT THE
FORMATION OF SUNSPOTS, USING MAGNETOGRAMS

J. Linke[1], J.H. Hernandez[2], V.L. Selivanov[3]

[1] IZMIRAN, 142092 Troitsk, Moscow region, USSR, and
Zentralinstitut für Astrophysik, 1561 Potsdam, DDR

[2] Pedagogical High School, Inst. of Geophysics and Astronomy, Havana, Cuba

[3] SibIZMIR, 664033 Irkutsk, USSR

ON THE EFFECT OF MAGNETIC FIELD BOUNDARIES IN COMPLEX SOLAR
ACTIVE REGIONS

V.N. Ishkov, J. Linke

IZMIRAN, 142092 Troitsk, Moscow region, USSR

VARIATIONS OF PHYSICAL PARAMETERS AND OSCILLATORY MOTIONS IN
SELECTED SUNSPOT GROUPS

Yu.A. Nagovitsin, G.F. Vyalshin

Main Astronomical Observatory Ac.Sc. USSR, Pulkovo, 196140 Leningrad, USSR

DIAGNOSTICS OF SUNSPOT UMBRAL FINE STRUCTURE BY OSCILLATIONS

J. Staude[1], Yu.D. Zhugzhda[2], V. Locans[3]
[1] Zentralinstitut für Astrophysik, DDR–1561 Potsdam
[2] IZMIRAN, 142092 Troitsk, Moscow region, USSR
[3] Radioastrophysical Observatory, 226524 Riga, USSR
Department of Physics, Ravishankar University, Raipur 492001, India

SEMI-EMPIRICAL MODELS OF UMBRAL LIGHT BRIDGES

M. Sobotka
Astronomical Institute of Czechoslovak Ac.Sc., 25165 Ondřejov, Czechoslovakia

MAGNETIC FIELD FRAGMENTATION IN FLUX TUBES IN SUBPHOTOSPHERIC LAYERS OF THE SUN

V.D. Kuznetsov
IZMIRAN, 142092 Troitsk, Moscow region, USSR

CURRENT SHEET MODEL OF A SUNSPOT

K. Jahn
Astron. Observatory of Warsaw University, Al. Ujazdowskie 4, 00–478 Warsaw, Poland

VERTICAL GRADIENTS OF SUNSPOT MAGNETIC FIELDS

A. Hofmann, J. Rendtel
Zentralinstitut für Astrophysik, Sonnenobservatorium Einsteinturm, DDR–1561 Potsdam

MAGNETIC FIELDS AND PROPER MOTIONS OF SUNSPOTS

M. Goshzhanov, V.I. Skazhenyuk
Physical-Technical Institute Ac.Sc. Tur.SSR, 744000 Ashkhabad, USSR

LARGE–SCALE PROPERTIES OF ACTIVE REGIONS

A. Hofmann
Zentralinstitut für Astrophysik, Sonnenobservatorium Einsteinturm, DDR–1561 Potsdam

INVESTIGATION OF THE INTEGRAL CHARACTERISTICS OF ACTIVE REGION MAGNETIC FIELDS

L.V. Ermakova
SibIZMIR, P.O. Box 4, 664033 Irkutsk, USSR

ON THE CONTINUUM INTENSITY-MAGNETIC FIELD RELATION DURING THE DECAY PHASE OF SUNSPOTS

V. Martinez-Pillet, M. Vazquez
Instituto de Astrofisica de Canarias, E–38200 La Laguna, Tenerife, Spain

THE EVOLUTION OF THE LONGITUDINAL MAGNETIC NEUTRAL LINE AND CHROMOSPHERIC MATTER FLOWS IN SOLAR ACTIVE REGIONS

Zhang Hongqi, Ai Guoxiang, Li Jing, Liu Jainqiang
Beijing Astronomical Observatory, Chinese Academy of Sciences

TEMPORAL AND SPATIAL PROPERTIES OF THE Ca AND Hα CHROMOSPHERIC NETWORK STRUCTURAL ELEMENTS

E.V. Kononovich, O.B. Smirnova
Sternberg State Astronomical Institute, Moscow University, 119899 Moscow, USSR

A STUDY OF DARK He-POINTS BY A TV METHOD

L.D. Parfinenko
Main Astronomical Observatory Ac.Sc. USSR, Pulkovo, 196140 Leningrad, USSR

SOLVING THE MAGNETOHYDRODYNAMIC EQUATIONS IN THE STRONG FIELD APPROXIMATION

V.S. Gorbachev, S.R. Kel'ner

Moscow Physical Engineering Institute, Kashirskoe s. 31, 115409 Moscow, USSR

FLUCTUATIONS OF STRONG MAGNETIC FIELDS AS A CAUSE FOR THE FORMATION OF PLASMA CONDENSATIONS

V.S. Gorbachev, S.R. Kel'ner

Moscow Physical Engineering Institute, Kashirskoe s. 31, 115409 Moscow, USSR

SOLAR FORCE-FREE MAGNETIC FIELDS IN AND ABOVE THE PHOTOSPHERE

Chen Zhen Cheng, Cheng Yang

Beijing Astronomical Observatory, Chinese Academy of Sciences

FLOW VELOCITIES ALONG A SOLAR Hα EMISSION LOOP

A.B. Delone, E.A. Makarova, G.A. Porfireva, E.M. Roschina, G.V. Yakunina

Sternberg State Astronomical Institute, Moscow University, 119899 Moscow, USSR

CORONAL PROLONGATION OF PHOTOSPHERIC STRUCTURES AND MAGNETIC FIELDS ACCORDING TO RADIO ASTRONOMY DATA

G.B. Gelfreikh

Main Astronomical Observatory Ac.Sc. USSR, Pulkovo, 196140 Leningrad, USSR

FINE STRUCTURE OF THE SOLAR ATMOSPHERE AND ITS CONVECTION INTO THE INTERPLANETARY MEDIUM

G.P. Lyubimov

Institute of Nuclear Physics, Moscow University, 119899 Moscow, USSR

PHOTOSPHERIC MAGNETIC FIELDS AND CORONAL TEMPERATURES FROM ECLIPSE OBSERVATIONS

A.B. Delone, E.A. Makarova, G.V. Yakunina

Sternberg State Astronomical Institute, Moscow University, 119899 Moscow, USSR

"PERESTROIKA" OF MAGNETIC FIELDS IN ACTIVE REGIONS BEFORE A GREAT FLARE AND ITS MANIFESTATION IN THE RADIO RANGE

V.N. Borovik[1], N.A. Drake[2], A.A. Golovko[3]

[1] Main Astronomical Observatory Ac.Sc. USSR, Pulkovo, 196140 Leningrad, USSR

[2] Astronomical Observatory, Leningrad University, Petrodvoretsk, 198904 Leningrad, USSR

[3] SibIZMIR, P.O. Box 4, 664033 Irkutsk, USSR

LARGE–SCALE CHANGES OF FILAMENT SYSTEMS CONNECTED WITH THE APPEARANCE OF CORONAL TRANSIENTS AND GAMMA–RAY BURSTS

A.B. Delone, E.A. Makarova, V.S. Prokudina, G.V. Yakunina

Sternberg State Astronomical Institute, Moscow University, 119899 Moscow, USSR

PRELIMINARY RESULTS OF NON-LTE MODELLING OF FLUORESCENT IRON LINES IN SOLAR FLARES

N.A. Sakhibullin, U.Sh. Bayazitov

Kazan State University, Kazan, USSR

EMISSION IN THE LINES OF Fe I AND Fe II IN SOLAR FLARES

E.A. Baranovsky[1], E.A. Kurochka[2]

[1] Crimean Astrophysical Observatory Ac.Sc. USSR, Nauchny, 334413 Crimea, USSR

[2] Kiev University, 252053 Kiev, USSR

NON–THERMAL HYDROGEN IONIZATION IN ELEMENTARY FLARE BURSTS

V.V. Zharkova, V.A. Kobylisky

Kiev University, 252053 Kiev, USSR

V. Large–scale Structure and Dynamics

VI. Generation of Solar Magnetic Fields

ON THE DIMENSION OF THE SOLAR ATTRACTOR

V.M. Ostryakov, I.G. Usoskin

Ioffe Physical-Technical Institute Ac.Sc. USSR, 194021 Leningrad, USSR

ON THE ORIGIN OF THE FINE-STRUCTURED SOLAR MAGNETIC FIELD

N.I. Kleeorin[1], A.A. Ruzmaikin[2], D.D. Sokoloff[3]

[1] Lenin State Pedagogical Institute, 119435 Moscow, USSR

[2] IZMIRAN, 142092 Troitsk, Moscow region, USSR

[3] Moscow University, 119899 Moscow, USSR

DYNAMO MECHANISM FOR ELECTRIC CURRENTS IN THE SOLAR ATMOSPHERE

N. Seehafer

Zentralinstitut für Astrophysik, 1591 Potsdam, DDR

MODERN CONCEPTS OF THE SOLAR CYCLE

Yu.I. Vitinsky[1], G.V. Kuklin[2], V.N. Obridko[3]

[1] Main Astronomical Observatory Ac.Sc. USSR, Pulkovo, 196140 Leningrad, USSR

[2] SibIZMIR, 664033 Irkutsk, USSR

[3] IZMIRAN, 142092 Troitsk, Moscow region, USSR

EFFECT OF NEGATIVE MAGNETIC PRESSURE IN THE TURBULENT SOLAR CONVECTIVE ZONE

N.I. Kleeorin[1], I.V. Rogachevsky[2], A.A. Ruzmaikin[3]

[1] Lenin State Pedagogical Institute, 119435 Moscow, USSR

[2] Fedorov Institute of Applied Geophysics, 107258 Moscow, USSR

[3] IZMIRAN, 142092 Troitsk, Moscow region, USSR

SOLAR CONVECTION: STRUCTURE AND ROLE IN SPOT FORMATION

A.V. Getling

Institute of Nuclear Physics, Moscow University, 119899 Moscow, USSR

TRANSFER OF SOLAR LARGE–SCALE MAGNETIC FIELDS BY THE INHOMOGENEITY OF THE CONVECTION ZONE MATTER DENSITY

V.N. Krivodubskij

Astronomical Observatory, Kiev University, 252053 Kiev, USSR

ANOMALIES IN THE SOLAR ACTIVITY CYCLE

V. Martinez-Pillet, M. Vazquez

Instituto de Astrofisica de Canarias, E–38200 La Laguna, Tenerife, Spain

CHANGES IN THE LENGTH OF SOLAR CYCLES DURING THE LAST FIVE CENTURIES

J. Strestik, I. Charvatova

Geophysical Institute Czechoslovak Ac.Sc., Bocni II, 14131 Praha, Czechoslovakia

THE SOLAR MAGNETIC FIELD ENERGY, CORONAL GREEN LINE EMISSION, AND SOME PROPERTIES OF THE CYCLE

V.P. Mikhailutsa, M.N. Gnevyshev

Main Astronomical Observatory Ac.Sc. USSR, Pulkovo, 196140 Leningrad, USSR

VII. Convection and Magnetic Fields in Solar-type Stars

STELLAR GRANULATION
D. Dravins[1], Å. Nordlund[2]
[1] Lund Observatory, S–221 00 Lund, Sweden
[2] Copenhagen University Observatory, DK–1350 Copenhagen, Denmark

MAGNETIC FIELDS AND OUTER ATMOSPHERES OF LATE-TYPE STARS
M.M. Katsova, M.A. Livshits
IZMIRAN, 142092 Troitsk, Moscow region, USSR

NON-LTE ANALYSIS OF THE CARBON SPECTRA OF THE SUN AND PROCYON
T.G. Shcherbina
Main Astronomical Observatory Ac.Sc. Ukr.SSR, Goloseevo, 252127 Kiev, USSR

OBSERVATION AND ANALYSIS OF THE NEUTRAL CALCIUM SPECTRUM OF THE SUN AND PROCYON
A.V. Perekhod, N.G. Stchukina
Main Astronomical Observatory Ac.Sc. Ukr.SSR, Goloseevo, 252127 Kiev, USSR

OVERIONIZATION OF NEUTRAL IRON IN THE ATMOSPHERES OF SOLAR-LIKE SUBDWARFS
I.F. Bikmaev[1], S. Bobritzkij[2], N.A. Sakhibullin[2]
[1] Special Astrophysical Observatory, 357147 Nizhnij Arkhyz, USSR
[2] Kazan University, 420008 Kazan, USSR

SOLAR PHOTOSPHERE

To:

President G. Bush

President M. Gorbachev

Dear President Bush,
Dear President Gorbachev,

We are deeply concerned by the fact that the Earth´s atmosphere
and its space environment are increasingly becoming an arena for
military activity, in particular involving nuclear weapons.

Atmospheric and space pollution with nuclear waste and
radiation threatens life on Planet Earth and will make it more
and more impossible to explore our Universe by astronomical
observations.

We approach you with an appeal to take every pertinent
measure to prevent that outer space will become a base for
placing nuclear or other weapons.

The above statement has been endorsed by acclamation by the
participants in the Internation Astronomical Union Symposium
No 138 in Kiev , May 15 - 20 , 1989. More than 200 scientists
from 24 countries took part in the symposium.

For the participants of the International Astronomical Union
Symposium No 138 :

J.O. Stenflo   E.A. Gurtovenko

Institute of Astronomy   Main Astronomical Observatory
ETH - Zentrum   Ac. of Sc. Ukr.S.S.R.
CH - 8092 Zurich   252127 Kiev, Goloseevo
Switzerland   U.S.S.R.
Chairman, IAU Symp.138   Chairman, IAU Symp. 138
Scientific Organizing Commitee   Local Organizing Commitee

Kiev, May 15-20, 1989

529

J. O. Stenflo (ed.), Solar Photosphere: Structure, Convection, and Magnetic Fields, 529–530.
© *1990 by the IAU.*

УВАЖАЕМЫЙ ПРЕЗИДЕНТ БУШ!
УВАЖАЕМЫЙ ПРЕЗИДЕНТ ГОРБАЧЕВ!

Мы глубоко обеспокоены тем, что земная атмосфера и околоземное космическое пространство все в большей и большей степени становится ареной военной активности с выносом в Космос также и лазерного оружия.

Загрязнение атмосферы и космического пространства ядерными отбросами и радиацией угрожает жизни на планете Земля и делает все более и более невозможными исследования нашей Вселенной при помощи астрономических наблюдений.

Мы обращаемся к Вам с призывом принять все необходимые меры для того, чтобы космическое пространство не было превращено в базу для размещения ядерного и других видов вооружения.

Это обращение получило полную поддержку участников Симпозиума № 138 Международного астрономического союза, состоявшегося в Киеве с 15 по 20 мая 1989 года. В работе Симпозиума приняли участие более чем 200 участников из 24 стран мира.

От имени участников Симпозиума № 138
Международного астрономического союза:

Дж.О.СТЕНФЛО

Астрономический институт
Швейцария
Цюрих
Председатель
Симп. № 138 МАС

Научный оргкомитет

Э.А.ГУРТОВЕНКО

Главная астрономическая
обсерватория АН УССР
Киев
Председатель
Симп. № 138 МАС

Местный оргкомитет

LIST OF PARTICIPANTS

AUSTRALIA
Durrant, C.J.   University of Sydney, N.S.W.

BULGARIA
Buyukliev, G.   Department of Astronomy, Sofia
Zhelyazkov, I.   Faculty of Physics, Sofia University

CHINA, PEOPLE'S REPUBLIC OF
Chen Zhen-cheng   Beijing Astronomical Observatory
Wang Jingxiu   Beijing Astronomical Observatory
Zhang Hongqi   Beijing Astronomical Observatory
Zou Yi-xin   Beijing Astronomical Observatory

CHINA, REPUBLIC OF
Dean-yi Chou   Tsing Hua University, Hsinchij, Taiwan

CZECHOSLOVAKIA
Antalová, A.   Astronomical Institute, Tatranská Lomnica
Bumba, V.   Astronomical Institute, Ondřejov
Charvatova, I.   Geophysical Institute, Praha–Spořilov
Sobotka, M.   Astronomical Institute, Ondřejov
Sykora, J.   Astronomical Institute, Tatranská Lomnica
Strestik, J.   Geophysical Institute, Praha–Spořilov

DENMARK
Nordlund, Å.   University Observatory, Copenhagen

FINLAND
Brandenburg, A.   University of Helsinki
Houvelin, J.   University of Helsinki
Tuominen, I.   University of Helsinki

531

FRANCE
Dollfus, A.	Observatoire de Paris, Meudon
Henoux, J.	Observatoire de Paris, Meudon
Koutchmy, S.	Institut d'Astrophysique, Paris
Laclare, F.	Observatoire de la Côte d'Azur, Grasse
Muller, R.	Observatoire de Pic-du-Midi, Bagnères-de-Bigorre
Rozelot, J.-P.	Observatoire des Alpes Maritimes, Grasse
Samain, D.	LPSP, Verrièrres-le-Buisson

GERMANY, D.R.
Hofmann, A.	Zentralinstitut für Astrophysik, Potsdam
Kochler, P.	Carl Zeiss Jena
Rüdiger, G.	Sternwarte Babelsberg, Potsdam
Seehafer, N.	Zenralinstitut für Astrophysik, Potsdam
Staude, J.	Zentralinstitut für Astrophysik, Potsdam

GERMANY, F.R.
Brandt, P.N.	Kiepenheuer–Institut für Sonnenphysik, Freiburg
Deubner, F.-L.	Institut für Astronomie u. Astrophysik, Würzburg
Schüssler, M.	Kiepenheuer–Institut für Sonnenphysik, Freiburg
Spruit, K.	MPI für Astrophysik, Garching
Steffen, M.	Institut f. Theoret. Physik und Sternwarte, Kiel

GREECE
Alissandrakis, C.E.	University of Athens

HUNGARY
Erdelyi, R.	Eötvös University, Budapest
Petrovay, K.	Eötvös University, Budapest

INDIA
Antia, H.M.	T.I.F.R., Bombay
Gokhale, M.H.	Indian Institute of Astrophysics, Bangalore
Hasan, S.S.	Indian Institute of Astrophysics, Bangalore
Krishan, V.	Indian Institute of Astrophysics, Bangalore
Pandey, S.K.	Ravishankar University, Raipur
Sivaraman, K.R.	Indian Institute of Astrophysics, Bangalore

ITALY
Belvedere, G.	Istituto di Astronomia, Catania
Marmolino, C.	Università di Napoli
Severino, G.	Osservatorio Astronomico di Capodimonte, Napoli
Smaldone, L.A.	Università di Napoli

JAPAN
    Hiei, E.   National Astron. Observatory, Mitaka, Tokyo

NETHERLANDS
    Bruls, J.H.M.J.   Astronomical Institute, Utrecht
    Hoyng, P.   Space Research Laboratory, Utrecht
    Rutten, R.   Astronomical Institute, Utrecht
    Uitenbroek, H.   Astronomical Institute, Utrecht
    van Geffen, J.H.G.M.   Space Research Laboratory, Utrecht
    Zwaan, C.   Astronomical Institute, Utrecht

NORWAY
    Engvold, O.   Inst. of Theoret. Astrophysics, Oslo

POLAND
    Jahn, K.   Astronomical Observatory, Warsaw University
    Sikorski, I.   Inst. of Theoret. Phys. and Astrophys., Gdansk

SPAIN
    Aballe Villero, M.A.   Instituto de Astrofisica de Canarias, La Laguna
    Jimenes, A.   Instituto de Astrofisica de Canarias, La Laguna
    Martinez-Pillet, V.   Instituto de Astrofisica de Canarias, La Laguna
    Rodriguez Hidalgo, I.   Instituto de Astrofisica de Canarias, La Laguna
    Sanchez Almeida, J.   Instituto de Astrofisica de Canarias, La Laguna

SWEDEN
    Dravins, D.   Lund Observatory, Lund

SWITZERLAND
    Keller, Ch.U.   Institute of Astronomy, ETH Zurich
    Steiner, O.   Institute of Astronomy, ETH Zurich
    Stenflo, J.O.   Institute of Astronomy, ETH Zurich

UNITED KINGDOM
    Jennings, R.   University of Newcastle upon Tyne
    Solanki, S.K.   University of St. Andrews, Scotland

USA
    Avrett, E.   Center for Astrophysics, Cambridge, MA
    Ayres, T.R.   Center f. Astrophys. & Space Astron., Boulder, CO

534

Kalkofen, W.	Center for Astrophysics, Cambridge, MA
Kwing, Lam Chan	NASA/Goddard S. F. C., Greenbelt, MD
Lawrence, J.K.	San Fernando Observatory, Northridge, CA
Martin, S.F.	California Institute of Technology, Pasadena, CA
Saar, S.	Center for Astrophysics, Cambridge, MA
Shine, R.	Lockheed Palo Alto Research Lab., Palo Alto, CA
Tarbell, T.D.	Lockheed Palo Alto Research Lab., Palo Alto, CA
Thomas, J.H.	University of Rochester, Rochester, NY
Title, A.	Lockheed Palo Alto Research Lab., Palo Alto, CA

USSR

Abramenko, V.I.	Crimean Astrophysical Observatory
Ajmanov, A.K.	Astrophysical Institute Kazakhstan, Alma-Ata
Amalskaja, S.I.	Main Astronomical Observatory, Leningrad
Atroshchenko, I.N.	Main Astronomical Observatory, Kiev
Baranov, A.V.	Ussurijsk Solar Station, Primorskij Kraj
Baranovsky, E.A.	Crimean Astrophysical Observatory
Bayazitov, U.Sh.	Shemachka Astrophys. Observatory, Azerbaijan
Berdichevskaja, V.S.	Sternberg State Astronomical Institute, Moscow
Bikmaev, I.F.	Special Astrophys. Observatory, Stavropol terr.
Burlov-Vasiljev, K.A.	Main Astronomical Observatory, Kiev
Chernobay, V.A.	Kishinev Astronomical Observatory, Kishinev
Chistyakov, V.F.	Ussurijsk Solar Station, Primorskij Kraj
Chumac, O.V.	Astrophysical Institute Kazakhstan, Alma-Ata
Demidov, M.L.	Sib.IZMIR, Irkutsk
Djalilov, N.S.	IZMIRAN, Troitsk, Moscow region
Dolginov, A.Z.	Ioffe Physical-Technical Institute, Leningrad
Dolgopolov, V.I.	Astronomical Observatory, Kiev University
Drake, N.A.	Astronomical Observatory, Leningrad
Dudnikova, G.I.	Inst. of Theoret. and Appl. Mechanics, Novosibirsk
Dudorov, A.E.	Cheljabinsk University, Cheljabinsk
Ermakova, L.V.	Sib.IZMIR, Irkutsk
Gadun, A.S.	Main Astronomical Observatory, Kiev
Gandzha, S.I.	Main Astronomical Observatory, Kiev
Gelfrejkh, G.B.	Main Astronomical Observatory, Leningrad
Getling, A.V.	Institute of Nuclear Physics, Moscow
Gigolashvili, M.Sh.	Abastumani Astrophys. Observatory, Abastumani
Gladushina, N.A.	Voroshilovgradsky Teacher Inst., Voroshilovgrad
Gnevyshev, M.N.	Kislovodsk Station of Pulkovo Observatory
Golovko, A.A.	Sib.IZMIR, Irkutsk
Golovkov, V.P.	IZMIRAN, Troitsk, Moscow region
Gopasyuk, S.I.	Crimean Astrophysical Observatory
Gorbachev, V.S.	Moscow Engineering Physical Inst., Moscow
Goshchdzhanov, M.	Inst.of Techn.Physics, TurkmenAc.Sc., Askhabad
Grigoryeva, S.A.	Sib.IZMIR, Irkutsk
Grigoryev, V.M.	Sib.IZMIR, Irkutsk
Gurtovenko, E.A.	Main Astronomical Observatory, Kiev

Ilchenko, A.G.	Poltava Gravimetric Observatory, Poltava
Ishkov, V.N.	IZMIRAN, Troitsk, Moscow region
Ivanchuk, V.I.	Astronomical Observatory, Kiev University
Juchimuk, A.K.	Main Astronomical Observatory, Kiev
Kandrashov, E.V.	Main Astronomical Observatory, Leningrad
Karpinsky, V.N.	Main Astronomical Observatory, Leningrad
Katsova, M.M.	Sternberg State Astron. Institute, Moscow
Kelner, S.R.	Moscow Engineering Physical Inst., Moscow
Kichatinov, L.L.	Sib.IZMIR, Irkutsk
Kim, I.S.	Sternberg State Astron. Institute, Moscow
Kleeorin, N.I.	Pedagogical Institute, Moscow
Klepikov, V.Yu.	IZMIRAN, Troitsk, Moscow region
Klyachkin, A.V.	Ioffe Physical-Technical Institute, Leningrad
Kononovich, E.V.	Sternberg State Astron. Institute, Moscow
Kontor, N.N.	Institute of Nuclear Physics, Moscow
Kootz, S.B.	Main Astronomical Observatory, Kiev
Kostik, R.I.	Main Astronomical Observatory, Kiev
Kotov, V.A.	Crimean Astrophysical Observatory
Kovalchuk, M.M.	Astronomical Observatory, Lvov
Kravchuk, P.F.	Astronomical Observatory, Kiev University
Krishtal, A.N.	Main Astronomical Observatory, Kiev
Krivodubskij, V.N.	Astronomical Observatory, Kiev University
Kuklin, G.V.	Sib.IZMIR, Irkutsk
Kulagin, E.S.	Main Astronomical Observatory, Leningrad
Kuli-zade, D.M.	Azerbaijan University, Baku
Kuznezov, V.D.	IZMIRAN, Troitsk, Moscow region
Latushko, S.M.	Sib.IZMIR, Irkutsk
Lejko, U.M.	Astronomical Observatory, Kiev University
Linke, J.	IZMIRAN, Troitsk, Moscow region
Livshits, M.A.	IZMIRAN, Troitsk, Moscow region
Lozitskij, V.G.	Astronomical Observatory, Kiev University
Magerramov, V.A.O.	Scient. and Industrial Assoc. of Cosmic Res., Baku
Makarov, V.I.	Kislovodsk Station of Pulkovo Observatory
Makarova, E.A.	Sternberg State Astron. Institute, Moscow
Malovichko, P.P.	Main Astronomical Observatory, Kiev
Markov, V.S.	Sib.IZMIR, Irkutsk
Matvejev, Yu.B.	Main Astronomical Observatory, Kiev
Minasynts, G.S.	Astrophysical Institute, Alama–Ata
Mogilevsky, E.I.	IZMIRAN, Troitsk, Moscow region
Mordvinov, A.V.	Sib.IZMIR, Irkutsk
Mozharovsky, S.G.	Ussurijsk Solar Station, Primorskij Kraj
Nagovitsin, Yu.A.	Main Astronomical Observatory, Leningrad
Obridko, V.N.	IZMIRAN, Troitsk, Moscow region
Ogir, M.B.	Crimean Astrophysical Observatory
Osipov, S.N.	Main Astronomical Observatory, Kiev
Ostryakov, V.M.	Ioffe Physical-Technical Institute, Leningrad
Parchevsky, K.V.	Crimean Astrophysical Observatory
Parfinenko, L.D.	Main Astronomical Observatory, Leningrad

Perekhod, A.V.	Main Astronomical Observatory, Kiev
Pevtsov, A.A.	Sib.IZMIR, Irkutsk
Pikalov, K.N.	Main Astronomical Observatory, Kiev
Plyusnina, L.A.	Sib.IZMIR, Irkutsk
Poljakov, E.V.	Main Astronomical Observatory, Leningrad
Ponyavin, D.I.	Institute of Physics, Leningrad University
Porfireva, G.A.	Sternberg State Astron. Institute, Moscow
Rogachevskij, I.V.	Institute of Appl. Geophysics, Moscow
Ruderman, M.S.	Institute for Problems in Mechanics, Moscow
Ruzmaikin, A.A.	IZMIRAN, Troitsk, Moscow region
Ryutova, M.P.	Institute of Nuclear Physics, Novosibirsk
Sakhibullin, N.A.	Kazan University, Kazan
Sattarov, I.S.	Astrophys. Institute Uzbekistan, Tashkent
Selivanov, V.L.	Sib.IZMIR, Irkutsk
Shaab, M.	Astronomical Observatory, Kiev University
Shcherbina, T.G.	Main Astronomical Observatory, Kiev
Shchukina, N.G.	Main Astronomical Observatory, Kiev
Sheminova, V.A.	Main Astronomical Observatory, Kiev
Skomarovsky, V.I.	Sib.IZMIR, Irkutsk
Smirnova, O.B.	Sternberg State Astron. Institute, Moscow
Sokoloff, D.	Moscow University
Solonsky, Yu.A.	Astronomical Observatory, Leningrad
Somov, B.V.	Lebedev Physical Institute, Moscow
Stepanian, N.N.	Crimean Astrophysical Observatory
Stepanov, A.I.	IZMIRAN, Troitsk, Moscow region
Stepanov, A.V.	Crimean Astrophysical Observatory
Stepanova, T.V.	Dept. of Astronomy, Moscow
Stoyanova, M.N.	Main Astronomical Observatory, Leningrad
Teplitskaya, R.B.	Sib.IZMIR, Irkutsk
Tsap, T.T.	Crimean Astrophysical Observatory
Vandakurov, Yu.V.	Ioffe Physical-Technical Institute, Leningrad
Vasiljeva, I.E.	Main Astronomical Observatory, Kiev
Vassilyeva, G.J.	Main Astronomical Observatory, Leningrad
Vitinsky, Yu.I.	Main Astronomical Observatory, Leningrad
Vojtenko, Yu.M.	Main Astronomical Observatory, Kiev
Yakunina, G.V.	Sternberg State Astron. Institute, Moscow
Yoshpa, B.A.	IZMIRAN, Troitsk, Moscow region
Zajtsev, V.V.	Institute of Applied Physics, Gorkij
Zharkova, V.V.	Astronomical Observatory, Kiev University
Zhugzhda, Yu.D.	IZMIRAN, Troitsk, Moscow region
Zlotnik, E.M.	Institute of Applied Physics, Gorkij

AUTHOR INDEX

538

SUBJECT INDEX